Inside IG Farben

In 1925 the three leading chemical firms in Germany – BASF, Bayer, and Hoechst – merged, together with some smaller firms, to become IG Farben. IG Farben became like no other firm synonymous for the participation of German industry in the most heinous crimes of the Nazi regime. This book deals in depth with one of IG Farben's leading factories, Hoechst, during the Third Reich. On the basis of long and meticulous archival research, including access to previously inaccessible company records, the author describes and analyzes the relationship between management and employees with the Nazi Party and its organizations. The author shows the exclusion and persecution of employees, particularly Jewish employees. He traces the extent of Hoechst's involvement in the exploitation of forced labor, and its active participation in human experiments in several concentration camps. Throughout, he tries to shed light on the motivations of those responsible for this conduct.

STEPHAN H. LINDNER is Professor of Interdependence of Technological and Social Change at the University of the Bundeswehr Munich. He formerly was a Lecturer of Economic History and History of Technology at the Technical University Munich. He published the German version of *Inside IG Farben* in 2005 to great reviews in German newspapers and academic journals.

Inside IG Farben

Hoechst During the Third Reich

STEPHAN H. LINDNER

University of the Bundeswehr Munich

English translation by Helen Schoop

CAMBRIDGE UNIVERSITY PRESS
Cambridge, New York, Melbourne, Madrid, Cape Town,
Singapore, São Paulo, Delhi, Tokyo, Mexico City

Cambridge University Press
32 Avenue of the Americas, New York, NY 10013-2473, USA

www.cambridge.org
Information on this title: www.cambridge.org/9780521178389

© Stephan H. Lindner 2008

First published 2008
First paperback edition 2011

A catalog record for this publication is available from the British Library

Library of Congress Cataloging in Publication data

Lindner, Stephan H.
[Hoechst. English]
Inside IG Farben : Hoechst during the Third Reich / Stephan H. Lindner;
English translation by Helen Schoop.
p. cm.
Includes bibliographical references and index.
ISBN 978-0-521-88766-3 (hardback : alk. paper)
1. Interessengemeinschaft Farbenindustrie Aktiengesellschaft – History.
2. Hoechst AG – History. 3. World War, 1939–1945 – Atrocities.
4. Chemical industry – Political aspects – Germany – History – 20th century.
I. Title.
HD9654.9.I5L5613 2008
338.7′660094109041 – dc22   2008003906

ISBN 978-0-521-88766-3 Hardback
ISBN 978-0-521-17838-9 Paperback

Gradually it was disclosed to me that the line separating good and evil passes not through states, nor between classes, nor between political parties either – but right through every human heart – and through all human hearts.

Alexander Solzhenitsyn

Contents

List of Illustrations   *page* ix

List of Abbreviations   xi

Foreword by Peter Hayes   xiii

Acknowledgments   xix

1   Introduction   1

2   From the Formation of IG Farben to the Great Depression   12

    2.1.   The End of Independence (1904–1925)   12

    2.2.   From the "Founders' Compromise" to First Reforms
        and Rationalization Measures   26

    2.3.   Forced Retirement of the "Old Hoechst Hierarchy"
        during the Great Depression   35

    2.4.   Hoechst's Development during the Great Depression
        (1929–1933/34)   52

3   Works Management, Workforce, and the National
    Socialist Party   59

    3.1.   From the Depression to the Nazi Seizure of Power   59

    3.2.   The New Rules in the Plant   70

    3.3.   The Works Management   76

        3.3.1.   Officer and "Leader" of the Plant: Ludwig Hermann   76

        3.3.2.   The Compromise Candidate: Carl
            Ludwig Lautenschläger   92

    3.4.   The NSDAP and Its Representatives in the Plant during
        the "Years of Peace"   108

    3.5.   "Nazification": Adjustment, Exclusion, and Political
        Persecution before the War   125

3.6. Not a "National Comrade": Jewish and Partly
      Jewish Employees   152

3.7. The Surrender of the Hoechst Management to the
      NSDAP during the War   184

3.8. The Rise of the "Crown Prince": Karl Winnacker   205

3.9. Not Part of the "Followers" in the War:
      The "Foreign Workers"   212

4   From Self-Sufficiency to War Production, Drugs, and
      Experiments on Human Beings   250

4.1. The Reorganization of the Plant under Ludwig Hermann
      and Friedrich Jähne   250

4.2. Research and Development   261

4.3. Production and Its Importance for Rearming and War   287

4.4. Drugs and Experiments on Human Subjects –
      The Pharmaceutical Department during the War   307

5   The Postwar Years: Dealing with the Past   337

5.1. The End of the War, Denazification, and the
      Nuremberg Trial   337

5.2. The Reestablishment of Hoechst and the "IG Family"   346

Bibliography   367
Index   381

Illustrations

Following Page 211

1. The Hoechst Works in August 1930 (Hoechst Archives)
2. Ludwig Hermann (Hoechst Archives)
3. Celebration of the 75th anniversary of Hoechst in January 1938 (Hoechst Archives)
4. Carl Ludwig Lautenschläger (Hoechst Archives)
5. Ludwig Retzinger (Bundesarchiv Berlin)
6. Hans Wagenheimer (Bundesarchiv Berlin)
7. Anniversary of Karl Ferdinand Blumrich in front of a "Hitler altar" (Hoechst Archives)
8. Franz Henle (Hoechst Archives)
9. Franz Henle's son Karl Henle (Hoechst Archives)
10. Hoechst's Pharmaceutical Research Laboratory in 1931 (Hoechst Archives)
11. Karl Winnacker (Hoechst Archives)
12. The arrival of female Eastern workers (*Ostarbeiterinnen*) at Hoechst during the War (Hoechst Archives)
13. The barracks of the Eastern workers – the so-called *Russenhof* (Hoechst Archives)
14. Georg Kränzlein (Hoechst Archives)
15. Hoechst's Pharmaceutical Bureau, headed by Julius Weber (Hoechst Archives)
16. Carl Ludwig Lautenschläger at the IG Farben Trial before the Nuremberg Military Tribunals (Hoechst Archives)

Abbreviations

AOG	Gesetz zur Ordnung der nationalen Arbeit 1934 (Law on the Organization of National Labor of 1934)
BA	Bundesarchiv (German Federal Archives)
BA-MA	Bundesarchiv-Militärarchiv (German Federal Archives – Military Branch)
BASF UA	BASF Unternehmensarchiv (company archives)
Bayer WA	Bayer Werksarchiv (business archives)
BDC	Berlin Document Center
Clariant WA	Clariant Werksarchiv (business archives)
HA	Hoechst Archives
HHStAW	Hessisches Hauptstaatsarchiv (Hesse State Archives)
IfZ	Institut für Zeitgeschichte, Munich
NI	Nuremberg document (industry)
PA	Personalakten/-ordner (personnel files)
PSW	(documents of) Personal- und Sozialwesen (Human Resources and Social Affairs)
TEA	(documents of the) Technical Committee of IG Farben
Wacker UA	Unternehmensarchiv (company archives) of Wacker-Chemie

Foreword

From 1925 to 1945 the I.G. Farbenindustrie AG was the largest non-state-owned corporation in Germany and by most indices the world's fourth largest such enterprise. Formed through the merger of the three preeminent chemicals firms on the European continent (BASF, Bayer, and Hoechst) with five smaller manufacturers, the company produced an immense array of goods, from dyes and pharmaceuticals to aluminum, fuel, and rubber, and its well-funded research operations added constantly to the total, achieving such lastingly valuable discoveries as sulfa drugs, magnetic tape, and a variety of synthetic fibers. But neither size nor inventiveness accounts for the firm's enduring name recognition. Some sixty years after the conquerors of the Third Reich ordered the dissolution of IG Farben, it remains infamously linked with Adolf Hitler's policies. In fact, it serves as the most common textbook example of the willingness of German big business to make common cause with barbarism.

More powerfully than any other industrial combine, IG Farben put its talents and capacities to the service of the Nazi program of armament, autarky, aggression, and annihilation. By 1943/44, the company and its numerous subsidiaries were supplying one-quarter of the artificial fibers, one-third of the fuel-from-coal, and all of the synthetic rubber (buna) on which the blockaded Reich's armed forces depended; most of the nation's munitions and the nitrogen they required; and virtually all of such indispensable substances as methanol, tetraethyl lead, and synthetic greases, along with the poison gases that Hitler kept in reserve. Foreign, often forced and sometimes enslaved, laborers made up one-half of the 333,000 personnel who provided these goods or worked on constructing new factories for that purpose. Among these were the approximately 30,000 concentration camp inmates killed while or after being compelled to help build IG Farben's installations and staff its coal mines in the vicinity of Auschwitz. Most sensationally, the combine owned 42.5 percent of the shares in the firm that owned the patents to Zyklon B, the pesticide used to murder approximately one million Jews at that death camp and at Majdanek. As a result, IG Farben seemed for decades – erroneously, we now know – to bear primary corporate responsibility for that product and its application.[1]

Of course, so notorious an enterprise has been the subject of many books, and until recently most of them were unscholarly. Indeed, the involvement of the combine's forerunners with chemical weapons production during World War I, as well as the firm's enormous market presence and putative power in and outside Germany during the 1920s, invited sensationalist accounts even before Hitler's accession. One of these presented IG Farben to American readers as "a monster camouflaged floating mine in the troubled sea of world peace"; another told Germans that the company constituted "the secret government of [their] republic."[2] Such conspiracy mongering acquired new impetus during World War II, especially in the United States, where the notion acquired currency that numerous cartel agreements between IG Farben and several large and vital American enterprises hamstrung military preparedness and mobilization.[3] Thus, an established image of the "Moloch IG" ensured that dismantling the corporation ranked high on the agenda of the victorious Allies in 1945. But neither this action nor the war crimes proceedings in 1947–48 against twenty-three leading executives of the firm stemmed the flow of publications about it. On the contrary, the split verdict of the American tribunal gave rise to a new, highly polarized round of writing. The judges voted 2 to 1 to acquit all the defendants on three of the five charges and ten on the remaining two as well; for the thirteen individuals convicted on one or two counts of the indictment, the maximum sentence came to eight years in prison, minus previous time in confinement, and all went free by 1951. As a result, during the ensuing thirty years, former prosecutors as well as spokespeople for communist East Germany, on the one hand, and erstwhile IG Farben employees, on the other hand, frequently retried the case in print, trading accusations and excuses in the guise of history.[4]

---

[1] On the true controlling corporate parent of Zyklon, see Peter Hayes, *From Cooperation to Complicity: Degussa in the Third Reich* (New York: Cambridge University Press, 2004), chapter 8.

[2] The quotations stem, respectively, from Victor Lefebure, *The Riddle of the Rhine* (New York: Chemical Foundation, 1923), p. 18 (of which 147,000 copies were in print by 1927), and Helmut Wickel, *I.-G. Deutschland* (Berlin: Der Bücherkreis, 1932), p. 5.

[3] See Joseph Borkin and Charles A. Welsh, *Germany's Master Plan* (New York: Duell, Sloan and Pearce, 1943), and, for summaries of these charges, Howard Ambruster, *Treason's Peace* (New York: Beechhurst Press, 1947), and Richard Sasuly, *IG Farben* (New York: Boni & Gaer, 1947).

[4] The principal works by former prosecutors are Josiah DuBois, *The Devil's Chemists* (Boston: Beacon Press, 1952), and Joseph Borkin, *The Crime and Punishment of I.G. Farben* (New York: Free Press, 1978). Representative of the East German publications are Willi Kling, *Kleine Geschichte der IG Farben – Der Großfabrikanten des Todes* (East Berlin: Dietz,

Only in recent decades has polemic given way to academic research on the nature, causes, and consequences of IG Farben's deeds during the Nazi era.[5] My own book on the subject argued that IG Farben pursued a largely reactive course toward Hitler's rise and rule.[6] Driven by long-standing dedication to massive investments in large-scale programs of chemical synthesis (initially dyes and nitrogen, later fuel and rubber from coal) and alternately bribed and bullied by the "carrot-and-stick" economic framework that Nazism created, IG Farben's leaders let themselves be transformed into agents of Hitler's expansionist and racist ideology. In the process, their firm's historic orientation around civilian and export markets gave way to concentration on domestic, primarily military, needs. The great growth and profitability that resulted between 1933 and 1945, and the attendant implication in spoliation of property and exploitation of human beings, reflected the workings of commercial and competitive calculations in a perverse context far more strongly than a putative identity of purposes between the giant combine and the NSDAP. According to my account, even the corporation's dreadful decisions to build a factory near Auschwitz, then to enlist slave laborers in that project, emerged from a chain of "rational" and somewhat defensive considerations, rather than from ideological zeal. That circumstance hardly mitigated the vicious effects of the firm's tunnel vision, however, or relieved its executives of responsibility for the horrors they worsened, including for the German people.

In 1990 Gottfried Plumpe published a *Habilitationsschrift* on the history of IG Farben, which adopted (though sometimes in distorted fashion and usually without acknowledgment) many of my findings but also departed from them in a way that proved both overdrawn and self-contradictory, yet fruitful nonetheless.[7] While insisting that the determinants of the firm's

1957), and Hans Radandt (ed.), *Fall 6: Ausgewählte Dokumente und Urteil des IG-Farben-Prozesses* (East Berlin: Deutscher Verlag der Wissenschaften, 1970). Among the defenses offered by former officials of Farben are Fritz ter Meer, *Die I.G. Farben Industrie Aktiengesellschaft* (Düsseldorf: Econ, 1953); August von Knieriem, *The Nuremberg Trials* (Chicago: Regnery, 1959) [*Nürnberg, rechtliche und menschliche Probleme* (Stuttgart: E. Klett, 1953)]; Curt Duisberg, *Nur ein Sohn* (Stuttgart: Seewald, 1981); and Heinrich Gattineau, *Durch die Klippen des 20. Jahrhunderts* (Stuttgart: Seewald, 1983). Also apologetic in tone are two biographies of Farben's founding fathers – Karl Holdermann, *Im Banne der Chemie. Carl Bosch, Leben und Werke* (Düsseldorf: Econ Verlag, 1953), and Hans-Joachim Flechtner, *Carl Duisberg – vom Chemiker zum Wirtschaftsführer* (Düsseldorf: Econ Verlag, 1959) – and Werner Reichelt, *Das Erbe der I.G. Farben* (Düsseldorf: Econ Verlag, 1956).

[5] The first, though unfortunately false, start in this direction was Hellmuth Tammen, *Die I.G. Farbenindustrie Aktiengesellschaft (1925–1933)* (Dissertation, Freie Universität Berlin, 1978).

[6] Peter Hayes, *Industry and Ideology: IG Farben in the Nazi Era* (New York: Cambridge University Press, 1987; new edition, 2001).

[7] Gottfried Plumpe, *Die I.G. Farbenindustrie AG: Wirtschaft, Technik und Politik 1904–1945* (Berlin: Duncker & Humblot, 1990). For a discussion of some of the errors and deficiencies of this book, see Peter Hayes, "Zur umstrittenen Geschichte der I.G. Farbenindustrie AG,"

behavior in Nazi Germany were almost exclusively internal and economic, rather than external and political, Plumpe also maintained that the Third Reich, despite the riches it brought IG Farben, retarded its long-term commercial position by discouraging research and development in such later rewarding fields as polymers for plastics and synthetic fibers. Logically inconsistent as these claims were, the latter one drew attention to a hitherto relatively underemphasized aspect of IG Farben's conduct between 1933 and 1945, namely, its disadvantages to the firm itself. Plumpe thus underlined the degree to which IG Farben's managers failed to defend their own and their stockholders' interests, as well as civilized norms.

The most recent contributions to the literature on IG Farben have moved away from examining the combine as a whole from the perspective of its boardroom in order to scrutinize in detail what happened within sectors of the mammoth enterprise: the plants of BASF, one of the predecessor and successor firms; the former Hüls AG, founded in 1938 as a subsidiary; the Monowitz factory near Auschwitz from 1941 to 1945; and, in the case of this book, the main installations of Hoechst AG, another of the entities that formed and emerged from the IG Farben concern. Nonetheless, like my and Plumpe's works, the studies of BASF and Hüls traced and analyzed managerial actions and overall trends but could not allot, because of gaps in the surviving archival sources, much attention to the biographies and attitudes of the individuals involved.[8] Bernd Christian Wagner's work concentrated more tightly on people, both those who ran and were subjected to Monowitz, but in a short time frame and under the most extreme conditions.[9] Despite the great merits of all three studies, none accomplished Stephan Lindner's feat of bringing readers, as his English-language title says, "inside IG Farben" during the entire span of its existence.

Thanks to Lindner's pathbreaking and painstaking research and the unusual circumstance among corporate archives that the files of the former Hoechst are richer in personal records and recollections than, for example, statistical data, readers can now discern more precisely than ever before how deeply Nazi ideology penetrated the thoughts and actions of

Geschichte und Gesellschaft 18 (1992), pp. 405–17, and the same author's review in *Business History Review* 65 (1991), pp. 1020–3.

[8] Raymond G. Stokes, "From the IG Farben Fusion to the Establishment of BASF AG (1925–1952)," in Werner Abelshauser, Wolfgang von Hippel, Jeffrey Allan Johnson, and Raymond G. Stokes, *German Industry and Global Enterprise. BASF: The History of a Company* (New York: Cambridge University Press, 2004), pp. 206–361, and Bernhard Lorentz and Paul Erker, *Chemie und Politik: Die Geschichte der Chemischen Werke Hüls 1938–1979* (Munich: C.H. Beck, 2003).

[9] Bernd Christian Wagner, *IG Auschwitz: Zwangsarbeit und Vernichtung von Häftlingen des Lagers Monowitz 1941–1945* (Munich: K. G. Saur, 2000).

Hoechst's leaders and molded life within its plants from 1933 to 1945. To be sure, Lindner's examination of the *tertius inter pares* of the founding firms of IG Farben devotes considerable attention to explaining why Hoechst occupied that status from 1914 to 1945 and how successive chief executives struggled to improve upon it. But his signal contributions to the literature on IG Farben lie in his exposure of the leading National Socialists within the Hoechst division, the extent to which its senior and mid-level managers shared or adopted and then acted upon the party's ideology, the depth and duration of corporate complicity in such crimes as forced labor and drug testing on concentration camp inmates by the SS, and the lengths to which the refounded firm's leaders went after the war to aid colleagues compromised by their deeds and ties to Nazism while showing scant human sympathy for either former slave laborers or persecuted employees. Over and over, Lindner confronts us with the worldview of the people who orchestrated Hoechst's collaboration with the Nazi regime – not only the chemists and managers, but also the party's representatives in the factory administration, and not just before 1945, but after, when such people, Lindner shows, repeatedly lied (the word is not too strong!) about their and the firm's earlier conduct. Moreover, he directs our attention to what these individuals recurrently ignored: the fates of Jewish and other personnel who were harmed by such actions. The result is one of the most penetrating and chilling accounts of the interaction of commerce and corruption in the Third Reich and its aftermath that has been published to date.

Dr. Lindner's findings cast dark shadows on many reputations, which makes the support of the former Hoechst AG for his book all the more worthy of note and praise. Deserving of special mention is the courageous commitment of Jürgen Dormann, first as chairman of the Managing Board of Hoechst AG, then chairman of the Managing Board of Aventis SA, the firm into which Hoechst merged in 1999, to the project. He assured the author – and myself as advisor to the work – of complete cooperation on the part of Hoechst's departments, full access to relevant records in the firm's possession, and total independence in preparing this text for publication, and, to the best of our knowledge, these promises have been kept. Also indispensable have been the inexhaustible efforts of Wolfgang Metternich, whose manifold responsibilities include managing and extending Hoechst's archive. He unearthed, gathered, and preserved many illuminating, previously uncataloged materials for this study; arranged several interviews with surviving family members and other witnesses to the events described here; and provided numerous stimulating suggestions for research. Without the aid of these two individuals, as well as countless other longtime associates of Hoechst AG, this book could not have come into being.

Stephan Lindner's unflinchingly forthright and honest account of Hoechst in the Nazi era will require students of IG Farben, including me, to reconsider and revise their interpretations. This book thus constitutes an important advance on our knowledge and a notable contribution to Germany's and German industry's continuing confrontation with its past.

Peter Hayes

Acknowledgments

When finishing a book one of the pleasant tasks is that of acknowledging the debts one owes to the people who gave their assistance on the long journey to completion. I have seldom received so much support as I did for this study, and listing all the people who helped me by name would far exceed the scope of these acknowledgments. I can therefore include only a few representative names of some of the people to whom I owe particular thanks.

My thanks go, in the first instance, to the former Hoechst AG, whose board agreed to finance the work for this study, granting free and full access to its documents, and renouncing any right of review over what I would write. I wish in particular to thank Jürgen Dormann, the former chairman of the board of Hoechst AG, later Aventis SA, for his great interest in this study and his unwavering support. I am also grateful to the Aventis Foundation for supporting the American edition of the book. My grateful thanks go to the staff of HistoCom who are responsible for the archives of Hoechst AG and who were always most supportive: Ursula Kotsch, Manuela Kuhl, Walter Molsberger, and Erwin Weishäupl. The head of the archives, Wolfgang Metternich, deserves special mention, since he supported the project and my work on it much more than one could expect from a business's archivist.

Peter Hayes was my most important advisor for this study right from the start, as a scholar and as a friend. His contribution to this book can hardly be overestimated; I am immensely indebted to him. I would also like to thank the other members of the research advisory committee: Gerald Feldman, Raymond Stokes, Jakob Tanner, and Ulrich Wengenroth. The book benefited enormously from their help and constructive criticisms. The reports of two anonymous referees were also extremely helpful, and I wish to thank them for their suggestions and the trouble they took. They commended the translation, which was carried out by Helen Schoop. She performed a difficult task very well – and deserves very special thanks. I also wish to express thanks to my editor at Cambridge University Press, Frank Smith, for the pains he took – and for his patience.

Finally, I owe thanks to Dieter Degreif and Hans-Hermann Pogarell and their colleagues in the other archives I consulted; they were all very generous and supportive. I was also permitted to make use of private papers, and here I wish to express my particular thanks to Anna Ercoli-Schnitzer, Oskar Henle, and Elisabeth Lautenschläger. Elisabeth Vaupel, Walter Wetzel, and my brother Michael Lindner very kindly assisted me in questions pertaining to chemistry, pharmaceutics, and medicine. I am also grateful for the support of my colleagues from my former workplace, the Institute for the History of Technology at the Technical University Munich.

Last but not least I wish to express my deep thanks to my wife, Sigrid, and our two children, Franca and Thomas, for their support and patience during the past few years.

I dedicate this book to the children of victims of the Nazi regime, in particular to the children of persecuted employees at the IG Farben plant at Hoechst. During my work on this study it became very obvious to me that they suffered no less from the persecution to which their parents were subjected; yet they are so often passed over in silence.

It goes without saying that I alone must be held responsible for any factual mistakes or misinterpretations in this book.

I

Introduction

At the end came disgrace. The history of the I.G. Farbenindustrie lasted only twenty years. It had been founded on December 9, 1925, with the recording of the merger contract between the German chemical companies Bayer, BASF, Agfa, Griesheim-Elektron, Weiler-ter Meer, and Hoechst. On the basis of the Allies' Control Commission Law No. 9 of November 30, 1945, the I.G. Farbenindustrie was seized and its dissolution planned. The decision taken by the victorious Allied powers stemmed not just from the company's entanglement in the crimes of the Nazi regime, but was a result of its active participation in these. "IG Farben" stood and stands for autarky, armament, exploitation, and – Auschwitz.

In the U.S. occupation zone, where the IG Farben plant Hoechst was located, the American military commander-in-chief had already taken over the management and control of IG Farben and its capital on July 5, 1945. Finally, at the end of July 1948, twenty-three former managers of IG Farben were tried before an American military tribunal in Nuremberg. All of them were found innocent on charges number one, four, and five (planning, preparation, initiation, and waging of wars of aggression; membership in the SS; and conspiracy). Ten managers also were acquitted of the two other charges. However, thirteen managers were sentenced to prison terms ranging from eighteen months to eight years (minus their time spent in custody), on charges number two (plunder and spoliation) and three (enslavement and murder of civilians, prisoners of war, and concentration camp inmates). The executives Otto Ambros and Walter Dürrfeld, who were directly involved in the construction of the IG Farben plant at Auschwitz, were both sentenced to the longest terms. Carl Ludwig Lautenschläger, the managing director of the IG Farben plant Hoechst, was one of the ten declared innocent on all counts. His deputy, Friedrich Jähne, was sentenced to eighteen months for plunder and spoliation.[1]

---

[1] See *Trials of War Criminals before the Nuremberg Military Tribunals under Control Council Law No. 10, vols. 7–8: The Farben Case* (Washington, DC: U.S. Government Printing Office, 1952–3).

Although the prosecutors in Nuremberg found the verdict of the American judges too mild, the defenders and the accused felt it was too harsh and considered the condemned men "victims" of "victors' justice," since the managers of IG Farben had only done their "patriotic duty." In March 1953, on the occasion of the first Extraordinary General Assembly of the newly founded Farbwerke Hoechst AG, the chairman of the board, Karl Winnacker, who had already been selected to manage the Hoechst works by the former management of the I.G. Farbenindustrie, expounded on this point, when he mentioned the Allies' decision to dissolve IG Farben: "Almost unnoticed by the majority of the people, who were preoccupied with the fight for their existence, this meant the destruction of the most important undertaking that German science and technology as well as German enterprise had ever developed." He continued: "In a trial fought with great bitterness, the responsible leaders of the I.G. Farbenindustrie AG, and thus our entire company, were able to rebut all the charges of war crimes, robbery and plunder levied against them."[2] Not a word was said about the conviction of thirteen leading managers by the Nuremberg judges, who, it should be emphasized, consistently followed the motto *"in dubio pro reo"* throughout the trial. Thus, Lautenschläger and his board colleagues had been cleared of participating in human experiments in concentration camps, since the tribunal pronounced that "where from credible evidence two reasonable inferences may be drawn, one of guilt and the other of innocence, the latter must prevail."[3]

Because of its entanglement in and involvement with the crimes of the National Socialist regime, the history of IG Farben during the Third Reich has been examined in numerous books and articles, but its role has been assessed in very different ways. Thus, a number of authors are convinced that IG Farben was much guiltier than the ruling of the Nuremberg Tribunal indicated. The most prominent of these is Joseph Borkin, who happened also to be one of the Nuremberg prosecutors. Accordingly, his book reads in parts like a public prosecutor's final address to the jury.[4] However, a number of other authors have come to a much more positive

[2] Karl Winnacker, quoted in Ernst Bäumler, *A Century of Chemistry* (Düsseldorf: Econ, 1963), p. 111.

[3] *Trials of War Criminals*, p. 1172; cf. in contrast the Dissenting Opinion of Judge Hebert who considered the accused managers guilty under count 3, pp. 1204–5 and 1307–25.

[4] Joseph Borkin, *The Crime and Punishment of the I.G. Farben* (New York: Free Press, 1978); also Josiah E. DuBois, *The Devil's Chemists. 24 Conspirators of the International Farben Cartel Who Manufacture Wars* (Boston: Beacon Press, 1952); cf. also historians of the (former) German Democratic Republic such as Dietrich Eichholtz, *Geschichte der deutschen Kriegswirtschaft 1939–1945*, vol. 1: *1939–41*, 3. durchgesehene Auflage (Berlin [East]: Akademie, 1984), pp. 59–61; idem, "Die IG-Farben-'Friedensplanung.' Schlüsseldokumente der faschistischen 'Neuordnung des europäischen Grossraums,'" *Jahrbuch für Wirtschaftsgeschichte 1966*, part 3, pp. 271–332.

assessment of IG Farben's role. Unsurprisingly, the most prominent representatives of this viewpoint are Fritz ter Meer, a former managing
director of IG Farben, and Gottfried Plumpe, a historian and executive of
the IG Farben successor firm Bayer. Plumpe considered the managers of IG
Farben as unhappily compromised through the conditions existing at the
time and by their professional dedication to their entrepreneurial and
managerial functions.[5] Finally, there is the balanced and carefully argued
account by the American historian Peter Hayes on IG Farben during the
Third Reich. It shows the responsibility, the guilt, and the entanglement of
IG Farben's management without extenuation. But it also highlights the
limited options of a company embedded in a state-regulated economy under
an increasingly totalitarian regime.[6]

IG Farben and its managers obviously did not determine the fundamental
political and economic conditions of Nazi Germany, but acted within them.
Moreover, IG Farben was neither influential in aiding Hitler's takeover
nor among his "stirrup holders." The "primacy of politics" was Nazi
Germany's hallmark, that is, political premises determined the course of the
economy.[7] However, as shown by Hayes, the managers of IG Farben were
only too ready to adapt to the new conditions and to take advantage of
them. The English historian Tim Mason, who decisively contributed to
launching the concept of the primacy of politics in Nazi Germany, therefore
reached the following conclusion about the relationship between the Nazi
state and the German economy: "The fact that numerous industrialists not
only passively co-operated in the 'Aryanization' of the economy, in the
confiscation of firms in occupied territory, in the enslavement of many
million people from Eastern Europe and in the employment of concentration camp prisoners, but indeed often took the initiative in these actions,
constitutes a damning judgement on the economic system whose essential
organizing principle (competition) gave rise to such conduct. But it cannot
be maintained that even these actions had an important formative influence

---

[5] Fritz ter Meer, *Die I.G. Farbenindustrie Aktiengesellschaft. Ihre Entstehung, Entwicklung und Bedeutung* (Düsseldorf: Econ, 1953); Gottfried Plumpe, *Die I.G. Farbenindustrie AG. Wirtschaft, Technik, Politik 1904–1945* (Berlin: Duncker & Humblot, 1990); Plumpe has been heavily criticized for abridging and doctoring a quotation on the erecting of IG Farben's plant in Auschwitz, and for not giving that plant the appropriate attention, as this added to the book's general apologetic tendencies: Peter Hayes, "Zur umstrittenen Geschichte der I.G. Farbenindustrie AG," *Geschichte und Gesellschaft* 18 (1992), pp. 405–17, and Bernd Wagner, *IG Auschwitz. Zwangsarbeit und Vernichtung von Häftlingen des Lagers Monowitz 1941–1945* (Munich: Saur, 2000), pp. 11–12.

[6] Peter Hayes, *Industry and Ideology. IG Farben in the Nazi Era*, 2nd ed. (Cambridge: Cambridge University Press, 2001), first published in 1987.

[7] Cf. Tim Mason, "Der Primat der Politik – Politik und Wirtschaft im Nationalsozialismus," *Das Argument* 8 (1966), pp. 473–94; Henry A. Turner, "Unternehmen unter dem Hakenkreuz," in *Unternehmen im Nationalsozialismus*, ed. by Lothar Gall and Manfred Pohl (Munich: Beck, 1998), pp. 15–23.

on the history of the Third Reich; they rather filled out in a barbaric manner a framework which was already given."[8]

In a totalitarian regime such as the National Socialist one, a company of any military importance had to choose between adapting or running the risk of being taken over by competitors or by the state. Given the internal logic and functioning of a company, this could hardly be the aim of its managers, and the latter option was not just an empty threat, as prewar confiscations in the German steel industry had proved.[9] One therefore has to question how the executives of a company could act, when the function and logic of a firm were thus challenged. As pointed out by the economist Edith Penrose, firms have a very specific function in a free economy. According to her definition, a firm is "a collection of resources bound together in an administrative framework," and the function of an industrial firm – hence also of IG Farben – is to acquire and organize these human and other resources "in order profitably to supply goods and services to the market."[10] Or to quote the historian Raymond Stokes: "Firms exist to make money."[11]

But even if the logic of entrepreneurial action and what historians Lutz Budraß and Manfred Grieger call the "morality of efficiency" is taken into consideration, one cannot ignore valid ethical norms.[12] However, as historians Dietmar Petzina and Werner Plumpe wrote, when important parts of the "common sense morality," which also operated during the National Socialist period, were called into question by a "politically determined and sanctioned legal morality" or even partially invalidated, then companies were caught up in "ethical dilemmas" that as "social systems embedded in specific environmental conditions," they could hardly hope to solve. The common strategy was therefore often to retain an individual "common

---

[8] Mason, "Primat der Politik," p. 490.

[9] Cf. Gerhard Th. Mollin, *Montankonzerne und 'Drittes Reich,' Der Gegensatz zwischen Monopolindustrie und Befehlswirtschaft in der deutschen Rüstung und Expansion 1936–1944* (Göttingen: Vandenhoeck & Ruprecht, 1988), pp. 102–9; for Mollin the "primacy of private property" is limited to the production only (p. 277); cf. also Rainer Zitelmann, *Hitler. Selbstverständnis eines Revolutionärs* (Hamburg: Berg, 1987), pp. 241–64.

[10] Edith Penrose, *The Theory of the Growth of the Firm*, 3rd ed. (Oxford: Oxford University Press, 1995), p. xi. The adverb "profitably" is missing in many definitions, though this is a basic prerequisite for the survival of companies: cf. Werner Plumpe, "Unternehmen," in Gerhard Ambrosius, *Moderne Wirtschaftsgeschichte. Eine Einführung für Historiker und Ökonomen* (Munich: Oldenbourg, 1996), p. 47, or Toni Pierenkemper, *Unternehmensgeschichte. Eine Einführung in ihre Methoden und Ergebnisse* (Stuttgart: Steiner, 2000), p. 18.

[11] Raymond G. Stokes, *Divide and Prosper. The Heirs of I.G. Farben under Allied Authority* (Berkeley: University of California Press, 1988), p. 25.

[12] Lutz Budrass and Manfred Grieger, "Die Moral der Effizienz. Die Beschäftigung von KZ-Häftlingen am Beispiel des Volkswagenwerks und der Henschel Flugzeug-Werke," *Jahrbuch für Wirtschaftsgeschichte* 1993, pp. 89–136.

sense morality" in the private realm and to respect the legally fixed standards in functional domains as a means of coming to terms with a paradoxical situation. This strategy seemed to be the appropriate path for many companies as well as for the social groups represented in them.[13] Still, it remains important to explore how far, while complying with the economic and political framework provided by the Nazi regime, entrepreneurs and managers tried to uphold "common sense morality" in the functional realm. For this stance was capable of being aligned with the "morality of efficiency," since, as Budraß and Grieger have suggested, it was double-edged and could contribute to either a better treatment of forced laborers or their "annihilating exploitation."[14] Thus the fundamental question is whether the Hoechst managers endeavored to preserve a "minimum of compassion" – to use the expression coined by the historian Avraham Barkai – or if they proved themselves to be "active and acquiescent aides" to the Nazi regime, while pursuing their entrepreneurial aims.[15]

A number of resistance fighters against National Socialism emerged in the military – an arena where obedience has always been an important, career-enhancing principle. Yet, amazingly little resistance arose within the economic realm, where "virtues such as freedom, initiative, and entrepreneurial daring were always held in high esteem," as historians Lothar Gall und Manfred Pohl put it.[16] This does not mean that there was no resistance among entrepreneurs and managers. Prominent examples include Oskar Schindler and Berthold Beitz, who both tried to save Jews, or Eduard Schulte, who broke the silence and exposed the Final Solution.[17] But these men were exceptions. To paraphrase Tim Mason's crucial question: Was it competition – as a central feature of the economic system – that made managers think only in business terms and repress any moral qualms?[18]

Between 1925 and 1945 IG Farben was one of the biggest firms in Europe, if not in the world, in terms of both employees and turnover. It had an influential role in Hitler's autarky and armament program, a role that paid off substantially. Indeed, between 1933 and 1945, IG Farben's net profit increased fivefold. In 1943 IG Farben owned no fewer than 334

---

[13] Dietmar Petzina and Werner Plumpe, "Unternehmensethik – Unternehmenskultur: Herausforderungen für die Unternehmensgeschichtsschreibung?" *Jahrbuch für Wirtschaftsgeschichte* 1993, pp. 12–13.

[14] Budrass and Grieger, "Moral der Effizienz," pp. 135–6.

[15] Avraham Barkai, "Die 'stillen Teilhaber' des NS-Regimes," in *Unternehmen*, pp. 117–20.

[16] Lothar Gall and Manfred Pohl, "Einleitung," in *Unternehmen*, p. 12.

[17] Hayes, *Industry and Ideology*, pp. 365–6; on Schindler, cf. Steven Spielberg's movie *Schindler's List*; on Schulte, cf. Walter Laqueur and Richard Breitman, *Breaking the Silence* (New York: Simon and Schuster, 1986); on Beitz, cf. Thomas Sandkühler, *'Endlösung' in Galizien. Der Judenmord in Ostpolen und die Rettungsinitiativen von Berthold Beitz* (Bonn: Dietz, 1996).

[18] Cf. Hayes, *Industry and Ideology*, pp. 379–81; Mark Spoerer, "Die Automobilindustrie im Dritten Reich," in *Unternehmen*, pp. 66–8.

factories in territories controlled by the Nazi regime, it had acquired shares
in almost 1,000 firms worldwide, and its balance sheets boasted assets
worth over 3 billion Reichsmark. A large part of this significant growth,
Peter Hayes wrote, "materialized during the war years at the expense of
victims of National Socialism." IG Farben bought Jewish and foreign
property in the "great economic sphere" (*Großwirtschaftsraum*) controlled
by Nazi Germany, and it employed an increasing number of foreign
workers and concentration camp inmates, who in 1944 made up at least
40 percent of its employees.[19]

During IG Farben's early years, the "founders' compromise" between the
three "big sisters" meant that all the boards of the individual firms were
consolidated into the IG Farben board. Moreover, the "founder firms"
were able to retain a vast degree of autonomy in the name of the rather
curious managerial concept of "decentralized centralization." Within IG
Farben, as Peter Hayes wrote, "it was known rather more aptly, with a nod
to another composite of pyramided dynastic unions, proud local magnates,
and heterogeneous dominions as the 'Habsburg model.'"[20] In my study,
reflecting this "Habsburg model," the main focus will not be on the IG
Farben headquarters, and IG Farben will not be discussed as a whole.
Instead, I will be examining a "province": the IG Farben plant Hoechst.
Looking beyond the structures of the whole company described and analyzed by Hayes and Plumpe, I will attempt to penetrate inside IG Farben
and to submit one of the plants of this mammoth corporation to a more
precise analysis.

Thus, this study is not strictly speaking a business history, but only the
history of a plant. The joint stock company "Farbwerke, vormals Meister
Lucius und Brüning," which had been independent until the end of 1925,
was no longer an independent company during the period under consideration. Instead, it was part of a large chemical corporation, the I.G.
Farbenindustrie AG. The reader can thus expect a history of the IG Farben
plant Hoechst, which in turn consisted of various components, like the Dye
or Acetic Acid Departments (which might also be called factories), as well
as laboratories. The reader should be aware that the concepts of "plant"
and "factory" are often used synonymously, but "factory" is also used for
certain departments within the plant. Furthermore, the concepts of "plant,"
"factory," "company," and "firm" are sometimes used synonymously, as,
for example, in the *Betriebsordnung* in the Third Reich, which can be
translated as "company regulations" or "plant regulations." Likewise, in

---

[19] Hayes, "Industrie," p. 124; according to Hayes, in 1944, 46 percent of the staff were
foreigners, prisoners of war, and concentration camp inmates; Plumpe, *I.G.
Farbenindustrie*, p. 629, gives a figure of a little under 40 percent for 1944.

[20] Hayes, *Industry and Ideology*, p. 19.

Nazi Germany the director of a company and/or a plant was also referred to as the *Führer des Betriebes*.[21]

Siegfried Kracauer divided the "structure of the historical universe" into two main groups, the micro- and macro-histories, with fuzzy boundaries between the two. According to Kracauer, readers do not learn enough about the past when a historian focuses on "the macro-units," or "big clusters of events." These cannot capture the entire "wealth" of history, which only presents itself in a "micro-dimension": "Therefore, attentive historians who aspire to a rich history favor a mutual permeation of macro- and micro-history."[22] It is the aim of this study to combine the macro-history of IG Farben during the Third Reich with the micro-history of the Hoechst plant. This attempt thus differs from most extant business studies focusing on the Third Reich in general and on IG Farben in particular, almost all of which have limited their investigations to the macro-level. The exceptions are Bernd Wagner's study on the IG Farben plant at Auschwitz and, to some extent, Raymond Stokes's chapter on BASF while it was part of IG Farben, which is embedded in a general history of BASF.[23]

Thus, the aim of this study is to provide a plant history of the IG Farben plant Hoechst on the basis of the available literature on IG Farben and the documents in the archives, while referring back to the whole corporation during the Third Reich. The main questions I attempt to answer are these: How far was Hoechst linked to the Nazi regime, its representatives and organizations, and to what extent was it entangled or even actively involved in their crimes? To paraphrase Tim Mason: How barbarically did

[21] See, for example, the term used within IG Farben for the respective combines affiliated to the three "big sisters," which were referred to as *Betriebsgemeinschaften*, which translates as "works groups" or "plant communities." For the Third Reich, cf. the contemporary legal publication of Felix Rössler, *Der Führer des Betriebes (insbesondere: Die Rechtsnatur der Betriebsgemeinschaft und des Führeramts)*, Schriften des Instituts für Wirtschaftsrecht 13 (Jena: G. Fischer, 1935).

[22] Siegfried Kracauer, *Geschichte – Vor den letzten Dingen*, Schriften 4 (Frankfurt am Main: Suhrkamp, 1971), pp. 103–32, especially pp. 103–4 and 115–19; cf. also Jürgen Schlumbohm "Mikrogeschichte – Makrogeschichte: Zur Eröffnung einer Debatte," in idem (ed.), *Mikrogeschichte – Makrogeschichte: komplementär oder inkommensurabel?*, 2nd ed. (Göttingen: Wallstein, 2000), pp. 7–32.

[23] Wagner, *IG Auschwitz*; Raymond G. Stokes, "From the IG Farben Fusion to the Establishment of BASF AG (1925–1952)," in Werner Abelshauser et al., *German Industry and Global Enterprise, BASF: The History of a Company* (Cambridge: Cambridge University Press, 2002), pp. 206–361: Stokes's discourse on the IG Farben period is limited, as his contribution had to integrate with the overall project on the history of BASF and its specific focus – cf. J. Steven Jones, "Review of Werner Abelshauser, Wolfgang von Hippel, Jeffrey Allan Johnson, and Raymond G. Stokes, German Industry and Global Enterprise BASF: The History of A Company," Economic History Services, May 14, 2004, http://www.eh.net/bookreviews/library/0785.shtml (last retrieved: November 20, 2005), and Raymond G. Stokes, "EH.Net Book Review: Author's Response," http://www.eh.net/bookreviews/response/Stokes.htm (last retrieved: November 20, 2005).

Hoechst comply with the framework provided by IG Farben and Nazi Germany? The focus is thus not just on individual entanglements or even the complicity of persons affiliated with the factory such as personnel or management. The institutional role of IG Farben and Hoechst is also discussed, and I therefore analyze its attitude toward various groups of employees, as well as toward production and research under the auspices of armaments production and war.

To understand the role of the IG Farben plant at Hoechst during the Third Reich, it is crucial to examine the position of the plant within IG Farben. A first step consists in showing Hoechst's path into and under the I.G. Farbenindustrie during the German Empire and the Weimar Republic. The main question centers around the role played by Hoechst in the merger and its position within IG Farben during the founding period in and after 1925. My aim is to show how Hoechst was integrated in IG Farben and the effects of the first reforms and rationalization measures undertaken during the 1920s, especially as a result of the Great Depression. Hoechst emerged much weaker from these events, as is particularly obvious from a review conducted by the plant manager in 1933/34, which revealed the problems that the plant had to contend with at the beginning of the Nazi regime.

In the main part of the study, I attempt to describe and analyze the interaction of the NSDAP, the plant management, and personnel. I therefore present the plant regulations that were enforced during the Third Reich, since they redefined the relations between plant management and personnel and made them both accountable to the new regime. From 1934 onwards plant directors enjoyed a particularly central role as "factory (or company) leaders" (*Führer des Betriebes*). I thus explore the principal values that governed their behavior and their work and the relationship they cultivated with National Socialism as well as with the Nazi regime. Special emphasis is placed on the nazification of the plant during the so-called peaceful years and the capitulation of the plant management to the NSDAP and its organizations during the war. In the process I will also describe and analyze the role and importance of the leading representatives of the NSDAP in the factory, as well as the adaptation and exclusion processes among employees – characteristically renamed "the followers." Moreover I will examine the persecution of employees who were Jews or defined as such by the Nazi regime, as well as the treatment of so-called foreign workers (*Fremdarbeiter*). In this context the increasingly totalitarian course of the Nazi regime often makes it difficult for a historian to distinguish between coercion and consent, that is, agreement with the prevailing system.[24] Nonetheless, the readiness to adapt to the Nazi system displayed

[24] See the subhead of Robert Gellately's *Backing Hitler. Consent and Coercion in Nazi Germany* (Oxford: Oxford University Press, 2001); cf. also Ralf Dahrendorf, *Gesellschaft und Demokratie in Deutschland* (Munich: R. Piper & Co., 1966), pp. 434–42.

by both the plant management and the executives in Hoechst was quite remarkable.

After this section, which mainly focuses on individuals and their social relationships, I examine research and development as well as production in the plant during the "peacetime years" and during World War II. On the basis of the already mentioned review of 1933/34, I attempt to elucidate the aims and actions of the plant management and how these contributed to making the company lucrative and strong again. These aims and actions include a wide range of activities, from construction operations to research goals. Moreover, I assess to what extent Hoechst's weakened position actually turned out to be an advantage by providing a spur to innovation. The research carried out in the "Process Engineering Department" and the close cooperation between industry and academia in the field of plastics were exemplary in this respect. I analyze the evolution of the most important areas of production in Hoechst during the entire Nazi period, while trying to assess the importance of Hoechst's role for the war effort. A special part of this chapter reviews the history of the Pharmaceutical Department, where R&D united both humaneness and monstrosity. As Hoechst worked on drugs to diminish individual suffering, it also took part in ghastly experiments and trials conducted on concentration camp inmates, even though the person responsible for this area in Hoechst played an active role in the Catholic resistance movement against the Nazi regime.

The final section covers how the Nazi past was dealt with in the first postwar years. Under the American administration, many "politically implicated" individuals were removed from their leading positions or fired in the course of denazification. Finally, the Nuremberg trial against the IG Farben management, during which both Hoechst's plant director and his deputy had to justify themselves, represented an acme in the confrontation of the plant with its Nazi past. In the reestablished Hoechst, such political measures ceased at the beginning of the 1950s. From that point on, Karl Winnacker, the new managing director, endeavored to subvert the outcome of the trials: his solidarity was clearly with the so-called victims of denazification, not with victims of the Nazi regime.

In recent years the secondary sources for a company or plant history of Hoechst during the Third Reich have increased significantly. A number of scholarly studies of companies involved in various fields during that period have described and analyzed the dealings of these companies, paying particular attention to their motives and the options open to them.[25] Moreover,

---

[25] For example, Stokes, "IG Farben Fusion"; Gerald D. Feldman, *Allianz and the German Insurance Business, 1933–1945* (Cambridge: Cambridge University Press, 2001); Peter Hayes, *From Cooperation to Complicity. Degussa in the Third Reich* (Cambridge: Cambridge University Press, 2004); Harold James, *The Deutsche Bank and the Nazi Economic War against the Jews: The Expropriation of Jewish-Owned Property*

recent studies on science and technology during the Third Reich make it easier to prepare a book on the chemical-pharmaceutical industry. In particular, several new publications on the pharmaceutical industry and on human experiments during the Third Reich have also explored the role of IG Farben and even of Hoechst.[26] Hoechst and some of its veterans have also published a number of works, none of them very self-critical.[27]

At the beginning of this project, the array of available primary sources that could illuminate the interpenetration of the macro- and micro-historical levels was not very satisfactory. Whereas the archives of the IG Farben trial in Nuremberg provided a good basis for describing events on the IG Farben level, as Hayes and Plumpe did, they offered little material for a detailed study at the plant level. Even the rather well-organized IG Farben files, which were subsequently allocated to the archives of the successor companies BASF, Hoechst, and Bayer, were not an adequate basis for an authentic plant history. However, in the course of my work, the source base improved significantly, a factor that also contributed to the length of time required for the project. Files from the management office of the former Hoechst AG on denazification in the factory as well as on foreign workers were transferred to the archives. Rich files from the Human Resources Department were found in Hoechst's basement. Among them were the employment books (*Diensteintrittsbücher*), which provide considerable information on the proportion of foreign workers, as well as details on their treatment.[28] The personnel files (*Personalakten*) of the so-called Department VI, which dealt with Hoechst's executives, were also discovered there. The perusal of these sources considerably broadened the available knowledge on the factory and its executives. Moreover, in these files I also found allusions to individuals whom I could then trace back to sources in other

(Cambridge: Cambridge University Press, 2001); Hans Mommsen and Manfred Grieger, *Das Volkswagenwerk und seine Arbeiter im Dritten Reich* (Düsseldorf: Econ, 1996).

[26] Among others, Wilhelm Bartmann, *Zwischen Tradition und Fortschritt. Aus der Geschichte der Pharmabereiche von Bayer, Hoechst und Schering von 1935 bis 1975* (Stuttgart: Steiner, 2003); Ute Deichmann, *Flüchten, Mitmachen, Vergessen. Chemiker und Biochemiker in der NS-Zeit* (Weinheim: Wiley-VCH, 2001); Ernst Klee, *Auschwitz, die NS-Medizin und ihre Opfer* (Frankfurt am Main: S. Fischer, 1997); Paul Weindling, *Epidemics and Genocide in Eastern Europe, 1890–1945* (Oxford: Oxford University Press, 2000).

[27] Karl Winnacker, *Challenging Years: My Life in Chemistry* (London: Sidgwick & Jackson, 1972); Ernst Bäumler, *Die Rotfabriker. Familiengeschichte eines Weltunternehmens* (Munich: Piper, 1988); Bäumler, *Century*.

[28] Previously the only available data on foreign workers were their number at the end of the war, as quoted, for example, in Bäumler, *Rotfabriker*, pp. 313–14. According to Dr. Wolfgang Metternich, director of the Hoechst Archives, the personnel files of the foreign workers were supposedly destroyed in the 1970s together with other files of the Department for Human Resources and Social Affairs (*Abteilung für Personal- und Sozialwesen, PSW*) to make room for a table-tennis table.

locations. The fact that the archives of the Human Resources Department are available in microfilm format made them usable, since it was impossible to work through all of the tens of thousands of individual employee files. Parallel references in the files from Department VI and in the denazification and compensation records in public archives made it possible to search for particular names in these microfilms. Other important files in the Hoechst archives were the personnel files, called "service files," on a number of important employees, as well as the large dossier on Carl Ludwig Lautenschläger, the plant director between 1938 and 1945 – a dossier that also contained several volumes of his memoirs. Files on the old IG Farben were also available at Hoechst and proved useful, especially those of the Technical Committee (*Technischer Ausschuss* or *TEA*) as well as the documents of the Nuremberg trial.

The archives of the old "IG Farben sisters" – that is, those of Bayer AG in Leverkusen and BASF in Ludwigshafen – were also consulted for this study. Additionally, I perused the files of Hoechst's former subsidiary plant in Gersthofen (near Augsburg) in the archives of Clariant in Gersthofen and the files of Wacker-Chemie in Burghausen, 50 percent of which belonged to IG Farben, or rather to Hoechst. The Hagley Museum in Delaware, New Jersey, where the files of the chemical corporation DuPont are stored, was also a useful source.

Besides the collections available in company archives, I drew upon documents in public archives. The German Federal Archives (*Bundesarchiv*) proved crucial with their collections on IG Farben, the Economic Group for Chemicals (*Wirtschaftsgruppe Chemie*), and the former Berlin Document Center. Other important locations were the Hesse State Archives (*Hessisches Hauptstaatsarchiv*) in Wiesbaden, where I was able to consult the denazification and compensation files, and the Institute for Contemporary History (*Institut für Zeitgeschichte*) in Munich, which has a microfilm version of the Nuremberg documents. Complementary material was found in the Frankfurt Municipal Archives (*Institut für Stadtgeschichte*), in the Bavarian State Archives (*Bayerisches Hauptstaatsarchiv*) in Munich, in the North Rhine-Westphalian State Archives (*Nordrhein-Westfälisches Hauptstaatsarchiv*) in Düsseldorf, in files of the district government (*Bezirksregierung*) of Düsseldorf, in the Swiss Federal Archives (*Schweizer Bundesarchiv*) in Bern, and in the National Archives in Washington, D.C.

I also had access to a number of private collections containing many valuable documents that contributed to the writing of the plant history. All the private and public archives named proved extremely open and accommodating. I was thus able to consult documents that, only a few years ago, would not have been accessible or would still have been classified. Finally, I am indebted to several contemporary witnesses for information obtained through our conversations.

2

From the Formation of IG Farben
to the Great Depression

2.1. THE END OF INDEPENDENCE (1904–1925)

In June 1920 the Supervisory Board of the Dyes Works (*Farbwerke*), formerly Meister Lucius & Brüning, in Hoechst upon Main granted the request of the Management Board to centralize the scattered buildings belonging to the Technical Department in a new structure. The modern building, based on plans by Peter Behrens, one of the foremost German architects of the time, was a demonstration of Hoechst's self-assurance, or, in the words of a celebratory volume published on the occasion of the company's seventy-fifth anniversary in 1938, an indication of its "strong will to mold the future."[1] The new administration building combined beauty with practicality, and its architecture was vaguely reminiscent of fortifications and castles. It was intended as a visible demonstration of the standing of one of the leading enterprises of the German chemical industry.[2]

Four years later, however, in June 1924 when the building was officially inaugurated, it had already lost its symbolic value. Because of the difficult situation in which Hoechst found itself, the company was already in the throes of negotiating a potential merger with the other two giants of the German chemical industry, the "Farbenfabriken, vorm. Friedrich Bayer & Co." in Elberfeld and Leverkusen (Bayer) and the "Badische Anilin- und Sodafabrik" in Ludwigshafen (BASF). The fusion of three major and a few minor chemical companies into the I.G. Farbenindustrie Aktiengesellschaft at the end of 1925 meant that the building no longer stood for a self-confident, independent company. Now it housed only the office of a member

---

[1] Hermann Pinnow, *75 Jahre Werksgeschichte Hoechst. Zur Erinnerung an die 75. Wiederkehr des Gründungstages der Farbwerke vorm. Meister Lucius & Brüning, 1863–1938* (Munich: Bruckmann, 1938), p. 133.

[2] Wolfgang Metternich, "Traditionsgebundene Baustrukturen," in Bernhard Buderath (ed.), *Peter Behrens. Umbautes Licht. Das Verwaltungsgebäude der Hoechst AG* (Frankfurt am Main: Prestel, 1999), p. 146.

of the IG Farben Management Board, who was both Hoechst's director and head of the Middle Rhine (later renamed Maingau) Works Group.[3]

The company, which manufactured aniline and aniline dyes, had been founded in Hoechst in January 1863 by two friends, the chemists Eugen Lucius and Adolf Brüning, together with August Müller and C. F. Wilhelm Meister, both merchants related to Lucius. In the trade register, the new company was first recorded as "Meister, Lucius & Co," then as "Meister, Lucius & Brüning." At the end of the nineteenth century the German dyes industry was booming, and so were the dyes works in Hoechst. On January 1, 1880, the founders and owners decided to transform the firm into a joint stock company to provide security for the future and to acquire new financial resources. Henceforth, the company's name was changed to "Farbwerke, vorm. (previously) Meister Lucius & Brüning." The Farbwerke continued to expand until the onset of World War I, not only in Hoechst itself, but also worldwide. A great variety of dyes were developed and produced in Hoechst, and a new Pharmaceutical Department was added. The management also established a second factory in Gersthofen, close to Augsburg. Additional factories were set up abroad – in England, France, and Russia. The Farbwerke also had sales offices and agencies all over the world. Thus, prior to the First World War, the Farbwerke in Hoechst together with Bayer and BASF played a significant role in the global dominance of the German dye industry.[4]

As early as 1904, that is, a whole decade before the beginning of the war, and despite the strong position of the German chemical industry, Carl Duisberg, a member of the Bayer Management Board, already attempted to initiate a merger of the country's most important chemical companies.[5] He described the purpose of this integration in a memorandum as "the most far-reaching reduction of production, administrative, and sales costs by the suppression of ruinous competition, with the aim of achieving the highest possible profits."[6] Duisberg admitted that such a merger was easier to organize in difficult times, since potential candidates for the merger tended to be more willing under such circumstances. Even if the chemical

---

[3] Ibid., p. 147; on the Works Groups, see Plumpe, *I.G. Farbenindustrie*, pp. 142 and 151.

[4] See the anniversary publication *Farbwerke, vorm. Meister Lucius & Brüning, 1863–1913* (Hoechst: Farbwerke Hoechst, [1913]); L.F. Haber, *The Chemical Industry 1900–1930: International Growth and Technological Change* (Oxford: Clarendon Press 1971), pp. 14–17; Carsten Reinhardt, *Forschung in der chemischen Industrie. Die Entwicklung synthetischer Farbstoffe bei BASF und Hoechst, 1863–1914* (Freiberg: TU Bergakademie, 1997), 245–7.

[5] See *Die Vorbereitung des Zusammenschlusses der IG-Farbenindustrie im Jahre 1904*, Dokumente aus Hoechster Archiven 9 (Frankfurt am Main–Hoechst: Farbwerke Hoechst AG, 1965).

[6] Carl Duisberg, "Denkschrift über die Vereinigung der deutschen Farbenfabriken, Januar 1904," in idem, *Abhandlungen, Vorträge und Reden aus den Jahren 1882–1921* (Berlin: Chemie, 1923), p. 344.

companies were by no means in difficulty, Duisberg nevertheless felt that the time for a fusion was right. The merger would prevent costly competition abroad, achieve better purchasing and sales prices, and offer higher yields from large-scale production, ensuring satisfactory returns for the years to come. Competition and legal restrictions meant that business was not always profitable – especially abroad. But the existing or potential difficulties were definitely surmountable: "No industry can boast as many virgin territories as the dye and pharmaceutical industries, territories where gold abounds." But it would be preferable to reach agreements "in order to eliminate the struggle which is the consequence of competition, without dispensing with the advantages to be obtained by this very competition." Duisberg saw a merger of the firms as a potential solution.[7] However, he believed it important to maintain an internal rivalry, since, for Duisberg, this constituted the basis of the impressive innovativeness and competitiveness of the German chemical industry. If competition between companies were to end, he feared "a stagnation of progress." He thus proposed to rationalize production as much as possible, without totally renouncing internal competition. Basically, production should be carried out at two different locations, and the chemists and engineers employed there were to "collaborate in a competitive spirit."[8]

However, at the time Hoechst felt itself to be in such a strong position that its managers declined the proposed merger, because they perceived more disadvantages than advantages arising from such a move. In his answer to Duisberg, Gustav von Brüning, a member of one of the founding families and on the Hoechst Management Board since 1899, wrote that Duisberg's assertion that the merger would lead to a decrease of production and sales costs as well as to a reduction of the "ruinous competition" between firms seemed "very seductive" and "quite feasible at first glance." But reservations had arisen on closer examination. For instance, the "trust" organization suggested by Duisberg seemed better suited to other forms of production than those of the chemical industry. Trusts like those binding the coal and coke associations could be dissolved at any time if the hopes generated by the merger did not materialize. After a disintegration of the trust, each coal mine could sell its own coal again, every coke factory could carry out its orders as previously, and every steel and iron company could compete independently. However, a merger of the big chemical companies would entail that some firms would have to leave a number of production processes to factories that worked more efficiently, personnel would be laid off, and sales would be centralized. But, according to Brüning, if the advantages of a merger did not materialize, then it would be "impossible to give the individual companies back their independence." They would have changed so much in the meantime that it would no longer

---

[7] Ibid., pp. 353–4.   [8] Ibid., pp. 354–6, 363, and 365.

be possible to restore them to their former state: "This move is a move into the unknown, with no possibility of turning back." Furthermore, the competition between German chemical plants significantly contributed to their impressive performance, and it remained doubtful whether it would be possible to remain similarly creative without competition. Hence Brüning came to the conclusion that, if an improvement of existing conditions was indeed desirable, it should not be carried out according to Duisberg's suggestions "because it would be irresponsible to endanger organizations and operations through too incautious and hasty a move, especially considering the many years and the extraordinary amounts of work, intelligence, and capital which have been necessary to actually bring about these conditions." At most, Brüning could imagine the "creation of a sales association or a similar institution."[9]

However, that same year Hoechst and the Cassella company in Mainkur, which had been producing dyes and pharmaceuticals under the management of the von Weinberg brothers, decided to merge – much to Duisberg's surprise. In 1907 this alliance was extended through the inclusion of "Kalle & Co." in Biebrich to form a so-called tripartite agreement (*Dreiverband*). This group then faced the "triple alliance" (*Dreibund*) of Bayer, BASF, and the "Aktiengesellschaft für Anilinfabrikation" (Agfa) in Berlin, formed in 1904. Although the two alliances were in competition with one another most of the time, they nevertheless concluded a few agreements, notably in the important field of indigo. Thus, at the time Hoechst still preferred to negotiate price-fixing agreements with the "big sisters"; in other words, in certain areas it favored collaboration over a real merger.[10]

During World War I the discussion arose once more as to whether German chemical companies should merge. Duisberg, who in the meantime had become chairman of Bayer's Management Board, once again took the initiative. In the summer of 1915 he prepared a revised version of his 1904 memorandum and emphasized two new reasons for a merger. On the one hand, the consequences of the war had to be taken into account: factories located in hostile countries had been sequestrated or placed under military control. It was not yet clear what would become of these and other assets in enemy territory after the war. But it was to be expected that "great losses"

[9] Gustav von Brüning, " 'Erwägungen' zu Carl Duisbergs Denkschrift 'Die Vereinigung der deutschen Farbenfabriken,' " *Tradition. Zeitschrift für Firmengeschichte und Unternehmerbiographie* 9 (1964), pp. 1–5.

[10] Tammen, Helmuth, *Die I.G. Farbenindustrie Aktiengesellschaft (1925–1933). Ein Chemiekonzern in der Weimarer Republik* (Berlin: H. Tammen, 1978), pp.10–11; Plumpe, *I.G. Farbenindustrie*, pp. 46–7; Pinnow, *75 Jahre Werksgeschichte*, pp. 91–2; Jeffrey Allan Johnson, "The Power of Synthesis (1900–1925)," in Werner Abelshauser et al., *German Industry and Global Enterprise, BASF: The History of a Company* (Cambridge: Cambridge University Press, 2002), pp. 127–30, 137–9; Bäumler, *Century*, pp. 79–90.

would be suffered. The war was also having dire consequences domestically as a result of restrictions or prohibitions on the use of particular raw materials, and exports were becoming a problem. Some works could switch to arms production, while others could manufacture intermediate products using machinery that would otherwise have stood idle. The war and its consequences affected the firms in various ways, but a merger would spread the negative consequences of the war more evenly over all shoulders.[11] Moreover, according to Duisberg, another important advantage of a merger would be the resulting better capacity to oppose labor organizations, for Duisberg considered unions to be the "train bearers of social democracy." The power of such a big company would be "so great that organized labor would not be capable of instigating rash strikes, and workers would not even dare to envisage them." Furthermore, in the event of a strike, the production sites not affected by the walkout would be able to help out with goods, and the group as a whole would thereby "victoriously withstand long-lasting strikes without incurring any significant damage."[12]

In contrast to 1904, Hoechst reacted positively to Duisberg's suggestion. But it wanted to be treated as an equal partner of Bayer and BASF, even though the balance sheet totals of the three companies as listed in the memoranda showed that Hoechst's relative standing had deteriorated. According to the 1902 balance sheet, quoted in the memorandum of 1904, Hoechst held an equivalent position to that of BASF and Bayer, while the figures for 1914, published in the memorandum of 1915, showed that was no longer the case. Indeed, in the years immediately preceding the outbreak of the war, Hoechst had not developed as well as BASF and Bayer.[13] This state of affairs may have contributed to Hoechst's willingness to form an alliance with the other two "big sisters," but Gustav von Brüning's death also seems to have affected the decision. In 1904 he had been the person responsible for turning down the proposed merger because he had feared a subsequent loss of independence. Since Brüning's death, however, Adolf Haeuser, a lawyer who was not descended from any of the founding families, was head of the Hoechst Management Board.[14] It is not necessary to concur with Duisberg, who did not consider lawyers competent to manage a firm.[15] But when comparing the leading personages of the three "big

[11] Carl Duisberg,"'Die Vereinigung der deutschen Farbenfabriken,' August 1915," in Treue, Wilhelm, "Carl Duisbergs Denkschrift von 1915 zur Gründung der 'kleinen I.G.,'" *Tradition. Zeitschrift für Firmengeschichte und Unternehmerbiographie* 8 (1963), pp. 205–6.

[12] Ibid., pp. 214–15.

[13] Ibid., pp. 224–5; Duisberg, "Denkschrift," pp. 346–7.

[14] Cf. Ernst Fischer, "Meister, Lucius und Brüning, die Gründer der Farbwerke Hoechst AG," *Tradition. Zeitschrift für Firmengeschichte und Unternehmerbiographie* 2 (1958), pp. 70–1.

[15] Cf. Karl Holdermann, *Im Banne der Chemie. Carl Bosch, Leben und Werk. Bearb. von Walter Greiling* (Düsseldorf: Econ, 1953), pp. 171–2: "Then Duisberg jumped up and said

sisters" and their professional achievements, it is worth noting that Duisberg planned and established a modern, efficient plant in Leverkusen, and Carl Bosch, the up and coming man at Ludwigshafen, successfully achieved the technical implementation of high-pressure synthesis. With the fertilizers obtained by this development, he was able to open up an important and lucrative field of trade for BASF.[16] Hoechst could hardly boast of similar successes but was clearly not prepared to admit it.

Duisberg wrote a telling memorandum for BASF about a discussion he had held with Haeuser in 1915, which highlighted this very point. According to Duisberg, the business reports of Bayer and BASF had aroused considerable surprise in Frankfurt – nobody at Hoechst "had expected such high figures." Nevertheless Haeuser had insisted on equal treatment. Duisberg apprised BASF of the fact that Hoechst "felt itself entitled to a status equal to that of Ludwigshafen and Leverkusen" and would start negotiating only "on this basis," since Haeuser, according to Duisberg, had emphasized that otherwise "Hoechst would lose for all time the public reputation as a company of equal standing it had enjoyed until then." In response to Duisberg's objection that negotiations about equality were currently possible only "if Cassella were ready to make considerable sacrifices, either directly or through the intervention of Hoechst," Haeuser reported that the Weinberg brothers were interested in reaching an agreement and ready to negotiate. To start negotiating on an equal basis with Hoechst, Duisberg demanded that Hoechst take over the Kalle company. Duisberg noted that a comparison of the balance sheets of the "triple alliance" (Bayer, BASF, and Agfa) and those of the "tripartite agreement" (Hoechst, Cassella, and Kalle) showed an assets ratio of 60 to 40, and a profits ratio of 66 to 34, while in the field of nitrogen for fertilizers there was no comparison whatsoever.[17]

After a year of negotiations, in August 1916, a community of interests contract (*Interessengemeinschafts-Vertrag*) was signed between Agfa, BASF, Bayer, Cassella, Hoechst, and Kalle, as well as the Chemische Fabrik Griesheim-Elektron in Frankfurt and the Chemische Fabrik formerly Weiler-ter Meer in Uerdingen upon Rhine. For the distribution of profits, the pooling of which was the main aim of the agreement, Hoechst and Kalle together were given the same weight as BASF and as Bayer, the other two "big sisters." An agreement had been reached, but the contract

---

heatedly that he did not care in the least about those statements. It was fortunate that the individual firms were managed by engineers and not by lawyers. The lawyers were mere assistants who had the duty of implementing decisions." Cf. also Johnson, "Power of Synthesis," pp. 185–6.

[16] Cf. Holdermann, *Im Banne der Chemie*, pp. 80–125; Hans-Joachim Flechtner, *Carl Duisberg. Vom Chemiker zum Wirtschaftsführer* (Düsseldorf: Econ, 1959), pp. 141–52.

[17] BASF UA, B4/558, report Duisberg's on discussion with Justizrat Haeuser on December 10, 1915.

ensured the continued independence of the allied companies: they kept their own organizations and were solely liable when trading with third parties. On the whole, despite a number of decisions that had to be made collectively, the member firms retained considerable autonomy.[18] The member companies were especially unwilling to give up business sectors in order to comply with the planned rationalization process within the new community of interests. This reluctance reflected the fact that the alliance had been concluded for a period of fifty years, but could be called off at any time. Accordingly, meetings of the Community Board (*Gemeinschaftsrat*), the highest executive body of the community of interests, were often unproductive. The potential to dissolve the community of interests at any given time was its Achilles' heel. Indeed, it ran counter to the aims Duisberg had striven to achieve with the merging of the chemical companies.[19]

Looking back in 1948, Richard Weidlich, a leading member of the Management Board of Hoechst and later of IG Farben, wrote that numerous individual interests persisted within the original community of interests of 1916. Because "it was necessary to anticipate the demise of the community of interests," it was always a difficult decision "when a firm was supposed to close down one of its less profitable branches and leave that area of production to another company – a move that would have resulted in a lasting disadvantage in the event that the community of interests ended." The biggest danger of a shift in the existing balance between the members lay in the BASF-owned nitrogen plants, which were subsequently placed under the collective control of the new organization. Nevertheless, a number of irresolvable problems remained, which were characteristic of "such an instable institution" as a community of interests. This state of affairs encouraged thoughts of an even closer cooperation that would do away with existing special interests and result in an indissoluable merger.[20]

But these visions did not materialize at the end of World War I. Hoechst's situation at the time was rather mixed. The Treaty of Versailles allowed the members of the Entente to confiscate German patents and foreign assets, and the German chemical industry was particularly hard hit.

[18] BASF UA, B4/2553, Bayer (Duisberg) to Agfa, BASF, Cassella, Griesheim-Elektron, Weiler-ter Meer, Hoechst, Kalle, June 14, 1916, including draft of the IG treaty; B4/2552, Interessengemeinschafts-Vertrag of August 18, 1916.
[19] Cf. Wolfram Fischer, "Dezentralisation oder Zentralisation – kollegiale oder autoritäre Führung? Die Auseinandersetzung um die Leitungsstruktur bei der Entstehung des I.G. Farben-Konzerns," in Norbert Horn and Jürgen Kocka (eds.), *Recht und Entwicklung der Grossunternehmen im 19. und frühen 20. Jahrhundert. Wirtschafts-, sozial- und rechtshistorische Untersuchungen zur Industrialisierung in Deutschland, Frankreich, England und den USA* (Göttingen: Vandenhoeck & Ruprecht, 1979), p. 479; Plumpe, *I.G. Farbenindustrie*, pp. 96–9.
[20] HA, PA Weidlich, Richard Weidlich, "Erinnerungen (aus dem Gedächtnis)," August 1948.

In Hoechst's case, numerous patents as well as three foreign factories located in England, France, and Russia were seized.[21] Moreover, during the war Hoechst had worked primarily for the German military, focusing on poison gases, smoke screens, hand grenades, and land mines. In 1918, according to one of Hoechst's company histories, about 70 percent of Hoechst's turnover was linked to the war production. The construction of armament factories had led to the expansion of the plant "without any strategic corporate planning and without taking a future peace-time economy into account. ... Shacks were built wherever there was space. In order to save room, various annexes were erected without taking account of the flow of traffic on the plant's streets. Thus, in the course of the war, Hoechst had been transformed into a motley construction with many technical disadvantages." Consequently, at the end of World War I, Hoechst found itself in a situation "which considerably hindered its conversion to a peace-time economy."[22]

During the initial postwar years, the chemical industry, including Hoechst, was able to profit from the inflation in Germany. Compared to countries boasting a stable currency, the German chemical industry could offer its products at cut prices, and was thus particularly competitive. In addition, the chemical industry, and hence also Hoechst, was awarded important reparations contracts by the government of the Republic.[23] However, Hoechst apparently did not properly exploit these advantages, probably also because of the problems outlined above. At any rate, in November 1920 Georg von Schnitzler, the member of the Management Board responsible for sales, complained that production was not keeping up with demand, especially in the field of dyes. The reserves originally meant for the Entente powers had already been used up to "more or less satisfy" customers' demands. Hoechst was not in a position to meet demand and, according to von Schnitzler, now had to suffer the consequences: "we are being edged out of our proprietary domains by the other IG companies." Haeuser, the managing director, was even blunter. He noted that, although the Hoechst plant was better supplied with coal than the Griesheim and Cassella works, it was less productive. He concluded that "we neither sufficiently exploit our coal nor our clerks and workers." According to Haeuser, this situation should be remedied quickly, "especially since our poor performance has led to a deep-seated negative attitude toward us

---

[21] Stephan H. Lindner, *Das Reichskommissariat für die Behandlung feindlichen Vermögens im Zweiten Weltkrieg. Eine Studie zur Verwaltungs-, Rechts-, und Wirtschaftsgeschichte des nationalsozialistischen Deutschlands* (Stuttgart: Steiner, 1991), pp. 18–21; Pinnow, *75 Jahre Werksgeschichte*, pp. 102–5; HA, 12 (1863–1963), Otto Hirschel, "Das Werk Hoechst im Verbande der I.G. (1916–1945)," May 1961, p. 6.

[22] HA, 12, Hirschel, "Das Werk Hoechst im Verbande der I.G. (1916–1945)," May 1961, p. 5; Plumpe, *I.G. Farbenindustrie*, p. 96.

[23] W. Fischer, "Dezentralisation," pp. 479–80.

within IG, and our current performance is earning us very little respect."[24] Nevertheless, Hoechst was apparently still making good money. In July 1920 von Schnitzler declared at a meeting of the Management Board that prices for dyes were "approximately 35 times higher than peace-time [i.e., prewar] prices." The company should take advantage of this situation for as long as possible: "individual IG firms have been planning to cut prices, but we contend that no reason justifies this move as yet. The other IG firms share our opinion."[25]

However, by 1923/24 Hoechst's situation had deteriorated even further. In January 1924, Paul Duden, a member of Hoechst's Management Board, reported that IG's business situation was generally worrying, with all sectors showing losses. There was no work for thousands of Bayer's employees and workers, and once the crisis had passed, only 60 percent at best of the current number of workers could hope for further employment. Hoechst faced the same situation. According to Duden, it would be necessary to "considerably limit the number of blue- and white-collar workers, which has multiplied far beyond the usual quota during the abnormal inflation period." In particular the number of sales clerks, office clerks, and craftsmen would have to be reduced.[26] Even though the first layoffs had already been initiated, in September 1924 Haeuser found that "approximately 9 thousand workers were employed in the factory, whereas, considering the current level of activity, approximately 5 thousand workers would suffice." He added: "Even considering all potential circumstances, a minimum of 1 thousand workers must be laid off, since it is impossible to explain how these extra 1 thousand workers are to be kept occupied."[27]

Faced with this difficult situation, with German dye plants operating at only 40 percent of capacity, the big chemical companies were ready to consider a closer association.[28] Erwin Selck, a member of Cassella's Management Board, analyzed their current situation. At the end of June 1924, he sent a letter to Oscar Michel, a member of BASF's Management Board, and included a memorandum with the prosaic title "Thoughts on the IG Contract." Selck noted that when the agreement had been concluded in 1916, the companies had still hoped to return to the world market after the war with something approaching their former position. The basic assumption underlying the association of 1916 was "to limit the autonomy [of member companies] as little as possible." As far as production and

---

[24] HA, TEA 95b, Sitzung des Direktoriums Hoechst, November 9, 1920.
[25] HA, TEA 95b, Sitzung des Direktoriums Hoechst, July 13, 1920.
[26] HA, TEA 95b, Sitzung des Direktoriums Hoechst, January 25, 1924.
[27] HA, TEA 95b, Direktionssitzung Hoechst, September 19, 1924.
[28] Plumpe, *I.G. Farbenindustrie*, p. 116; the debate on the merger is thoroughly discussed in W. Fischer, "Dezentralisation." For the negotiations on the merger, especially from Duisberg's point of view, see Flechtner, *Carl Duisberg*, pp. 335–53.

research were concerned, the companies remained largely independent, and only pooled their experience and suggested improvements. A form of " 'moderate competition' (the expression was used repeatedly) should exist" in the commercial field. These assumptions had been justified in 1916, and a phase characterized by a community of interests, consisting of largely independent member companies, had been "genetically necessary." But circumstances had changed: the war had "ended unfavorably," and the period of inflation, which had been "initially advantageous" for the export industry, had changed, "turning into its exact opposite." Large areas of the market had been lost, and there was now a general monetary and financial crisis. The currently high production costs for dyes were endangering exports, and those costs could be "effectively counteracted only if material and human equipment as well as market possibilities are ruthlessly streamlined." But these problems could not be solved, because a community of interests "is a loose construct which can be dissolved for many different reasons." A radical structural change and a true fusion of member companies were the only feasible options.[29]

Once again it was Carl Duisberg who, at the end of October 1924, aired the suggestion of a closer alliance. After the "inflation year of 1923 with no payment of dividends," the year 1924 was a year "of unemployment and downsizing." The situation of the chemical industry was particularly uncertain, especially in the field of dyes, since it had become obvious that shrinking markets were not a mere temporary phenomenon. Increasing competition and the "pauperization of Europe" meant that a lasting decline in sales was to be expected, accompanied by a decrease in production amounting to around 40 to 50 percent. "The beautiful years of carefree development and the continuous expansion of our community of interests' factories are thus over." Everything should be done to cut production costs, simplify the organization, and organize sales more efficiently. But, Duisberg noted, "this is no longer possible if we continue on our previous path and allow the individual companies to retain full independence." Duisberg believed the best solution would be to adapt the community of interests' contract to the new circumstances, but he anticipated immediate difficulties if the rights of individual companies were restricted. A merger might be the solution, but many versions were possible. One possibility would be to incorporate all the companies into a large joint stock company that would be a new creation; another possibility was a takeover of all the firms by a single company. The latter solution, however, would mean the regrettable loss of established names such as BASF, Hoechst, Bayer, and Agfa – the names should therefore be retained in the names of subsidiaries. Duisberg anticipated further problems and costs from the resulting large number of members of Management and Supervisory Boards, as well as potential

[29] BASF UA B4/634, Selck to Michel, June 26, 1924, enclosed: "Gedanken zum I.-G.-Vertrag."

tax disadvantages, and the danger of "foreign infiltration," if preferred shares were eliminated. Thus it was necessary to take another route: "We believe that we have found one in the guise of a holding company." If a holding company did not necessarily solve the problems of the previous organizational form, it could at least minimize them and, above all, avoid the "greatest evil": decentralized sales and the resultant chaotic competition.[30]

However, following a discussion of the memorandum on October 29 at a meeting of the Community Board (*Gemeinschaftsrat*), the majority of firms declined the solution of a holding company and requested an "immediate and complete merger." Only the representatives of Bayer and Hoechst favored Duisberg's ideas and demanded at least a transition period prior to a complete merger. During the meeting Duisberg was surprised to discover "a sharp difference" of opinions between Carl Bosch, the chairman of the BASF Management Board, and himself over this "important and significant question of organization." Haeuser apparently suggested a "temporary merger," a move that Duisberg found preferable to immediate fusion.[31]

At the beginning of November, in the aftermath of this meeting, Duisberg submitted a new memorandum, in which he commented on Bosch's visions of a merger and its organization. Duisberg believed it would be impossible to include all the board members of the various allied firms in the board of the new corporation – as BASF had suggested – because the latter would be too big to take important decisions, such as those concerning personnel. Duisberg strongly objected to "autocratic" structures that might seem successful for a time but were generally tailored too strongly to individual personalities and, moreover, tended to treat "all the people active in the organization as if they were lifeless tools without will or soul." He knew from personal experience "the importance and necessity of an organization" in which responsibilities were clearly recognizable and structured, "and in which nevertheless everyone can freely think and act." He proposed "a collaborative consultation and decision making" of all responsible individuals, and not just by those who just happened to be present, to deal with "all important questions." Duisberg felt that Bosch wanted "friends" in important positions, who would be allowed great freedom in their decision making; but this "indirect autocratic system" was too dependent on individual personalities, without whom the system would fail. Duisberg wanted a "democratic system" for the new corporation. He preferred to "start with the foundations" and build up "the organization from the bottom up, objectively and logically." At the time the

---

[30] Bayer WA, 4 C 1, Carl Duisberg, "Vorschläge zur Vereinfachung, Verbilligung und Befestigung der Interessengemeinschaften der deutschen Teerfarbenfabriken und zur Beseitigung der noch bestehenden Mängel und Schwierigkeiten," October 26, 1924.

[31] Ibid., part II, November 4, 1924, with reference to the meeting of October 29, 1924.

IG consisted of thirty-three chemical plants, a coal mine, eight lignite pits, three limestone factories, and two gypsum works – and employed more than 90,000 people. To organize these production sites and individuals with a minimum of bureaucracy and maintain an "ideal amount of necessary competition," it was important to avoid "egalitarianism" and to "set up largely autonomous territorial plant communities" that would encourage the scientific, technical, and commercial areas to function to the best of their ability. Therefore, all the IG factories should be organized in four large groups, but in the process "current circumstances" were to be taken into account "as much as possible." A number of essential decisions were to be left to the future "works groups" (*Betriebsgemeinschaften*) and sales partnerships (*Verkaufsgemeinschaften*). A few leading personalities from these plant and sales communities would become members of the board of the community of interests (*Interessengemeinschaftsrat*). The board would thus consist of between twenty-eight and thirty-six members, most of them chemists, businessmen, lawyers, and engineers. Moreover, Duisberg suggested the establishment of a Community Board (*Gemeinschaftsrat*), consisting of ten to twelve leading personalities of the IG board. "The overall control of the various plants and sales communities including their equipment" would be assigned to the chairman of the Community Board or, depending on the division of labor, to his deputy and the other members. While Duisberg was fully aware of how important individual personalities could be, he still wanted this particular organizational blueprint to be extensively implemented.[32]

But at that point Hoechst stepped in and offered its own contribution to the debate. In a memorandum written in December 1924, Haeuser noted that the former IG had been organized around the tenet that collaboration between the companies in technical and commercial fields would take place in an atmosphere of trust, while the companies remained largely independent. But these principles had fostered conflicts from the very beginning, since a rational collaboration between allied companies sometimes demanded sacrifices. Conflicts had arisen, especially in the commercial sector, and had escalated because doubts had persisted concerning the indissolubility of the contracted alliance: "The ongoing debate on the question of how IG's existence should be secured necessarily led to a strengthening of the existing egoistic tendencies of individual firms. It must be said this has now gone so far that, particularly in the commercial field, a positive collaboration is only imaginable if the special interests of individual firms are set aside. The only way to achieve this is by a merger." Haeuser admitted that the takeover of all the firms by one company would be "the easier" solution and would probably prove the strategy favored by most. Yet he emphasized his regret that IG "would no longer be organized" along

---

[32] Ibid., part III, November 8, 1924.

existing lines. He considered the establishment of a big trust as a "not very promising" solution and would have preferred a slower development. But he recognized that, "in a sense," the companies were acting under "compulsion." Having stated this, he felt that the current structure of IG "with all its advantages" should be kept on and its disadvantages should be accepted, because every company in the alliance was a "valuable historical entity, embodying a particular spirit." From a technical perspective, this diversity encouraged progress more than excessive standardization would. "To secure technical progress will require the preservation of the individual firms to a large extent." Haeuser was of the opinion that this aspect should also be considered in the future and categorically proscribed measures aiming at a "uniform organization of the corporation, because of an exaggerated emphasis on saving costs." As to the management of the newly merged corporation, Haeuser suggested incorporating all the members of Managing and Supervisory Boards of the allied companies into a new Management and Supervisory Board, at least for a transitional period. Following Duisberg's suggestion, the Management Board should then form an Executive Committee in which members from the three "big sisters" would be dominant. Thereafter, the new firm should be organized in a "top-down" fashion by the Executive Committee – a suggestion that was diametrically opposed to Duisberg's concept.[33]

At the end of February 1925, Carl Bosch finally took a hand in the debate on the organization of the merger, since the necessity of a merger was by now universally accepted. Bosch emphasized that the main consideration underlying his proposals was to create operational units that could take action immediately. His idea was to "quietly and continuously develop the existing entities and to limit new forms of organization to what was absolutely unavoidable." Bosch proposed a "smaller board" as the responsible body, while the Supervisory Board should primarily control the corporation and nominate the members of the Management Board, but would not be involved in the daily management of the company. The smaller board would consist of all full members of the Management Board. "Because of their personalities," some of its members were bound to have more influence than others. This would not contradict the principle of full equality among the members of the Management Board. The smaller board would represent "the real management" of the corporation and would be "in a sense, omnipotent." However, since its size would still often make it unwieldy, a committee of a few members should be selected from among its midst, who would be responsible for preparing bigger projects and taking particularly important decisions such as those regarding personnel or "business of a discrete nature." The Management Board should mainly "take note of important issues and take responsibility." The basic work

[33] Bayer WA 4 C 1, Haeuser to Bosch, Duisberg, Weinberg et al., December 13, 1924.

would be carried out in advance by different groups of specialists. To a large extent, work at IG would therefore continue as before: "What distinguishes my suggestion from the others is its striving toward a greater degree of simplification in the management of the company; bodies with more or less overstretched powers should be eliminated and the overall responsibility transferred to the smaller board, and finally, individual departments should remain as independent as possible." Bosch thus opposed the establishment of independent works groups. Nor did he see the necessity for a continued competition within IG to maintain its competitive potential: "A capable work force does not need to be egged on. The knowledge that good work will be honored, even if it is outperformed by another department, is more important. But it is even more crucial that progress is not hampered by sluggish and narrow-minded managers. To my mind, this last point is the most important one." Bosch also wished to do away with meetings of the Community Board and IG meetings, since a lot of time was spent in "unproductive talk" or in smoothing over antagonisms between allied firms. And finally, the name "community of interests" (*Interessengemeinschaft*) should be eliminated, since after a merger, the company would become a corporation or association, and the old name would be redolent of the "darker side of German dye manufacturers."[34]

The controversial discussions and significant disagreements made it difficult to imagine that the parties involved would reach a consensus rapidly. But Bosch had rightly emphasized their shared aims, even if opinions differed as to the means of achieving them. On July 10, 1925, Duisberg and Bosch wrote a letter to the other members of the IG Community Board, stating that they had discussed the "question of reorganization" extensively. They had reached the conclusion that "the best and simplest means of resolving all divisions and implementing the planned merger was by all members of the previous Community Board, with the exception of Bosch, joining the Supervisory Board on the day of the fusion, under conditions yet to be defined."[35] The new blueprint for the planned merger, a "gentlemen's agreement," as it was termed in the letter, was the so-called "founders' compromise." To a large extent, Duisberg had prevailed in the questions respecting the name of the company and its organization. After all, neither Bosch nor Duisberg had strongly favored centralization. The historian Wolfram Fischer comments that both Duisberg and Bosch had worked toward creating "a combination of centralized and decentralized elements in the structure of the new corporation." Bosch, however, had largely managed to push through his own views in the matter of staff, and as the chairman of the Management Board, he became the leading figure in the

[34] Bayer WA, 4 C 1, BASF to Gemeinschaftsrat, February 2, 1925, enclosed: Niederschrift Bosch's of the same date on the question of the merger.
[35] Bayer WA, 4 C 1, Duisberg and Bosch to Gemeinschaftsrat, Leverkusen, July 10, 1925.

new corporation – all the other members of the former Community Board, including Duisberg, became members of the Supervisory Board, a change that was apparently lavishly recompensed.[36] As for Hoechst, it had succeeded in its aim of being treated as one of the three "big sisters." This was Haeuser's achievement, and he was accordingly praised for it in Hoechst: "It is owing to *Geheimrat* Haeuser and his skilful negotiations and representation of Hoechst's interests that the Hoechst Dye Works were able to retain their status as an equal [among the big sisters]."[37]

2.2. FROM THE "FOUNDERS' COMPROMISE" TO FIRST REFORMS AND RATIONALIZATION MEASURES

In his study of IG Farben, the historian Gottfried Plumpe wrote that the company's organizational development, as well as that of its decisional structures, could be divided into three phases: first, the age of the "founders' compromise" and "decentralized centralization"; then a period characterized by the development of a divisional structure to overcome the problems engendered by the initial compromises and the Great Depression; and finally, an era of "polycentrism," as a result of the extraordinary expansion in the 1930s.[38]

In the mid-1920s Carl Bosch was questioned by a Reichstag Committee of Inquiry into Production and Market Conditions in the German Economy (*Reichstagsausschuss zur Untersuchung der Erzeugungs- und Absatzbedingungen der deutschen Wirtschaft*) about the changes in economic organizations. He described the process that IG Farben had gone through, with its move from "decentralized centralization" toward a closer cooperation between individual plants, clear agreements, and rationalization measures. He began by emphasizing that, in the aftermath of World War I, which had resulted in the loss of many patents after the Versailles Treaty, and because of widespread protectionist policies, IG Farben had lost half of its markets – and there was no sign of an upswing. Bosch stated that under these circumstances there was no alternative except to close down half of the plants, to try to reduce production costs, and to form joint purchasing and sales organizations. The former community of interests had not been able to implement these measures and thus, "once considerable resistance and reservations were overcome, a complete merger was decided on."[39]

---

[36] W. Fischer, "Dezentralisation," pp. 483–5; see also Plumpe, *I.G. Farbenindustrie*, pp. 136–44.

[37] HA, PA Haeuser, Festreden zum 70. Geburtstag des Herrn Geheimrat Dr. Haeuser, speech of Weidlich, p. 12.

[38] Plumpe, *I.G. Farbenindustrie*, p. 163.

[39] Carl Bosch on IG Farben, in Ausschuss zur Untersuchung der Erzeugungs- und Absatzbedingungen der deutschen Wirtschaft, *Verhandlungen und Berichte des Unterausschusses für allgemeine Wirtschaftsstruktur (I. Unterausschuss), 3. Arbeitsgruppe: Wandlungen in den wirtschaftlichen Organisationsformen. Erster Teil: Wandlungen in den*

Once the fusion had been concluded, Bosch's task had been mainly to "draw the necessary conclusions from this merger." He started with production, insisting that the manufacture of most products should now be carried out in a single plant. The only exceptions, where manufacturing continued to be carried out in two locations, were a few major products for which the capacities of a single plant did not suffice. This was followed by the rationalization and reduction of the range of dyes produced by IG Farben "in order to eliminate all unnecessary products, which previously competed against one another." Even the streamlining of the administration and sales organizations had finally been completed. However, this had proved to be a "rather lengthy process," since the individuals involved had had to be convinced of the necessity of "acting radically." As Bosch recalled, "During the initial exchange of ideas leading up to the merger, we made a few concessions in order to accommodate some people. At first, dyes continued to be sold from four different sites, in order to avoid completely depopulating some of the offices. We divided up the sale of dyes according to countries. This proved to be a fundamental mistake. Therefore, we have now decided to completely centralize this branch in Frankfurt." It was to be hoped that a definitive solution had now been found. Bosch drew the following conclusion from these experiences: "A merger can never be implemented too radically. With hindsight, all half measures turn out to be wrong."[40] Thus, the management had recently decided to centralize sales of pharmaceutical products, which until then had been divided up according to countries between the Leverkusen and Hoechst plants.[41]

Bosch then analyzed the new organization. The integration of all members of the Management Board of the allied firms into the IG Farben Management Board had led to the subsequent creation of a Working Committee (*Arbeitsauschuss*) of twenty-six members, which he chaired. It met every three to four weeks in Frankfurt and was responsible for fundamental decisions affecting the company: "The other board members are now managers of departments. They no longer attend the general sessions." The Supervisory Board had undergone the same process, since all members of the allied firms had initially been taken on. However, only ten of the former board members from the original companies taken into the Supervisory Board were appointed to the Administrative Council (*Verwaltungsrat*). And only that body was responsible for important issues like controlling the Management Board, nominating directors, and validating contracts.[42] However, Bosch emphasized that every member of the Management Board had the right to consult the minutes of the Working

Rechtsformen der Einzelunternehmungen und Konzerne (Berlin: Mittler, 1928), pp. 436–51, p. 437 for the quotation.

[40] Ibid., pp. 437–8.   [41] Ibid., p. 441.   [42] Ibid., pp. 439–40.

Committee and could raise objections at any time against its decisions. In addition, there were a number of technical committees and specialist committees, for example for dyes and pharmaceuticals. According to Bosch, the actual work was done by these committees: "The Working Committee is not expected to deal with topics about which no decisions can yet be taken and considers only matters affecting our general policies." Bosch's decided views on this matter were illustrated by a further comment: "I refuse to deal with matters on which my colleagues in their specific fields cannot agree." Thus, the other members of the Management Board had more power than was first apparent. When questioned by the Reichstag Committee about competition within IG Farben and opportunities for employees' personal development, Bosch answered: "Based on my observations, the energy and activity in the firm is a lot stronger and more stimulating than ever before, because people no longer have to deal with unnecessary matters. Previously, they wrote an infinite number of useless letters; that has now ceased. I have always stressed that every head of a department is fully responsible for his own department, and should only present matters which have already been settled to us, that is, to the Management Board."[43]

What did these changes described by Bosch mean for Hoechst? What was its position within IG Farben? As Plumpe emphasizes in his study, membership in the Working Committee, which was responsible for all strategic decisions, was determined not only by the fact that all important domains of IG Farben's activity had to be represented, but also by "provenance from one of the founding companies."[44]

After Haeuser's transfer to the Supervisory Board and the Administrative Council, Paul Duden, Hoechst's new plant manager, became a full member of IG Farben's Management Board, as well as director of the Middle Rhine (later Maingau) Works Group and a member of the Working Committee. Duden had worked in Hoechst for twenty years. He studied chemistry in Marburg, Geneva, and Würzburg and obtained his university lecturer qualification (*Habilitation*) in Jena under the supervision of Professor Ludwig Knorr, becoming associate professor (*Extraordinarius*) there. In 1905 Paul Duden accepted Hoechst's offer and moved into industry, where he became head of the Central Laboratory. During the First World War he pioneered the development of acetylene chemistry: acetone, a crucial ingredient in the manufacturing of explosives, was produced through the catalytic oxidation of acetylene into acetaldehyde and acetic acid.[45] If Duden proved a very successful chemist, the same could not be said of his

---

[43] Ibid., pp. 441–3.   [44] Plumpe, *I.G. Farbenindustrie*, pp. 144–5.

[45] HA, PA Paul Duden, article on the occasion of Duden's 25th anniversary at IG Farben on January 2, 1930; Jens Ulrich Heine, *Verstand & Schicksal. Die Männer der I.G. Farbenindustrie A.G. (1925–1945) in 161 Kurzbiographien* (Weinheim: VCH, 1990), pp. 82–3.

managerial skills. He did not display the toughness required by his duties as plant director, particularly in Hoechst, where a number of small fiefdoms had developed. These fiefdoms assiduously, even pettily, endeavored to protect their competences and products from one another. As Karl Winnacker euphemistically wrote in his memoirs, "a deeply-entrenched departmental federalism flourished in Hoechst."[46]

Thus, when Albrecht Schmidt, one of Hoechst's most notorious "minor princelings," praised Duden for his equable manner, this was not a particularly good sign for the plant. In his laudatory speech for Duden on the occasion of Duden's twenty-fifth anniversary of joining the company, Schmidt, who often complained to Duden about putative or actual humiliations, stressed that "Duden had and has no enemies!" This was because Duden always behaved decently, even in the most difficult situations, and his word could be relied on. With his conciliatory manner, Duden was able to achieve "complete pacification" of even violent quarrels. Duden's "integrity and reliability" coulded be "absolutely counted on." Schmidt was unable to refrain from mentioning that some persons – clearly including himself – had perceived Duden as an "intruder," who "seemed effortlessly to succeed, where others had tried in vain for decades." But Schmidt emphasized that Duden's equable manner had made him acceptable to everyone, including himself.[47] But this diplomatic and diffident form of management proved increasingly problematic for Hoechst, since his propensity for appeasement meant that many necessary decisions were not implemented.

The other full members on the IG Farben Management Board and the Working Committee from Hoechst were Alfred Ammelburg, the head of the Pharmaceutical Department, Wilfrid Greif and Georg von Schnitzler from Sales, and Richard Weidlich from the Legal Department. Schmidt, who has been mentioned above, was deputy technical manager and a full member of the Management Board, but he was not a member of the Working Committee.

Two additional committees, the Commercial Committee and the Technical Committee, helped the Working Committee in preparing, implementing, and monitoring decisions. All important commercial and economic questions and problems were discussed in the Commercial Committee, which typically included only men with commercial training and lawyers. The Technical Committee, composed of chemists and engineers, mainly debated investment policies and technical questions, which ranged from research and development to production. Karl Krekeler, manager of the Lower Rhine Works Group, headed the Technical Committee until the

---

[46] Winnacker, *Challenging Years*, p. 65.
[47] HA, PA Duden, vol. 6: Persönlich, speech of Albrecht Schmidt, in: 1905/1930 – zum 25-jährigen Jubiläum von Duden, January 11, 1930.

end of 1932; he was succeeded by Fritz ter Meer, who chaired it until 1945. At the end of 1932, the IG Farben management dissolved the Commercial Committee and transferred its duties to the Working Committee; the expansion of trade led to its revival in 1937 with Georg von Schnitzler as its chairman. In addition to these two committees, a number of specialized committees existed for various important operational areas. As IG Farben's structure was still strongly decentralized in the 1920s, there was no consistent planning and control of investments and research, despite all the committees and councils – largely because of members' commitments to previously independent plants.[48] Reminiscing about this period, Fritz ter Meer wrote: "The term 'decentralized centralization' was used to describe the organization of the I.G. Farbenindustrie Aktiengesellschaft. It was an unlovely term, but nevertheless proved accurate. It was particularly applicable to those plants that had largely managed to preserve their independence in spite of losing their sales departments, and it was accepted that this decentralization of all technical activities was not the most economical solution."[49]

The rationalization of dye production in the 1920s is a good case in point of how reforms and rationalization measures were implemented under those circumstances, and what the resulting problems were. One of the major problems facing IG Farben's management after the merger was the excessive production of dyes, a problem that Bosch had outlined to the Reichstag Committee. Only a fraction of capacity was actually utilized, and the range of dyes remained far too large. In 1926, after the Coloristic Committee eliminated 60 percent of all dye types, the IG Farben plants were still producing 32,000 commercial dyes. Furthermore, a number of dyes were produced at multiple locations, and production costs differed widely. Discussions within IG Farben about which plants were profitable and which should be rationalized often lasted for months and sometimes led to vehement disputes.[50] They provide a good illustration of the issues Bosch reported to the Reichstag Committee and retrospectively described by ter Meer. The correspondence and the minutes of the various commissions and subcommissions for the rationalization and restructuring of the production of dyes and intermediate products convey a reliable picture of the acrimony with which some of the disputes concerning particular dyes and manufacturing departments were waged. The two following examples illustrate Hoechst's position and its weakness in negotiations within IG Farben.

At the end of July 1926, the IG Farben plant at Mainkur (Cassella) wrote a letter to the other IG Farben plants stating that the solution envisaged at the end of June 1926, namely, to cut production of Azo dyes at the Mainkur

[48] Plumpe, *I.G. Farbenindustrie*, pp. 145–7. [49] Ter Meer, *I.G. Farbenindustrie*, p. 29.
[50] Bäumler, *Century*, p. 100.

plant, was unacceptable under any terms. In its production and sales of Azo dyes, Mainkur was second only to Leverkusen. Furthermore, the plant had already suffered great losses in its range of intermediate products and triphenylmethane dyes.[51] By fall, still no compromise had been achieved, and Mainkur once again made it clear that it "totally disagreed with the current plan of restructuring and rationalization." The plant had always been very cooperative and willing to compromise to accommodate the interests of IG Farben, but this time it was not prepared to give way. Mainkur had enough of sacrificing its own interests: for once, it demanded that one of its own areas be strengthened. A revision of the current plan was indispensable "since the proposed adjustments are contrary to the interests of both IG and Mainkur and clearly do not offer Mainkur, which manufactured only dyes, a solid basis."[52]

Mainkur felt – and this impression was not entirely unjustified – that it was being sacrificed by Hoechst, its former partner in the tripartite agreement.[53] For its part, Hoechst had good reasons to cling to the plan contested by Mainkur, since in May 1926 Leverkusen demanded that Hoechst "generally cease the manufacture of Azo dyes."[54] In the ensuing discussions, Hoechst was able to improve its situation considerably, so it was not surprising that it did not want the results jeopardized. Hoechst argued that, after the decision of the Technical Committee taken in August at Bosch's instigation, "peace [had] at last returned": "If decisions prove temporary, no plant will know what it can expect in the future, no plant can properly deal with the question of customers, because no plant can afford to abandon any lines of production." Hoechst insisted that huge sacrifices had been made "on all sides" after "long and difficult" battles. It emphasized that, as a result of Mainkur's pressure, significant concessions had been granted with regard to blue and green sulfur dyes, which had been ceded to Mainkur and Leverkusen in return for a small compromise respecting sulfur black. Hoechst argued that it had given up production of many dyes despite having "the most modern equipment": in all, some 232 types of dye and seventeen intermediate products. If Mainkur was prepared to list its sacrifices, then Hoechst would do the same; furthermore Mainkur had not been particularly generous to Hoechst. Hoechst had always sought to achieve a general agreement for all plants, while Mainkur pursued individual arrangements with other plants, bypassing Hoechst: "Considering the

---

[51] HA, TEA 61, Mainkur to Ludwigshafen, Leverkusen, Hoechst, Elberfeld, and Wolfen, July 24, 1926. Cf. also Besprechung über Zusammenlegung der Azofarben und der wichtigsten Zwischenprodukte für Azofarben, June 30 and July 1, 1926.

[52] HA, TEA 61, Mainkur to I.G. Werke and TEA, September 8, 1926 (Abschrift).

[53] Cf. HA, TEA 61, Mainkur to Ludwigshafen, Leverkusen, Hoechst, and Wolfen, September 21, 1926.

[54] HA, TEA 61, Besprechung über Zusammenlegung der Fabrikationen von Leverkusen-Mainkur, May 27, 1926.

enormous sacrifices made by all the plants and the generally applauded resumption of productive work after struggles that lasted for months and consumed all energies, we are confident that, like the other plants, Frkft. [Mainkur] will be able to adapt to the smaller scale proposed."[55]

Conflicts about other dyes also continued over many months, and disputes were fought using all available means. As soon as an agreement was reached in one area, new disagreements flared up in another. Thus, at the beginning of August 1926, Elberfeld wrote to the other plants that the Berlin (Agfa), Biebrich (Kalle), Elberfeld (Bayer), Griesheim, Mainkur, Ludwigshafen, and Uerdingen plants had reached an agreement respecting the restructuring and rationalization of the production of multicolored sulfur dyes. Mainkur was to keep the blue dyes, Elberfeld the green and reddish-brown ones together with most of the food colorings, and Berlin would manufacture the rest. The letter continued: "Unfortunately, up to now, the Hoechst plant has agreed with this arrangement only insofar as it is prepared to surrender the blue and green thiogene dyes to Mainkur." Elberfeld demanded Hoechst's approval of the existing arrangement, but added: "we have heard that Hoechst is not willing to cede its food colorings to Berlin and Elberfeld because it is not prepared to recognize the advantages of the Berlin and Elberfeld food coloring facilities over its own. Hoechst is the only plant to hold this opinion, and further negotiations appear pointless." It would be in IG Farben's interest to limit the manufacturing of food colorings to two locations; this had already been proved by the transfer of products from Mainkur, which resulted in a much better utilization of capacity and limited the number of manufacturing brands: "For the sake of IG's interests, we would find it a great pity if Hoechst were to prove unrelenting and continue its comparatively small production of food colorings, which, in our opinion, could never be organized profitably in the current circumstances."[56]

Hoechst defended itself against these accusations. It continued to resist the transfer of other product lines, since it had already agreed to give up the blue and green thiogene dyes, which entailed the shutdown of "a large number of excellently arranged modern facilities." This had been "a momentous step taken in the interests of IG" and Hoechst was not prepared to make any further concessions to either Mainkur or Elberfeld with respect to food colorings. Both plants should finally accept that, for purely economic reasons, Hoechst was unable to comply. Hoechst had repeatedly and clearly explained why it could not give up the food colorings, as these constituted a growing group of products and included important brand names. Not only were the food colorings important for Hoechst, but the production facilities were very efficient and its profitability was "absolutely

---

[55] HA, TEA 61, Hoechst (A. Schmidt) to Werke der I.G. and TEA, September 21, 1926.
[56] HA, TEA 66, Elberfeld to Berlin, Hoechst, Mainkur, and Ludwigshafen, August 6, 1926.

equal" to that of Elberfeld and much superior to that of Berlin. Finally, Hoechst suggested awaiting the results of the already implemented extensive rationalization and restructuring measures before undertaking new measures.[57]

But the matter was still not settled. Elberfeld insisted on taking over more dyes from Hoechst: "The main goal of restructuring and rationalization, apart from improving the utilization of equipment, is to limit the number of brand names. But neither purpose is attained, if one of the plants refuses to cooperate and retains its former production, even though this represents only a fraction of the output of the new manufacturing site." In the case of food colorings, Elberfeld suggested that Hoechst pay a visit to ascertain the outstanding quality and economic viability of Elberfeld's facilities.[58] Hoechst coldly responded that it did not need to view the facilities, since they were well known. Hoechst also demanded that Hoechst – and not Wolfen – should be chosen as a second production site for food colorings, because its equipment was as profitable as Elberfeld's and much more so than Wolfen's. If one of the plants was obliged to yield, in the interest of IG Farben, it should be Wolfen.[59]

Elberfeld replied sharply, noting "regretfully" that Hoechst had not attempted to disprove "our objective report," but tried to lecture the other plants on the "IG position," taking the absurd line that the new allocation of products should be decided on the basis of prior production. "Obviously, we cannot agree with this totally erroneous point of view." The only decisive factor was the "profitability of the facilities" and there was "no doubt as to where the best equipment was to be found." Moreover, the figures quoted by Hoechst were misleading, since they could not be subjected to comparisons.[60] Wolfen's reaction to the figures quoted in support of the efficiency of the various locations was even sharper. In response to Hoechst's claim that its facilities were equal to those of Elberfeld and superior to Wolfen's, Wolfen replied that "as the previous correspondence shows, Hoechst initially did not want to communicate its figures." In contrast, the Elberfeld, Mainkur, and Wolfen plants already exchanged data in February and sent the figures to Hoechst: "It was only at that point that Hoechst was prepared to participate and sent its figures on March 8, 1926. We were particularly struck by the fact that in Hoechst's letter dated March 12, its costs were always lower than those of Wolfen, whereas in the first exchange in 1925 they were approx. 30% higher. We were also struck by the fact that Hoechst tried to pass off its primitive equipment as equal to

[57] HA, TEA 66, Hoechst to Mitglieder der Unterkommission für bunte Schwefelfarben und schwarze Schwefelfarben, August 19, 1926 (copy).

[58] HA, TEA 66, Elberfeld to Mainkur, Hoechst, and Wolfen, September 1, 1926.

[59] HA, TEA 66, Hoechst to Mainkur, Wolfen, and Elberfeld, September 11, 1926 (copy).

[60] HA, TEA 66, Elberfeld to Mainkur, Hoechst, and Wolfen, September 18, 1926.

the Elberfeld Frederking facilities. Given Hoechst's views, we felt it would be pointless to respond to the letter, since the figures of the 3 different systems had already been evaluated by Elberfeld." Wolfen added that Hoechst's assumption that its equipment was on a par to that of Elberfeld was "probably not shared by anyone."[61] The negotiations with Hoechst on rationalizing and restructuring the food colorings sector had "reached a dead end."[62] Elberfeld still attempted to impose its solution and enlisted the support of Krekeler, chairman of the Technical Committee: Hoechst should surrender its production of food colorings to Elberfeld and Wolfen, since their equipment was more modern and more efficient. This would uphold the principle that dyes should be produced only in the two locations with the most profitable facilities.[63]

The disputes briefly outlined above establish three important points. First, in both cases Hoechst was isolated, since the other parties managed to reach an agreement. Hoechst was either bypassed deliberately, or it stood apart and attempted to retain products by questionable means, as the the example of the belated submission of its figures for the food colorings shows. Second, Hoechst had to fight for recognition within IG Farben, in particular against frequently expressed and often legitimate accusations of low profitability and antiquated facilities. Third, the discussion shows that the plants still needed to find a common ground and that rationalization measures were difficult to enforce. All in all, it is clear that Hoechst's position within IG Farben was difficult, partly because of its outdated production facilities and partly because of the inability of its managers to form coalitions. As a result, Hoechst faced coalitions forged between the other plants.

Hoechst lost only 2.5 percent of its dye production after the first measures by IG Farben. However, production at Leverkusen increased by 27.4 percent, and at Ludwigshafen by 6.2 percent. Hoechst's Azo Department was the hardest hit: its production sank by approximately 20 percent. As the commemorative volume published on the occasion of the company's hundredth anniversary commented, "For the first time, therefore, the equilibrium of the former big three was upset. Until then, all three were of approximately equal strength in the dyestuffs field. But now, after the great sacrifices on the I.G. altar, Hoechst was reduced to third place."[64] But, to a large extent, Hoechst could only blame itself for the situation it found itself in.

Hoechst's unfavorable circumstances in the years preceding the Great Depression became clear in the summer of 1926, when Carl Bosch sent Fritz

[61] HA, TEA 66, Wolfen to Mainkur, Hoechst, and Elberfeld, October 28, 1926.
[62] HA, TEA 66, Elberfeld to Struss (TEA), September 29, 1926 (handwritten).
[63] HA, TEA 66, Elberfeld to Krekeler and Leverkusen, October 19, 1926.
[64] Bäumler, *Century*, p. 101.

Gajewski, a trusted colleague, to the plant to investigate investment plans and equipment. During this "inspection," violent quarrels arose between Hoechst's management and Gajewski.[65] In retrospect, one of Schmidt's closest confederates at the time even claimed that Bosch had wanted to assess whether it was still worthwhile investing in the Hoechst plant.[66] This was probably a somewhat exaggerated interpretation of Gajewski's mission, but it nevertheless highlighted the extent to which the "old Hoechst hierarchy," its data, and its statements had become discredited. In his memoirs Karl Winnacker, later Chairman of the Management Board of the Farbwerke Hoechst, elaborated on the plant's problems within IG Farben. He remembered from his student days at the Technical University Darmstadt "that the papers had then carried reports that the works of the Maingau were to be closed down altogether."[67] He recalled Fritz ter Meer's particularly negative opinion of Hoechst, even though Winnacker commented that ter Meer could "be very unjust in his opinions." Thus, much later, the two men met up on the evening before the Hoechst anniversary of 1963: "Ter Meer had never been very fond of the old Hoechst hierarchy or its politics and even now, over a glass of wine, he repeated his charge that all had not been well during the negotiations preceding the I.G. merger. He probably meant that Hoechst had glossed over its real situation."[68]

During the years of the Great Depression, the "old Hoechst hierarchy," as ter Meer called them, suffered the consequences of the IG Farben management's understandable mistrust. They became increasingly isolated within IG Farben and were gradually forced into retirement. Moreover, in the course of the Depression, Hoechst had to suffer significant cuts in both production and sales.

2.3. FORCED RETIREMENT OF THE "OLD HOECHST HIERARCHY" DURING THE GREAT DEPRESSION

The rationalization measures in IG Farben continued apace during the Great Depression, and the rule of retaining at least two production sites for every important product was abandoned.[69] Ter Meer's statement that the individual plants remained "strong and self-contained organisms" was adhered to until the Depression.[70] But in 1929, given the mounting pressure of the economic slump, on Duisberg's initiative the Working Committee set up a committee to determine potential cost reductions. With this aim in mind, IG Farben's various areas of activity were grouped into three major

---

[65] See HA, Hoe 24, visit of Gajewski in the summer of 1926.
[66] HA, Otto Hirschel, "Aus meiner I.G.-Zeit. Erinnerungen," November 1967, pp. 10–11.
[67] Winnacker, *Challenging Years*, p. 70.   [68] Ibid., p. 103.
[69] Bäumler, *Century*, p. 101.   [70] Ter Meer, *I.G. Farbenindustrie*, p. 30.

Product Divisions (*Sparten*). Carl Bosch commissioned three managers he considered particularly competent to monitor the investments in these divisions. Division 1 under Carl Krauch dealt with nitrogen and fuels. Fritz ter Meer was responsible for Division 2, which included dyes, chemicals, intermediate products, solvents, pharmaceuticals, and pesticides. Fritz Gajewski became the head of Division 3 consisting of photography products and synthetic fibers. To begin with, the three men were supposed to only implement economy measures and control investments. But the company's management soon realized that this new divisional organization was also suitable for the management of the company according to technical requirements. The new structure meant that the plants lost a large part of the autonomy they had formerly enjoyed, since from then on the divisions had a direct say in their management.[71]

In 1930 the IG Farben Central Committee, chaired by Bosch, was created to replace the Working Committee as the top executive body in all important questions, including those on personnel and "confidential matters."[72] Until well into the 1930s, the Central Committee was IG Farben's "top management" and was responsible for strategic decisions. Paul Duden and Georg von Schnitzler were the only Hoechst members on this Committee, the former until his retirement in 1932, and the latter from 1930 to 1945, first in his position as head of the Dyes Sales Group (*Verkaufsgemeinschaft Farben*), then from 1937 as chairman of the Commercial Committee.[73]

Following Duisberg's death in 1935, Bosch joined the Supervisory Board, and the Central Committee consequently declined in importance, so that it dealt mostly with decisions relating to personnel. This development was also linked to the changes in joint stock corporation law that took place in 1938 and resulted in the dissolution of the Working Committee. The entire Management Board subsequently became the topmost decision-making body. The differentiation between full and deputy Management Board members also was abrogated. The Management Board initially included twenty-seven members, but during the war years this figure was reduced to around twenty-one. Thus, ultimately, the entire Board was approximately the size of the former Working Committee.[74]

After the war Fritz ter Meer wrote that board meetings had been like "a discussion between good friends." Of course, there had been divergent opinions but "never fights." Board members were well aware that each member's considerable discretionary powers, and the existing distribution

---

[71] Hayes, *Industry and Ideology*, pp. 19–21; Plumpe, *I.G. Farbenindustrie*, pp. 147–9; cf. ter Meer, *I.G. Farbenindustrie*, pp. 31–2.

[72] Ter Meer, *I.G. Farbenindustrie*, pp. 56–7.

[73] Hayes, *Industry and Ideology*, p. 23; Plumpe, *I.G. Farbenindustrie*, 149–52.

[74] Hayes, *Industry and Ideology*, pp. 23–4; Plumpe, *I.G. Farbenindustrie*, pp. 156–8.

of tasks meant that each "had to be the most knowledgeable in his field" and to act accordingly.[75] After 1938 each member of the Management Board saw "only that portion of the minutes and IG's balance sheets related to his sphere of operations." As Peter Hayes put it, IG Farben's "spiraling growth" in the 1930s "only strengthened centrifugal forces within the firm."[76]

Hoechst had to make many sacrifices after the fusion and often stood alone against the other IG Farben plants. Most crucially, Hoechst could not argue convincingly for its own interests if the figures it presented did not add up and the other plants' equipment was more modern and operated more profitably. Thus, the rationalization measures implemented during the Great Depression affected Hoechst with particular severity, as it lost many products and even entire product lines. The company suffered from the fact that its management had not taken appropriate measures in the 1920s to cut the costs of production and compete with the other IG Farben plants. Thus, the plant was not only technically outdated, its organization was also rather inefficient.[77]

As mentioned before, a number of small "fiefdoms" had formed within the company. Accordingly certain product groups were manufactured simultaneously at several different sites in the plant, and the management failed to centralize these production sites, a move that might have led to greater economies of scale and better results by bringing together the expertise scattered in different locations. Vat dyes, for example, were produced in no fewer than three departments both before and after World War I: in the Alizarin Dyes Department, in the department headed by Albrecht Schmidt, and in the Azo Dyes Department. The fragmented production was for "historical" reasons, since Hoechst followed the principle of allowing new products to be manufactured by the department where they had been invented or developed.[78] This also applied to the so-called Helidone dyes. At first little attention was paid to thioindigo dyes, but once their quality was assured, they were developed further at two different locations, Schmidt's Patents Laboratory and Wilhelm Roser's Central Laboratory: "In both laboratories, a great number of thioindigo dyes were developed which were later manufactured under the 'Helidone' brand. The dyes created in the Central Laboratory were manufactured in the Azo Department, while those from the Patents Laboratory were produced by Schmidt's department."[79] It is surely unnecessary to comment on the decline in profitability resulting from such idiosyncrasies. A centralization

[75] Ter Meer, *I.G. Farbenindustrie*, pp. 54–5.

[76] Hayes, *Industry and Ideology*, p. 24; cf. Plumpe, *I.G. Farbenindustrie*, p. 158.

[77] Winnacker, *Challenging Years*, pp. 69–71; Bäumler, *Century*, pp. 100–2.

[78] HA, 12, Hirschel, "Das Werk Hoechst im Verbande der I.G. (1916–1945)," May 1961, p. 154.

[79] Ibid., p. 158 for the quotation; Bäumler, *Century*, pp. 100–1.

of production and combination of skills and strengths would have led to better outputs and a more efficient utilization of capacity – thus certainly ensuring a higher profitability. Hoechst continued to retain old habits that it could no longer afford, and Duden's conciliatory management style was incapable of ending these practices in favor of higher economic viability and competitiveness.

Albert Schmidt, mentioned above, was one of the "old Hoechst hierarchy" who clung to his privileges and categorically insisted on maintaining "historically established" customs, for vat dying and Helidone dyes as well. He was considered one of IG Farben's important researchers and chemists, not least because of his successful self-promotion. His memoirs and the publication in 1934 of *Industrial Chemistry and Its Significance for the View of the World* (*Die industrielle Chemie und ihre Bedeutung im Weltbild*), a book imbued with Nazi ideology, contributed to his reputation.[80] A committed member of the NSDAP and the SS, he retired from Hoechst at the end of 1931, after which he officiated as a consultant to the *Gauleitung* of Hesse-Nassau on the Four-Year Plan and on university matters. Himmler, for whom he had written some reports, conferred a high honorary rank in the SS on him. In 1944 Schmidt even became *SS Brigadeführer*.[81]

On the occasion of the festivities held to mark Schmidt's retirement, Paul Duden used the following metaphor to describe him: "If I may be so daring as to compare you with a chemical compound, to which group do you belong? It is quite obvious that you are not part of the aliphatic series, but neither do you belong to the aromatic compounds. Rather, you should be counted among the highly unsaturated compounds, which, with their strong primary and secondary valences, attract everything within close range. But you are not one of those chemical systems that can, without difficulty, be rotated around their own axis. No, you resemble much more the rigid, nay, spiny systems, charged with inner tension and energy, and which are thus particularly interesting for chemists, since they discharge their energy and tension toward the outside in the form of interesting chemical reactions. Perhaps there is a certain affinity with explosives, the power of which is also measured in terms of energy content."[82] This chemical metaphor was a polite way of describing a very ambitious person who craved for recognition and continually believed himself slighted and insufficiently appreciated, in both material and immaterial respects.

---

[80] Cf. Christopher Kobrak, *National Cultures and International Competition. The Experience of Schering AG, 1851–1959* (Cambridge: Cambridge University Press, 2001), pp. 35–8; Heine, *Verstand*, pp. 126–8; Reinhardt, *Forschung*, pp. 304–9.

[81] See Heine, *Verstand*, pp. 126–8; HA, PA Schmidt, vol. 2, Biographisches and article from *Hoechster Kreisblatt* of July 3, 1939; see also BA, BDC file Schmidt and NS 19/3875, Schmidt's correspondence with Himmler and other top ranks of the SS in 1941 and 1942.

[82] HA, PA Schmidt, vol. 2, speech on Duden in Worte anläßlich der Feier des Ausscheidens von Schmidt und Tiedtke, February 16, 1932.

After studying chemistry, Schmidt's first permanent position was with the Schering pharmaceutical company in Berlin, where he was head of the "Scientific Research Laboratory" from 1888 to 1898. Even then, he considered himself to be undervalued and complained accordingly.[83] He joined Hoechst in 1898, established the Patents Laboratory, and worked primarily in the field of dyes. During the First World War he developed a smoke bomb and apparently worked successfully on a combat gas. In 1910 he became an authorized signatory (*Prokurist*) for Hoechst, and in 1916 a deputy member of its Management Board, before becoming a full member of the IG Farben Management Board in 1925.[84] His greatest success in research, which he never tired of mentioning and which Duden referred to in his speech on his retirement, was the invention of artificial fog used by the German navy during World War I. He considered this invention as a personal, private matter that he had naively signed over to Hoechst at a time when he was not in possession of all his faculties due to illness. The company had made enormous profits from it and from other achievements of his. Instead of being given 5 percent of the profits, he had been fobbed off with a single lump-sum payment. His anger surfaced in a series of detailed letters written between 1917 and 1918. Hoechst had awarded him 50,000 RM, a sum he found far from adequate. He felt that Hoechst had "sacrificed" his "incentive to work" in favor of a rigid principle, while he had expended "immense" efforts "over many years and in his spare time" to work on "purely military inventions."[85]

In addition to these material questions, he was also concerned about his status within Hoechst's hierarchy, believing that, as had previously occurred at Schering, Hoechst did not sufficiently appreciate him. Thus in December 1920 he wrote to Haeuser, stating that he could not accept what he and others perceived "to be objectively an incredible humiliation": "For me, the formal affront is the central issue, while the proposed material trade-off in my area of work is really in the interest of the company which will be able to make better use of my skills and experience than has previously been the case." In a letter he presented what he considered an acceptable solution: "the official announcement of my appointment as the representative of Prof. Duden in technical matters." Furthermore, when Max Epting – then a member of the board – resigned, Schmidt should become a full member of the Management Board.[86] Haeuser replied in mid-January 1921, stating bluntly that Schmidt had no reason to feel passed

[83] Kobrak, *National Cultures*, pp. 35–8, see especially n. 88 on pp. 35–6.
[84] HA, PA Schmidt, vol. 2, Biographisches, Zeittafel.
[85] HA, PA Abteilung VI, Schmidt, Schmidt to vom Rath, May 31, 1917, and further letters with similar, slightly varying arguments of October 20, 1917, May 29, 1918, and June 23, 1918, where also the profits for Hoechst and the military importance of the invention were stressed.
[86] Ibid., Schmidt to Haeuser, December 29, 1920, see also letter dated January 10, 1921.

over. The important persons in Hoechst "all agreed that the appointment of Dr. Weidlich as a full member of the Management Board cannot be interpreted as a measure aimed to exclude or even disparage you." The reasons prompting Weidlich's appointment, which Schmidt perceived as an insult, were of a strictly "administrative nature." Haeuser added that Schmidt's views of the significant difference between full and deputy members of the Management Board were not generally shared, nor that "the promotion of a deputy board member to full board member was a slight to the other deputy board members." If that were the case, then full board membership would only be linked to the number of service years, and that would be truly undesirable. Consequently the Hoechst Management Board refused to commit itself to proposing to the Supervisory Board that Schmidt be promoted when a member of the Management Board resigned. However, the board was prepared to organize Schmidt's area of activities in accordance with his wishes. Moreover, as he was the oldest member of the Management Board responsible for technical matters, he was entitled to "represent Prof. Duden at technical conferences, etc."[87] The result of these negotiations can be found in the form of a handwritten, nondated declaration in Schmidt's employee records signed with the characteristic "S" Schmidt used for his signature. The document is worth quoting in full: "In connection with his promotion to full member of the Management Board and representative of Prof. Duden in technical matters, S. freely and in all due legal form, as sealed by a handshake in good faith, states that his activities on the board are and will remain of a strictly technical nature. He agrees that all administrative matters are and will remain outside his sphere of authority and interest. Furthermore, he agrees – without perceiving it as an affront – that his inclusion in or exclusion from particular sessions with the full members of the board is and will remain subject to the judgment of the chairman of the board."[88] Thus Schmidt's demands for formal recognition were met, but this was not linked to an increase in either authority or influence.

In December 1925 Schmidt at last received the official promotion to the Management Board. However, probably on the basis of the above-cited statement, he did not become a member of the important committees, that is, the Working and Technical Committees. His role was nevertheless not insignificant, since he often – in his typical, rather arrogant way – represented Duden when dealing with technical matters outside the plant and thus had some external impact. If Schmidt's own statements are to be believed, his work was very advantageous for Hoechst. This may sometimes have been the case, but his presence and behavior probably contributed

---

[87] Ibid., Haeuser to Schmidt, January 14, 1921.
[88] Ibid., handwritten statement of Schmidt, signed only with paraph, undated, supposedly written between 1920 and 1925.

more to Hoechst's isolation than to improving its position within IG Farben. And since it was generally known in the company that the nature of his authority was formal rather than actual – not least because he was not an official member of either the Technical or the Working Committee – to his annoyance, he saw himself treated accordingly.[89]

He often complained to his superior Duden. In September 1928, for example, he wrote that he had been "severely snubbed" by Duden, but did not want to let "any ill will" develop; he only insisted on setting the matter straight. He believed that it was crucial not to air quarrels "in the presence of subalterns," especially since he was convinced that Duden, when he had time to think things over calmly, would agree with him. The conflict between the two men must have escalated considerably, since Duden, usually a quiet and equable man, appears to have reacted violently to Schmidt's words and threatened to take over the Alizarin Department, which Schmidt headed at the time. Schmidt complained that Duden would never have dared do that with another member of the Management Board, such as Ammelburg, and that it was not appropriate to confront a full member of the board, holding the same rank, in such a way. In his forty years as an executive in industry, thirty of which he had spent at Hoechst, he had occasionally held views that differed from those of others. However, "the company's success has always vindicated the views I held, even though I have reaped only anger, ungratefulness, and often personal attacks." Without his intervention, which took its toll on his nerves, many products would have been lost to Hoechst. Yet he experienced nothing but ingratitude, as Duden's remarks showed. Not "lust for power" but objective reasons had prompted him to take over the Alizarin Department, since the struggle with Ludwigshafen and Leverkusen and the department's situation demanded a unified management. Furthermore, his "old colleague, Dr. Kränzlein," also needed "permanent technical etc. leadership to reveal his considerable talents." Schmidt continued: "When I think of the frequent scenes occurring in connection with the excellent Dr. Kränzlein, and consider how today he has appealed to the older management, your statement offends me – and I am not going to broach the subject of your granting Dr. Kränzlein the chemists I had good reasons for refusing him. Do not misunderstand me: doubtlessly we owe our current sound basis largely to the excellent Dr. Kränzlein. But if he and Commission IX are not treated in an authoritarian fashion, his labors will not be successful." Schmidt also reminded Duden of how he had managed to retain the production of a number of dyes for Hoechst in the teeth of opposition from Ludwigshafen's Kurt Hans Meyer, a member of IG Farben's Management Board. Yet this achievement only resulted in Schmidt being reproached for extremely

[89] Cf. HA, Hirschel, "Aus meiner I.G.-Zeit. Erinnerungen," November 1967, pp. 9–12, on the time when he was a close colleague of Schmidt.

inappropriate behavior, which "negatively affected my relationship to Mr. Bosch." But now to reap ingratitude from Hoechst was too much: "Moreover, although I am aware it is not your fault, there is my miserable position within IG: I am neither a member of the T.C. nor of the W.C.; I am barely tolerated in the first committee, while in the second I am not accorded the recognition given to younger people, such as Dr. Greif." This made his "struggles extremely difficult," and therefore Duden should in future avoid "such hurtful treatment within the company." Furthermore, Schmidt asked Duden to remember that he was only acting in obedience to "a hereditary, unavoidable sense of duty" by striving "to achieve what is best for the plant and thus finding my own inner satisfaction." Of course, it was impossible for them to always agree, since Duden held an important position and had to take more account of the interests of IG Farben than Schmidt did. Schmidt felt that his own duty lay in "enforcing Hoechst's interests," in contrast to Duden, who was "persona gratissima whereas I am persona ingrata, but this does not affect me very much." Immediately afterwards, he contradicted this statement by once again refering to his role within IG Farben and the various committees: "I have not been given a single mandate in the Technical Committee to permanently legitimize my position. I am thus dependent on my good conduct, on my being tolerated in the Technical Committee."[90]

Schmidt was convinced of his own superiority and considered everyone else as not particularly competent, even going so far as to make comments in a similar vein about Fritz ter Meer. Schmidt remarked that ter Meer was easily dazzled by appearances and had a limited viewpoint: "despite his overall capabilities" ter Meer saw "everything through the lens of the small Uerdingen plant. He has not yet dealt with more complex situations, at least not to any great extent."[91] Schmidt's letter to Duden on the production of rhodamine has a similar tenor: "Contrary to the opinion held by Dr. ter Meer, there can be no doubt that our facilities are absolutely modern. I firmly demand that Dr. ter Meer examine this matter once more. ... There is nothing to reproach Hoe[chst] with, apart from earlier, mistaken calculations, which have now been completely revised."[92] In other circumstances Schmidt had more faith in ter Meer, particularly when he believed that they held the same views. Schmidt's colleagues were no doubt aware how inaccurate Schmidt's perception of his own position was and how imprudently he could act – and this did not cast a positive light on Hoechst. Ter Meer's visit to Hoechst in April 1930 provides another good example of Schmidt's behavior. Schmidt reported to Duden that, in the course of their extended discussion, he had received the impression that

[90] HA, PA Schmidt, vol. 1, Schmidt to Duden, September 12, 1928.
[91] Ibid., Schmidt to Duden, September 13, 1929 (altogether four letters).
[92] Ibid., Notiz Schmidt for Duden, July 4, 1930.

ter Meer was willing "to find out the truth," indeed, Schmidt felt that ter Meer was "gradually siding with us." Schmidt had decided not to invite Ludwig Hermann to take part in these very important discussions "in order not to disturb what was a private conversation." In his talk with ter Meer, Schmidt nevertheless reported on "all the functions Dr. Herrmann has taken over and what is being planned for him."[93]

This was doubly strange in view of the fact that Ludwig Herrmann, who had been manager of the IG Farben plant in Gersthofen until the end of 1929 and joined Hoechst at the beginning of 1930 as its future manager, had been chosen by Carl Bosch and was not the candidate of choice of the "old Hoechst hierarchy." During the forty-eighth session of the Working Committee chaired by Bosch on November 22, 1929, in Leverkusen, the "reorganization of the Middle Rhine and Central Germany Works Groups" was discussed under point 11b of the agenda. Paul Duden's appointment as manager of the Middle Rhine Works Group was confirmed, and Ludwig Herrmann was designated his deputy and successor.[94]

However, shortly after Hermann's appointment, Hoechst's situation deteriorated dramatically in the winter of 1929/30. Carl Ludwig Lautenschläger, who later succeeded Hermann as Hoechst's manager, wrote in his personal memoirs that the plant suffered extensively from the rationalization and restructuring measures carried out in IG Farben between 1926 and 1932, when many of its production sites were closed down: "Moreover, a number of unsuitable arrangements made during the inflation years by a few members of our directorate and other employees proved to be irregular, so that the good and established reputation of our company was strongly discredited. Thus, during the first years after the merger, it appeared as if the future existence of the former Hoechst Farbwerke was in doubt."[95] Indeed, no sooner had Duden's successor been confirmed than a scandal involving members of the Hoechst management broke out because of these "irregularities." It lastingly damaged the reputation of the entire management outside the firm, within IG Farben, and particularly in Hoechst itself. At the root of the scandal was apparently the purchase of the Grüneburg site for the new IG Farben headquarters in Frankfurt and/or some dubious financial speculations. Unfortunately the existing sources do not allow the scandal and its causes to be fully reconstructed. Suffice it to say that it was significant enough to find mention in the memoirs Lautenschläger compiled for his children.

[93] Ibid., Notiz Schmidt for Duden, April 14, 1930, on ter Meer's presence on April 9, 1930.
[94] HA, TEA 106, Niederschrift über 48. Sitzung AA, November 22, 1929; Winnacker, *Challenging Years*, pp. 74–5; PA Hermann, speech of Fritz ter Meer on the occasion of Hermann's funeral.
[95] HA, PA Lautenschläger, Lautenschläger, "Erinnerungen," vol. 2 (unpaginated).

The principal actors of the scandal were the Management Board members Richard Weidlich and Carl Ref, together with the authorized signatory Jakob Safran – all originally from Hoechst. Subsequently all three men were obliged to resign. The minutes of the fiftieth session of the Working Committee, held on February 28, 1930, laconically state: "*Geheimrat* Bosch then reports that Dr. Weidlich has resigned from the Management Board, as has Mr Ref (Hoechst). The Working Committee has duly taken note of this."[96] It was an unusual turn of phrase to describe the departure of members of the Management Board. During the forty-eighth session at the end of November 1929, which also saw the appointment of Hermann, Bosch had announced that several Management Board members would be resigning at the end of the year. He thanked them in the name of the Working Committee for "their tireless efforts in promoting IG's interests" and wished them "the very best for the years to come."[97] There was no mention of any such recognition of services in the fiftieth session.

Weidlich was Haeuser's protégé and close colleague, and in Hoechst he had apparently been considered Duden's successor.[98] Like Haeuser, he had studied law and chemistry. After joining Hoechst in 1903, he had a brilliant career – to Schmidt's great displeasure. In 1911 he became an authorized signatory for Hoechst; in 1916 he was taken onto the Hoechst Management Board as a deputy member, and he became a full member in 1920. As one of Hoechst's representatives, he played a pivotal role during the negotiations prior to the fusion. In 1925 he became a full member of the IG Farben Management Board and a member of the Working Committee. He also held important positions within the corporation, as director of the Patents Commission, head of the pension fund for white-collar workers, and the company's representative in a number of employer organizations and industrial associations.[99] In 1928 he was still feted on the occasion of his twenty-fifth anniversary of joining IG Farben, but one year later he was transferred – apparently not entirely voluntarily – to Agfa in Berlin, where his task was to reorganize some of the administration according to instructions from Weinberg, Bosch, and Duisberg. At the time of his forced resignation early in 1930, Weidlich was only fifty-two years old.[100]

[96] HA, TEA 106, Niederschrift über 50. Sitzung AA, February 28, 1930.

[97] Ibid., Niederschrift über 48. Sitzung AA, November 22, 1929.

[98] HA, PA Haeuser – Weidlich's and Haeuser's speeches on the occasion of Haeuser's seventieth birthday allude to the subsequent arrangements.

[99] HA, PA Weidlich, Biographisches, article on Weidlich from *Die Farbenpost*, October 31, 1962.

[100] HA, PA Abteilung VI, Weidlich, Weidlich to Office of the U.S. High Commissioner for Germany, IG Farben Control Office, March 7, 1950. Cf. also Heine, *Verstand*, pp. 152–3; Heine, usually very well informed, writes on Weidlich: "1929 relocation to Berlin NW 7. 1930 retirement. 1930 attorney-at-law at the Superior Court of Justice of Berlin, activity in the field of intellectual property law in Berlin and Zurich. 1945 retirement in Baden-Baden."

Carl Ref, the second member of the Management Board obliged to resign against his will, began working for Hoechst in 1893. He became an authorized signatory for the firm in 1905. In 1920 he was promoted to deputy member of Hoechst's Management Board and subsequently became a member of the Management Board. Ref's letter of dismissal, dated February 28, 1930, curtly states that "herewith your appointment as a member of the Management Board of our company" is revoked. A single sentence follows: "The settlement of your contractual relationship with our company is subject to further decision."[101] There is no mention of gratitude for particular achievements or of best wishes for his retirement. Ref was sixty-four at the time and died only a few months later, on December 3, 1930. But he did not receive an official pension, and his employee records mention only that he was granted a "voluntary allowance." After his sudden death, payment of this allowance to his widow was continued, but it was emphasized that this was voluntary and thus could be revoked at any time.[102] These proceedings were even more unusual than the very unceremonious and thankless dismissal by the Working Committee – they even smacked of a threat.

Jakob Safran, one of Weidlich's closest confidants and colleagues, was retired at the same time, although he was granted a pension. He had been a Pallottine monk prior to joining Hoechst in 1910, where he quickly rose to become an authorized signatory. From 1916 on he worked in the Real Estate Department.[103]

The reason for these forced retirements cannot be ascertained from the available archival documents. The IG Farben management apparently did everything to preserve the greatest secrecy, as the treatment of Ref and his widow and the payment of the "voluntary" retirement allowance shows. Given the available primary sources, it is difficult to interpret it other than that there must have been financial misconduct, probably in connection with real estate transactions.

Interestingly, an important clue that appears to bear out this interpretation is provided by a collection of newspaper clippings on Weidlich's forced retirement in his employee records.[104] Among them are the following items, taken primarily from the local press and publications linked to IG Farben. The Frankfurt *General-Anzeiger*, dated March 7, 1930, states that

---

[101] Bayer WA, 271/2.1 PA Ref, Duisberg to Ref, February 28, 1930.

[102] HA, PA Abteilung VI, Ref, see Vermerk Orth, June 11, 1930 re. support of Ref and Flach, Abteilung VI (Hoechst) to Nobbe, IG-Werk Leverkusen, March 22, 1939. On Ref Heine only states: "1926 deputy board member of IG Farben, member of the Accounting Commission and the Finance Commission. In 1930 retired ten months before he died" (Heine, *Verstand*, p. 123).

[103] HA, PA Safran.

[104] HA, PA Weidlich, the newspaper articles and reports cited below – collected, clipped, and pasted – can be found there.

Weidlich was fired. But in the *IG Rundschau* from the same day, Weidlich is said to have resigned from the Management Board "on the basis of a friendly mutual agreement." On March 8, "Berlin NW 7," IG Farben's press bureau, announced: "As to the rumors that have been circulating in Hoechst upon Main about IG Farben's Hoechst plant, we have been informed by IG's management that some of the transactions carried out by the plant's Real Estate Department were not approved by the Management Board and that the latter has accordingly taken the necessary measures. The company expressly emphasizes that it has not suffered any pecuniary damages." On the same day, the *Rot-Fabriker* reported that it was becoming more and more evident "that huge frauds have been uncovered, involving sums that probably amount to several millions." On March 8 the *Gross Frankfurter Volksstimme* wrote about a "mystery" at IG Farben. The authors of the articles wondered about Ref's and Safran's suddenly deteriorating health and Weidlich's unexpected resignation – since he was an "extremely healthy man." Although there was talk of a "huge fraud," it was hoped that this type of rumor was only a "malicious report." But it was strange that no clear denial was forthcoming and that IG Farben persisted in its "hushing-up tactics," since this only lent plausibility to the rumors. On March 9, mirroring the announcement given by IG Farben in Berlin, the city section of the *Frankfurter Zeitung* wrote that certain real estate transactions had not been approved by the Management Board, which had subsequently undertaken steps to deal with the matter accordingly. However, the article added that Weidlich, who had previously been the director of the department, had been asked to resign "even though he apparently was not involved in the above-mentioned affair." On March 12, 1930, the *Freie Presse* reported that it was now known that "a thorough investigation had become indisputably necessary. .... The final result of this inquiry has been to hold those employees accountable who were negligent in their management of the particular transactions and went against the business regulations of their superiors. Apparently, these circumstances have compelled Messrs Safran and Ref to resign. Furthermore, the Management Board member Dr. Weidlich, who is said not to be involved in this affair in any way and who bears only formal responsibility, has nevertheless drawn the consequences and resigned from the Management Board in agreement with the firm." However, IG Farben had not incurred any damages. In contrast, on March 14, *Die Fackel* reported "frauds amounting to millions" at Hoechst. As in the Favag scandal,[105] "highly respected directors" were embroiled in these crimes "at a time, when most people in Germany are starving," yet such men pocketed "indecently high salaries." But on March 18 the city section of the *Frankfurter Zeitung* reported the explanations given by IG Farben's management and Weidlich concerning the latter's

[105] See Feldman, *Allianz*, pp. 39–45.

resignation. Apparently, there had been nothing more than "massive disagreements between Dr. Weidlich and the gentlemen of IG's Management Board concerning the scope of activity and competences of the Real Estate Department under Dr. Weidlich, as well as about his supervisory duties." Moreover, it was emphasized that IG Farben had not suffered any pecuniary damages and that Weidlich's personal integrity was not in question. An article in the *Deutsche Bergwerks-Zeitung* from March 22 summed up the debate. It deplored the "loathsome sensation mongering" of the press that had resulted in the fabrication of "the worst possible manner" of legends: "At any rate, we are in full possession of all information and can confirm that this affair has nothing to do with what could be described as scandal, embezzlement or the like." Instead, the top management had merely intervened "to prevent a subordinate from acting in a way that could be construed as discrediting. Thereupon, the Management Board member responsible for supervision voluntarily drew the consequences."

A letter from Duisberg, dated December 4, 1929, to all the members of the Management Board of IG Farben, which – significantly – remains in Ref's employee records, supports this latter version of events. It states that a number of Supervisory Board members had been told, particularly by shareholders, that "the salaries earned by members of our Management Board were probably too high, since both they and members of their family publicly indulge in forms of excessive consumption, which are not commensurate with the current economic situation." Therefore, at the end of November 1929, the Administrative Council decided that, "in the interest of our firm's reputation," Duisberg should urgently entreat all members of the Management Board "to adapt their standard of living and their conduct to the rather unpropitious situation of our economy, in order to avoid eliciting disparaging comments or criticism directed against individual members of the board." Duisberg added that recipients of the letter "should take note of the provision in their employment contracts, stating that all members of the board are bound 'to refrain from stock exchange transactions or other speculative ventures that could somehow bring discredit on their position as Management Board members.'"[106] Was this what had actually happened? Had members of the board engaged in "discrediting ventures" in these difficult times? Or was something rather more serious involved?

A few letters on real estate transactions can be found in the files. At the beginning of August 1927, Gustav Orth from Hoechst reported to Weidlich, who was spending a holiday in Karlsbad, that Safran had just left on vacation: "He seemed very satisfied about the outcome of the last negotiations in connection with Grüneburg Park. Apart a few negligible plots, he has bought all the necessary sites and houses. He ended up having

---

[106] Bayer WA, 271/2.1, PA Ref, Duisberg to ordentliche und stellvertretende Mitglieder des Vorstandes der I.G. Farbenindustrie Aktiengesellschaft, December 4, 1929.

to pay a bit more than he had originally anticipated for some of the houses. But the final accounts will probably prove quite advantageous."[107] In October 1929 Orth reported to Weidlich in Berlin, the latter's new place of work, that a first discussion about a planned real estate transaction had been held with Müller, the former mayor of Hoechst, who became a Frankfurt alderman after the incorporation of the municipality of Hoechst into the city of Frankfurt: "The negotiations completely exhausted Mr. Safran. At the end, he was almost apathetic, probably due to a breakdown he suffered during the previous night." Safran was now convalescing.[108] It must be added that at the end of May 1929, Safran was involved in a train accident that left him suffering from nervous disorders and depression. It is difficult to assess how far Safran was involved in the affair, but he apparently received his normal pension.[109]

In many respects, these accounts still leave us in the dark. Unfortunately Weidlich's memoirs and those of his widow, both of which have survived in Weidlich's employee records, do not shed much more light on the matter. In August 1948 Weidlich penned reminiscences about his time at IG Farben, in which he enlarged on his own role for the company. He deplores the massive discrimination he felt Hoechst had suffered after the fusion, repeatedly stresses his conflicts with Duisberg, and eventually mentions the purchase of the Grüneburg site, the premises for the new IG Farben administration building. Hoechst had been entrusted with finding a suitable location, and the site looked relatively affordable. Although IG Farben had not taken a decision yet, he had argued in favor of taking the risk and immediately seizing the opportunity. At first, the company had rejected the purchase, since the Griesheim plant had already acquired a plot of land for the construction of a skyscraper. Thereupon, according to Weidlich, Hoechst began negotiating to sell the land. While these negotiations were taking place, Bosch again expressed his interest in the Grüneburg site because the Griesheim site near the Main was too small. Hoechst therefore transferred the Grüneburg site to IG Farben. Yet Weidlich reaped very little gratitude for his efforts. On the contrary, Duisberg accused Hoechst, and especially Weidlich, of "speculating at the expense of IG." The Hoechst members of the Supervisory Board, Walther vom Rath and Adolf Haeuser, as well as Hoechst's director, Paul Duden, vehemently contradicted the accusation and clarified the state of affairs: "It goes without saying that we did not demand or obtain anything from this transaction, but withdrew in favor of IG as soon as it was clear that it wanted the site (a matter I do not wish to omit, since there have been rumors that I made huge profits from the purchase and sale of the Grüneburg site to IG)." Weidlich then added: "I do not want to elaborate here on my activities on Agfa's Management

---

[107] HA, PA Weidlich, Orth to Weidlich, August 4, 1927.
[108] Ibid., Orth to Weidlich, October 12, 1929.   [109] HA, PA Abteilung VI, Safran.

Board in 1929 and 1930, or on my resignation from IG. I have done this elsewhere."[110] Unfortunately the document he refers to has not survived, but a report by Weidlich's widow exists.

On October 31, 1962, an article on Weidlich was published in the *Farbenpost*, Hoechst's company magazine. The plant's archivist at the time had researched the affair and contacted Weidlich's second wife Marianne (Weidlich's first marriage ended in divorce in 1926). In July 1962 she replied to Fischer: "In my opinion, the resignation of my husband is of no relevance to an examination of his activities in expanding the Farbwerke and IG. It need not even be mentioned."[111] This wish was gallantly respected by the *Farbenpost*. It mentioned only that Weidlich had celebrated his twenty-fifth anniversary in 1928 and on that occasion had been awarded an honorary doctorate from Braunschweig Technical University: "A few years later, Weidlich resigned from IG Farben and worked in Zurich in the field of industrial property rights and as a lawyer in Berlin, where he worked at the *Kammergericht* until the end of the war."[112] However, in her letter Marianne Weidlich provided her version of the affair that she did not wish to see published. According to her, the reason for her husband's forced resignation lay in the merger and the power struggles it had unleashed within IG Farben. Weidlich was "in effect the only person" who stood up for Hoechst's interests, especially against Duisberg and had thus often quarrelled with him: "After a key discussion with Haeuser, my husband had the impression that, despite 25 years of working together, he was not getting sufficient support, so without further ado he decided to resign from the Management Board and avoid further discussions. Thus, personal rivalry and resentment were the only reasons that played a role." Erwin Selck, her husband's "only good friend" on the Management Board, had apparently played a particularly disgraceful part. Not only had he supported the attacks against Weidlich very early on, but subsequently, together with Bernhard Buhl, he instigated a press campaign to prevent Weidlich from rejoining IG Farben. However, later Weidlich worked for the company as a lawyer.[113]

This version may have its merits, but the reproaches against Buhl and Selck seem unjustified, since the affair was settled by oral agreements, concluded at the highest level, and treated as strictly confidential. A report for Duisberg, who was vacationing in Cannes at the time, shows how the matter was settled. Fritz Nobbe, the director of the Finance Committee, wrote to Duisberg: "I returned from Frankfurt yesterday morning. Dr. vom Rath as well as *Geheimrat* Haeuser and *Geheimrat* Bosch entrusted me with

---

[110] HA, PA Weidlich, Weidlich, "Erinnerungen (aus dem Gedächtnis)," August 1948.
[111] Ibid., Marianne Weidlich to Fischer, July 2, 1962.
[112] Ibid., article "Männer machen Geschichte. Dr. jur. Dr. rer. nat. Richard Weidlich (1878–1960)" from *Die Farbenpost*, October 31, 1962.
[113] HA, PA Weidlich, Marianne Weidlich to Fischer, July 2, 1962.

informing you that they had settled the matter concerning W. and R. in such a way as to protect IG's interests. They request that you save any details for a verbal discussion with them." The affair thus not only required the direct intervention of vom Rath and Haeuser, the most important Hoechst representatives on the IG Farben Supervisory Board, it even involved Bosch, the chairman of IG Farben. It was settled by oral agreement, and the persons involved remained anonymous as only the abbreviations W. for Weidlich and R. for Ref were used. Significantly the letter continues: "On the instructions of the above-mentioned gentlemen, I have not carried out further investigations, since I have been entrusted only with the task of drawing up the W. and R. accounts and settling matters with R. directly. Dr. Buhl has been instructed to negotiate with Mr. W. in Berlin. I have delivered my reports to *Geheimrat* Haeuser."[114] This supports the assumption that IG Farben made use of all means at its disposal to solve the matter internally and confidentially, and avoid any leaks.[115] However, these efforts can only be described as naive, since the affair was seized on by the press, and the policy of maintaining secrecy only fueled the rumors.

But Buhl appears to have had nothing to do with the rumors. On the contrary, he attempted to obtain an advantageous settlement for Weidlich. For in December 1930 Buhl arranged on behalf of IG Farben for Weidlich to move to Zurich, where he would be responsible for the company's interests in the area of industrial property rights, if he so wished. Weidlich could continue representing IG Farben's interests even after December 31, 1933, "in accordance with the provisions of an employment contract … and on condition that the pension due to you from January 1, 1934 will substitute for a retainer." But "this commitment to further cooperation becomes invalid if, after December 31, 1933, you decide to definitively stop working as a consultant in the field of industrial property rights." In January 1931 Weidlich was nevertheless obliged to confirm that this arrangement did not make him an employee of IG Farben. Buhl had to fight hard for this agreement, for the Administrative Council objected to it on the grounds that "this indicated a reemployment of Dr. Weidlich, a move that it wanted to avoid at all costs." Buhl contested this interpretation and consequently wrote the January letter. Weidlich decided to accept, and Duisberg finally agreed.[116] Selck also remained a good friend to Weidlich.

[114] Bayer WA, 271/2.1 PA Ref, Nobbe to Duisberg, March 10, 1930.

[115] HA, PA Gustav Orth. On Orth's main activities in the previous two years a Vermerk (Entwurf) of July 19, 1932, among other things mentions "criminal case versus Frohmader-Bernheim" or "Morschbach bribery affair," but then "settlement (*Abwickelung*) Dr. W. und R." – where they felt obliged to anonymize names by using initials.

[116] HA, PA Weidlich, Buhl to Orth, Januar 27, 1931, including in transcript: Buhl to Weidlich, December 17, 1930, Buhl to Weidlich, January 14, 1931, and Buhl to Mitglieder der Rechtsabteilung IG, January 15, 1931.

When the publishing company, which Weidlich acquired after his forced resignation, went bankrupt, Weidlich found himself in serious financial straint, as a lien was placed on all payments from IG Farben. Selck reacted by collecting the sum of 15,000 RM – approximately five times the annual salary of a chemist at the time – among Weidlich's friends and volunteering to be the guarantor of the loan.[117]

From these various sources it appears that Weidlich had indeed been responsible for the department. However, a direct involvement in the financial or property scam seems highly unlikely, since he would surely no longer have been trusted by the management of IG Farben. The company's behavior toward Ref and his widow in the matter of his pension seems at least to indicate an involvement in dubious transactions, if not financial irregularities.[118]

Lautenschläger's memoirs indicate what impression this affair made at Hoechst. Ludwig Hermann's daughter also clearly recollected the affair, for in the course of a conversation about the plant, she immediately referred to the "awful mess back then" that had caused her father so much worry and led to such ill feeling at Hoechst.[119] The negative repercussions within the company can be surmised if one remembers that, at the same time, many employees had lost their jobs and quite a number were reduced to abject poverty. Rationalization measures carried out within IG Farben, and the poor economic situation resulted in a decrease of the number of blue-collar workers at Hoechst from 8,981 in 1929 to 5,615 in the summer of 1930.[120]

Nevertheless, the affair and its consequences and Schmidt's retirement did not result in the complete departure of the "old hierarchy" in Hoechst. Some of the full members and deputy members from Hoechst on the Management Board found attractive posts in other IG Farben locations: Georg von Schnitzler became head of the Dye Sales Department in the new "Grüneburg" headquarters; Hans Eduard Wolff also transferred to Grüneburg, while Wilfrid Greif moved to New York and William Weber to China. On January 1, 1932, Richard Tiedtke, the chief engineer, took early retirement. Only two days later, Hermann Wagner committed suicide,

---

[117] Bayer WA, Personalia 271/2, Weidlich, s.v. "Treuhänderkonto W," Selck to Weidlich, February 27, 1933 (*"von Deinem Selck"*). For the bankruptcy of Hackebeil-Verlag, see HA, PA Weidlich, press review from *Die Fackel* of January 16, 1931 (Bruno Elkin, Was geht im Hackebeil-Verlag vor?), *Frankfurter Zeitung* of January 27, 1931; Abendblatt der *Frankfurter Zeitung* of 26 June,1931; *Freie Presse* of March 1, 1932; *Hoechster Kreisblatt* of March 16, 1932.

[118] On Ref's odd business practices, cf. also HA, PA Abteilung VI, Hübner, Hübner to Duden, October 20, 1932, with a report enclosed.

[119] Personal communication, Lore Wittmer, Ludwig Hermann's daughter.

[120] See Bäumler, *Century*, p. 101.

apparently because of despair at the losses incurred by his department at Hoechst.[121]

Thus the number of IG Farben Management Board members in and from Hoechst dropped dramatically, and the Great Depression and the consequent rationalization measures within IG Farben led to substantial losses in production and sales for Hoechst.

2.4. HOECHST'S DEVELOPMENT DURING THE GREAT DEPRESSION (1929–1933/34)

Early in 1934, one year after taking over the management of the plant, Ludwig Hermann had a statistical report compiled on Hoechst's development during the Great Depression. The report made it painfully clear how much the company had suffered from the Depression and the mismanagement of the "old Hoechst hierarchy." Hermann wanted to show the top representatives of IG Farben in the Technical Committee the serious impact that both the merger and the economic slump had had on Hoechst, as the Technical Committee was responsible for all company investments. The statistical figures used to compare individual plants during different periods were based on the so-called volume of costs: "These include all costs connected with production, that is, personnel costs, energy consumption, maintenance, capital costs (amortization and interest), internal costs of transportation, costs of analyses and dying according to utilization, as well as the pro-rata costs for social security contributions and office expenditures (Accounting Department, Personnel Department, Wages Department)." But it did not include the expenses for research or new buildings.[122] Hermann presented the results of the statistical report to the Technical Committee on February 2, 1934.[123]

Hermann's introduction emphasized that the rationalization and restructuring of various production processes as a result of the merger and the economic slump, and the shrinkage that this had necessarily entailed for a number of products, such as indigo, were "unfortunate in more ways than one" and had proved "a real headache." This was a euphemistic depiction of the developments over the past four years, since the entire volume of costs, which amounted to 31.2 million RM in 1929, fell to a mere 19.2 million RM in 1933. This was a decrease of approximately 38 percent – despite the fact that the volume of costs had already begun to drop prior to

[121] See Pinnow, *75 Jahre Werksgeschichte*, pp. 192–3; on Wagner's suicide, see HA, PA Wagner, "In memoriam Dr. Hermann Ludwig Wagner" and the obituary in "Chemiker-Zeitung," February 13, 1932.

[122] HA, 12, Hirschel, "Das Werk Hoechst im Verbande der I.G. (1916–1945)," May 1961, p. 27; there, also a summary of Hermann's report.

[123] HA, PA Hermann, Endgültige Fassung des Referats Dr. Hermann vor dem TEA, Hoechst, February 2, 1934.

1929. The decline affected individual departments differently. The Nitrogen Department was among those hit hardest, and its costs fell from 7.3 million to 3.4 million RM, that is, by 53 percent. This was due in a large part to the economic downturn, which affected the demand for nitrogen, especially abroad, and worked to Hoechst's particular disadvantage: "The guiding principle behind the allocation of the shrunken volume of production was to concentrate ammonia processing in those plants responsible for the primary manufacturing of ammonia – that is, ammonia synthesis. Thus, in the aftermath of the redistribution, Hoechst lost the production of calcium nitrate in 1929 and of potassium nitrate in 1931." The plant had hoped to find a substitute with a new crystalline saltpeter process that was developed at Hoechst: "But at the end of large-scale testing, it turned out the process could be carried out in the currently idle ammonium sulfate factory at Oppau, without requiring any considerable investment. This production line was thus also lost for Hoechst."

The Inorganics Department was also badly affected. Its volume of costs dropped from approximately 4.5 million RM in 1929 to around 1.7 million RM in 1933, a decrease of about 61 percent. But despite the enormous decline in output, production costs had developed "comparatively favorably." The sector for organic intermediate products looked more promising; the volume of costs decreased only from 2.5 million RM in 1929 to around 2.3 million RM in 1933. However, the production of aniline dropped sharply, from 3,648 tons in 1929 to 929 tons in 1933, notwithstanding the fact that Hoechst's own consumption of aniline in 1933 stood at around 1,345 tons. This meant that the "basis for inorganic and organic intermediate products used for the production of dyes in Hoechst has become extremely unfavorable. The company is dependent on the supply of numerous preliminary products, both inorganic and organic, for its dyes production. And Hoechst's production of many preliminary products is limited to satisfying its own needs, so that it cannot profit from sales." The volume of costs for solvents dropped from 3 million RM in 1929 to approximately 1.7 million RM, a decrease of around 45 percent – but production costs had also fallen.

In the factories exclusively devoted to the production of dyes, that is, excluding any affiliated facilities manufacturing product groups such as synthetic resins, the volume of costs fell by 30 percent between 1929 and 1933, from 8.9 million to 6.2 million RM – and this despite the fact that dyes had already suffered considerable losses prior to 1929. Between 1929 and 1933 the volume of costs in the Azo Department sank from 2.2 million to 891,000 RM, and production fell from 2,168 to 1,094 tons – in other words, the initial figures were approximately halved. Results for the aniline and sulfur dyes were more favorable since the volume of costs dropped by only about 19 percent, from roughly 3 million to around 2.4 million RM. While the triphenylmethane dyes held up fairly well and sulfur black more

TABLE 2.1. *Shipments at Hoechst of Dyes and Dyeing Assistants (in Tons)*

Shipment	1929	1930	1933
Indigo	8,635	5,867	2,385
Sulfur dyes	3,238	2,149	1,966
Dyeing assistants	827	709	1,792
All other dyes	6,490	5,771	3,066
TOTAL	19,190	14,496	9,209

Source: HA, PA Hermann, Endgültige Fassung des Referats Dr. Hermann vor dem TEA, Hoechst, February 2, 1934.

or less remained the same, the development of the new textile auxiliaries Igepon A and T was perceived as very promising, since production of these auxiliaries actually rose from 388 tons in 1929 to 626 tons in 1933. The Indigo Department suffered most. Between 1929 and 1933 its volume of costs sank from 2.8 million to 1 million RM, a decrease of almost two-thirds, and production fell by about 71 percent, from 3,204 to 942 tons. Despite the massive fall in production, the company nevertheless managed to reduce the production costs: in 1929 the production of 100 kg indigo (100 percent) cost 219.07 RM, whereas in 1933 the figure was only 170.50 RM. According to Hermann, the Alizarin Department performed best, since its volume of costs between 1929 and 1933 increased from about 2.2 million to approximately 3 million RM – a growth of around 33 percent – while production rose by about 20 percent, from 248 tons in 1930 to 295 tons in 1933. Taken altogether, the Dye Departments at Hoechst "with the exception of the alizarin and triphenylmethane dyes" did badly. Hoechst had been severely affected by the two big mergers of different production locations within IG Farben, particularly by the merging of facilities carried out in 1930. But the hopes placed in the dyes Hoechst had managed to retain, in particular the two biggest dyes, indigo and sulfur black, were dashed, for their turnover declined sharply.

Table 2.1 shows the strong fall in dyes production and the growing importance of dyeing assistants between 1929 and 1933. Altogether, shipments sank by 52 percent. In 1932, according to Hermann, Hoechst's share of IG Farben's entire dye business was only 18 percent, calculated on the basis of the gross profit sales for dyes. In contrast, the shares of Leverkusen and Ludwigshafen were 32 and 28 percent, respectively.

Compared with the dyes sector, the Pharmaceutical Department was Hoechst's pride and joy: between 1929 and 1933 its volume of costs declined by only 12 percent, that is, from around 3.7 million to 3.3 million RM. The Pharmaceutical Department manufactured popular drugs that included painkillers and antipyretic medications such as Pyramidon and Antipyrin, the anti-infectious drug Melubrin, and Salvarsan, which was

TABLE 2.2. *Reductions in the Hoechst Workforce, Excluding Sales, 1930–1933*

Group	January 1, 1930	January 1, 1934
Members of the Management Board	8	2
Directors	15	10
Authorized signatories (*Prokuristen*)	23	14
Employees with university degree	451	365
Technicians including colorists	165	115
Master craftsmen	618	438
Commercial clerks	1,397	904
Blue-collar employees	7,074	5,571
TOTAL	9,751	7,419

Source: HA, PA Hermann, Endgültige Fassung des Referats Dr. Hermann vor dem TEA, Hoechst, February 2, 1934.

used against syphilis, as well as hormones and other preparations, including solvents such as acetoacetate. In the spring of 1931, Hoechst's serum production was transferred to the Behring-Werke in Marburg, but the manufacture of vaccines was definitively earmarked to remain at Hoechst. Between 1929 and 1933, the volume of costs for sera including vaccines dropped by about 60 percent, from 522,000 to 208,000 RM. In 1932 Hoechst's share in the turnover of IG Farben's pharmaceutical division (pharmaceuticals and pesticides) was 42.6 percent, and its share of net profits even reached 46.4 percent. Thus, this was an area in which Hoechst was particularly profitable.

From 1930 onwards, however, the plant management massively tried to reduce costs through layoffs and pay cuts in order "to adjust a suit of clothes that has become much too large for Hoechst's skinny body," as Hermann put it. Thus, between 1930 and 1933 the number of employees decreased from 9,751 to 7,419 (Table 2.2), the salaries of white-collar employees dropped from 17.5 million RM (100 percent) to 11.6 million RM (66 percent), and the wages of blue-collar employees declined from 14.47 million RM (100 percent) to 9.8 million RM (68 percent)(Table 2.3).

But reductions in costs alone were not enough to ensure the future viability of the plant. Hermann made this clear to the members of the Technical Committee: "Gentlemen, thanks to the substantial amounts you have granted us to renovate our factories and auxiliary plants, we have improved many things and want to accomplish even more." Between 1930 and 1934, about 20 million RM were invested in new facilities and in repairs. Simultaneously, just over 22 million RM were raised through amortizations: "With this [money], we have primarily undertaken the tasks of constructing modern factories and merging factories and their auxiliary plants to achieve a necessary reduction in the costs of production while at

TABLE 2.3. *Overheads at Hoechst, 1930 and 1933*

Cost Category	1930 (million RM)	1933 (million RM)	Decline (%)
General shop costs	2,400.0	1,201.0	50.0
Welfare costs	4,950.0	3,143.0	36.5
Fire protection costs	0.5	0.3	33.6
Office costs (excluding technical departments and purchasing)	2,812.0	2,042.0	27.3
Purchasing (a) chemical	0.3	0.2	27.2
Purchasing (b) technical	0.6	0.4	34.1
TOTAL	11,514.0	7,299.0	36.6

Source: HA, PA Hermann, Endgültige Fassung des Referats Dr. Hermann vor dem TEA, Hoechst, February 2, 1934.

the same time providing more light, air, and cleanliness for the plant and improving traffic conditions."

Not many new investments were carried out in the Nitrogen Department due to the decline in its production. However, two 30-ton Wedge furnaces for the manufacture of SO_3 were installed in the Inorganic Department, and the NaCl solubilizing station was rebuilt. New and expensive buildings were also erected for the Intermediate Products, Solvents, and Azo Dyes (lacquer paints) Departments. The Pharmaceutical Department also came in for its share of improvements: the Pharmaceutical Laboratory was completed, applications to set up new buildings for the production of Antipyrin and Pyramidon were granted, and work was started on a new building for the hormones department.

On the whole, "despite the unfavorable conditions described," Hermann reported that the general state of the Hoechst plant could "be described as tolerable." While Hoechst hoped for better days, it had set itself certain targets, since hope was not enough: "Work in our scientific laboratories will take precedence, and their mission will be to open up new areas. For many valuable products have emerged from their labors in the past." Nevertheless, at present, none of the work in progress was expected to result in an increase in production, so that other paths would have to be tried: "Hoechst can only improve its situation by making better use of its facilities and incorporating more production lines." Further cost-saving measures would still be necessary, but at the current volumes of production and under the existing circumstances, even with the best will in the world Hoechst would not be able to achieve the cost prices of Leverkusen.

The figures and their presentation were an indication of the overall critical situation of the plant in 1933/34. Three days later, at a meeting of the Management Board and the technical directors in Hoechst on February 5, 1934, Hermann gave an account of his report to the Technical Committee and emphasized that the committee had listened attentively. He thought that the presentation had led to "a better understanding of Hoechst's difficult situation." However, even the senior employees of the plant were taken aback by the figures, as none of them had been aware of the full extent of the problems at Hoechst.[124]

The relinquishing of Duisberg's suggestion that every important product be manufactured in at least two different sites had had particularly harsh repercussions for Hoechst. The plant lost the better part of its share in the production of dyes and intermediate products, and the remaining production of Azo dyes had to be partially transferred to other plants; the only two exceptions were the paint and pyrazolone dyes. In some areas, however, Hoechst had been strengthened, since it became the sole producer of patent blue, methylene blue, saffranine, and rosalinine blue.[125] The plant was given an important boost in the field of pharmaceuticals, when it took over the production of certain pharmaceuticals from the Mainkur (Cassella) and Biebrich (Kalle) plants. To compensate for the takeover, serum production was transferred to the IG Farben–affiliated Behring Werke in Marburg, but Hoechst continued to manufacture vaccines.[126]

Between 1925 and 1932, however, the plant suffered not only losses in production, but also in the sales of its dyes and pharmaceuticals. In 1948 Weidlich retrospectively wrote that the famous "founders' compromise" was soon revised. Thus, the sale of dyes was centralized because this move seemed to promise greater efficiency and lower costs. At the time Duisberg had argued vigorously with Bosch in an attempt to save the Leverkusen Sales Department, "because one could not expect him to look out from his house onto empty office windows." Finally Bosch and Duisberg reached an agreement, but "at the expense of Hoechst." Duisberg agreed to a centralization of dyes sales "in return for Leverkusen receiving the sales of pharmaceuticals." Sales of inorganics, acetic acid, and solvents were then relocated to Griesheim. As a result, Hoechst "became solely a place of production and completely lost its position in the market."[127]

In his memoirs Lautenschläger, the former manager of the Pharmaceutical Department, also refers to problems that seemed to indicate that "the very survival of the Hoechst Farbwerke was in doubt." Lautenschläger felt that

---

[124] HA, TEA 95c, Niederschrift der Vorstands- und technischen Direktionssitzung on February 5, 1934, in Hoechst.
[125] Bäumler, *Century*, p. 101.   [126] Pinnow, *75 Jahre Werksgeschichte*, pp. 175–6.
[127] HA, PA Weidlich, Weidlich, "Erinnerungen (aus dem Gedächtnis)," August 1948.

the danger had been averted by "the firm hand of a few leading personalities." Thus, Ludwig Herrmann and his colleagues had managed to reinstate Hoechst, so that it once again became "a strong pillar" within IG Farben: "At the time, more than in any other plant, the fight to ensure the survival of our Hoechst plant was carried out with tireless perseverance and steely determination – by means of purposive research, rational reorganization and the redesigning of old and new production departments."[128]

Indeed, after the forced retirement of the "old Hoechst hierarchy," the new management was no longer bound by old traditions and habits. Despite the initially negative consequences of the Weidlich and Ref affair for Hoechst and its reputation, these events proved positive in the long term, since they offered an opportunity for a fresh start under Ludwig Hermann, the new manager of the plant – and the candidate of Bosch and Duisberg and not of the "old Hoechst hierarchy."

[128] HA, PA Lautenschläger, Lautenschläger, "Erinnerungen," vol. 2, [unpaginated].

3

Works Management, Workforce, and the National Socialist Party

3.1. FROM THE DEPRESSION TO THE NAZI SEIZURE OF POWER

Hoechst was no Nazi stronghold prior to 1933, neither the works itself nor the area of the same name in the city of Frankfurt; after April 1928 Hoechst had ceased to be an independent communality and was incorporated into the city of Frankfurt.[1] In the local elections of May 1928, the NSDAP received only 12,526 votes out of a total of 280,563 valid votes in Frankfurt, which amounted to a share of around 4.5 percent. In the district of Hoechst the NSDAP received only a bare 2 percent of the votes or 312 of 15,812 valid votes. In contrast, the Social Democrats, the Communists, the DVP (German People's Party), and the Catholic Center Party did better in Hoechst than they did in Frankfurt as a whole. The Center Party did particularly well in Hoechst, where it obtained 20.8 percent of the votes, almost double the share it received in Frankfurt, where it stood at 10.8 percent. The decisive factor here was the greater number of Catholic voters in Hoechst compared to the whole of Frankfurt – mainly due to the fact that the communality of Hoechst had previously formed part of the Catholic bishopric of Mainz while Frankfurt had been an imperial free city with a predominantly Protestant population. However, by the time of the local elections of November 1929 the situation had changed: while the NSDAP got 9.9 percent of the votes in Frankfurt and only 5.4 percent in

---

[1] The following election results of Frankfurt and the district of Frankfurt-Hoechst are based on Institut für Stadtgeschichte, photocopies from *Städtisches Anzeigeblatt Frankfurt am Main* (issue 21 of May 26, 1928, issue 49 of November 25, 1929, and issue 12 of 1933). The results for Frankfurt are also published in Dieter Rebentisch, "Zwei Beiträge zur Vorgeschichte und Machtergreifung des Nationalsozialismus in Frankfurt: Von der Splittergruppe zur Massenpartei," in Eike Hennnig (ed.), *Hessen unterm Hakenkreuz. Studien zur Durchsetzung der NSDAP in Hessen*, 2nd ed. (Frankfurt am Main: Insel, 1984), p. 289; see also Josef Fenzl, *Aus der Geschichte der Hoechster Kaserne 1920–1945* (Frankfurt am Main: Kramer, 1998), p. 66.

TABLE 3.1. *Local Elections 1928–1933 in Frankfurt am Main and in the Administrative District of Frankfurt-Hoechst (%)*

Party	May 20, 1928		November 17, 1929		March 12, 1933	
	Frankfurt	District of Hoechst	Frankfurt	District of Hoechst	Frankfurt	District of Hoechst
Total number of valid votes	280,563	15,812	272,918	15,221	331,613	19,451
NSDAP	4.5	2.0	9.9	5.4	47.9	40.6
DVP	10.3	11.8	12.9	16.0	2.5	3.2
Center	10.8	20.8	12.2	22.7	11.5	18.6
SPD	32.5	35.1	27.6	25.1	19.1	18.5
KPD	12.8	14.0	13.0	16.3	9.7	11.8

Source: Institut für Stadtgeschichte, photocopies from *Städtisches Anzeigeblatt Frankfurt am Main* (issue 21 of May 26, 1928, issue 49 of November 25, 1929, and issue 12 of 1933).

Hoechst itself, the percentage of votes for the Communist Party rose to 13 percent in Frankfurt and to 16.3 percent in Hoechst. The Center Party also made some gains, while the Social Democrats suffered losses, falling to 27.6 percent in Frankfurt and 25.1 percent in Hoechst. At the end of 1929 there was no visible indication that the Nazis would assume power at a municipal level in Frankfurt, and in the district of Hoechst this appeared even less likely.

Even in the local elections in Frankfurt on March 12, 1933, which could no longer be considered truly free elections and which saw the mobilization of many people who had not voted previously, the NSDAP did not perform as well in Hoechst as in Frankfurt as a whole. Compared to the November elections of 1929 the number of valid votes cast for the NSDAP in Frankfurt increased from 272,918 to 331,613 and in Hoechst from 15,221 to 19,451. In Frankfurt the NSDAP achieved a clear victory with 47.9 percent of the votes – while in Hoechst the share of the votes given to the NSDAP came to 40.6 percent. After the NSDAP the best result was achieved by the Center Party with 18.6 percent in Hoechst but only 11.5 percent in Frankfurt. In Frankfurt the Social Democrats fell to 19.1 percent and in Hoechst to 18.5 percent. In contrast, the Communists were still at 11.8 percent in Hoechst and 9.7 percent in Frankfurt. Compared to the previous results of 1929, the formerly strong German People's Party (liberal) melted away, plummeting from 12.9 to 2.5 percent in Frankfurt and from 16 to 3.2 percent in Hoechst (Table 3.1).

Bernhard Schacke, an "old fighter" of the NSDAP who worked for Hoechst as a chemist, referred to the "red-black population" of Hoechst,[2] an assessment that was borne out by the results of the municipal elections. Yet according to the historian Josef Fenzl, the National Socialists experienced their greatest difficulties in "proletarian environments" such as in the Frankfurt district of Nied adjoining that of Hoechst, while in two other districts also adjacent to Hoechst, the districts of Zeilsheim and Unterliederbach, the outcome of the voting varied greatly from location to location. In Hoechst itself, according to Fenzl, the "small-town, bourgeois and craftsmen milieu" had predominated.[3] Therefore when we look at the voting behavior both in Hoechst and in the municipalities or townships around Hoechst where a large part of the workers and employees of the Hoechst Works lived, it becomes evident that up until the end of the 1920s little pointed to the NSDAP's enjoying a strong support among the population of Hoechst or its immediate neighbors.

Just how weak the NSDAP was in the IG Farben plant Hoechst itself is borne out by the results of the elections to the Works Council at the beginning of the 1930s (Table 3.2). Even in 1931, at the height of the Great Depression and despite large-scale layoffs in the plant, the NSDAP did not have a strong position. The National Socialists were far from massing a majority of blue- and white-collar workers behind them. In the elections in 1930 and 1931 they won 6.6 and 8.2 percent of workers' votes, while the unions that supported the "Weimar parties" – in other words, the Social Democrats, the Center, and the Liberals – stood at over 50 percent. The Communists too were strongly represented in the plant. In March 1931 they received more than one-third of the votes. However, the elections for the Blue-Collar Employees Committee (*Arbeiterrat*) in March 1931 in Hoechst had to be repeated, as the works management sacked all Communist members of the Works Council together with a large number of their deputies, accusing them of industrial espionage. The Communists lost heavily in the repeat elections in May. They were unable to mobilize their supporters as they had done in March – this can be argued on the basis of a

---

[2] BA, BDC file on Bernhard Schacke, Schacke to RuSHA, June 21, 1934, and Schacke to Himmler, January 26, 1935. Schacke was a member of the NSDAP since February 1, 1928 (membership number 75,698), and cofounder of the local SS. On the Hoechst Party branch and its internal conflicts see Fenzl, *Aus der Geschichte*, pp. 126–36, especially pp. 185–9 (document Fritz Weber dated June 6, 1933).

[3] Fenzl, *Aus der Geschichte*, pp. 59–60; Nied, a formerly independent municipality, was incorporated into Frankfurt in 1928 and was subsequently administered by the "district administration for the western suburbs"; cf. Adalbert Vollert, *Nied – wie es einmal war. Historische Notizen eines Frankfurter Stadtteils* (Frankfurt am Main–Nied: Heimat- u. Geschichtsverein, 1989).

TABLE 3.2. *Results of the Elections to the Hoechst Works Council in 1930 and 1931 – Blue-Collar Employees Committee*

	1930		March 1931		May 1931	
Group	Number	Percentage	Number	Percentage	Number	Percentage
Workers eligible to vote	6,504	100	5,516	100	5,703	100
Voter turnout	5,323	81.8	4,694	85.1	3,739	65.6
Free trade unions	1,989	37.4	1,885	40.2	1,648	44.1
Christian trade unions	861	16.2	751	16.0	688	18.4
Communists	1,937	36.4	1,663	35.4	1,030	27.5
National Socialists	435	8.2	308	6.6	276	7.4
Invalid votes	101	1.9	87	1.8	97	2.6

Sources: HA, PSW 657, Betriebsratswahl 1931 and Bekanntmachung der Wahl vom 29. Mai 1931, dated May 30, 1931. See also Zollitsch, *Arbeiter*, p. 198, Table 33. The percentages occasionally vary slightly; the figures given here were all recalculated. A rounding error gives a final result of 100.1% for the figures of the 1930 election.

generally lower voter turnout in May.[4] Compared to other IG Farben works the Communists were relatively strong in Hoechst and the National Socialists fairly weak. Thus in the Leverkusen works the NSDAP received 13.2 percent of the votes in August 1931. However, it should be noted that at the same time in the Krupp plants the NSDAP received only 4.5 percent of the votes.[5]

An examination of the election results of the White-Collar Employees Committee (*Angestelltenrat*) in the works council elections of the Hoechst works shows that the NSDAP enjoyed a distinctly higher popularity among

[4] HA, PSW 657, press clipping from *Hoechster Kreisblatt*, issue 105 of May 6, 1931: "Arbeiterratswahl bei der I.G., Werk Hoechst. Infolge der Werkspionage-Entlassungen." Cf. Wolfgang Zollitsch, *Arbeiter zwischen Weltwirtschaftskrise und Nationalsozialismus. Ein Beitrag zur Sozialgeschichte der Jahre 1928 bis 1936* (Göttingen: Vandenhoeck & Ruprecht, 1990), pp. 198–9.

[5] Zollitsch, *Arbeiter*, pp. 199–201; Hayes, *Industry and Ideology*, p. 53, in contrast emphasizes the radicalization in the Hoechst plant, resulting in 42 percent of the vote for KPD and NSDAP combined at the elections of March 1931.

TABLE 3.3. *Results of the Elections to the Hoechst Works Council in 1930 and 1931 – White-Collar Employees Committee*

Group	1930		1931	
	Number	Percentage	Number	Percentage
Employees eligible to vote	2,614	100.0	1,997	100.0
Voter turnout	2,260	86.5	1,809	90.5
Gedag associations	1,132	50.1	756	41.8
GDA associations	343	15.2	828	45.8
AFA associations	482	21.3		
National Socialists	281	12,4	208	11.5
Invalid votes	22	1,0	17	0.9

Source: HA, PSW 657, Betriebsratswahl 1931 Werk Hoechst – Angestelltenrat.

this group than among blue-collar workers (Table 3.3). In 1930 the NSDAP received 12.4 percent of the votes and in 1931, 11.5 percent. But here too the National Socialist employees' organizations did not hold the majority; in fact, in 1931 the organizations with ties to the "Weimar parties" were even able to increase their share of the votes. The Christian and nationalist *Gesamtverband deutscher Angestelltengewerkschaft* (Gedag; General Association of German White-Collar Employees Unions) was particularly strong with almost half of the votes in 1930 and still over 40 percent in 1931. In 1931 the liberal, nationalist *Gewerkschaftsbund der Angestellten* (GDA; Federation of White-Collar Employees Trade Unions) joined together with the *Allgemeinen freien Angestelltenverband* (Afa; General Free White-Collar Employees Association), which had close ties to the Social Democrats, to form a joint list; together they were able to gain almost 46 percent of the valid votes.

Thus, after the elections in 1931 only two out of twenty-three members of the works council were from the NSDAP – one in the Blue-Collar Employees Committee and one in the White-Collar Employees Committee. These two National Socialists faced seven representatives from unions with ties to the Social Democrats, two from the Christian unions, apparently six Communists (although this is somewhat unclear because of the special elections held in May), and three representatives each from the Gedag and the GDA/Afa lists. Prior to the seizure of power on January 30, 1933, nothing pointed to any particular Nazi inclinations on the part of Hoechst employees. As a caveat it should be added that the the results of elections to the works council depended in no small measure on factors specific to the company and on personal factors and therefore did not necessarily reflect political voting behavior.[6]

All in all, the elections to the works council in the final phase of the Weimar Republic do show, as the historian Wolfgang Zollitsch wrote, that "there could be no question of a tense revolutionary situation." This also applied to companies such as Hoechst in which the Communists were strongly represented, as "the fear of losing one's job, the threat of complete destitution, curbed the willingness of employed workers to carry out radical protests in practice. During the depression pressures to adapt were stronger."[7]

Like the majority of its workforce, the works management showed no particular sympathies for the National Socialists prior to 1933. The top managers of IG Farben – and also those of Hoechst – tended to sympathize with the bourgeois-liberal centrist parties, the liberal DDP (German Democratic Party), and above all the nationalist-liberal DVP.[8] In the presidential elections of 1932 Carl Duisberg – speaking for the whole of the IG Farben management – strongly supported the reelection of Paul von Hindenburg; he expressed mistrust of the NSDAP and Hitler, primarily because of the similarity of their economic program to that of "Marxism."[9] While there certainly were some National Socialists among the managers of IG Farben, Wilhelm Rudolf Mann was the only Nazi on the board of directors. In these circles National Socialism was still generally rejected in 1932, and hopes were pinned on the cabinets of Franz von Papen and Kurt von Schleicher.[10] In Hoechst, for example, an employee was sharply reprimanded in the summer of 1932 for having distributed leaflets in support of the NSDAP on factory premises during office hours despite the interdict against such activities. The works management felt it necessary to censure him severely: "We wish to express our severe disapproval on this occasion and wish to expressly draw your attention to the fact that in the event of a repeat offence this will result in the most serious consequences on our part."[11] As late as mid-January 1933 the manager of the local NSDAP

[6] See Matthias Frese, *Betriebspolitik im "Dritten Reich." Deutsche Arbeitsfront, Unternehmer und Staatsbürokratie in der westdeutschen Grossindustrie 1933–1939* (Paderborn: Schöningh, 1991), pp. 30–1.

[7] Zollitsch, *Arbeiter*, pp. 208–9.

[8] Plumpe, *I.G. Farbenindustrie*, pp. 527–9. Hayes, *Industry and Ideology*, pp. 48–54; in Hoechst, before 1933 chief engineer Friedrich Jähne and Georg Kränzlein, head of organics research, were members of the DVP.

[9] Plumpe, *I.G. Farbenindustrie*, pp. 535–6.

[10] Peter Hayes, "Die I.G.-Farbenindustrie," in *Unternehmen*, p. 109; Plumpe, *I.G. Farbenindustrie*, pp. 539–40.

[11] HA, PA Schmidt (microfilm), Hoechst management (Hermann and Schwamborn) to Schmidt, April 21, 1932.

branch, a former employee who had been dismissed during the Great Depression, was refused when he requested his reemployment in Hoechst citing his miserable economic circumstances.[12]

With the seizure of power the *Nationalsozialistische Betriebszellen-Organisation* (NSBO) or National Socialist Workers' Cell Organization, the Nazi representation of blue- and white-collar employees in companies, hoped that its position would be strengthened in the IG Farben works – and in Hoechst. While the NSBO, the "SA of the factories," had attacked both socialism and communism in its publications and propaganda leaflets, it had also vilified capitalism. According to Hitler, the NSBO were the "ideological combat troops" of the NSDAP in the factories.[13] The number of NSBO members increased enormously between 1930 and 1933 – from around 4,000 employees at the beginning of 1931 to around 260,000 in January 1933. But the influence of the NSBO in factories and businesses prior to the seizure of power had not lived up to the expectations of its leaders.[14] The question was whether that would now change.

The historian Peter Hayes writes that at the time the IG Farben leadership was attempting to walk a fine line between satisfying the demands of NSDAP activists, retaining managerial control of the company, and preserving "the morale and the efficiency of the numerous workers who continued to loathe Nazism."[15] A letter from the Economic Department of IG Farben dated April 7, 1933, to the heads of the works and sales groups made it clear that the management had no intention of caving in to the demands of the NSBO. On the instructions of the chairman of the board, Carl Bosch, the NSBO were informed of a "decree of the Chancellor of the Reich enclosed": this was an appeal by Adolf Hitler dated March 10, 1933, to the members of the NSDAP, SA, and SS that ordered them to desist from carrying out individual acts as they saw fit. "As of today the national government has the executive powers for all of Germany in its hands. This means that in future the national invigoration will be guided from above and be more systematical." It continued: "the harassment of individual persons, obstructions of cars or disturbances of business must categorically cease."[16] The management considered the last part of the sentence to be particularly important; no matter how open-minded the IG Farben management might be toward the new National Socialist government, it had every intention of being and remaining "master in its own house."

[12] HA, PA Zeh (microfilm), Note Landmann dated January 17, 1933.
[13] Michael Schneider, *Unterm Hakenkreuz. Arbeiter und Arbeiterbewegung 1933 bis 1939*, Geschichte der Arbeiter und der Arbeiterbewegung in Deutschland seit dem Ende des 18. Jahrhunderts 12 (Bonn: Dietz, 1999), pp. 159–61; Zollitsch, *Arbeiter*, pp. 212–13.
[14] Schneider, *Unterm Hakenkreuz*, pp. 163–4. [15] Hayes, *Industry and Ideology*, p. 108.
[16] HA, Pol 6, I.G. Berlin, wirtschaftspolit. Abt. to Leiter der Betriebs- und Verkaufsgemeinschaften, Betreff: NSBO, dated April 7, 1933, in the appendix proclamation of Hitler, published in *Völkischer Beobachter*, issue 70 of March 11, 1933.

The attitude described by Hayes is very noticeable on the occasion of the first celebration of the "Day of National Labor" held in the Hoechst plant. One of the first measures of new regime had been to proclaim May 1, the traditional day of the international workers movement, a "Day of National Labor."[17] In the Hoechst plant this was elaborately celebrated in 1933 with addresses by the NSDAP county leader, the new plant manager Ludwig Hermann, and Walter Hirschelmann, the NSBO member of the Works Council, who had been acting as the "representative of the workforce" since the end of April 1933.[18] While the speeches of Hirschelmann and the county leader were only briefly summarized in the commemorative brochure specially published by Hoechst, Hermann's speech was printed in full. In his speech Hermann spoke of the "feeling of loyal comradeship" that now filled the entire plant and its members and that would unite them with all German people. The times when "barriers" had existed between works management and workforce were now forgotten: "Mindful of the fact that in every single one of our employees we see a fully enfranchised member of the factory, the state and the national community, we wish to fill our house with the spirit which inspired our whole nation in August 1914 and to cultivate the same loyal comradeship which proved its worth many thousand times over on the field of battle." He ended his speech, significantly enough, not simply with a reference to Hitler but with the ringing words "Our beloved fatherland and its leaders, General Fieldmarshal von Hindenburg and the Chancellor of the Reich Adolf Hitler, Heil, Heil, Heil!"[19]

The head of the NSBO in Hoechst, the laboratory assistant Walter Hirschelmann, had become the "representative of the workforce" already prior to the smashing of the unions that took place the following day on May 2, 1933 – a step that Hermann, as he later emphasized, expressly welcomed.[20] The previous chairman of the Works Council, the worker Leonhard Roth, and his deputy, the locksmith Johann Weber, both from the free trade unions, had been reelected to the Blue-Collar Employees Committee at the beginning of April 1933. Roth was apparently initially confirmed in office. However, already in April Hirschelmann increasingly took over his functions as representative of the workforce – despite the results of the April election to the Blue-Collar Employees Committee, the position of the NSBO had been strengthened, not least because of its recognition by the works management as the representative of the

[17] See Zollitsch, *Arbeiter*, pp. 212–13.
[18] On the NSBO and the other trade unions in the plants between January and June 1933, see Frese, *Betriebspolitik*, pp. 42–52; for Hirschelmann's role in the Hoechst plant, see *Hoechster Kreisblatt* of April 26 and 29, 1933.
[19] HA, Pol 6, Brochure of I.G. Farbenindustrie Aktiengesellschaft Frankfurt a.M.–Hoechst, "Zur Erinnerung an den 1. Mai 1933."
[20] HA, PA Hermann, Typescript of Hermann's speech on Labor Day (May 1), 1934; see also Plumpe, *I.G. Farbenindustrie*, p. 622.

workforce on May 1.[21] Roth and Weber, the two last freely elected chairmen of the Works Council, were both forced to resign from office in mid-June 1933, apparently "in the course of the *Gleichschaltung* (consolidation) of works councils." However, neither of them was dismissed, and both continued to work in the Hoechst plant.[22]

Wolfgang Zollitsch noted that after 1933 "consolidation" at the level of businesses met with "almost no resistance" – and this also applied to Hoechst. "The employees maintained an attitude of resignation and clung to their jobs."[23] And even the workers' representatives appear to have resigned. In the autumn of 1933 the Blue-Collar Employees Committee constituted itself anew – without elections. Representatives of the old unions in the Blue-Collar and White-Collar Employees Committees had been replaced, as in the case of Roth and Weber, or had adapted.[24]

After Hirschelmann had voluntarily left the plant in August 1933, Ludwig Retzinger, a young worker and member of the NSBO, took over his position as representative of the workforce.[25] However, the enhanced position of National Socialists in the plant was not solely due to the proscription of the unions. A number of employees had newly joined the NSDAP, the so-called *Märzgefallene* or "turncoats of March," who became "Party comrades" in the months from March to May 1933. As a result of the wave of applications for membership the NSDAP even ordered a moratorium on new admissions that lasted until 1937/38 – and there were few exceptions to this policy. The number of NSDAP members in the Hoechst plant also increased because of preferential hiring of previously unemployed "old fighters" of the NSDAP.[26]

[21] HA, PSW 665, On the 1929/30 list of members of the Blue-Collar Employees Committee Weber is in second place and is referred to as the "1st member of the Works Council," Roth is listed first and described as "Betr.Vorsitz" (*Betriebsratsvorsitzender* = chairman of the Works Council). Elsewhere (for example, in Archiv der Friedrich-Ebert-Stiftung, Personalia: Johann Weber, Mein Lebenslauf, dated September 25, 1945) Weber is described as the chairman of the Works Council at the time.

[22] HA, PA Weber (microfilm), PA Roth (microfilm), PSW 665, Records on Blue-Collar and White-Collar Employees Committee 1919–1933/34; on Weber see also Archiv der Friedrich-Ebert-Stiftung, Personalia: Johann Weber, Mein Lebenslauf, dated September 25, 1945; cf. also Frese, *Betriebspolitik*, pp. 42–4, 65–73.

[23] Zollitsch, *Arbeiter*, p. 212.

[24] HA, PSW 665, one instance from the White-Collar Employees Committee: the works foreman H. Schenkelberg had apparently joined the NSBO but was dismissed from the White-Collar Employees Committee and the Works Council at the end of October 1933 because of his previous membership in the SPD and the *Reichsbanner*.

[25] HA, PA Hirschelmann (microfilm), personal data sheet and memo "Sozialabteilung für Alizarinfabrik zu Freistellung Hirschelmanns für den Betriebs-Ausschuss nach Wahl Betriebsrat," dated April 19, 1933.

[26] HA, TEA 95c, Report by Schwamborn on "Arbeiter- und Angestelltenfragen" for the directors' meeting of July 26, 1933; cf. Frese, *Betriebspolitik*, p. 50: In the Krupp plants approximately 25 percent of new employees were members of Nazi organizations,

Overall the works management of Hoechst continued their "tightrope act": they did not make the giving of the "German salute" mandatory in the plant, but they warned against any disrespect being shown to the salute. In the summer of 1933 collections for the NSDAP were generally banned in the plant, but this was coupled with the decision to give a donation to the SS. At the beginning of 1934 the works management decided to continue to forbid collections in the plant while reserving the right to make exceptions to this ban. They obliged Nazi activists by enacting generous regulations for Nazi paramilitary exercises: "In consultation with the *Wehrverbände* [paramilitary associations] an arrangement concerning temporary leave and wages for persons taking part in military athletics camps has been agreed on in writing."[27]

But the question of hiring new staff remained tricky. On March 8, 1933, the head of the Personnel Department of Hoechst, director Wilhelm Schwamborn, reported that since the lowpoint on October 1, 1932, the number of blue-collar workers and, more recently, of white-collar employees had increased, albeit with fewer increases in the latter number. However, he advised "exercising restraint" when hiring new staff, as the situation remained uncertain, and suggested that if there was a "need for engineers, chemists, or economic and administrative staff" it would be possible to use "supernumerary employees who were still present in some departments."[28] Nor did Hoechst intend to give up its policy of restraint for some time. In the Maingau directors' meeting held in mid-May 1933 Hermann reported: "Unfortunately, the request of representatives of the National Socialist organization that we take on more workers cannot be complied with, as Hoechst currently has 150 superfluous workers, in the next month there will be no work for another 100 persons and at present it is almost impossible for the works management to find emergency work for these 250 men."[29] It had to be admitted that in Hoechst the levels of employment in the plant lagged behind that of the plants Ludwigshafen and Leverkusen.[30] Although on May 1, 1933, Hermann had emphasized how much he was affected by the circumstances of former employees, now unemployed, he could not and would not employ persons whom he did not need – and in this matter he had no intention of being told what to do by the NSDAP and its representatives in the plant.

---

compared to around 80 percent in the IG Farben plant at Leuna. There are no exact figures for Hoechst.

[27] Hayes, *Industry and Ideology*, p. 108; NI-5857, Niederschrift über Vorstandssitzung am 14. August 1933 in Hoechst; HA, TEA 95c, Niederschrift der Vorstands- und technischen Direktionssitzung, March 5, 1934 [=NI 5873].

[28] HA, TEA 95c, Niederschrift der Maingau-Direktionssitzung, March 8, 1933.

[29] HA, TEA 95c, Niederschrift der Maingau-Direktionssitzung, May 17, 1933.

[30] HA, TEA 95c, Niederschrift der Maingau-Direktionssitzung, June 21, 1933.

At the end of July 1933 Schwamborn was finally able to inform the directors' meeting of IG Farben's intended stance toward demands by the NSDAP, the NSBO, and the *Deutsche Arbeitsfront* (DAF; German Labor Front) respecting issues affecting blue- and white-collar employees – and that stance was characteristic for IG Farben and Hoechst. Schwamborn reported that the chairman of the works council of the Kalle plant, acting for the "Central Committee of IG Works Councils" had presented three demands: First, there should be no hirings or layoffs without consulting the members of the works councils; second, all punitive measures taken against National Socialists should be repealed, and third, all Communists should be dismissed. These demands were accompanied by the assertation that they had already been accepted by the IG Farben Management Board – however, this was not the case; the board had rejected them four weeks previously. It was agreed that the first demand should be "strictly rejected" out of hand, as employment offices and employers were solely responsible for hiring people or for their recruitment and employers alone must be responsible for dismissals – "in compliance with current legal regulations." The second demand should also be rejected, as none of the punitive measures against National Socialists had been undertaken "for political reasons," and with respect to the hiring of National Socialists the "special instructions" were still held to apply, namely, that "NSDAP members with membership numbers lower than 200,000 and members of recognized paramilitary associations were to receive preferential treatment." Compliance with the third demand went without saying; in fact, "on express request" the factories had convinced "the currently solely responsible District Personnel Department of the NSDAP to designate those persons who were still considered to be Communists, as otherwise dangerous mistakes would inevitably occur." According to Schwamborn, that department had also confirmed that it would be possible to refuse to reinstate employees who "had committed serious offences as former members of the German Communist Party or other antinational parties" and who had been castigated for this: "this refusal [to reemploy them] will continue in force even if the persons in question have since become members of the NSDAP or one of its affiliated organizations."[31]

The management's conduct toward former members of left-wing parties – the phrase "former members of the German Communist Party or other antinational parties" could be used to refer to all left-wing parties including the Social Democrats – was therefore even more intransigent than that of the National Socialists. In general, the management had no intention of doing their work – unless the plant itself would benefit. Thus, a request presented by the Blue-Collar Employees Committee "to help it determine

---

[31] HA, TEA 95c, Report by Schwamborn on "Arbeiter- und Angestelltenfragen" for Hoechst's directors' meeting of July 26, 1933.

those persons who were, as yet, not members of any organization by the distribution of a questionnaire to all workers" was turned down by Hoechst. After consultation with the Federation of Employers and the NSDAP District Personnel Department such a request was considered to be "precipitate." "Caution was called for" when "lower administrative levels of the NSDAP" made such requests. According to Schwamborn, any "cooperation" on the part of the company was "only permissible" if – and the following formulation clearly expresses the management's cost-benefit calculations – such a request was "a general order that had been undoubtedly issued by the NSDAP headquarters, and if the company's cooperation was essential or cooperation by the works management was expedient to carry it out."[32]

Thus, during the phase in which the Nazi regime began to consolidate its power the works management of Hoechst acted no differently from IG Farben as a whole. Hoechst attempted as far as possible to advance its own interests and only to grant the new regime those concessions, which the plant considered necessary – particularly if these coincided with the management's own interests. Legally the relationship between works management and workforce was regulated by the *Gesetz zur Ordnung der nationalen Arbeit* (AOG; Law on the Organization of National Labor), which was passed on January 20, 1934, and the Factory Code of Rules prescribed therein – and here again, the specific implementation of the regulations remained a matter of dispute between NSDAP organizations and the management.

3.2. THE NEW RULES IN THE PLANT

According to the Law on the Organization of National Labor passed on January 20, 1934, all persons employed in a company or plant became a *Betriebsgemeinschaft* or "plant community," the head of the company or plant was now referred to as the *Betriebsführer* or "plant leader," and the workforce now constituted a *Gefolgschaft*, which translates approximately as "followers." A legal commentary on this law and on the role of the plant "leader" in particular – the book was part of the library of works manager Hermann[33] – stated that the law aimed at "ending conflicts within companies," as was indicated by the fact that the manager was referred to as the "leader" and the employees as his "followers." "The mutual obligation to be loyal, which binds together the leader and his followers, rules out any divisive conflicts." Putting aside all self-interest management and employees were expected to "work together for the common good of the people and state" in the *Betriebsgemeinschaft*. To accomplish this goal

---

[32] Ibid.
[33] HA, Pol 20, including Rössler's work, bearing the stamp "Bibliothek Dr. Hermann."

the law placed the manager at the head of the organization and expected all employees, irrespective of whether they were white- or blue-collar employees, to obey. However, the law did not aim to create merely "one person who gives orders and many who obey them, but companions in work and destiny, bound together by feelings of comradeship." The "leader" of a company and the employees should carry out their duties in a spirit of mutual "personal attachment." "However, as in larger companies the leader is not able to make personal contact with every one of his followers, he will be assisted by members of a Council of Trust [*Vertrauensrat*] who will represent the followers and offer advice."[34]

Despite the power accorded the "plant leader," pains were taken to point out that he was by no means an absolute and free "master in his own house"; rather he was under obligation to the National Socialist state, which would also exercise "a close scrutiny over life in the plant community."[35] This was repeatedly emphasized. "Within the scope of his power to make decisions" the "leader" of a factory was not "bound by any rights of participation of others." And he was permitted to make his decisions "at his own discretion," even if the Council of Trust, "which served him in a purely advisory capacity," did not approve them.[36] But the powers of the "leader" of a factory were also curbed: "That the plant leader's almost all-embracing power to make decisions may not be exercised in an arbitrary fashion, but must be circumscribed by the national and social spirit and by the spirit of community, has already been emphasized. His decisions must serve the purposes of the factory and not merely the entrepreneur's pursuit of profit, and the decisions must be made for the general advantage of the people and the state while also fulfilling the leader's obligation to care for his followers."[37]

The AOG also made it mandatory for all companies with more than twenty employees to have a Factory Code of Rules, which in contrast to the work rules previously in effect, applied to both blue- and white-collar employees.[38] The procedure to create a new factory code of rules was as follows. Certain items of the code were prescribed by the DAF. These were passed on from the Reich Trustee of Labor to the *Betriebsobmann* (shop steward) via the District Administrator and the County Administration of the DAF. The specific items for a particular factory were then inserted into or added to this predetermined "framework" and then discussed and finalized with the management of the company, the respective Legal

---

[34] Rössler, *Führer des Betriebes*, pp. 20–1.

[35] Ibid., pp. 29–33, see also pp. 42–6 elaborating on the relationship of plant leader and state.

[36] Ibid., p. 35.

[37] Ibid., p. 37; Schneider, *Unterm Hakenkreuz*, pp. 293–300, provides a concise summary of the AOG; on the genesis of the AOG, see Frese, *Betriebspolitik*, pp. 93–113; on AOG and IG, cf. Plumpe, *I.G. Farbenindustrie*, pp. 620–2.

[38] Frese, *Betriebspolitik*, p. 138.

Department, and the Council of Trust as the representative of the work-force. This version was then returned to the Trustee of Labor, who had to approve it.[39]

On September 27, 1934, a new Factory Code became effective for the entire Maingau Works Group of IG Farben, the plants at Hoechst, Griesheim, Mainkur (Cassella), Marburg (Behring-Werke), and Offenbach.[40] The preamble, bearing the sober heading "Scope and Commencement," stated that the "leaders" and "followers" of these plants now constituted a "factory community sharing a common destiny," who were to work together "in faithful comradeship and mutual trust" – for the good of the factory, for Germany and its economy and according to the principle of "common weal before private gain." Every employee should carry out his allotted tasks conscientiously and in a disciplined manner, showing a "joyful commitment to the National Socialist state and its *Führer*" as well as "unwavering loyalty toward the plant community, the *Volksgemeinschaft* [national community] and the German fatherland." The following para-graphs then outlined the regulations concerning hiring, working hours, work habits, prevention of accidents, wages, vacation, infringements of the Factory Code, and generalities. Certain points of the new Factory Code deserve particular mention.

A compromise had been achieved between Nazi representatives and the works management in the question of the hiring of employees. Ultimately the plant leader had the final say in the hiring of a particular person, but the decision had to be made "with due observance of the legal regulations and in consultation with the Council of Trust and the *Betriebszellen-Obmann* [shop steward]." The hiring of new staff also depended on their "personal and professional suitability" – a phrasing that was capable of meaning many things in a specific case.[41] For Hermann, who wanted to be sure that conflicts would be settled within the plant, the point "Generalities" was particularly important, as it included the following paragraph. "All matters pertaining to the plant must in principle be settled within the plant. In the event of differences an attempt must first be made to settle them together with the head of the factory and then – if necessary – to call on the Council of Trust. Only if no understanding can be reached should other instances outside the factory be appealed to such as legal aid offices, labor courts, or Trustees."[42] According to the Factory Code only gatherings that had been agreed on between the plant leader and the Council of Trust were permit-ted, while "all agitation which will endanger the industrial peace,

[39] HA, PSW 648, Note from the company archive dated April 28, 1978, on an interview with Retzinger of April 26, 1978.

[40] Wolfgang Hromadka, *Die Arbeitsordnung im Wandel der Zeit: am Beispiel der Hoechst AG* (Cologne: Heymann, 1979); the Factory Code of 1934 can be found on pp. 115–25.

[41] Ibid., p. 115. [42] Ibid., p. 118.

furthermore the bringing, distribution or affixing of unauthorized pamphlets and placards," was forbidden.[43] This gave the plant management and the Nazi representatives in the plant a wide leeway for interpretation – and they were to make use of it. A list of sanctions, graded according to the seriousness of the infringement against the regulations of the Factory Code, was also included. In addition to oral or written cautions sanctions included fines, hearings in front of a disciplinary committee, and, the highest punishment, immediate dismissal. The last punishment could be effected for a number of reasons, which included repeated drunkenness during working hours or the handling of stolen goods to the detriment of the factory. But above all, this was a sanction that could be used to punish "subversive behavior" and "if an employee endangered the industrial peace in the factory through malicious incitement of the followers or consciously presumed to intervene in the management of the factory in an inadmissible manner or continued to maliciously disrupt the community spirit within the factory community."[44] While the last sentence could also apply to members of the Council of Trust or even the *Betriebszellen-Obmann*, as in Hoechst – in the first instance it targeted opponents of the Nazi regime.

The Factory Code of 1934 already had a strongly political bent, but its successor, a code enacted in January 1939 that applied solely to the workforce of the IG Farben plant at Hoechst, was even more political and included even more details.[45] It can serve as an indication of how polycentric the structure of IG Farben became in the course of its colossal expansion during the 1930s, for the individual plants of IG Farben operated increasingly independently in the prewar years, until finally the members of IG Farben's Management Board were informed only of matters for which they were directly responsible and not more.[46] But it is also interpretable – and that would not be erroneous – as an indication of the decreasing importance of the Maingau Works Group and of its manager, Carl Ludwig Lautenschläger.

Now that full employment had been achieved in Germany, the preamble of the Factory Code of January 1939 placed an even greater emphasis than that of 1934 on the duties of employees. Thus the "leader" and the "followers" constituted a "plant community, to which everyone was subordinate." Everyone was expected to work together as comrades in a "National Socialist spirit" for the "common good of the people and the state and for the furtherance of the aims of the factory." "The National Socialist state has made the right to work a reality. Out of this arises the duty of every productive person to work." All employees were united "by

---

[43] Ibid., p. 119. [44] Ibid., pp. 122–5, quotation on p. 124.
[45] Ibid., the Factory Code of 1939 can be found on pp. 126–49.
[46] On this topic, see Hayes, *Industry and Ideology*, p. 24, also pp. 185 and 206 on the "atomization" of IG Farben.

the same honor of work," which they must not offend lest they be called to account for harming the community. In return for the loyalty and trust shown to the plant community and its leader, the leader promised "leadership and care which would extend beyond the workplace and the hours of work." What this meant was already outlined in the first paragraph, which addressed the question of "company care." This comprehensively listed the social benefits provided by the factory, from warm meals to a maternity home and to housing estates for employees.[47]

The second paragraph was devoted to the functions of the Council of Trust, the members of which were no longer elected by the workforce after 1935. The Council was described as the "organ of the plant community" in which "the social questions of the factory entrusted to it in accordance with the Law on the Organization of National Labor" should be deliberated. These included industrial safety and general working conditions, the solution of conflicts and a strengthening of the community spirit: "In the fulfillment of his tasks the leader of the factory shall be assisted by the members of the Council of Trust who will act in an advisory capacity." In addition to the Council of Trust, which was limited to the works, an advisory committee, the *Unternehmensbeirat*, was also set up for all of IG Farben to improve the connections between individual factories and offer advice on general social questions. Paragraph 3 briefly set forth the task of the *Betriebsobmann* (shop steward), as he was now called: "The *Betriebsobmann* as the representative of the German Labor Front, shall give his support to strengthening and reinforcing the community spirit."[48]

While the prominent position in the Factory Code accorded Nazi representatives in the plant already indicated their importance, the pivotal position of the Party and its organizations was emphasized in two further paragraphs. Paragraph 4 addressed the topic of "admission into the factory." The first sentence categorically stated: "Only those persons may be admitted into the factory community, who are members of the DAF [German Labor Front] and who expressly undertake to comport themselves loyally and honestly toward the German constitution and the leaders of the state." Hoechst had finally acceded to the demand of the DAF; it must be added, though, that in Hoechst the *Betriebsobmann* had been pressuring employees to join the DAF since 1934.[49] Furthermore, when hiring new staff, preference was to be given to previous employees who had been made redundant through no fault of their own "if they showed a sufficient aptitude" as well as "those deserving fighters who had contributed toward the national rebirth"; in both groups "war veterans" were to be given

---

[47] Hromadka, *Die Arbeitsordnung*, pp. 126–8.   [48] Ibid., pp. 128–9.
[49] Ibid., p. 129; HA, Pol 6, Letter from Retzinger, dated April 9, 1934, with the demand that "works comrades" should "now join the German Labor Front in their own interest"; on the demand of the DAF, see Frese, *Betriebspolitik*, pp. 162–8.

precedence.[50] In addition to this increased pressure to join one of the Nazi organizations, the Factory Code also included a separate paragraph on the "advancement of youth," fourteen-to-eighteen-year-olds. This stipulated that they should receive particular "support and care," but also that they must "endeavor at all times to have a National Socialist point of view" and be members of either the Hitler Youth or the League of German Girls.[51]

Other clauses of the Factory Code determined in extensive detail the working hours, absences from work, maternity leave and maternity protection, vacations, and behavior during working hours. As in 1934 the regulation of behavior included the prohibition of any gatherings that had not been specifically approved as well as "all agitatorial activities" that would disturb "the industrial peace." The obligation to maintain secrecy and the "fight against industrial espionage" were particularly emphasized. Employees were explicitly reminded of their duty not to disclose information; they were additionally expected to report "every sign of treason and high treason including subversive propaganda, espionage, and sabotage" immediately to the factory leader or his deputy. This exhortation amounted to an invitation to denunciations – which, however, had begun even prior to the new Factory Code.[52]

The so-called offences against social honor were treated in detail; such offences were to be punished by the disciplinary courts already provided for in the AOG. Detailed examples of such offences were given, most of which dealt with people holding office. An employer was held to have violated the "social honor" if he misused his position of authority and "maliciously" took advantage of employees or offended against "their honor." Employees were at fault if they endangered industrial peace "by malicious incitement of the followers." This also applied to members of the Council of Trust if they "consciously presumed to intervene in the management of the factory in an inadmissible manner or maliciously disrupted the community spirit within the factory community" or illicitly passed on confidential information which they had gained through their position. The range of possible sanctions by the factory was listed in detail; the severest punishment was dismissal. Summary dismissal was possible "in the event of serious offenses against the community spirit" and offenses against safety regulations, likewise for theft, the receiving of stolen goods or breaches of trust. The first point, that is, "serious offenses against the community spirit," was explained in more detail: thus, summary dismissal was possible "for legally ascertained subversive activities" but also for defamation or slandering a "works comrade"; misuse of one's position as a superior was sanctioned in the same way as the "malicious" incitement of employees or "malicious" disruption of the "community spirit."[53]

[50] Hromadka, *Die Arbeitsordnung*, p. 129.   [51] Ibid., pp. 142–3.   [52] Ibid., pp. 133–6.
[53] Ibid., pp. 145–9.

The Factory Code and its provisons not only made the integration into certain Nazi organizations mandatory, but also, as will be shown later, poisoned the oft-invoked industrial peace. The open and honest comradeship called for by the Factory Code was simply not possible if every employee was constantly aware that frank or even unconsidered words could create great difficulties.

The academic literature is nearly unanimous in holding that the AOG once more made employers "masters in their own houses," at least toward their employees.[54] The historian Martin Broszat even referred to the AOG as "the sociopolitical Basic Law of the Third Reich," which cut back workers' rights in favor of state regimentation and limited social partnership within factories in favor of a "relationship between leader and followers."[55] In contrast, Wolfgang Zollitsch has pointed out: "If one examines the practices at an operational level, a more differentiated picture appears. The members of the Council of Trust often did not content themselves with the narrow scope accorded them but took the activities of the former works council as their model. The tensions that this created within this body show that sweeping condemnations do not do justice to the behavior and the aims of many Councils of Trust members."[56] What was it like in Hoechst, then?

3.3. THE WORKS MANAGEMENT

3.3.1. Officer and "Leader" of the Plant: Ludwig Hermann

On January 1, 1933, Ludwig Hermann was appointed head of the IG Farben plant at Hoechst and of the Maingau Works Group. As mentioned above, Hermann had already been chosen by Carl Bosch at the end of 1929 for the works management of Hoechst – against the wishes of the "old Hoechst hierarchy." This was despite the fact that Hermann was a "Hoechst man" himself, even though he had spent most of his time in the Gersthofen works near Augsburg. This had been founded by Hoechst in 1902 as a branch plant, and Hermann rebuilt and enlarged it in the years 1919–29, turning it into a very profitable installation. Bosch and others from IG Farben's management were confident that Hermann, together with the new technical director and chief engineer Friedrich Jähne, would be

---

[54] See Zollitsch, *Arbeiter*, p. 228; Schneider, *Unterm Hakenkreuz*, p. 294; David Schoenbaum, *Hitler's Social Revolution: Class and Status in Nazi Germany, 1933–1939* (New York: Norton, 1980), p. 92.

[55] Martin Broszat, *Der Staat Hitlers. Grundlegung und Entwicklung seiner inneren Verfassung*, 11. Aufl. (Munich: dtv, 1986), p. 196.

[56] Zollitsch, *Arbeiter*, p. 228.

able to turn the Hoechst plant around and make it fit for future bigger tasks.[57]

Hermann was born on April 10, 1882, in Memmingen, where his father owned a brewery. Hermann went to school in Memmingen and subsequently attended an industrial college in Augsburg. In 1900 he switched to the Technical University of Munich, where he also became a member of the prestigious Corps Vitruvia fraternity. After passing his preliminary examinations in 1902 he served his one year of military duty in the Bavarian army in a field artillery regiment. At the end of 1905 he completed his study of chemistry with a doctorate in engineering. Prior to completing his studies Hermann had already begun working as a chemist in the potassium works in Aschersleben, where together with the Viennese professor Jean Billiter he set up the chlorine electrolysis facilities. In 1911 he moved to Hoechst, where he also introduced the process of the electrolytic generation of chlorine and developed a process for the compression and liquefaction of chlorine.[58]

At the beginning of the First World War, which he was later to look on as the most formative period of his life, Hermann, a convinced monarchist, enlisted as a volunteer.[59] Apparently on the basis of his knowledge of chlorine, at the beginning of 1915 he became part of an inner circle preparing the use of chemical weapons. Colonel Max Peterson had been given the assignment to set up a first company to work with poison gas, which was disguised as a "disinfection unit." His staff at the time consisted only of his adjutant Otto Lummitsch, the scientist and spiritual architect of chemical warfare, Professor Fritz Haber, and the first lieutenant of the reserve Ludwig Hermann.[60]

In the following years Hermann fought as an officer in the 35th Württemberg Regiment of Engineers, the "Gas Regiment" headed by Colonel Peterson, on both the Western and the Eastern fronts; as of the summer of 1916 he held the rank of a captain. The Gas Regiment used a so-called blowing process for the gas, which was extremely effective due to the thickness of the cloud of poison gas it created, but dependent on the prevailing direction of the winds. If the wind turned, the cloud of poison gas

[57] Cf. Winnacker, *Challenging Years*, pp. 74–5; Heine, *Verstand*, pp. 89–90 on Hermann; HA, PA Hermann, eulogies given at the memorial service for Hermann on June 2, 1938, here, speech of Fritz ter Meer.

[58] HA, PA Abteilung VI, Hermann, letter of application from Hermann dated January 20, 1911; PA Hermann, article "Männer, die Geschichte machen: Dr. Ludwig Hermann (1882–1938)" from *Die Farben-Post*, December 21, 1961.

[59] HA, PA Hermann, Speech script of Hermann, "Privat, zum Pioniertreffen in Stuttgart," May 4, 1935; oral communication by Lore Wittmer, Ludwig Hermann's daughter.

[60] Margit Szöllösi-Janze, *Fritz Haber 1868–1934. Eine Biographie* (Munich: Beck, 1998), pp. 327–9 and p. 780, n. 351; Dietrich Stoltzenberg, *Fritz Haber. Chemiker, Nobelpreisträger, Deutscher, Jude* (Weinheim: VCH, 1994), pp. 244–7, reports a slightly different version, given by another colleague of Haber's named Richardt, who does not mention Hermann. However, Richardt joined the group only at a later date.

might be pushed back again, inflicting losses on their own troops. The assembly of the gas cylinders for the blowing process was time-consuming and could be dangerous for the gas troop under enemy artillery fire. Because of these problems the gas troops repeatedly suffered heavy losses, as happened in May 1916 to the gas company commanded by Hermann in the Champagne area.[61]

Hermann never appears to have had a problem with the use of chemical warfare agents. His colleague and later friend, the Swiss Johannes Hess from Wacker Chemie, wrote to Hermann in August 1916 that he had been "somewhat anxious" whether Hermann had "got through the last offensive well." Hess then went on to mention a mutual acquaintance, who was now fighting against the Italians "also with chlorine gas," adding: "and yet it is a terrible weapon. But let us take help from wherever it comes!"[62] It is not known that Hermann voiced even such moderate criticism; he was apparently convinced of the value of the weapon.[63]

As an officer at the front, Hermann demonstrated both excellent professional skills and leadership qualities. His capabilities as well as probably his lack of scruples marked him out for higher tasks in chemical warfare. Thus, in the autumn of 1917 the army command put him in charge of the camp in Breloh in the Lüneburg Heath, which produced poison gas cylinders for the artillery. Writing to Hess at the beginning of October 1917 Hermann commented: "Currently I have a lot to do and am always on the 'qui vive.' The production of the artillery gas munition is now being carried out in Germany – some of it in plants which have been completed, some of it in plants which are still under construction. As the current head has been

[61] BA-MA Freiburg N 102/10, Heber memoirs: "At the end of April 1916 on the Western Front – the gas from the German gas pioneers rolled back in thick patches of cloud over the parapets of our own trenches in such high concentrations that the gas penetrated the masks and led to high losses"; see also Rudolf Hanslian (ed.), *Der chemische Krieg, vol. 1: Militärischer Teil*, 3. völlig neubearb. Aufl. (Berlin: Mittler, 1937), pp. 23, 77–106 on the blowing process and the German deployment; HA, PA Hermann, speech script of Hermann's: "Privat, zum Pioniertreffen in Stuttgart," May 4, 1935; the Hauptstaatsarchiv Stuttgart and the Bundesarchiv-Militärarchiv Freiburg have reported that no further documents exist on the 35th Pioneer Regiment.

[62] Wacker UA, 1. A. 24, Hess to Hermann, August 5, 1916.

[63] See also Gerit von Leitner, *Der Fall Clara Immerwahr. Leben für eine humane Wissenschaft*, 2. durchgesehene und verbesserte Auflage (Munich: Beck, 1994), pp. 201–2: In her description of the first weapons testing of gas in 1915 near Cologne Leitner writes about Hermann as follows: "The director from Farben Industries is completely possessed by this great patriotic feat of chemistry and does not admit any criticism. He is working on the smooth transportation and removal of the reusable standard steel flasks and keeps making new calculations in order to gauge the numbers – which depend on the tested distances of the chlorine gas flasks – per kilometer of the front line and to coordinate the quantity of chlorine gas to be filled and the filling with the plant in Leverkusen." On Hermann's opinion on the use of gas as a weapon see also Otto Hahn, *My Life*, trans. Ernst Kaiser and Eithne Wilkins (London: Macdonald, 1970), pp. 118–29.

'ousted,' I will probably be appointed commander. On the one hand I dislike leaving the war at the front, while on the other hand I am pleased that I will be placed in charge of something that big, because I do believe that I am strong enough to straighten things out."[64] It was this attitude, his ambition, and his will to master even the most difficult problems that were to qualify him later on for a job in industry as a top manager. Other qualities that later distinguished and characterized him were already manifest: Hermann was a perfectionist who paid attention to even the most minute details; likewise he had an awareness of potential dangers and he drew up corresponding instructions, sometimes in his own hand. As the production of the chemical weapon mustard gas remained risky, this attitude was not unimportant.[65]

Hermann was still in Breloh when the war ended. The outcome – defeat and revolution – did not fill him with enthusiasm. Just how much he disliked the revolution was still apparent in the 1920s. A fellow officer from Breloh, a certain Dr. Frantz, who had sided with the revolution in November 1918 was harshly attacked for this by the former military commander in Breloh, Colonel von Wangenheim. As his professional position suffered, Frantz asked several fellow officers who had served together with him in Breloh to write testimonials for him. Hermann heard about this from one of the officers who had received a letter from Frantz but decided to ignore his appeal. Hermann supported this attitude and clearly expressed his revulsion toward Frantz: "You were quite right not to react to his importunities. The fellow has not just dragged the honor of German officers through the mud but demonstrates by his unqualified behavior that in my opinion he is not worthy of being treated as a professional and a colleague."[66] The story had a sequel not untypical for the times: Frantz became mixed up in an affair of honor with another officer from Breloh who had publicly insulted him. The Council of Honor of the League of German Officers demanded proof of the alleged dishonorable behavior of Frantz, who was accused among other things of having spoken out against officers at a meeting of soldiers in November 1918. Typically Hermann offered himself as a witness for the prosecution: "On the basis of the notes in my diary I can offer a faithful picture of the events which occurred in Breloh from November 7–15, 1918, and which should hopefully be of some

---

[64] Wacker UA, 1. A. 24, Hermann to Hess, October 8, 1917.

[65] On this topic, see Stoltzenberg, *Fritz Haber*, pp. 291–5; Dieter Martinetz, *Der Gaskrieg 1914/ 18. Entwicklung, Herstellung und Einsatz chemischer Kampfstoffe. Das Zusammenwirken von militärischer Führung, Wissenschaft und Industrie* (Bonn: Bernard & Graefe, 1996), pp. 71–2; private files Stoltzenberg, letters from Stoltzenberg to his wife, dated February 17, 1918, March 3, 1918, August 14, 1918; handwritten order from Hermann to Stoltzenberg, April 4, 1918; order to 2. Füllkompanie, April 26, 1918.

[66] Clariant WA, file 1546, Otto Haehnel to Hermann, June 16, 1921, response from Hermann, June 21, 1921.

help to Hebler. Your memory of how much Frantz offended our Colonel at the time by the ostentatious wearing of the red ribbon in conferences with the old gentleman is quite right, because from that point on von Wangenheim broke off every connection to Frantz and, following his lead, all of us did likewise. Frantz was a rather dubious gentleman, and it would do him good to be pilloried."[67]

After the war Hermann returned to Hoechst and took over the management of the branch plant in Gersthofen near Augsburg. Hermann, as has already become clear, strongly opposed the revolution of 1918. This is evident in his choice of words on the Bavarian Soviet Republic. In April 1919 he reported on strikes and discussions with workers about wages and working conditions to the works management in Hoechst – and made his annoyance about the developments clear. At the beginning of April 1919 he wrote: "people are saying that a Bavarian Soviet Republic has been declared, but officially we know nothing about that as yet. In any case the mess is great."[68] By mid-April he had calmed down again: "The atmosphere is somewhat freer of the taint of Bolshevist germs since the workforce in Augsburg voted against the Soviet Republic on Sunday in a mass meeting."[69] Even after the Bavarian Soviet Republic had been toppled through the intervention of the national government Hermann never became a supporter of the republic.

In the years 1919–29 he succeeded in completely modernizing the Gersthofen plant and in introducing a number of new and lucrative products that made the plant profitable. The products of chlor-alkali electrolysis were used to produce caustic soda, bleaching liquor, and hydrochloric acid – all of which found a good market in southern Germany. Hermann also set up a profitable production of waxes and camphor in Gersthofen. Over the years he created a network of relationships, which included former fellow students and officers as well as his fraternity brothers, other industrialists, and managers. He also accepted a number of honorary posts, including that of representative of the employers' association, a position in the company's health insurance scheme and a patron of the arts – which probably contributed to his being awarded the honorary title of commercial councilor (*Kommerzienrat*) in Bavaria in 1927.[70]

[67] Clariant WA, file 1546, Wolfes (Hermann's adjutant in Breloh) to Hermann, February 10, 1924, response from Hermann, February 13, 1924: the Hebler mentioned here probably refers to Karl Heber, who served in the 35th Pioneer Regiment, see Martinetz, *Der Gaskrieg*, pp. 167–71, members' list of the Officers' Association of the former gas troops. Hermann's diary quoted here has unfortunately been lost.

[68] Clariant WA, file 1546, report of Hermann from Gersthofen, April 8, 1919.

[69] Clariant WA, file 1546, report of Hermann from Gersthofen, April 15, 1919.

[70] HA, PA Hermann, Article "Männer, die Geschichte machen: Dr. Ludwig Hermann (1882–1938)" from *Die Farben-Post*, December 21, 1961; see also eulogies given at the memorial service for Hermann on June 3, 1938, especially Fritz ter Meer's speech.

Through his connections he was able to intervene in a timely fashion when he wanted to ward off competitors or sought cooperation. In such cases he shied away from no expense and few means. Thus he also played a "considerable role" in the acquisition of 50 percent of the share capital of Wacker Chemie by Hoechst.[71]

His perfectionism was as much in evidence in Gersthofen as it had been in Breloh; he had no sympathy for inefficiency. Thus he once wrote personally to Duden, then head of Hoechst, when the wife of one of the directors of Wacker Chemie, which was now affiliated with Hoechst, had fallen ill with typhoid and urgently required antityphoid lymph to inoculate himself and his children. As the vaccine was not available in Munich, Hermann requested a certain Dr. Joseph in Hoechst to dispatch the remedy produced there to the address of the affected director as an express delivery, and Dr. Joseph promised to do so. However, Hermann wrote, the drug had never arrived, which had very much disturbed the director: "I am very sure that Dr. Joseph reacted promptly; it appears to me, however, that the Hoechst mailing office did not." As, according to Hermann, this was the second time something like this had occurred, it was "possibly in the interests of the reputation of our company to look into the matter." It was necessary to investigate where the "embarrassing delay" had occurred. And after he had apologized to Duden for troubling him in "such a matter," he added that the director in question, whose wife had by now died, had "urgently requested" them to refrain "from any form of condolences etc." In this way Hermann made it clear how much he considered the reputation of the company to be at risk if such incidents of unreliability were not redressed.[72]

The successes of Hermann in Gersthofen, along with his energetic manner and willingness to take charge, made him an ideal candidate in the eyes of the IG Farben management to succeed Duden in Hoechst. At the funeral service for Hermann in 1938 Fritz ter Meer thanked his "friend Hermann" for all he had achieved. As plant manager in Gersthofen he had shown his great capabilities as a technician and chemist, built up the plant, and brought it to a high technical standard: "Early on he matured into a versatile industrialist, who accepted general obligations within the industry and public duties in addition to his work in the plant. The confidence of his great teacher and mentor Carl Bosch then called him to the plant in Hoechst, whose management he took over in 1933, together with the management of the remaining factories of the Middle Rhine Works

---

[71] Clariant WA, file 1546, Hermann to Eduard von Reuter, Ministerialdirektor im Ministerium des Innern, Vorstand der obersten Baubehörde in München, dated February 17, 1921. Cf. also HA, 12, Hirschel, "Das Werk Hoechst im Verbande der I.G.," including "Verflechtung der Wackergesellschaft mit den Farbwerken Hoechst nach dem I. Weltkrieg."

[72] Clariant WA, file 1546, Hermann to Duden, February 5, 1924.

Group." There Hermann's task had been to put the Hoechst plant "on a
new, healthy footing, to reorganize its technical facilities and to fill the
factories with a fresh spirit and new life." As a "true German man" he had
been proud to be able to demonstrate in his position that he had understood
"the spirit of the times" and had "heard the call which our great leader
Adolf Hitler" had addressed "also to him." All would recall his achievements in Hoechst and IG Farben, "his strong personality, his simple, direct
spirit, his loyal, sincere character, and his gift, which few others shared, to
clearly grasp every question, whether it was of an objective or a personal
nature, and solve it with the same clarity and simplicity." Ter Meer continued: "And I am convinced that his open and warm manner, combined
with a soldierly severity, created a bond between him and every one of his
followers, just as he became a kind friend and valued colleague of ours in
the Board of Management."[73]

All in all Hermann was already a mature and experienced personality,
a "doer," when Carl Bosch picked him for the succession in Hoechst in
1929 – where indeed many things required doing. Strictly speaking, the new
task should have tempted him; for, as he himself once wrote, "such
Sisyphean tasks have always had a certain attraction for me."[74] Nevertheless he went to Hoechst with a fair amount of skepticism. He wrote to a
colleague in Ludwigshafen at the end of 1929 that he was obliged to leave
"Gersthofen of which I have become rather fond" at the beginning of the
year 1930 to take over the Indigo Department in Hoechst: "What other
tasks will follow I cannot as yet guess. I only know that I am obliged to take
on a task at an unpropitious moment in time, which will be no little
headache. But there is no help for it: if one is called, one cannot refuse; in
such cases personal reasons must be pushed aside." Yet he added: "It will be
difficult, but I hope that I will be able to manage."[75] Here again his sense of
duty toward the company is clear, obliging him to shelve his own wishes
and if necessary his own convictions.

After Hermann had familiarized himself with his new sphere of activity
in Hoechst from 1930 to 1932, he took on the management of the IG
Farben plant at Hoechst and the Maingau Works Group on January 1,
1933. With the National Socialist seizure of power on January 30, 1933 –
Hermann had been in his new position for only one month – his duties now
took on additional political dimension, which had not existed to the same
degree in the Weimar Republic. Hermann now did not just have to manage
the crisis in the Hoechst plant, he also had to come to grips with the new
political demands.

[73] HA, PA Hermann, eulogies given at the memorial service for Hermann on June 3, 1938,
   here speech of Fritz ter Meer.
[74] Clariant WA, file 1546, Hermann to Stoltzenberg, June 2, 1921.
[75] Clariant WA, file 1547, Hermann to Dr. P. Müller, Ludwigshafen, November 20, 1929.

As was the case with a number of supporters of the Nazi regime – among whom Hermann must undoubtedly be counted – Hermann's relationship to National Socialism was somewhat complicated. Hermann was a nationalist, a conservative, a reactionary. In the Weimar republic he probably supported or voted for the nationalist-liberal DVP or the more nationalist, even reactionary DNVP (German National People's Party), but not for the NSDAP. After 1933 he did not just welcome the establishment and consolidation of the Nazi regime but even became an enthusiastic supporter of Hitler and his economic and foreign policy. On August 1, 1935, Hermann also became a member of the NSDAP; he was one of only three top managers of IG Farben who were permitted to join the NSDAP in the period when there was a general ban on new members.[76] According to his daughter he received his membership as a "present" from *Gauleiter* Jakob Sprenger, which would appear to be borne out both by the date on which he joined – for after May 1933 there was an official policy of refusing to admit new members to the NSDAP – as well as the fact that he apparently never paid membership dues.[77] But Hermann was no mere "fellow traveller" out of opportunism. Handwritten additions to speeches added by Hermann testify to his genuine enthusiasm for Hitler, even in speeches held on semiprivate occasions. The few surviving letters also speak for themselves, even if it is necessary to add that the general climate of spying and persecution in the Third Reich did not lead just to a decrease in the number of letters but also to a certain caution on the part of most letter writers in their choice of phrases.

While Hermann had not cared for the Weimar Republic, he was now very taken with the Germany led by Hindenburg and Hitler, for that is how he perceived it. Here he believed to have found the "soldierly spirit" of the First World War again. As "plant leader," to use the new term for the heads of departments and works according to the new Law on the Organization of National Labor from January 20, 1934, he perceived himself as an officer and the "followers" as his soldiers. Hermann recognized the community spirit of the trenches of the First World War again in the "national community" (*Volksgemeinschaft*) of the Nazi regime – as did other industrialists and managers.[78] In an article on Hermann published in 1961 it was stated – correctly – that in the Nazi celebrations on May 1, he

---

[76] Hayes, *Industry and Ideology*, p. 102: the other two were Heinrich Hörlein and Heinrich Gattineau.

[77] BA, BDC file on Hermann, membership number 3,698,808, under the heading "contributions paid" there is a handwritten annotation "none"; oral communication by Lore Wittmer, Ludwig Hermann's daughter.

[78] Cf., e.g., Hermann Schlosser at Degussa (Hayes, *Cooperation*, pp. 49–53) or Hans Constantin Paulssen (Cornelia Rauh-Kühne, "Hans Constantin Paulssen: Sozialpartnerschaft aus dem Geiste der Kriegskameradschaft," in Paul Erker and Toni Pierenkemper (eds.), *Deutsche Unternehmer zwischen Kriegswirtschaft und Wiederaufbau. Studien zur Erfahrungsbildung*

had seen the "materialization of his ideal of comradeship and allegiance to the plant."[79]

The manuscript of a talk held by Hermann on May 4, 1935, bearing the annotation "private, for the meeting of the Sappers [Regiment] in Stuttgart," illustrates his ideas particularly well, as it gives a clear exposition of his convictions and values.[80] Hermann held a talk in front of the assembled sappers, addressing them as his "dear wartime comrades" and "friends," for whom he had "always felt a deep bond of great love and devotion," and all of whom "are and will remain my best friends throughout my whole life." He reminded them of the times "in which we were vouchsafed the opportunity to protect hearth and home from the clutches of the enemies surrounding our Fatherland." According to Hermann, the memory of the "bloody offering of millions" of the First World War was both "moving and inspiring" for all of them: "moving because with those millions of brave soldiers never before had so many German values for the future fallen victim to a militarily inadequate past. Inspiring, because we know today that our brave brothers did not fall in vain." For "from the spirit of our fallen soldiers arose the soldierly principle which in its purified and elevated state we now call the spirit of the people [*Volkstum*]. Our present state and our Reich are built on their sacrifices." In the war one had "been truly connected as a people," even if wearing the military gray uniform one had held a "different rank": "Do you remember our Christmas celebration in 1915 in Boltelager near Montfaucon? Was that not similar to the organization that we today call '*Kraft durch Freude*' [Strength through Joy]?" Hermann also reminded his hearers of the dark side of the war, of a day on which the dead and wounded in his company had clearly been particularly numerous: "It was the 16th of May 1916 in the Champagne area. In all the years which have since passed that day has never left my memory, its impact on the company emotionally affected me to an extent like no other event during my whole life. These sad events welded us together even more, if that were still possible."[81] Hermann ended his talk with quasi-historical philosophical reflections, in which he described Hitler as a gift from Providence. For hundreds of years, Hermann

von Industrie-Eliten (Munich: Oldenbourg, 1999), pp. 109–92 – the subhead reads "social partnership born out of the spirit of the war comradeship).

[79] HA, PA Hermann, Article "Männer, die Geschichte machen: Dr. Ludwig Hermann (1882–1938)" from *Die Farben-Post*, December 21, 1961; cf. also Pol 6, Brochure of I.G. Farbenindustrie Aktiengesellschaft Frankfurt a.M.-Hoechst, "Zur Erinnerung an den 1. Mai 1933."

[80] HA, PA Hermann, Speech script of Hermann, "Privat, zum Pioniertreffen in Stuttgart," May 4, 1935, typewritten, end of speech handwritten.

[81] It was not possible to find out more about the day referred to here, May 16, 1916, in Champagne, particularly as no more documents on the 35th Pioneer Regiment have been found, not even in various diaries of the war on the battles on the Western Front at this time.

went on, German history had vacillated between "peaks and troughs," but had risen up "again and again from the deepest depths to the heights." In 1844 the writer Gottfried Keller had stood "at the 'old and large grave of Germany' mourning the great corpse of the German people" – yet "a grave containing the will to be resurrected." Now, for the last two years, the German people were locked "in a tremendous struggle to prepare the way for its anewed ascent." Keller's predictions were coming true, thanks to Hitler, whose name, however, Hermann now did not mention. In the manuscript the end of his speech was annotated by hand, so that it now read: "If today staunch German brothers come together, they fondly remember the simple soldier for whom the experiences at the front were the original leaven to create a living national community out of his people. The resurgence of the Fatherland will be achieved from within this national community. He leads the way and serves as the best example. He gives us the spirit of his tremendous creativity. Indeed, he gives us more, he gives us his whole being and life. Let us help him, every one of us, by our devotion and faithful confidence in him, let us follow him to wherever he leads us." In this speech Hitler is envisioned as a cross between Moses and Christ; the ending resonates like a prayer. Hermann's speech reveals his devout, honest, and serious humility toward Hitler. Hermann appears to have completely accepted Hitler's self-stylization – despite the murder of many innocent people during the so-called Röhm coup and the behavior of the Nazi regime toward former comrades who were Jewish or were counted as Jewish.[82]

But Hermann's belief in Hitler and his regime, his willingness to support both the man and the regime "as a soldier" with all the force he could muster, was also evident on many other occasions. There are phrases that grab one's attention, phrases that would not have been necessary even during public functions. In the Maingau directors' meeting held on December 22, 1933, Hermann expressed his pleasure that "for the first time since a long time" the German people could once again celebrate "a German Christmas." Without touching on the issue of the persecution of political dissidents or the anti-Semitic excesses of the year 1933, he found only positive words to comment on the developments of the past year: "Much has changed within one year; a new courage and a new joy in life have entered many German hearts. Millions of our brothers now have work and bread. Progress is unmistakably apparent in many economic areas. The *Winterhilfswerk* [Winter Relief Effort] has done its bit so that nobody needs to suffer cold or hunger anymore, thanks to the generous care which is given to all."[83] But, of course, "care" was not given to all, and the persecution of

---

[82] Cf. especially Peter Fritzsche, *Germans into Nazis* (Cambridge, MA: Harvard University Press, 1998).

[83] HA, PA Abteilung VI, Albert Beil, Vortrag Hermann – Abschiedsworte für Beil and Dornhecker in der Maingau-Direktionssitzung am 22.12.33.

political dissidents and "racially inferior persons" had already begun. For Hermann this took a back seat when set against the decrease in unemployment and the continually improving economic mood.

In the public sphere his jubilation about the achievements of the regime was even more obvious, his veneration of Hitler almost religious. In March 1934 on the "great battle day of the campaign for work" Hermann praised the government for its work creation schemes. Directly after the First World War, particularly after the beginning of the Great Depression, the situation for young people who had completed their vocational training had become intolerable, Hermann commented, as they often either lost their jobs or never found a job in the first place. Nobody should be surprised that they had begun to doubt the purpose of the economy and looked to someone who could give them a belief in a better future: "And then the apostle came to the front ranks, who had already traveled up and down the country for many years but had been heard only by few. With the greatest urgency and fervor he pointed to new paths, showing us how to emerge out of the hopeless confusion and save our people from the fall into the abyss." And, according to Hermann: "at the last minute the kind fates decided in his favor: the apostle became the leader of the German Reich!" Since that time, Hitler and his government had filled "almost all of the people and thereby the German economy with new hope."[84]

Hermann perceived the achievements of Hitler's government not just as they affected the general economy but also in his own company. Hermann had apparently much regretted having to dismiss many employees of Hoechst during the Great Depression and seeing them plunged into the distress of unemployment. The reduction in the numbers of unemployed since Hitler's seizure of power had therefore impressed him.[85] But, above all, he appreciated the smashing of the unions and the 1934 Law on the Organization of National Labor. He emphasized this in his speech held on May 1, 1934. On this day the members of the Council of Trust, the committee that the law had created to advise the "plant leader" on the problems of the "followers," took their oaths of office. Hermann celebrated the end of "class struggle" and emphasized the "feeling of loyal comradeship held by all employees for one another." He praised Hitler's policies, which had given work and bread to two and a half million "fellow Germans suffering want." Hermann also welcomed the help given to those who were still unemployed: "To help those still sentenced to inactivity over the worst times during the winter months, an appeal was made to the willingness of

---

[84] HA, PA Hermann, speech script of Hermann's for the "great battle day of the campaign for work" (*Grosskampftag der Arbeitsschlacht*), March 21, 1934.

[85] Ibid., typescript of Hermann's speech on Labor Day (May 1), 1934; oral communication by Lore Wittmer, Ludwig Hermann's daughter.

the whole German people to make sacrifices. The people outdid themselves." The other course that Hitler had traced, according to Hermann, had been "born out of the 1st May 1933." "It was and is directed at our most important economic assets, the persons who are active in the economy. Not even 12 hours after the *Führer's* great speech at the Tempelhofer Feld the unions were brought into line." And on May 19 Hitler had declared at the Congress of the German Labor Front that "nothing filled him with so much pride as being able to say some day: I have won over the German workers for the German Reich!" Hermann was of the opinion that Hitler had found the solution to the difficult question of replacing the unions and the employers' associations. He had clearly seen "where and how he could incorporate the working German people so that the specter of class struggle would be eternally banished and the focus would be on the people themselves." Hitler had brought back "work rules" to the economic setting, which offered the greatest opportunities of overcoming differences that would continue to exist: "the works plant, where the condition required for a personal community is easiest to achieve." This had been brought to fruition in the Law on the Organization of National Labor. The law was based on three fundamental realizations: "leadership and responsibility" belonged together, followers and "leaders" were bound together "come what may," and all agreements and directives were meaningful only if they were concluded "in a spirit of decency and honor." He was filled with joyful pride, Hermann said to his "followers," "that he had been called on to be their leader." He looked on the responsibility not as a burden, but approached the task with great joy: "I will carry out my office with justice, respect, and love toward every single individual, with clemency where it is appropriate, with severity where that is necessary. Above all shall stand the dedication to the plant and the willingness to make sacrifices for the good of the plant as a whole and beyond that for all of our people. Day and night my greatest endeavor shall be to provide you, my followers, with permanent employment and additionally to do everything to incorporate new employees in your ranks." Both the "followers" and he himself as their "leader" were accountable to the *Führer*. The members of the Council of Trust stood between the workforce and himself as the plant leader as "honest intermediaries": "You are the bridge to the entirety of your workmates. You perceive the 'soul of the plant' and therefore will and can always create a true picture of the current frame of mind of the followers for the management." The members of the Council of Trust were a "selected host of fighters," who would march together with the followers and leaders, with unreserved candor, in loyalty and obedience – as the "connecting link" between followers and leadership. Hermann concluded his speech with a fervent hymn to Hitler: "Let us all look up in awe, admiration, love, and gratitude to the brilliant man, whose purest wishes and thoughts, whose courage and energy and confidence, whose farsighted gaze have placed a wealth of inconceivable successes in the lap of our people."[86]

Of course, one has to ask how much of this was simply the rhetoric of the Third Reich, how much was purely lip service – but all in all at least the enthusiasm for Hitler appears to mirror Hermann's own convictions, as such comments have also been recorded as uttered on semiprivate occasions such as the reunion of the Sappers Regiment. Hermann apparently did reject the anti-Semitism of the Nazis. He put in a word for the daughters of the well-known Frankfurt professor Gustav Embden, who were friends of his own daughters. Professor Gustav Embden, a Jew, suffered so much from the massive disruptions of his lectures by his students that he took his life in the summer of 1933. The Pharmaceutical Department of Hoechst had worked together with Embden for many years. Likewise, one of Hermann's daughters, a keen horsewoman, spent almost every day at the famous stud farm of Carl von Weinberg, who held Hermann in high esteem.[87] While Hermann's dislike of anti-Semitism appears quite believable, it must be noted that despite his idealization of the old wartime comradeship he accepted that former comrades-in-arms who were Jewish or descended from Jewish families found themselves increasingly excluded in Nazi Germany, including comrades from the gas troops of the First World War. They were no longer "fellow members of the community" (*Volksgenossen*) and therefore no longer "comrades." Even if Hermann apparently did try to help some of them, nevertheless he placed a greater value on the new "national community" in the Nazi state than on the old feelings of comradeship between brothers-in-arms.

Hermann did have serious reservations about the attempts to subordinate the Lutheran church to the influence of the National Socialists and even involved himself in attempts to prevent it. He had no wish to mix religion and ideology. Hermann was a committed Protestant who had let himself be persuaded by one of his schoolfriends, the minister Oscar Daumiller, who later held an important position in the Bavarian Protestant regional church after 1933, to finance the education in the seminary at Memmingen for the son of a friend who had fallen in the First World War.[88] Quite early on Hermann contributed to endowments for persons in need. In the 1930s he actively supported the building of a Protestant church in Gersthofen. In the spring of 1934 he wrote to Karl Weber, his successor as plant manager in Gersthofen, that IG Farben had agreed "on my urgent

---

[86] HA, PA Hermann, typescript of Hermann's speech on Labor Day (May 1) 1934; cf. Plumpe, *I. G. Farbenindustrie*, p. 622.

[87] Oral communication by Lore Wittmer, Ludwig Hermann's daughter; private files of Hermann family, very personal letter from Carl v. Weinberg to Lore Wittmer on Hermann's death dated June 2, 1938.

[88] Clariant WA, file 1547, Daumiller to Hermann, December 8, 1927; response from Hermann, December 10, 1927.

request" to donate 20,000 RM for the building of a Protestant church; that was around 80 percent of the required sum. However, Hermann added: "The regrettable quarrel within our Protestant church fills every believing member of the church with serious concern." Referring to a statement of Hitler's that one should differentiate between the organization of an ideology and that of a belief, he ordered that Weber must take pains to see that "the church in Gersthofen did not fall in the hands of the German Christian Movement." "If this were the case or if there is even a likelihood of this occurring, then the sum should be considered as not having been approved." Indeed, one should ask oneself whether there was any purpose in building a church as long as "such a gruesome spectacle" continued to be enacted within the Protestant church. Weber should talk this over with his friend Daumiller, now a member of the church synod. Weber traveled to Munich and informed Hermann of his talks with representatives of the Protestant regional church, the gist of which was that the Bavarian regional church supported "their Bishop Meiser" and "wanted to have nothing to do with the new movement." Weber did not share this optimism and wrote to Hermann that in Bavaria it would not be possible to offer "resistance to a ruthless decree, even if the decree is unconstitutional, as it was in Prussia." And as it was not yet clear which line of thought was likely to be victorious in the regional church, Weber was of the opinion that the money should be kept back until everything had been settled. "We should not disregard the fact when building a Protestant church in Gersthofen that with the new, large housing estate, which I already told you about, many families completely loyal to the Party and thereby also to the new religious movement might move into this area, which could lead to unpleasant complications."[89]

Yet criticism of certain aspects of the Nazi ideology or of some NSDAP members did not make Hermann into a critic of Hitler or of the Nazi regime. It was possible to be frank with him, as Weber's letter shows – but Hermann himself was not very critical of the new regime despite its measures against Jews and the Protestant church. On May 1, 1935, he praised Hitler and the regime for what had been achieved during the past two years: prior to the seizure of power the nation had lain under a pall of "paralyzing pessimism, tired resignation, gray despair," Hermann stated. "Was it therefore surprising that millions of our best brothers, particularly those condemned to inactivity through no fault of their own, despaired at finding any purpose in life?" No one dared "to believe in the intervention of a

[89] Clariant WA, file 1547, Weber to Hermann, December 4, 1933; Hermann to Weber, March 15, 1934; Weber to Hermann, March 19, 1934. On the Bavarian "church struggle" against the German Christians and on Daumiller's role, see Helmut Baier, *Die Deutschen Christen Bayerns im Rahmen des bayerischen Kirchenkampfes* (Nuremberg: Verein für Bayerische Kirchengeschichte, 1968), especially pp. 146–9, 425–7; in detail on this topic, see Doris L. Bergen, *Twisted Cross. The German Christian Movement in the Third Reich* (Chapel Hill: University of North Carolina Press, 1996).

tremendous personality who would bring about a turning point." Yet this personality had already been there, "even if he was still unrecognized or even resisted." Since Hitler had assumed the reigns of government, things had begun to move "forward and upward in quick succession." Hermann rejoiced: "With the mobilization of all energies it was possible to break the paralysis of the economy and with state help to thaw the frozen despair. Confidence and trust, the mainstays on which the success of an economic rebirth rest, were restored." Unfortunately this economic revival remained largely limited to the domestic market, while problems persisted in export markets "which can be attributed in no small measure to malice and ill-will in the world." Everything that had been achieved depended on Hitler's personality, and Hermann found particular words of praise for the enactment of the *Wehrgesetz* (Law concerning Armed Forces): "Let us give him that which even the poorest fellow German is capable of giving: our firm will to follow his leadership in faithful confidence and devoted loyalty wherever he may lead. In this way we serve the nation and by honoring the *Führer* we honor ourselves."[90]

From being someone who might have been described as conservative, reactionary, and nationalist, Hermann had become a convinced supporter of Hitler and his policies. Hitler's foreign policy met with Hermann's unqualified approval. At the end of December 1933 he stated contentedly during a meeting: "The League of Nations is dead, the disarmament conference is done for, the world economic conference has failed. The disgraceful treaty of Versailles, that work of the devil, is beginning to totter, you can clearly feel it, something new is coming." Hermann did, however, hope that the foreign policy would be peaceful: "I hope that German and French negotiators will soon be sitting together at a table to come to a political and economic understanding and arrangement for both countries which will not have hatred and war as its aims but peace and prosperity for both peoples." But this would require Germany being given "justice and equality."[91] The return of the Saar area in 1935 and the military reoccupation of the Rhineland in contravention of international law in 1936 left Hermann jubilant. He commissioned a statue, "the liberated Saar," from the well-known sculptor Scheibe for the Hoechst plant, and a small copy of the statue to put on his desk.[92]

Just how positively Hermann continued to view the government was once again shown in his speech on the occasion of May 1, 1937. In the meantime

---

[90] HA, PA Hermann, typescript of Hermann's speech on Labor Day (May 1), 1935.

[91] HA, PA Abteilung VI, Albert Beil, Vortrag Hermann – Abschiedsworte für Beil and Dornhecker in der Maingau-Direktionssitzung am 22.12.33.

[92] HA, article "Eine Saar-Statue im Werk Hoechst" from "Von Werk zu Werk. Der Maingau," March 1936; oral communication by Dr. Wolfgang Metternich, director of Hoechst Archives.

the Nuremberg Laws had deprived Jewish citizens and citizens descended from Jewish families of their rights and persecuted them; even during the empire, law had been held in greater respect. But Hermann praised what had been achieved and believed that "a new great time for Germany has dawned." For much had been achieved in the years of the National Socialist regime: "The technical achievements were in the vanguard during this rebuilding. Great difficulties, caused by our lack of raw materials, by agitation and boycotts, had to be overcome. Despite these dangers the National Socialist government has managed to bring work and bread to an army of unemployed, several millions strong, and given the severely shaken German economy a new impetus." All this was achieved "by the German people, with their own efforts and without any foreign help," and all of this, particularly the economic upsurge, was due only to Hitler: "industry and economy once again have been given tasks to complete and are able to plan for the longer term." But Hermann did not consider that this was all. Looking beyond one's own sphere of work, other successes were apparent: "Germany once again has a strong national army. With the occupation of the garrison towns of the Rhineland by soldiers on March 7, in the name of the people the *Führer* has restored full sovereignty to Germany after years of the deepest national [*völkisch*] humiliation." He added: "We are proud of the decision of our people on March 29, the day on which it accorded its *Führer* a mark of confidence in a celebratory act of voting which was unequalled in history. German unity, established by Adolf Hitler, is now after more than three years of National Socialist rule unshakeably established."[93]

Just how seriously Hermann took the National Socialist "plant community" is evident from his activities in Hoechst. He established, apparently on the suggestion of the *Betriebsobmann*, annual celebrations of employees' anniversaries. He seems to have gotten on very well with Ludwig Retzinger, who was *Betriebsobmann* at Hoechst during most of Hermann's time as works manager. On the occasion of Hermann's funeral service Retzinger said that with Hermann's death they had lost not just a plant leader and an "exemplary workmate," but also a "fatherly friend, in whom we could place our complete trust at all times." Decades later Retzinger still spoke highly of Hermann, calling him "outstanding" with "a great understanding of social problems."[94]

Hermann's receptiveness to the NSDAP and its organizations in Hoechst fit well with his interests. The new factory code ensured that he would be "master in his house"; he even had a decisive influence on the choice of

[93] HA, PA Hermann, typescript of Hermann's speech on Labor Day (May 1), 1937.
[94] HA, PA Hermann, eulogies given at the memorial service for Hermann on June 3, 1938, Retzinger's speech as *Betriebsobmann;* PSW 648, note from the company archive dated April 28, 1978, on an interview with Retzinger of April 26, 1978; see also Bäumler, *Rotfabriker*, pp. 283–4.

candidates for the Council of Trust. He was therefore correspondingly prepared to work together with them and show himself to be a "good negotiating partner" vis-à-vis the National Socialist representatives in the plant and the *Gau*. Under Hermann there already was a separate contact man with the *Gauleitung* and the local Gestapo in the plant, Hans Wagenheimer, although Wagenheimer did not yet play the role he was to play under Hermann's successor.[95]

As plant manager Hermann was able to modernize and reorganize the plant with the help of his chief engineer, Friedrich Jähne. Around 1936 Hermann developed the first symptoms of cancer of the throat. He spent only short periods in clinics or sanatoriums – he wanted to return as soon as possible to his work at the plant, a wish that also accorded with his sense of duty. Yet he was increasingly forced to direct the fortunes of the plant from afar. Letters he received from his own management office or from senior employees were annotated by him in the margins with the corresponding instructions.[96] At the celebrations of firm's seventy-fifth anniversary he was barely able to speak. But he continued to come into the plant up until a very short time before his death on May 31, 1938. In his memoirs Karl Winnacker recounted that he had visited Hermann in his study only a few days before his death. Even then Hermann had demonstrated his energy and openness toward all things new, for he immediately took up and implemented Winnacker's suggestion to purchase an electron microscope for the plant's Process Engineering Department.[97]

Up until his death Hermann showed his dedication to Hoechst. He is perhaps best described as an "officer," which is how he saw himself. He had always been leader with a great sense of both responsibility and authority, but he also always remained someone who followed orders – and he executed orders, as he had done in the war in the matter of poison gas, without any caveats. The soldierly virtues of an officer, who looks after his "followers" and seriously cares for his plant, were closely bound up for him with the other aspect of this profession, namely, obedience. As works manager Hermann was both "leader" and led – and he accepted both positions with great dedication, without scruples and with no consciousness of wrongdoing.

3.3.2. The Compromise Candidate: Carl Ludwig Lautenschläger

In January 1938 Hoechst celebrated its seventy-fifth anniversary, although it was impossible to overlook Hoechst's loss of importance within IG

---

[95] HA, PA Hermann, letter from Hermann's office (Bormann) to Hermann, February 1, 1938; HA, Hirschel, "Aus meiner I.G. Zeit," p. 25.

[96] HA, PA Hermann, Letters from Hermann's office (Bormann) and from von Brüning to Hermann, 1937 and 1938.

[97] Winnacker, *Challenging Years*, pp. 87–8.

Farben even in the anniversary publication. The products of Hoechst's respected Pharmaceutical Department now left the works bearing the Bayer cross, for the double emblem "Bayer-Meister Lucius," which had been introduced in 1927, had disappeared in 1934.[98] More depressing for Hoechst was the fact that Hermann was now so ill that he could say only a few words of greeting and the anniversary speech was given by the grandson of one of the founders of the company, Gustav von Brüning, who now, according to Winnacker's memoirs, appeared to be "the coming man."[99]

Another potential successor, the director Karl Staib, had died unexpectedly in the plant just a short time before, at the end of December 1937, at the age of only forty-three. After completing his chemistry studies at the Technical College in Stuttgart, Staib served as a highly decorated artillery officer in the First World War. He was severely wounded in the war and was classified as 40 percent war-disabled. After the war he completed his dissertation in 1920 in Stuttgart, followed by a job as assistant at the Inorganic Chemistry Institute of the University of Breslau. In 1921 he left the university to work in industry at the chemical plant at Griesheim-Elektron. After its fusion with IG Farben a number of factories were closed, and Staib moved to Bitterfeld, where he gave a very good account of himself. In 1931 he was entrusted with the management of the small IG Farben plant at Rheinfelden; in 1934 Hermann appointed him head of the Inorganic and Nitrogen Departments at Hoechst. Here Staib once again showed his talent, and as early as May 1936 the works management put in an application for him to be appointed a director, which was approved by the Central Committee of IG Farben in June.[100] How highly Hermann rated Staib is indicated by the fact that Hermann did not let his own advanced illness deter him from attending Staib's funeral, where he gave the funeral address and spoke of his esteem and affection. Hermann did not just find words of praise for Staib's professional achievements, he also praised his leadership qualities. With his exemplary discipline, his hard work, and his cooperative and gregarious manner he had managed to forge a close bond with his "workmates": "a new and good spirit took up residence in his department when he came. Everywhere buildings were being demolished, renovated, rebuilt, improved; stimulating discussions paved the way for new goals, clear instructions were given: in short, everything was movement, liveliness, health." Staib had been "almost at the zenith of his life"

---

[98] See *Chronik der Hoechst AG 1863–1988* (Frankfurt am Main: Hoechst-Aktiengesellschaft, 1990), p. 159; Pinnow, *75 Jahre Werksgeschichte*, p. 72: production of Panflavin pills.

[99] Winnacker, *Challenging Years*, p. 87.

[100] On this point, see HA, PA Abteilung VI, Staib; PA Staib, including obituary of Staib, dated January 14, 1938, and article on Staib's death from "Von Werk zu Werk, Ausgabe Maingau," February 1938.

and "many a great deed had been expected" from him in the future.[101] As these words show, Staib was a man very much to Hermann's liking. A not unimportant point for a potential successor to Hermann was that he had begun work in Griesheim and therefore stemmed from the Maingau Works Group. It appears plausible that Staib had been considered as a potential successor – but now he was dead.[102]

In Winnacker's memoirs and in Bäumler's book *Die Rotfabriker*, published in 1988, the discussions about the succession are described as follows: after the death of Staib and of Hermann a "certain amount of confusion" prevailed among the management of IG Farben. According to Winnacker, the view seemed to be that management recruits should come from within the works groups, and the founding families continued to play a role far beyond the importance of their shareholdings. Finally in August 1938 Gustav von Brüning, who stemmed from one of the founding families and who had taken over the Inorganic Department after Staib's death, called Winnacker into his room and disclosed to him that he would be entrusted with the management of Hoechst "within a year." However, Brüning, who according to Winnacker was well prepared for this job, took his own life on October 4, 1938, "in a fit of depression." Thereafter there was a "long interregnum" in Hoechst until finally, "to everyone's surprise," Lautenschläger was appointed successor.[103]

However, this depiction of Hermann's succession is false, which is somewhat surprising as in the history of the plant written by Otto Hirschel, which served Bäumler as a basis for his book *A Century of Chemistry*, the chronology is given correctly.[104] On June 4, 1938, the "Secretary's Office of Dr. Ludwig Hermann" in Hoechst asked the IG Central Office in Ludwigshafen to send all contracts that had previously been addressed to Hermann to Lautenschläger, "who, as the deputy of Dr. Hermann up to now, will continue to carry out his functions."[105] According to the resolution of the IG Farben Supervisory Board, dated May 28, 1938, the former deputy members of the Management Board were appointed full members of

---

[101] HA, PA Staib, funeral speech from Hermann, December 31, 1937; Winnacker, *Challenging Years*, p. 87, places the funeral service as taking place sometime during the first days of January; and writes that the illness had already visibly left its mark on Hermann.

[102] Winnacker, *Challenging Years*, p. 87: "Staib had enjoyed everybody's respect because he was endowed with the progressive intellect that was so characteristic of life in the modern central German plants. He had probably been quietly selected as Hermann's successor"; cf. also HA, Hirschel, "Aus meiner I.G. Zeit," p. 23.

[103] Winnacker, *Challenging Years*, pp. 88–90; Bäumler, *Rotfabriker*, pp. 299–300, is also incorrect in his description of Hermann's succession.

[104] See HA, 12, Hirschel, "Das Werk Hoechst im Verbande der I.G.," p. 31.

[105] BASF UA, B 4/1915, Hoechst (Sekretariat Hermann) to Ludwigshafen (Zentralstelle für Verträge), June 4, 1938; Ludwigshafen (Zentralstelle für Verträge) to Hoechst, June 7, 1938: Hoechst was requested to pass on a letter that had been addressed "Middle Rhine Works Group for the attention of Dr. v. Brüning" to Lautenschläger.

the board; in Hoechst this affected Lautenschläger and Chief Engineer Friedrich Jähne.[106] On June 30, 1938, Lautenschläger announced in a circular that he had been appointed "plant leader of the Hoechst works as well as leader of the Middle Rhine Works Group" in a resolution of the Central Committee passed on June 17. Further on the circular continues: "At the same time Dr. Gustav von Brüning has been appointed director with instructions to take over the technical management of the production facilities and to merge the laboratories with the exception of the Pharmaceutical Laboratories."[107] Brüning thus occupied a very prominent position in the plant, for he was given priority above all other directors in the plant unless they belonged to the Pharmaceutical Department; he was virtually a primus inter pares. Yet he was appointed a director of Hoechst only after having been entrusted with new duties.

In his personal memoirs Hirschel wrote that after Staib's death and the interregnum of Lautenschläger Brüning had been expected to take over the works management: "As the successor to Dr. Hermann the IG Working Committee placed Professor Dr. Carl Ludwig Lautenschläger in charge of the Hoechst plant and the Maingau Works Group; Dr. von Brüning was given the job of overseeing technical management of the factories and the task of unifying the research laboratories with the exception of pharmaceutical research while remaining manager of the inorganics plant. After a period of settling in and getting to know the ropes he was expected to succeed Professor Dr. Lautenschläger as manager of the Hoechst plant and the Maingau Works Group." But Brüning's suicide put paid to that.[108]

That Brüning had indeed originally been envisaged as the future works manager of Hoechst is borne out by an exceedingly positive testimonial by Hermann, dated February 1934, concerning Brüning, which was probably written on the occasion of Brüning's being appointed an authorized signatory in April of the same year. Hermann praised Brüning's professional abilities, who, Hermann said, had successfully completed several difficult tasks. At the end of 1930 he had taken over the Indigo Department as Hermann's deputy and managed it independently in the following two years, proving that he was able to "direct a large, troubled factory without letting it go to rack and ruin but on the contrary maintaining it in a quite good condition." In January 1933 Hermann therefore had appointed him a member of the newly created "Directorate T," the central coordination and testing office in the plant, as one of his personal staff. Immediately afterwards Brüning was given the job of investigating the problem of the very high production costs in the Alizarin Department and was able to report a successful outcome. As of

[106] Hoechst Human Resources, PA Abteilung VI, Lautenschläger, memo by Flach for Direktoren und Prokuristen im Werk Hoechst dated June 4, 1938.
[107] HA, PA Abteilung VI, Gustav von Brüning, circular dated June 30, 1938.
[108] HA, Hirschel, "Aus meiner I.G. Zeit," p. 23.

March 1933, Brüning reported to Hermann "in all general technical matters" and was responsible for communication between the works management and technical operations in the individual factories. All in all, Hermann came to the conclusion that "Dr. v. B. is a very knowledgable, hard-working, and farsighted employee, who with his calm, matter-of-fact, well-considered handling of the tasks entrusted him has not just earned our satisfacation in the highest degree, but also enjoys the full confidence of the entire work force (white-collar employees and blue-collar employees)." He added: "In summary it is necessary to attest that in addition to his general suitability v. B. does have those leadership qualities which would qualify him later to become works manager."[109]

In the years that followed Brüning was therefore increasingly employed for special tasks inside IG Farben and – as the potential works manager – was sent to gain experience outside of the main plant. In 1935 the IG Farben management entrusted him with the delicate job of setting up an office in Berlin that would "establish a contact between all IG plants on the one hand and the offices of various authorities on the other hand to deal with confidential matters." Once again, to quote from the testimonial of March 1938 on his appointment to deputy director, Brüning handled this task, "with great finesse by creating the Liaison Office W (*Wehrmacht*), which he headed until the spring of 1937."[110] Here Brüning successfully worked together with Carl Krauch, later head of IG's Supervisory Board, in this highly political and extremely important position for IG Farben as the representative of Division 2.[111] Brüning then returned again to Hoechst, where after Staib's death he was put in charge of the inorganics plants and the Nitrogen Department.

What is interesting about the testimonial written by Hermann about Brüning in 1938 is that this merely ends with the words: "Dr. von Brüning is a very knowledgable, dedicated, farsighted employee, with an irreproachable character, who with his calm, matter-of-fact, well-considered handling of the tasks entrusted him has not just earned our satisfaction in the highest degree, but also enjoys the full confidence of his co-workers."[112] The paragraph that had been added in 1934 of his suitability as a works

[109] HA, PA Abteilung VI, Gustav von Brüning, handwritten expert opinion by Hermann dated February 10, 1934.

[110] HA, PA Abteilung VI, Gustav von Brüning, expert opinion by Hermann dated March 23, 1938; BA, R 8128/A 139, personal data sheet of von Brüning, "joining the company": Berlin Liaison Office, October 15, 1936.

[111] On the Liaison Office ("Vermittlungsstelle W") see Hayes, *Industry and Ideology*, pp. 142–3; on Brüning see HA, Firmengründerfamilie von Brüning, note on the life of von Brüning dated June 10, 1942: the note states that he was in Berlin in the Liaison Office from September 1935 until December 1936.

[112] HA, PA Abteilung VI, Gustav von Brüning, expert opinion by Hermann dated March 23, 1938.

manager is missing; instead there is a brief comment on Brüning's irreproachable character. The sources do not indicate why Brüning apparently was no longer considered as a future works manager for Hoechst, particularly after he had evidently proven his worth in the delicate and political assignment of creating the "Liaison Office W," which retained its importance during the Second World War. Nor can it have been due to his lacking the proper political attitude, as Brüning had been a member of the NSDAP since April 1933.[113] The reasons behind the rejection of Brüning as well as for his suicide cannot be reconstructed on the basis of the existing records. There were rumors that Brüning was a homosexual, which would have cost him the works management and might have driven him to suicide, and 1938 had seen an increase in the persecution of homosexuals in Nazi Germany. But there is no concrete proof of this reading of events; the official version was that he had died from a heart attack.[114]

We can only speculate what the reasons were for the disavowal of Brüning and his suicide. Yet his death was important for Hoechst and was to have significant consequences, as Lautenschläger lost his most important employee, whom he urgently needed for running the plant and the works group. For Lautenschläger was a completely different person from Hermann, and a compromise candidate. Lautenschläger had not put himself out to get the position, although he had functioned as Hermann's deputy since 1935. Bormann, Hermann's office manager, wrote him a letter, sending it to where he was on vacation, to inform him that on January 30, 1935, *Gauleiter* Sprenger would be giving a speech in the works as part of a demonstration of the German Labor Front, and Erwin Selck, a member of IG Farben's Supervisory Board, would be in attendance. Bormann added: "and thus Prof. Lautenschläger will be spared the task of giving the welcome speech, which will certainly not upset him."[115] As

---

[113] Cf. on this point, BA, R 8128/A 139, personal data sheet of von Brüning: member of the NSDAP since April 30, 1933, membership number 2,399,434. Interestingly the BDC does not have a file on him, however, one should remember that the BDC has the records of only about 80–90 percent of all NSDAP members. Brüning was related to Oswald Pohl, a high-ranking member of the SS.

[114] Independently of one another, both Ludwig Hermann's daughter, Lore Wittmer, and Carl Ludwig Lautenschläger's widow mentioned Brüning's alleged homosexuality. This at least would indicate that in 1938 and even later this was a common reading of the events. On the increased persecution of homosexuals in general cf. Gellately, *Backing Hitler*, pp. 113–15. Bäumler, *Rotfabriker*, p. 299, refers to his death as follows: "Officially Brüning was held to have died from heart failure. In fact, he had committed suicide during a bout of depression. The reasons for this were only known to his immediate family." See also HA, Firmengründer Familie von Brüning: Hoechst to Werke der I.G., October 4, 1938, Lautenschläger to Abteilungsleiter in Hoechst, October 4, 1938; both letters refer to his death "from heart failure"; cf. also PA Abteilung VI, Gustav von Brüning, obituary by the IG in *Frankfurter Volksblatt*, October 5, 1938.

[115] HA, PA Hermann, letter [Bormann] to Hermann, January 29, 1935.

Hermann's illness grew steadily worse, Lautenschläger was forced to take over more and more of his functions inside IG Farben and the plant as well as to represent him outside the works, particularly in maintaining relations with the NSDAP. Lautenschläger did this reluctantly, particularly as Hermann evidently did not delegate any responsibilities to Lautenschläger but only allowed him to play the role of works leader to outsiders.

Bormann, whom Hermann apparently trusted, reported to Hermann at the end of January 1938 that ter Meer had given Lautenschläger the choice of attending a meeting of the Technical Committee in Hermann's stead. Lautenschläger accepted the prepared files and cost estimates but had then telephoned Hermann's office the following morning and said that Friedrich Jähne should do it as in the short time available he had not been able to familiarize himself sufficiently with the subject such that he would be capable of giving a proper account to the Committee. Bormann wrote that he had the impression that Lautenschläger "felt somewhat burdened by the responsibility, and that he would prefer someone else to be in his position, who would also be able to function as your representative toward the outside." Lautenschläger was of the opinion, Bormann reported, that as a doctor of medicine he did not have enough knowledge to be able to fill the role. If he had to represent Hermann, then he would also have to familiarize himself with the other departments, "which was absolutely not his thing." Lautenschläger had rejected Bormann's cavil, namely, that his role was primarily of a merely representative nature. Lautenschläger's view, according to Bormann, was that as the representative of the works manager he should "also have the necessary knowledge, so as not to make a fool of himself." Bormann added that he was "convinced" that Lautenschläger would "prefer to continue pursuing his own medical and scientific concerns and would prefer someone else, even if he were younger, to be entrusted as your representative."[116]

Sometime around 1942 in the unpublished memoirs he prepared for his children Lautenschläger wrote about his becoming works manager and Brüning's suicide: he had been obliged more and more to stand in to represent the gravely ill Hermann. "In the last days of his sufferings I often sat at his bedside to discuss the succession, because I confidently hoped that the difficult position of plant leader would not devolve on me at the present time and that maybe a technician from one of the other big IG plants could be appointed manager of the Middle Rhine Group." But the Central Committee had selected him; "fate" had called on him to take on "the scientific, technical, and sociopolitical leadership of our Maingau plants in the new National Socialist spirit." He "felt that it would have been cowardice" to evade this call. Yet he had found it hard to take over the works management, and it had not been made easy for him. "Difficult and

---

[116] HA, PA Hermann, letter [Bormann] to Hermann, January 27, 1938.

momentous months followed my appointment to the new position; much unwarranted resistance on the part of my closest co-workers had to be overcome, and many unfriendly checks to my serious work and difficulties in my path had to be removed." Lautenschläger counted the suicide of Brüning as one of these difficulties: "only a few months later I suffered another big disappointment; Dr. von Brüning, whom I had included in the works management as my closest employee, shot himself. Previously I had never understood that life should be a battle; I was always of the opinion that loyalty and comradeship, openness and diligence would prevail." But in the course of his professional life, particularly since he had become works manager, he had "unfortunately all too often" learnt to understand some of Hitler's words "in their full truth and meaning: 'he who wishes to live must also fight, and he who does not wish to fight in this world of eternal struggle does not deserve life!' "[117]

The suicide of Brüning, from whom he had hoped to receive much help and an easing of his burden in the works management, led to more than just the harsh judgment of Lautenschläger. On the same day, on October 4, 1938, Lautenschläger also released Hans Wagenheimer, who had been serving as the contact with the *Gauleitung* and the local Gestapo under Hermann because of his close connections to *Gauleiter* Sprenger, from all his other duties in the plant and appointed him personal intermediary between the works management and the *Gauleitung*.[118]

There are extensive sources available on Lautenschläger that inform us rather well about his actions and thinking: numerous letters, his office files, the files of the Nuremberg trial, the files of the denazification tribunal. Then there are the multivolume memoirs already referred to, which he wrote for his children in installments, the first in 1936, then in 1942, and finally in the early 1950s. Yet the insights that these give into his way of thinking should not be allowed to obscure his activities as works manager.

Carl Ludwig Lautenschläger was born on February 27, 1888, in Karlsruhe into a middle-class family; his father was an architect, his mother the daughter of a publisher in Stuttgart with a strongly Pietist background. Lautenschläger left high school without a diploma and began training to become a druggist. In 1907 he completed the first stage of his training, the first state examinations for druggists, in which he achieved very good marks. During his subsequent training as an assistant he began preparations to take his high school diploma examinations as an external pupil, to be able to gain admission to a university. In October 1908 he enrolled as a student at the Technical College of Karlsruhe to study pharmacy, which at that time was still possible without a high-school diploma. He attached

---

[117] HA, PA Lautenschläger, Lautenschläger, "Erinnerungen," vol. 3, [pp. 49–50].
[118] HA, PA Abteilung VI, Wagenheimer, draft of a letter from Lautenschläger to Kränzlein, dated October 4, 1938, with handwritten note "resolved by word of mouth."

great importance to supporting himself and earned a living by working as a substitute in different pharmacies. In 1910 he completed his studies by taking the state examination in which he achieved the mark "very good." This was followed by the offer of a position as assistant to Professor Carl Engler, the director of the Chemistry Institute in Karlsruhe; the only condition was that he would need to sit for his high-school diploma (*Abitur*), otherwise he would not be permitted to sit for any other exams. Lautenschläger passed his *Abitur* as an external pupil and the diploma examinations for chemistry in 1912. At the end of 1913 he completed his dissertation on the relationship between autooxidation and polymerization of various unsaturated hydrocarbons with Engler as his supervisor. Although he then turned his attention to synthetic chemistry and biochemistry, he continued to pursue his pharmaceutical interests and worked toward acquiring a license as a pharmacist. Around 1913/14 he began to entertain the idea of an academic career.[119]

In the First World War he volunteered but spent only a short time at the front as he was discharged as unfit for service in 1915. In contrast to Hermann, Lautenschläger had not enjoyed the war, and he was later to write that the "period of military service and the weeks in the army were the most unsatisfying days of my life till then, not just with respect to my activities but also with respect to the people I met." But the war did change his life, as in February 1915 his older brother Erwin died from an accident on the Western Front. Erwin had studied medicine and had been working in 1914 at the outbreak of the war as a doctor at the Laryngological Clinic of the University of Frankfurt, where he intended to pursue his *Habilitation*. Erwin had begun carrying out animal and clinical experiments with chemotherapeutic agents and had suggested that they work together: "As with the death of my brother this joint experimental therapeutic research work had abruptly come to an end and a suitable trustworthy medical co-worker was difficult to find in this field, I decided to also study medicine in order to be able to unite within myself the work of a chemist and a medical doctor."[120] Just how deeply the death of his brother affected him is evident in the chapter devoted to him in his memoirs. "My brother Erwin was the only intimate playmate and in later years my only friend, with whom I shared the joys and sorrows of my life until his tragic death." But he was more than that, he was an example whom Lautenschläger ardently admired and sought to emulate. He had a keen sense of his own imperfections compared to his brother. For his brother had not just been affable, respected, and popular, but had also been a talented, highly gifted man – in

[119] HA, PA Lautenschläger, Lautenschläger, "Erinnerungen," vol. 1 [unpaginated]: *Kinderjahre* (childhood), *Schuljahre* (school years), *Lehrjahre und Conditionsjahre als Apotheker* (learning years as a pharmacist), *Studienjahre* (years of study).
[120] Ibid., vol. 1: *Studienjahre* (years of study).

painting and music and as a doctor. Lautenschläger himself was "a stolid Alemannian, with few gifts or talents." He had achieved his goals only through diligence and hard work.[121]

Lautenschläger studied medicine in Heidelberg and Würzburg, and soon was working as an assistant for physiology at the University of Heidelberg under the Nobel Prize laureate Professor Albrecht Kossel. As he wished to continue his pharmaceutical education, Lautenschläger switched to the University of Freiburg at the end of 1917, where he worked as an assistant at the Pharmacological Institute under Professor Walther Straub. His work included cooperating with the laboratory for gas burns in Cambrai on the treatment of gas phlegmon. To complete his medical training Lautenschläger went to Erlangen in 1919, where he sat only for his state medical examination. Then he began his practical clinical training at the Municipal Hospital in Karlsruhe. He submitted his *Habilitation* in Karlsruhe and completed his medical dissertation in Freiburg in 1919, receiving the grade *summa cum laude*. Even before he had completed his *Habilitation* the Prussian Ministry of Education and Cultural Affairs offered him the chair of pharmaceutical chemistry at the University of Greifswald in 1919, which he accepted after completing his *Habilitation*.[122]

He found little to like in his work in Greifswald, which he considered a "transitory passage," a "spring board" to a bigger university. The only goal of many young professors was to get a place at a larger university. Anybody who stayed in Greifswald for any length of time had, according to Lautenschläger, "missed the boat." Such professors were either annoyed at their lack of success or they had resigned themselves to their fate and lived "as though they were already semi-retired." In his memoirs Lautenschläger depicted life in Greifswald as extremely boring and monotonous. He found it particularly noteworthy that everyone met up on Sunday in the university church as most of the lecturers were Protestants: "Only a few members of the university were Catholics, and in my time Greifswald only harbored a single Jew, the professor for mathematics; maybe he was necessary because calculations are best learnt from Jews." His research work in Greifswald satisfied him, but he found teaching unedifying as the "student material" did not amount to much. He attempted to initiate a transfer to Göttingen, a wish that was turned down by the Ministry. But Lautenschläger wanted to leave Greifswald, so the offer from Hoechst came at the right time.[123]

In December 1919 he received an inquiry from Hoechst asking him whether he would be interested in accepting the position of head of the Scientific Department for Medicinal Products, which had recently become vacant. After a first meeting with Haeuser and Ammelburg in Frankfurt in December 1919, both sides reached an agreement that Lautenschläger

[121] Ibid., vol. 2: *Mein Bruder* (my brother).   [122] Ibid., vol. 1: *Studienjahre* (years of study).
[123] Ibid., vol. 2: *Greifswald*.

would join Hoechst in October 1920. As he wished to continue his research at the university, Hoechst allowed him to also work at the University of Frankfurt.[124]

While Lautenschläger liked the plant and its proximity to Frankfurt, he disliked what he perceived as the prevailing Jewish influence and the "discontent of left-wing extremists." Class snobbery guided his pen when he wrote in his memoirs about the "strange customs" of the Hoechst workforce. There had been a Blue-Collar Employees Committee that had demanded equality of social circumstances, worker participation, and profit sharing: "strikes, wage negotiations, and other complications were the order of the day, and the unreasonableness and conflicts gradually resulted in deep discord between employer and employee." In his memoirs Lautenschläger described one of the "strange customs" in Hoechst: "[T]hus at the time of my joining the Dyes Works it was ordered that blue-collar employees, white-collar employees, university graduates, and directors should take their meals in the casino together; one had to join the queue in the order in which one appeared in the casino hall and eat the sometimes rather unappetizingly prepared stew. Accordingly, the brief lunch break offered neither mental nor physical refreshment."[125] Such a comment would have been unthinkable from Hermann and demonstrated the class snobbery that Lautenschläger cultivated – even later, when the Nazi regime propagated its "national community."

At first, Lautenschläger found little to enjoy in his work in Hoechst, as he was mainly occupied with office work. He was responsible for maintaining contacts between Hoechst and various clinics, doctors, and research institutes, for preparing products created in the laboratory for clinical testing, and for advertising Hoechst products. In 1922 he was made an authorized signatory; at the same time he was put in charge of all pharmaceutical research laboratories, which up to then had been attached to the pharmaceutical factories. One "small island" of research under Professor Wilhelm Roser in the so-called Central Laboratory eluded his influence until Roser's death in 1923. While initially research within Hoechst was writ small and important research results and new substances were mainly purchased from outside, Lautenschläger managed to gradually build up research in Hoechst.[126]

At the beginning Lautenschläger was directly involved in research and set up his own laboratory. Increasingly, however, the management of the

[124] Hoechst Human Resources, PA Abteilung VI, Lautenschläger, Lautenschläger to Ammelburg, February 27, 1920; Hoechst to Lautenschläger, March 1, 1920; HA, PA Lautenschläger, Lautenschläger, Erinnerungen, vol. 2: *"Greifswald"* and *"In den Hoechster Farbwerken und in der I.G. Farbenindustrie A.G."* (at Hoechst and IG Farben).
[125] HA, PA Lautenschläger, Lautenschläger, "Erinnerungen," vol. 2: *"In den Hoechster Farbwerken und in der I.G. Farbenindustrie A.G."* (at Hoechst and IG Farben).
[126] Cf. Bartmann, *Tradition*, pp. 80–4, 119–21.

whole department took up more and more of his time, and after about 1927 – the year in which he became a member of the board of directors of the plant – he was no longer able to carry out research himself. Yet he succeeded in pushing forward research and development in the fields of pharmaceuticals and drugs, setting up new laboratories and providing new substances. As it turned out, he had a particular affinity for biological and biochemical research – which is understandable, given his training as a pharmacist, and which allowed him to remain open to such research even at a time when work on chemical synthesis completely predominated. As a result of his encouragement of biochemical research – a new laboratory was built for biochemical and biological research in 1929 – in the 1930s Hoechst was one of the leading and most important producers of insulin in Germany.[127]

Probably on the basis of his achievements Lautenschläger also became head of the Department for Crop Protection and Pesticides after Schmidt retired. At the end of 1930 when his direct superior, Ammelburg, retired after a lengthy illness, Lautenschläger took over the management of the entire Pharmaceutical Department. In 1931 he became a member of the Management Board of IG Farben. In the same year Lautenschläger also got the opportunity to intervene directly in the operations of the pharmaceutical factories after their previous manager, Franz Scholl, retired. Alfred Fehrle henceforth managed the factories under Lautenschläger's direction. However, in 1930 Hoechst had to hand over the serum division to the Behring Works in Marburg, which IG Farben had taken over; in return Lautenschläger became a member of Behring's Supervisory Board in 1934.

In June 1938 the Central Committee of IG Farben appointed him plant manager, a duty that, as previsouly mentioned, he had not sought. Lautenschläger's original statements in Nuremberg, which he later retracted, were in general pretty close to the truth – they also testify to his firm belief that he had nothing to reproach himself for. He stated at the time that *Gauleiter* Sprenger had volunteered to confirm his appointment as plant manager if he became a member of the NSDAP, which he duly did after consulting with Friedrich Jähne and "various other senior staff of the plant," as at the time "there was no other suitable man who was contemplated for the management of the plant." This had prevented any "Party henchman of the *Gauleiter* employed in the plant or other employee of the *Gauleitung* from being installed as plant manager."[128] This is quite plausible because Retzinger, who was *Betriebsobmann* of Hoechst at the time,

---

[127] Ibid., p. 127; HA, PA Lautenschläger, Lautenschläger, "Erinnerungen," vol. 2: "*In den Hoechster Farbwerken und in der I.G. Farbenindustrie A.G.*" (at Hoechst and IG Farben).

[128] BA, Film 44839, affidavit by Lautenschläger given to Benvenuto von Halle on March 23, 1947, in Nuremberg.

declared in 1978 that the fanatical Nazi Georg Kränzlein, a "self-assured man" and "careerist," was not appointed Hermann's successor "as he had hoped." In Retzinger's opinion "the top management of IG Farben had no objection to people being Party members, but persons who were too politically active were considered undesirable in managerial positions."[129] As it was expected that the plant manager would come from the Works Group, the only obvious compromise candidate as Hermann's successor was Lautenschläger, who was a member of the Management Board and already deputy manager. The chief engineer of the plant, Friedrich Jähne, also a member of the Management Board, had apparently refused to stand as a candidate because he did not come from Hoechst and was not a chemist.[130] However, it is not clear whether the Management Board of IG Farben ever considered Jähne as a possible successor to Hermann.

Lautenschläger applied for membership in the NSDAP on April 29, 1938, and was accepted retroactively from May 1, 1937.[131] He was in good company. After admission to the NSDAP was once more generally possible, pressure to join was apparently brought to bear on managers and businessmen. Thus, almost all of the top managers of IG Farben, including Friedrich Jähne, became Party members; only Hermann Schmitz, the chairman of the Board of Directors, never became a member.[132]

But how would Lautenschläger act now that he was works manager? Otto Hirschel, one of his closest members of staff, praised his conscientiousness and sense of duty as works manager, but added that Lautenschläger was not just a "loner" but also did not have much understanding of engineering. "[H]e was not what one imagines a works manager to be. He was a scientist who had done a lot for the Pharmaceutical Division." Hirschel added: "I often had to defend him when critical voices rejected him. If I had not done that, I would have disappointed him so badly that I would have lost his confidence."[133] Lautenschläger's successor, Karl Winnacker, who rose to become the "crown prince" during the war, found far more negative words to describe Lautenschläger. Winnacker acknowledged that Lautenschläger "had had a wide education as apothecary, chemist, and doctor," but noted that he "was the embodiment of the professor type engrossed in his work" who as works manager had "always seemed rather unhappy when the complexity of the organizational and administrative tasks weighed down upon him." Winnacker viewed Lautenschläger's achievements as head of the

---

[129] HA, PSW 648, note from the company archive dated April 28, 1978, on an interview with Retzinger of April 26, 1978.

[130] Winnacker, *Challenging Years*, pp. 88–90.

[131] BA, BDC file on Lautenschläger, membership number 6,086,459; cf. also HHStAW, Abt. 520 F (A-Z), Lautenschläger, decision of the Spruchkammer Frankfurt dated June 17, 1949.

[132] BA, BDC file on Jähne; cf. Hayes, *Industry and Ideology*, p. 200.

[133] HA, Hirschel, "Aus meiner I.G. Zeit," pp. 25–6.

Pharmaceutical Division extremely critically because he had been unable to prevail against Leverkusen or build up his position in the works: "Lautenschläger, who was very introverted and loath to get involved in any kind of conflict, had never been able to cope satisfactorily with this situation and was certainly not prepared to face up to it. However, then as now, such a large complex simply had no room for men afraid to fight for their position." To emphasize this statement, Winnacker added: "I am sure that I am not being unjust to Carl Ludwig Lautenschläger but I do not think that he was very happy."[134]

Whether this comment was correct must be left unresolved. For Lautenschläger did appear to enjoy the authority and status associated with his position, but he remained a mystery to many of his contemporaries. Retzinger, who was *Betriebsobmann* at the time, remembered Lautenschläger as "not equal to his task," but also thought that he had "not been easy to read."[135] It surely would be false to characterize him as incompetent. He was a well-known scientist with a clearly stupendous knowledge of medicine, chemistry, and biology. He had managed to gather a competent team around himself in the Pharmaceutical Department, including a Jew, whom Lautenschläger retained in a senior position until the summer of 1938. Nor had he pushed for the position. But on the whole, his performance as works manager was anything but glorious. Due to his personal difficulties in communicating with others he remained excluded from important decisions within IG Farben. Even within the plant his authority was often called into question, and he did not cut a good figure. Lautenschläger preferred research and enjoyed lecturing at the university – up until the middle of the war he gave lectures at the University of Frankfurt on interdisciplinary subjects from the fields of medicine, chemistry, and biology. But he was torn between the prestige of his position as works manager, which he clearly relished, and the duties that went with it, which gave him little pleasure or even exceeded his capacities.

Like Hermann, Lautenschläger was a Protestant – but of a different type. Lautenschläger had been very influenced by his Pietist mother and, in contrast to the sociable Hermann, preferred to enjoy his free hours in peace: "For the the human heart to be awakened, enlightened, and inspired, the Lord is pleased to take us into a place of quiet, so that His spirit may speak more clearly to us and His work in us be done undisturbed." During his first years at Hoechst Lautenschläger rarely took any vacation; the only relaxation he allowed himself was to spend time over Christmas and the New Year and the Easter period with his mother in a house of the Inner Mission

---

[134] Winnacker, *Challenging Years*, pp. 89–91.
[135] HA, PSW 648, Note from the company archive dated April 28, 1978, on an interview with Retzinger of April 26, 1978.

in Langensteinbach, the "rest home Bethanien." Relaxing in peace and taking part in "profound devotions" gave him the strength he needed for his professional life. Lautenschläger around 1935 even wrote that "the days spent there belong to the most rich and beautiful hours of my life to date."[136] He tended to place individual years and particular events under the motto of a psalm. He married late, in the spring of 1929, a young lady of twenty-three from "a good family" in his hometown of Karlsruhe. They had apparently become acquainted through his mother.[137]

There is more to report concerning his time as head of the Pharmaceutical Department than just defeats at the hands of Leverkusen. The department began to prosper under his management. Yet, while he was able to attract and employ good staff and create good working conditions for them, he found it difficult to defend the interests of the department against the competition from Leverkusen/Elberfeld, which managed to take over the sales of the pharmaceutical products. Even afterwards he needed to speak for the interests of Hoechst, particularly in the regular monthly conferences in the Coblenzer Hof in Koblenz, a meeting point halfway between the works. There representatives from Leverkusen and Hoechst swapped information on the scientific and technical concerns of the Pharmaceutical Departments. Lautenschläger later wrote à propos of these meetings: "I have retained very few pleasant memories of these oftentimes impassioned sessions. Sometimes I had very serious altercations with colleagues from the other camp, in which I fought for the honor and the rights of my department, and again and again I needed some time to recover from these battles at home."[138] Accordingly, he found it hard going after becoming works manager.

But he was aware of his weaknesses and his strengths and looked for appropriate support – first from Gustav von Brüning, and, after Brüning's suicide, from Hans Wagenheimer, who was the personal contact man with the *Gauleitung* and the local Gestapo and whom he trusted. Lautenschläger was clearly anxious to avoid mistakes and encouraged Wagenheimer to put out his feelers in all important political matters to protect himself and avoid problems as far as possible. When *Gauleiter* Sprenger ordered an investigation of the political and ideological affiliations of the economic elite at the end of 1941 in his *Gau* Hesse-Nassau, no one had anything bad to say about Lautenschläger. The local group leader wrote that Lautenschläger was generally "described as a quiet person" with a "good" character, his ideological outlook was "firm," and his "attitude to the national

---

[136] HA, PA Lautenschläger, Lautenschläger, "Erinnerungen," vol. 2: "*In den Hoechster Farbwerken und in der I.G. Farbenindustrie A.G.*" (at Hoechst and IG Farben).

[137] Ibid., "*Meine Frau und ihre Familie*" (my wife and her family).

[138] Ibid., "*In den Hoechster Farbwerken und in der I.G. Farbenindustrie A.G.*" (at Hoechst and IG Farben).

community" was "good."[139] Other economic leaders in the *Gau* did not get off so lightly and, in particular, with so few negative comments. The gist of the reports on two other leading IG Farben men in the *Gau*, Friedrich Jähne and Constantin Jakobi, works manager in Griesheim, was far more negative; both were reproached for keeping their distance from the National Socialist regime.[140]

Despite his previously mentioned class snobbery, Lautenschläger had become a convinced follower of National Socialism and Hitler, even if after the war he presented himself in a different light. He was also strongly anti-Semitic.[141] This is not contradicted by the fact that he allowed a professing Jew to stay on in a senior position in his Pharmaceutical Department until the summer of 1938. For this co-worker was an extremely successful scientist, and Lautenschläger benefited from his continued presence, particularly in view of the competition with Leverkusen. At least until the end of 1942 Lautenschläger continued to believe in the "final victory," and even in his personal memoirs, which he penned for the eyes of his children alone, he echoed Goebbels's slogans: "Today we face the last phase of the great battle, the final battle, which will decide whether all the sacrifices made since 1914, the privations and the burdens and all the dead of this war have been sacrificed in vain or not! We have no other choice. The law of war is a hard one and commands: he who fails will perish! We have not yet reached our goal. But in inexpressible pride we look on what has been achieved, as this offers us a guarantee for a happy and victorious outcome in this struggle between nations."[142] It was only during the course of 1943 that he appears to have begun to distance himself from the regime, and his confidence in the "final victory" began noticeably to fade.

It is difficult to do justice to Lautenschläger's personality. His thinking and his actions point to a certain vacillation, an indecisiveness – and this resulted in an inability to assert himself. He seems to have been a rather weak character, animated by class snobbery and above all by the consciousness of always doing the right thing.[143] Under Lautenschläger's management the Hoechst plant had become very compliant with the regime and its representatives. It was not just with respect to politics or in his manner toward representatives of the regime that Lautenschläger reacted rather than acted. He was no different within IG Farben or even within the plant itself. Under his management the position of Hoechst within IG

---

[139] HHStAW, Abt. 483, Nr. 10542, Lautenschläger.

[140] HHStAW, Abt. 483, Nr. 10530, Jakobi; Nr. 10531, Jähne.

[141] HA, PA Lautenschläger, Lautenschläger, "Erinnerungen," vol. 6: the worst antisemitic comments are in this volume, which was written around 1952; cf. also Bartmann, *Tradition*, pp. 83, 199.

[142] HA, PA Lautenschläger, Lautenschläger, "Erinnerungen," vol. 3 [unpaginated, toward the end of the book].

[143] Cf. Bartmann, *Tradition*, pp. 199–200.

Farben and even his own position became so weak that in 1941 the IG Farben management began grooming Karl Winnacker as a successor and appointed him to a senior position in the plant – this, despite the fact that at the time Lautenschläger was only in his early fifties.[144]

3.4. THE NSDAP AND ITS REPRESENTATIVES IN THE PLANT DURING THE "YEARS OF PEACE"

The NSBO did not succeed in taking over from the unions and being entrusted with the task of transforming them into a unified organization for blue- and white-collar employees. This task was assigned to the German Labor Front (*Deutsche Arbeitsfront*, DAF), founded on May 10, 1933. Hitler assigned control of the DAF to the chemist and former IG Farben employee Robert Ley.[145] The most important representatives of the workforce in the plant were the NSBO *Betriebszellenobmann* and after 1934 the *Betriebsobmann* of the DAF as the "leading NS official in the plant"[146] together with the Council of Trust. Until 1935 the *Betriebsobmann*, the DAF, and the plant management jointly nominated members for the Council of Trust, who then were put up for election; after 1935 members to the Council were appointed directly. The NSBO began to lose its influence as of the summer of 1934; it was only retained in the NSDAP organization book for reasons "of piety," as the historian Martin Broszat wrote, while its tasks were now undertaken by the DAF.[147]

With the exception of an interval of about one year the worker Ludwig Retzinger was *Betriebsobmann* of Hoechst from mid-1933 until the beginning of 1940. After Walter Hirschelmann, the NSBO works council member, left the plant voluntarily in August 1933, Retzinger was appointed *Betriebsobmann* by the county administration of the DAF, probably on Hirschelmann's suggestion; at the time the office of *Betriebsobmann* was still affiliated with the NSBO. Retzinger, who was only twenty-four years old, was very young for this position; by the end of 1934 he had barely attained the minimum age of twenty-five required to become a member of the Council of Trust. But sometime around the end of 1934 or the beginning of 1935 he was relieved of his office as *Betriebsobmann* and demoted to deputy *Betriebsobmann*. In his

[144] Winnacker, *Challenging Years*, pp. 98–106.
[145] Frese, *Betriebspolitik*, p. 74; for a full account on Ley, see Ronald Smelser, *Robert Ley: Hitler's Labor Front Leader* (Oxford: Berg, 1988).
[146] Frese, *Betriebspolitik*, pp. 218–19, especially n. 169: Originally the NSBO *Betriebszellenobmann* was mentioned in the AOG as someone who "would be replaced by the DAF *Betriebswalter* during the elimination of the NSBO": "after the ranks of the DAF were reorganized, the coexistence of two different shop stewards was abandoned. The top Nazi official in the plant was now the '*Betriebsobmann*.'"
[147] Cf. Broszat, *Der Staat Hitlers*, pp. 196–7; Frese, *Betriebspolitik*, pp. 90, 112.

denazification proceedings Retzinger wrote about his appointment in 1933 and subsequent demotion as follows: "I was appointed *Betriebsobmann* for IG Farben in mid-1933, by the county administration [*Kreiswaltung*] of the DAF. My objections that I was too young and had too little political experience were rejected by the *Kreisobmann* of the DAF. However, in 1935 I was removed from office as B.O. [*Betriebsobmann*] and appointed deputy B.O. for these very reasons and was additionally accused of holding Marxist views."[148] This statement may be correct as Retzinger came from the NSBO, which was increasingly losing its influence and moreover had the reputation of being "leftist." Nevertheless, Retzinger did retain the office of deputy *Betriebsobmann*.

After Retzinger's demotion, the chemist and "old fighter" of the NSDAP Bernhard Schacke became the new *Betriebsobmann*. Schacke had been a member of the NSDAP since February 1, 1928, and a member of the SS since January 1932. He had worked his way up within the NSDAP in the 1920s, but by the beginning of the 1930s he had begun to withdraw from Party work, becoming increasingly involved in the SS, where he became an *Oberscharführer* around the same time he became *Betriebsobmann* and worked as an SS indoctrination officer. However, in 1934 Schacke was facing serious problems both in the NSDAP and in the SS. While the SS refused to grant him a marriage permit due to a hereditary disease in his family, not that of his fiancée, in the NSDAP Schacke also faced so-called "Uschla proceedings," that is, proceedings in front of the Investigation and Arbitration Committee of the NSDAP because of alleged homosexuality.[149] Considering the atmosphere of intrigue that prevailed at the time in the NSDAP in Hoechst, the truth of these accusations remains doubtful.[150]

Schacke had a number of friends and supporters in the NSDAP and the SS but also a number of detractors, not least because of his craving for importance. Whatever the reasons may have been to forbid his marriage because of alleged "hereditary health" risks, the SS could have hardly been referring to the deafness that ran in his family. Schacke, who was clearly under strong pressure because of the rumors surrounding his alleged homosexual inclinations, married despite the prohibition and was immediately punished by being thrown out of the SS.[151] In a letter to Himmler requesting his reinstatement it is clear that for Schacke the NSDAP and, even more, the SS represented a kind of substitute for family and friends.

[148] HHStAW, Abt. 520 F (A-Z), Ludwig Retzinger, Retzinger's petition on denazification dated June 16, 1946.

[149] BA, BDC file on Schacke, letter from Schacke dated June 21, 1934, SS-Oberabschnitt Rhein to RFSS Himmler, September 15, 1934.

[150] Cf. Fenzl, *Aus der Geschichte*, pp. 126–35, on conflicts inside the NSDAP and between NSDAP and SA.

[151] BA, BDC file on Schacke, Brandt to Schacke, January 16, 1935; Schacke to RFSS Himmler, January 26, 1935.

Schacke's craving for recognition is also revealed in his description of his "career in the NSDAP": after becoming a member and involving himself in Party work while living and working among people who were predominantly "red and black," that is, Socialist/Communist or Catholic, he had been exposed "to the most furious attacks." Schacke claimed to have been hated by his colleagues in the Hoechst plant, but he had also suffered from intrigues against him within the NSDAP. The "Red Front" of Hoechst had once beaten him up, while another other time "I put 20 men to flight who attacked me in a bar in Weilbach." Schacke continued: "I had to take legal action 20 times on behalf of myself and other PGn. [*Parteigenossen*, Party comrades] who were being persecuted by the [Weimar] system, and I won all lawsuits with the exception of the last one when I was unable to provide proof of the activities of a Communist informer." However, these activities for the NSDAP had exhausted him "financially and mentally" so that he had resigned all his positions. In January 1932 he had joined the SS "to continue fighting the enemy at the front." He clearly stated what the SS meant to him: "My friends were the old fighters of the SS. [I] would lose them if I were thrown out. After my disappointment with the P.O. [Party organization] to experience the same disappointment in the SS would depress me very much."[152]

Himmler did not reinstate him, but Schacke apparently was not thrown out of the NSDAP. His dismissal from the SS and the rumors about his possible homosexual inclinations, rumors moreover that the SS Race Office for the SS Upper Section Rhine had described to the Central SS Office for Race and Resettlement as "quite credible,"[153] had to affect his work as *Betriebsobmann*. In a directors' meeting in Hoechst in March 1935 Hermann briefly informed the persons present that "Schacke, the previous *Betriebszellen-Obmann*, has been removed from his position; Grosch will be acting as provisional shop steward until new elections are held."[154] At the time Heinrich Grosch was *Betriebsobmann* for the administrative building of IG Farben in Frankfurt-Grüneburg and retained the position as provisional *Betriebsobmann* of Hoechst until the end of 1935 or beginning of 1936.[155]

However, Schacke fought against his demotion in the NSDAP, the SS, and the plant. Suspecting his enemies were members of the Council of Trust, he began circulating extensive accusations against them, only to suffer another defeat. He was obliged to give a written declaration that he

---

[152] Ibid., Schacke to RFSS Himmler, January 26, 1935.

[153] Ibid., Himmler to Schacke, March 12, 1935; Rassereferent beim SS-Oberabschnitt Rhein to RFSS-RuSA, September 15, 1934.

[154] HA, TEA 95c, Niederschrift über die Maingau-Direktionssitzung in Hoechst, March 13, 1935. This information was included only in the internal minutes – there were always two versions of the minutes, an internal one and the version that was sent to other plants or works groups.

[155] HA, PSW 648, note from the company archive dated April 28, 1978, on an interview with Retzinger of April 26, 1978.

"withdrew his accusations" against the Council of Trust and the individual members "in their entirety" and was obliged to "express his regret" as these accusations had been "completely untrue." Further on in the letter he stated: "I realize the full extent of the wrong I have done and particularly promise herewith that in future I will engage in no more activities against the Council of Trust of the Hoechst plant."[156]

After this short and inglorious intermezzo with an "old fighter" of the NSDAP and the interim stewardship of Grosch, the DAF turned once again to Ludwig Retzinger.[157] Retzinger, born the son of a worker in Kostheim on March 24, 1909, had attended primary school in Hoechst and Zeilsheim. Retzinger had worked in the laboratory in Hoechst since 1926. From 1926 to 1931 he was a member of the free labor unions affiliated with the Social Democrats and of the left-wing *freie Naturfreunde* (Free Friends of Nature). In 1931 he resigned from the free labor unions and joined the NSBO, allegedly after being recruited by his supervisor, Hans Schlichenmaier, who had been a member of the NSDAP since 1931. As was emphasized during his "denazification" proceedings, Retzinger came to the NSDAP through his work as a representative of the employees. According to Retzinger's own statement, he was then recruited to join the Party and the SA in the same year by Hirschelmann. Retzinger remained in the NSDAP until the end of the war in 1945, but he left the SA in 1933.[158]

Yet Retzinger's position as *Betriebsobmann* was never uncontroversial. As mentioned before, he was relieved of office and demoted to deputy *Betriebsobmann* at the end of 1934 or the beginning of 1935. He remained a member of the Council of Trust, but came last out of all candidates in the elections to the Council in 1935.[159] In his denazification proceedings he wrote about his reappointment: "in 1936 I was reinstated and was once more removed from office in February 1940. Once again I stood accused of incompetence. In reality I had uncovered the machinations of an SD [Sicherheitsdienst, Security Service of the Party] man (Pg. [Party comrade] Wagenheimer) and I was removed because I spoke out against him and other older Party comrades."[160]

[156] HA, PA Schacke (microfilm), statement by Schacke dated June 28, 1935.
[157] HA, PA Retzinger (microfilm). In the questionnaire he filled out with his job application for the post of commercial clerk dated December 6, 1938, Retzinger wrote, "and I have been *Betriebsobmann* of our plant community with only a brief interruption since January 1, 1933."
[158] HHStAW, Abt. 520 F (A-Z), Retzinger, Meldebogen, April 26, 1946, Retzinger's petition on denazification dated June 16, 1946; BA, BDC file on Retzinger, NSDAP membership number 760,632.
[159] HA, PSW 657, Vertrauensratwahl am 12.4.1935, Werk: Hoechst – Ergebnisse, Sozialabteilung; notice on the results of the election to the Council of Trust dated April 15, 1935.
[160] HHStAW, Abt. 520 F (A-Z), Retzinger, Retzinger's petition on denazification dated June 16, 1946.

His professional advancement was closely linked with his position as ombudsman in the office of the Council of Trust and as *Betriebsobmann.* Retzinger had joined the plant as an unskilled laborer with a mere eight years' schooling. At the end of 1938 he applied for a job as a commercial clerk, although he had had no further education nor had he attended any training courses. Nevertheless Hoechst took him on as a commercial clerk, employing him in this position retroactively from December 1, 1938; in fact, he continued to work in the office of the Council of Trust. After his second removal from his position as *Betriebsobmann* at the beginning of 1940 he was employed in the Materials and Building Materials Department; Lautenschläger reported at the beginning of March that Retzinger "had been removed from his position as *Betriebsobmann.*"[161] Thus, in the years following the seizure of power Retzinger had benefited professionally and considerably increased his income; he remained a member of the Council of Trust until he was drafted into the army in 1941.

After the war Retzinger managed to be classified more or less as a fellow traveler in his denazification proceedings, since after his hearing the denazification tribunal simply dropped the proceedings against him under the "Christmas amnesty."[162] A little later Retzinger was even rehired by Hoechst. In his book *Die Rotfabriker,* published in 1988, Ernst Bäumler devoted much attention to Retzinger, giving him an overall positive, even too positive, assessment. The book erroneously states that Retzinger had "given up his office as first *Betriebsobmann* anyway in 1939." But at least the statement "he wanted to get on and advance from being a worker to being a white-collar employee. And he succeeded in this aim" was correct.[163] Retzinger had begun, as Bäumler wrote, as "a boy rinsing out vessels" in the laboratory in Hoechst where his father and his brother-in-law also worked. Bäumler described how Retzinger joined the NSDAP, which attracted him because of its ideal of a "national community" (*Volksgemeinschaft*) in which everyone held the same rank. Moreover,

[161] HA, PA Retzinger (microfilm), Bewerbungs-Fragebogen für kaufmännische Angestellte, dated December 6, 1938; reference from Hoechst (Pensel), dated September 12, 1946; TEA 95e, Niederschrift über die technische Direktions-Sitzung in Frankfurt-Hoechst, March 11, 1940. The minutes state that "Lautenschläger announces the resignation of Retzinger from his position as *Betriebsobmann.*" The quotation in the text is from TEA 0265, Niederschrift über die 10. technische Abteilungsleiter-Besprechung in Frankfurt-Hoechst, March 18, 1940, in which Lautenschläger refers to his being dismissed from office, which would also be in accordance with Retzinger's statement quoted above during his denazification proceedings.

[162] HHStAW, Abt. 520 F (A-Z), Retzinger, Protokoll der öffentlichen Sitzung, and decision of the Spruchkammer, dated September 1, 1948: *Weihnachtsamnestie* (Christmas amnesty).

[163] Bäumler, *Rotfabriker,* pp. 301–2; HA, PSW 648, note from the company archive dated April 28, 1978, on an interview with Retzinger of April 26, 1978. This goes on to state "he retired from office at his own request in 1939 at the age of 30 in order to pursue his career."

"Retzinger, who was keen on sports, liked the parades, the flags, the physical training."[164] After being appointed *Betriebsobmann* and elected to the Council of Trust in 1934 he was released from his duties in the laboratory to work full time for the Council. Bäumler did find some critical words to say about him: "of course, Retzinger was dedicated and enthusiastic. He believed in the Party and in Hitler and saw only the positive side everywhere." The struggle against unemployment and the social and political ideas of the DAF would have met with his particular approval.[165] He got on well with works manager Hermann, whom Retzinger did not consider a "reactionary." Nor did Retzinger have any objection to the reemployment of Hans Bassing, a worker and dedicated Social Democrat who was to become chairman of the Works Council of Hoechst after the war, but who was briefly taken into "protective custody" in the Frankfurt-Preungesheim jail because of his political convictions in 1933. Bassing's wife, who sewed and repaired shirts for Retzinger's family, had sounded him out. This is not unimportant, for Retzinger, as Bäumler wrote, was now not just a prosperous man – as this story of Bassing's wife shows, who now sewed for the former co-worker – but had also become a "powerful man." As Bäumler noted: "very few decisions could be taken, particularly on matters concerning employees, without his consent being sought." Bäumler added: "To Retzinger's honor we can say now that he did not attempt to use any brutal Nazi methods in Hoechst," and at Retzinger's denazification proceedings "no witness who reproached him for anything bad" appeared.[166]

Yet Josef Fenzl has highlighted Retzinger's participation in certain brutal SA measures taken against political opponents and commented that Bäumler's "very positive assessment" of Retzinger can no longer be upheld.[167] Altogether, Retzinger's denazification requires some explanation. Reading the files of the denazification tribunal, one should look at not only *what* was said about Retzinger, but *by whom* it was said. Who were the people who helped Retzinger during his denazification proceedings? During the trial it was said that he had also helped persons who were ideologically opposed to him and that he had behaved "decently." If one

[164] Bäumler, *Rotfabriker*, pp. 270–5.
[165] Ibid., p. 283; HA, PSW 648, Note from the company archive dated April 28, 1978, on an interview with Retzinger of April 26, 1978: At the time Retzinger said: "The 3rd Reich initially achieved much that was good for the German people, but the importance of this was completely overlaid and destroyed by the later developments." What "much that was good" implied for political dissenters and Jewish employees or employees considered to be Jewish in the years between 1933 and 1938 will be discussed in more detail below.
[166] Bäumler, *Rotfabriker*, pp. 282–5, misdates the proceedings and gives 1950 as the date of the trial; on Bassing's imprisonment in Preungesheim in June 1933, see HHStAW, Abt. 486, Gestapo-Kartei.
[167] Fenzl, *Aus der Geschichte*, p. 93.

reads the corresponding statements exactly, then it is clear that these statements did indeed come from people who had ties to the Communist Party (*Kommunistische Partei Deutschlands*, KPD) or were members of the SPD – as far as that went, it is indeed true that Retzinger helped political dissidents. But it was also the case that these people had been known to him for many years. The witness with links to the KPD and the member of the SPD were acquainted with Retzinger because they had been all members of the same sports club for many years. Thus Retzinger had not helped them despite their ideological differences but because he knew them well and they were friends of long standing.[168] Retzinger was not only a convinced National Socialist, he was also a keen athlete and sportsman, and he lived both loyalties simultaneously, just as many others did at the time. Retzinger trusted his sports buddies, and at the same time he had faith in Hitler and the NSDAP. For him and for many others this does not appear to have been mutually exclusive. We must assume that in 1935 Retzinger was simply doing a favor to an old school friend or sports buddy whom he might even have hoped to win over to the NSDAP. It did not make him less of a Nazi. It does not make him a "resistance fighter" or even particularly "decent"; in his environment he was simply behaving in conformity with the group. Retzinger did not have friends just among NSDAP members as Schacke did, for whom dismissal from the SS and attacks from within the NSDAP had been a personal catastrophe, but he also had other and older connections that he continued to maintain.

For Retzinger was also a committed Nazi, who not only officiated at the first National Socialist "marriage consecration" ceremony celebrated by the *Werkschar* of the Hoechst plant.[169] He also supported the severe punishment meted out to an employee by the DAF in 1938. Karl Lang had spoken contemptuously of the "*Führer's* reconstruction work," the *Wehrmacht*, and the economic measures of the Nazi regime. He had criticized the buildings in Nuremberg as "a disgrace" as long as workers had "nothing to eat." He had also daily made fun of the "German salute" by calling out "Heil Hitler" and forming a "Harlequin figure with his two raised hands." For these and other remarks he was reported for "behavior hostile to the state." Although already seriously ill, Hermann personally interrogated Lang in the presence of representatives from the Council of Trust and came to the conclusion that Lang had "oftentimes in conversations with other co-workers during working hours commented on everyday events in the worst manner": "This behavior must be described as downright irresponsible and has disturbed the industrial peace in the most serious manner." Based on the material against him and the declarations of

---

[168] HHStAW, Abt. 520 F (A-Z), *Retzinger*, Franz [Wolf] to Retzinger, dated July 5, 1946; statement of Franz Wolf (undated) and statement of Karl Bacher (dated July 10, 1946).
[169] Ibid., Document "Erste Eheweihe am 19.6.38 in Ffm.-Zeilsheim."

witnesses during the proceedings, Lang would have "deserved being summarily dismissed as the only correct form of punishment." However, "as a practical application of the plant leader's obligation to provide for the welfare of his employees" Hermann refrained from dismissing Lang in order to give him "a final opportunity to become a fully useful member of our plant community again." He therefore expressed "the severest disapproval" to Lang "on his irresponsible behavior," gave him a "final caution," and sentenced him to pay a fine of 20 RM. To Retzinger this punishment was too mild, and he asked the DAF to proceed against Lang: "As in this case it is a question of exceedingly grave political lapses and because the plant leader cannot be permitted to pass judgment on offenses of a political nature, I hereby request that the German Labor Front hold disciplinary proceedings."[170] So Retzinger did consider his task to be political and correspondingly proceeded against plant employees. With his important position in the plant, he contributed substantially to the functioning of the Nazi regime and to the oppression of dissidents.

After the war Retzinger described his own role and his relationship to the Gestapo to the denazification tribunal as follows: "I can assure you that during my whole time in office as *Betriebsobmann* I never had anything to do with the Gestapo. Everything to do with the Gestapo was in the hands of the works management. I particularly emphasize that I was not questioned in any way by the Gestapo. If arrests happened, the gentlemen went to the office of the Defense Commissar. Those people who were to be arrested were sent for and went there. I was only informed by telephone that the member ... had been arrested for political reasons. I was neither asked nor consulted in any form. That is how matters were carried out in our plant."[171] This was clearly exaggerated because, as Bäumler had noted, in matters pertaining to employees and staff, even if these concerned the Gestapo, Retzinger was usually consulted as *Betriebsobmann* and member of the Council of Trust. And he could even take the initiative himself, as the case of Lang shows. However, it is also correct that the *Betriebsobmann* was not as influential as people thought. In addition to the Defense Commissar, who served "as the long arm" of the Gestapo,[172] there was also Hans Wagenheimer, who enjoyed the confidence of the works management and who functioned as the direct contact with the *Gauleitung* and the local Gestapo in Hoechst. However, before examining his role and

[170] Ibid., Retzinger to DAF, Betriebsgemeinschaft I.G. Frankfurt-Grüneburg, March 17, 1938, in the appendix complaint against Karl Lang filed by *Blockwalter* Georg Roth and by NSDAP *Amtsleiter* Curt Heuer, undated; letter from plant leader Hermann to Karl Lang, Filmabteilung, March 14, 1938 (copy).

[171] Ibid., Protokoll der öffentlichen Sitzung, July 7, 1948.

[172] Lotfi, Gabriele, *KZ der Gestapo. Arbeitserziehungslager im Dritten Reich* (Stuttgart: DVA, 2000), p. 42.

that of the Defense Commissar, it will be necessary to discuss the Council of Trust.

As mentioned before, the Council of Trust acted only in an advisory capacity. The comment of Felix Rössler's quoted above emphasizes its limited role: "for there to be actual mutual trust between the plant leader and his followers, the leader must always be informed of their interests and wishes." This was best achieved through direct contacts, although in larger businesses this was not possible, so that in companies with at least twenty employees the plant leader would be advised by "men of trust," who would function as a "link" between him and the followers. "This purpose of the Council of Trust also made its sole task plain: that of deepening the mutual confidence within the plant community." The plant leader needed to be a member of the Council of Trust and should also chair the Council, since ultimately he was the one who took the decisions; the Council of Trust was complete only if it also included the plant leader. Every year in March the plant leader made out a list of members for the Council of Trust and their deputies "in agreement with the representative of the National Socialist *Betriebszellenorganisation*" and according to the legal regulations and held an election to allow the "followers" to vote on the list. When the members of the Council had been confirmed, the plant leader had "to give them the information necessary for them to fulfill their tasks" since they would be unable to form any judgment on questions that concerned them without sufficient information. "However, the Council of Trust has no say in the final decision but acts merely in an advisory function. The individual Council members can express their opinions. It is up to the leader whether he wants to take them into consideration or not. He alone decides what is right and he makes his decision independently, according to his own judgment and as he feels himself to be in duty bound. His decision is absolutely authoritative. A majority of Council members can only appeal to the Trustee of Labor and only if the decision concerns general working conditions and they are not compatible with the economic and social conditions in the plant."[173]

The plant leader was constrained only if he was guilty of misdemeanors toward his "followers" or infringed the "public good" as defined by the Nazi state. The only area in which the Council of Trust was allowed to operate was when dealing with social or political matters; the Council had virtually no say in the management of the plant. In his study on Nazi business policies the historian Matthias Frese describes the Council as follows: "The AOG allowed the Council of Trust very little leeway to influence working conditions or company policy. The Council of Trust was not even allowed to act as a collectivity."[174] But the actual power of the Council of Trust tended to depend on the specific persons in office.

[173] Rössler, *Führer des Betriebes*, pp. 39–41. [174] Frese, *Betriebspolitik*, p. 169.

To be appointed a member of the Council, it was necessary to satisfy various criteria. Before the first election to the Council in 1934 Hermann had a brief summary put together: under the terms of §9 of the AOG, in March of every year the "leader" of the plant should put together a list of Council members and their deputies "in agreement with the representative of the National Socialist *Betriebszellen-Organization*": "The followers give their views on the list in a vote by secret ballot." If the "leader" and the *Betriebszellenobmann* were unable to come to agree about the list, or if the workforce did not approve the list, then the Council members would be appointed by the Trustee of Labor. Both blue- and white-collar employees should be "adequately" represented in the choice of Council members. Only persons who were twenty-five years of age and older, who had worked in the plant for at least one year, and who had worked in their job or in a similar profession for at least two years were eligible to become Council members. However, these regulations were not imperative. But there were also binding regulations: a Council member had to have his civil rights, to be distinguished by his "exemplary human qualities," be a member of the DAF, and to support "the national state at all times." Every employee who was older than twenty-one and was in possession of his civil rights was allowed to vote on the Council members. Voters had the option of either agreeing to the list or rejecting individual persons by crossing out their names.[175]

In the elections to the Council of Trust held on April 12, 1934, Hermann as the works manager of the Maingau plant community and the *Betriebszellenobmänner* for the works in Hoechst, Griesheim, Mainkur, Marburg, and Offenbach put together a joint list for all plants of the works group, which consisted of ten Council members and ten deputy members. However, it appears that the still incumbent National Socialist works council in Hoechst chaired by Ludwig Retzinger[176] had taken the initiative. It was noted during the meeting of the Board of Directors and of the technical directorate on March 5, 1934, that "currently the lists for the members of the Council of Trust have been put together by the works council. After the lists have been submitted the plant management will comment on them; the plant management places special importance on the position of the representative for university graduates. The persons present should pass on their suggestions to Hermann in person."[177]

Six candidates were finally nominated from Hoechst, together with one candidate from each of the other works. The voter turnout in April 1934

[175] HA, PSW 657, note for Hermann on the election to the Council of Trust dated March 15, 1934.
[176] HA, Pol 6, Retzinger to *Arbeitskameraden* (work comrades), dated April 9, 1934 (in his position as chairman of the Hoechst Works Council on DAF membership).
[177] HA, TEA 95c, Niederschrift der Vorstands- und technischen Direktions-Sitzung in Hoechst, March 5, 1934.

was about 92 percent; 86 percent of the ballot papers handed in were unaltered, about 14 percent had been crossed out either completely or in part. Thus only a relatively small minority had rejected the list. In Hoechst 658 employees had rejected the candidate Ludwig Retzinger, while 5,843 agreed to his election. Hans Schlichenmaier, who had allegedly recruited Retzinger into the NSBO, became a Council member as the representative for university graduates with 5,825 persons voting for him and only 676 employees rejecting him.[178] In his speech on May 1, 1934, Hermann stressed the achievements of the new regime and described the Council of Trust as a "link" between "followers" and plant management.[179] In the course of the celebration the Council members with works manager Hermann as their chairman gave a solemn pledge: "We do solemnly promise during our administration to consider only the well-being of our plant and the community of all *Volksgenossen* and to put our own selfish interests last and by our conduct and in the fulfillment of our duty to serve as an example for all employees."[180]

Although Retzinger was replaced by Schacke as *Betriebsobmann* sometime in 1934 or at the beginning of 1935, his position within the Council of Trust remained unchallenged. Nevertheless the results in 1935, held on April 12, were considerably worse than in the previous election. This time the candidates were only from Hoechst, and every plant community now apparently had its own Council of Trust. According to the official election results 7,586 employees in Hoechst had the right to vote. A total of 6,977 envelopes were handed in, which corresponds to a voter participation of 92 percent. Of the votes cast, seventy-eight were invalid, leaving 6,886 valid votes. Of these, 5,588 ballot papers had been handed in unchanged; 1,298 ballot papers were crossed out or partly crossed out. Out of all valid votes the list of Council members was accepted by around 85 percent, while 15 percent turned it down. Retzinger had the worst result with 1,083 rejections. He was followed by Ernst Schmidt with 1,074 rejections, and Alfred Eckelmann as the representative for the university graduates in the plant and the successor to Schlichenmaier with 1,053 rejections. The most prominent candidates would appear to have come off worst, with results ranging between 84.3 and 84.7 percent; the best result was 85.5 percent.[181] Altogether, however, consent to the list of Council members was very high (Table 3.4).

---

[178] HA, PSW 657, Bekanntmachung, Betr. Abstimmung über die Liste der Vertrauensmänner 1934/35, dated March 28, 1934; election to the Council of Trust on April 12, 1934, list of results.

[179] HA, PA Hermann, typescript of Hermann's speech on Labor Day (May 1), 1934.

[180] HA, PSW 657, Das feierliche Gelöbnis des Vertrauensrates zum 1. Mai 1934.

[181] HA, PSW 657, Vertrauensratwahl am 12.4.1935, Werk: Hoechst – Ergebnisse, Sozialabteilung; notice on the results of the election to the Council of Trust dated April 15, 1935.

TABLE 3.4. *List of Proposed Members for the Hoechst Council of Trust in 1935–1936*

		Council Members	
1	Retzinger, Ludwig	Laboratory worker	Council of Trust office
2	Schmidt, Ernst	Commercial clerk	Council of Trust office
3	Eckelmann, Dr. Alfred	Chemist	Patents Laboratory Ch 76
4	Hofmann, Johann	Worker	Council of Trust office
5	Schörg, Karl	Graduate engineer	Electrical Department
6	Krug, Wilhelm	Locksmith	Council of Trust office
7	Oesterling, Heinrich	Fireman	Safety
8	Schaar, Ernst	Lathe operator foreman	Apprentices Workshop S 62
9	Blum, Ernst	Clerk	Packing Department Ch 83
10	Dinges, Anton	Worker	Dyes Warehouse F 38
		Deputy Members	
1	Trinkaus, Heinrich	Tailor	Technical Warehouse A 26
2	Grimminger, August	Commercial clerk	Coloristic Department
3	Schlichenmaier, Dr. Hans	Chemist	Alizarin Main Laboratory A 11
4	Krause, Hermann	Worker	Council of Trust office
5	Zeh, Hermann	Graduate engineer	Works Department A 70
6	Fischer, Georg	Electric fitter	Eltwe Ch 70
7	Hartwig, Lorenz	Master craftsman	Indigo Warehouse Ch 54
8	Schäfer, Fritz	Locksmith	Railroad Workshop S 62a
9	Kilp, Wilhelm	Junior clerk	Salary Department
10	Ressl, Michael	Electric fitter	Eltwe A 1

Source: HA, PSW 657, Bekanntmachung, Betr. Abstimmung über die Vorschlagsliste der Vertrauensmänner 1935/36, dated March 26, 1935.

Wolfgang Zollitsch noted that these election results should be compared with those of the plebiscites held in the Third Reich, which had approval rates of almost 100 percent. This made it "clear that within companies it was possible to stand firm against the National Socialist compulsory elections and that approximately one-third of the workforce was not afraid of expressing their discontent and rejection." However, Zollitsch did note that there was "a high degree of consent to the Council of Trust lists submitted" in the IG Farben works. While the results in Leverkusen had already been good, the approval rate in the Hoechst plant was particularly high.[182] However, Zollitsch also mentions the "Reports on Germany" of the exiled Social Democratic Party, which describe the election in 1935 in the IG Farben plant in Hoechst: "Workforce 5 to 6,000 persons. 30 percent invalid ballots. The election was held in the canteen. Massive terror tactics."[183] The canteen was only one of several places where votes were registered. Just how the official result was achieved and how great the pressure on the employees in Hoechst may have been cannot be reconstructed on the basis of the existing documents. If strong pressure was indeed brought to bear in a place like the canteen, this may have affected voting behavior in other voting locations and contributed to the high approval ratings.[184]

But it is necessary to point out that the position of the Council of Trust was not particularly important, neither in the AOG nor for daily work in Hoechst. According to Retzinger, the Council of Trust "never had any grave differences of opinion from the factory management."[185] This led Hermann to decide at the end of 1934 that "in light of the good working relationship with the Council of Trust" no persons should be employed without consulting the Council.[186] What this meant in practice is revealed by examining the personnel files. For example, when Reinhard Böker came to Hoechst at the beginning of 1935 from Professor Fries in Brunswick, Council member Retzinger, and the then *Betriebszellenobmann* Schacke, were only informed that Böker would be taken on – they were informed but had no say over or control in the matter.[187]

---

[182] Zollitsch, *Arbeiter*, pp. 221, 224.

[183] *Deutschland-Berichte der Sopade (Sozialdemokratischen Partei Deutschlands), Zweiter Jahrgang 1935* (Frankfurt am Main: Zweitausendeins, 1980), pp. 546–7.

[184] Cf. Zollitsch, *Arbeiter*, pp. 225–6.

[185] HA, PSW 648, Note from the company archive dated April 28, 1978, on an interview with Retzinger of April 26, 1978.

[186] HA, TEA 95c, Notiz für Herrn Bormann, Direktionsabteilung T – Nachtrag für Direktionssitzungsprotokoll, dated November 29, 1934.

[187] HA, PA Böker, communication to Retzinger and Schacke on the employment of Böker dated February 14, 1935. After the war it was sometimes said that the Council of Trust had held an important position in the plant – but no convincing evidence was found for such statements. Cf., for example, HA, PA Abteilung VI, Julius Weber, written testimony of Dr. Niessen, who "took over the management of the scientific department of the Bayer office in Frankfurt in 1941," dated June 27, 1945. He talks of an "omnipotent Council of Trust" in Hoechst. See also HHStAW, Abt. 520 F (A-Z), Weber, According to the written statement of Dr. Emil Meyer, dated March 30, 1946, the works council in 1935 demanded that Weber be dismissed because of his contacts with Jewish employees. But this demand was dropped apparently on the intervention of Hermann. At the same time Hermann apparently stated that it would no longer be possible to make Weber a director. On the same occasion Hermann "admitted that he would be obliged to dismiss any member of the plant community whose dismissal was demanded by the Labor Front for political reasons." This statement is somewhat contradictory as Hermann had just prevented Weber from being dismissed despite the previous demand for Weber's dismissal.

After 1936 there were no more elections to the Council of Trust; the term of office of the Council members was legally extended by one year, and when someone left office, the DAF and the plant management jointly appointed a successor. This decreased the legitimization of Council members even more – and the Council became increasingly unimportant. Although Zollitsch writes that Council members could "develop a form of independent existence, this was only possible if the member was supported by authorities outside the company – by the Party or the DAF." According to Zollitsch, in the second half of the 1930s "the emphasis shifted noticeably from the Council members to the 'administrators' of the DAF" in the works, the *Betriebsobmänner:* "They had the backing of a large external organization and were able to bring the political weight of the DAF to bear on their activities."[188] They needed the support. While the DAF and its *Betriebsobmann* did have an important role to play in the plant, which was strengthened by regular visits from important DAF members, yet at least until 1941 the role of the unofficial Party comrade who served as a direct contact with the *Gauleitung* and the local Gestapo was of greater, perhaps of the greatest, importance.

While Hermann apparently had no problems in dealing with Retzinger and the Council of Trust, after Brüning's suicide Lautenschläger tried to protect himself from the intrigues in the plant by creating a special position for Hans Wagenheimer. Wagenheimer's task as of October 1938 was to serve as a direct contact with the *Gauleitung* under *Gauleiter* Jakob Sprenger and with the local Gestapo. This gave Wagenheimer an extremely important position within the plant. After the war Retzinger claimed in front of the denazification tribunal that the works management had been responsible for maintaining ties to the Gestapo.[189] This task was shared by Hans Pöhn, the Defense Commissar in the plant after 1935,[190] and Hans Wagenheimer, whom Retzinger described during the denazification proceedings as an "SD man." So who was Wagenheimer, the man given the nickname "the milkman" after the war, when he once again climbed his way up in Hoechst?

On February 13, 1934, the Department for Job Creation of the Personnel Bureau of the NSDAP for the district of Hesse Nassau asked IG Farben in Frankfurt whether it would be possible to take on the "Pg. [Party comrade] and SS man Hans Wagenheimer" with NSDAP membership number

---

[188] Zollitsch, *Arbeiter*, pp. 224, 233–4.

[189] HHStAW, Abt. 520 F (A-Z), Retzinger, Protokoll der öffentlichen Sitzung, dated July 7, 1948.

[190] BA, Film 44839, according to Lautenschläger's testimony to Cooper dated March 20, 1947, Kränzlein and Hirschel were the Defense Commissaries. However, this is false. Cf. HA, PA Abteilung VI, Pöhn: Lautenschläger and Jähne to Pöhn, November 1, 1944; HHStAW, Abt. 520 F (A-Z), Nr. 4815, Protokoll der öffentlichen Sitzung, dated January 12, 1949: on Zeh – Pöhn's own statement.

215,305 as a trainee in the "Chemo Technology" Department: "Pg.
Wagenheimer is 22 years old, is interested in the profession, but was unable
to do any training as he became unemployed due to his long-time membership in the N.S.D.A.P. Enclosed please find a letter from
Obersturmführes [*sic*] Mages of the SS who warmly recommends him; Pg.
Wagenheimer also enjoys the full support of the *Gauleitung.*"[191] IG Farben
in Frankfurt passed the letter on to Hoechst, and the Personnel Department
asked director Georg Kränzlein, an active NSDAP and SS man, whether
Wagenheimer's application should be granted. On April 10, 1934, the
NSDAP Job Creation Department for the district increased its pressure and
explicitly reminded Hoechst of Wagenheimer's application. Only two days
later, on April 12, Wagenheimer was hired as a worker by Kränzlein's
Alizarin Department.[192] Wagenheimer was apparently in financial difficulties since there was an order of attachment on his property; moreover he
was keen to get ahead as quickly as possible. In August 1934 he already
requested an increase in pay or a transfer to another department where he
would have "a quicker opportunity for advancement."[193]

Initially, Wagenheimer toyed with the idea of making his career in the
Coloristic Department, but it was pointed out to him that a long period of
training would be necessary. He therefore took up Kränzlein's suggestion of
pursuing a "career as a master craftsman"; the Social Department then
pointed out that this too would take some years to complete. At the
beginning of 1935 "at the request of Dir. Dr. Kränzlein" he was finally
given a job as a junior clerk and after a short probationary period was given
a permanent position in Hoechst in May 1935. This was not the end of his
career: on October 1, 1935, Wagenheimer became a commercial clerk.
Hermann personally made the decision at the suggestion of the head of the
Personnel Department Schwamborn and of Kränzlein and – according to a
note in the personnel files – "without additional assessment."[194]

Wagenheimer had started work in April 1934 at about 80 RM per
month; now he was already earning 240 RM per month – in less than one
year. In 1938 on a single day, April 30, his income, which had amounted to
315 RM since 1937, was increased to 340 and then to 400 RM. By January

[191] HA, PA Abteilung VI, Wagenheimer, Personal Amt der NSDAP Gau Hessen-Nassau,
Abteilung Arbeitsbeschaffung, to I.G. Farben Frankfurt, February 13, 1934.
[192] Ibid., I.G. to NSDAP Kreisleitung Gross-Frankfurt, February 19, 1934, forwarded to
Hoechst; NSDAP Kreisleitung to Director Schwamborn (Hoechst), April 10, 1934;
Personalblatt, Auszug aus den Personalakten (Strictly Confidential).
[193] Ibid., Arrestpfändung, July 17, 1934; Aktennotiz, August 24, 1934.
[194] Ibid., Aktennotiz, August 24, 1934; petition of Wagenheimer, dated September 2, 1934;
note by Nüsslein dated September 20, 1934; note by Schwamborn dated October 29,
1934; Schwamborn to Wagenheimer, February 6, 1935; note dated May 6, 1935; note by
Schwamborn dated October 8, 1935; Schwamborn to Wagenheimer, October 10, 1935;
Auszug aus den Personalakten (Strictly Confidential).

1, 1940, his monthly income amounted to 700 RM. Wagenheimer owed the massive increase in income between 1938 and 1940 to the new position of trust conferred on him by Lautenschläger on the day of Brüning's suicide on October 4, 1938, when Wagenheimer left Kränzlein's department and became directly responsible to Lautenschläger "as a contact man between the plant work and the *Gauleitung,* etc."[195] He was to remain in this position until 1941.

For someone born in 1911, who had attended only primary school and only worked in his parents' dairy shop, this was an unusual and steep career trajectory, which, as he was well aware, he owed only to his political involvement. He was not even an "old fighter." Although he had joined the NSDAP in March 1930, he had left it in October of the same year and rejoined only on December 1, 1932[196] – yet the Party strongly supported him, as he himself admitted. In a résumé written in January 1935 he wrote that after the death of his father his parents' business had fallen off badly, and he was obliged to work on the farm of Prince Leopold of Prussia. When his mother fell ill, he returned to Frankfurt and continued the business, which, however, was worth less and less: "As I had gained some insight into large corporations [when supplying them] early on I already wished to carry out my vocational training in such a company. Thanks to the *Gauleitung* I was able to be employed by IG Farben."[197]

As a contact man he appears to have been very successful. In his personal recollections Hirschel wrote: "Wagenheimer – already during Dr. Hermann's time – was responsible for contacts with the Party. Professor Dr. Lautenschläger maintained this arrangement as one highly beneficial to Hoechst."[198] This was an understatement. While Hermann had already made use of his services, Wagenheimer's influence increased even more under Lautenschläger. It was Wagenheimer's job to help the plant with all problems, whether these were with Nazi organizations or were due to shortages of materials. If employees had problems with the Gestapo or other Nazi organizations Wagenheimer had to coordinate the position taken by the plant with that of the *Gauleitung* and the Gestapo or find a solution in advance. Together with Director Schwamborn Wagenheimer took care of all difficult matters concerning employees for Lautenschläger, who was generally interested only in receiving clear directions from the Nazi organizations.

---

[195] Ibid., Auszug aus den Personalakten (Strictly Confidential); draft of letter from Lautenschläger to Kränzlein, October 4, 1938.

[196] BA, BDC file on Wagenheimer (NSDAP Zentralkartei and Ortsgruppenkartei).

[197] HA, PA Abteilung VI, Wagenheimer, handwritten resumé of Wagenheimer, January 25, 1935.

[198] HA, Hirschel, "Erinnerungen aus meiner I.G. Zeit," p. 25.

Wagenheimer's role in the plant was viewed very differently. A senior employee in Hoechst who ran into difficulties with the Gestapo during the war later said of Wagenheimer that although he had been a member of the NSDAP he had nevertheless "been generally known" to be the "good genius of victims of political persecution." Wagenheimer had "in all respects justified the confidence placed in him" and saved the man from being persecuted by the Gestapo by admitting him into the NSDAP.[199] After the war a member of the works council interpreted this story and Wagenheimer's role completely differently: "Wagenheimer joined IG at the request of the SS and was an intermediary between IG and the SD [*Sicherheitsdienst*]. If really important people from IG did not join the Party willingly, Wagenheimer raised hell at the Gestapo, and after that the person in question had no choice but to join the NSDAP. This is what must have happened in this case as well."[200] No matter how one interprets this story, at the time Wagenheimer's intervention meant for the persons concerned protection from the Gestapo – and for the Hoechst plant the avoidance of potential difficulties with the Nazi authorities because of a senior employee.

So in Hoechst, until the first years of the war, in addition to the *Betriebsobmann* and the Council of Trust as the official representatives of the regime there was Wagenheimer as an informal representative, who wielded more power and influence in matters concerning employees and general staff than the two official representatives together. Matthias Frese wrote that the aim of the AOG was to ensure that all conflicts "at the operative level" would be worked out exclusively "between employer and employees." He added: "If conflicts nevertheless developed which could not be settled internally, e.g. strikes, then in addition to the Trustees of Labor this was also the responsibility of the Gestapo."[201] Before any conflicts developed that would have required the intervention of the Trustee of Labor, Hoechst wanted them to be cleared up with the help of the "SD man" Wagenheimer, as Retzinger called him.

However, Wagenheimer's position did not just diminish the role of the *Betriebsobmann* and the Council of Trust. It also limited the role of the Defense Commissar in the work, Hans Pöhn, who was also head of the factory administration and the air warden of the plant.[202] According to a shop steward after the war, as Defense Commissar Pöhn was

[199] HHStAW, Abt. 520 F (A-Z), Julius Weber, Weber to the Mayor of Frankfurt via IHK, December 20, 194[5] and Protokoll der öffentlichen Sitzung der Spruchkammer, January 10, 1949 (Weber's testimony).

[200] Ibid., Expert testimony by Trost, Protokoll der öffentlichen Sitzung der Spruchkammer, January 10, 1949, on Weber.

[201] Frese, *Betriebspolitik*, p. 95.

[202] HA, PA Abteilung VI, Pöhn, Direktion Hoechst to Pöhn, November 1, 1944.

responsible for preparing "the interrogations of factory employees by the Gestapo."[203] In contrast to this statement Pöhn was keen to emphasize after the war that his position did not make him a political official but that he had been working for the military authorities. He had not collaborated with the Gestapo but had rather tried to exonerate or even gain the release of works employees. He had joined the NSDAP only in 1941 and had done this primarily to avert possible problems for the plant.[204]

Indeed, Pöhn does not seem to have played the role in Hoechst that the newer literature generally attributes to the Defense Commissars.[205] But this seems to be due more to the particular circumstances of the Hoechst plant, where Wagenheimer already functioned as an "extended arm" of the Gestapo. In this context it is important to mention that during his interrogation in Nuremberg in answer to the question as to who had been Defense Commissar in Hoechst, Lautenschläger did not mention either Pöhn or Wagenheimer, but two senior employees, Georg Kränzlein and Otto Hirschel, of whom more will be said later.[206]

3.5. "NAZIFICATION": ADJUSTMENT, EXCLUSION, AND POLITICAL PERSECUTION BEFORE THE WAR

The historian Robert Gellately has demonstrated the distinction between "coercion" and "consent" under the Nazi regime and has emphasized the extent of voluntary support for the regime. It is his main thesis that consent and force are inseparably intermingled in the history of the Third Reich, as terror and compulsion were used selectively against individuals, minorities, or social groups that enjoyed little popularity among the general population: "Coercion and terror were highly selective, and certainly did not rain down universally on the heads of the German people." According to Gellately, terror alone was insufficient to explain why the National Socialists came to and stayed in power.[207] However, as his study also goes on to show, it is often very difficult for historians to distinguish between compulsion and consent in a totalitarian system such as that of the National

---

[203] HHStAW, Abt. 520 F (A-Z), Pöhn, information as to Pöhn, provided by the Betriebsvertretung (Trost), dated November 13, 1947.

[204] Ibid., Pöhn to Spruchkammer, November 29, 1947, with appendices including 1 and particularly 1a, dated June 6, 1946, on a conversation with Wagenheimer in which it was stated that Pöhn's joining the NSDAP had prevented his being replaced by an SS man. In BA there are no BDC records on Pöhn.

[205] Lotfi, *KZ der Gestapo*, p. 42.

[206] BA, Film 44839, Lautenschläger's testimony to Cooper dated March 20, 1947.

[207] Gellately, *Backing Hitler*, p. 2.

Socialist regime.[208] One wonders often where consent predominated or compulsion dictated behavior. This is difficult or even impossible to answer with any certainty in retrospect or on the basis of what was passed down. The return of the Saar to Germany in 1935 or the military victory over France in 1940 surely met with great general approval and even brought the regime further support. But this should not be interpreted as indicating that people did not see and fear the exclusions, persecutions, and denunciations – and looked for and found suitably unobtrusive means of nonagreement or even rejection. As Gellately also wrote, the readiness of citizens to inform the police or Party authorities about their suspicions had fatal consequences for every form of resistance: "Because many ordinary people served as the eyes and ears of the police, those who might have wished to resist could not gather to organize or to form solidarities. Those who still wanted to say 'no' had to swim against the tide, and were driven to individual acts of defiance that were important for them as moral individuals, but in the short run not threatening to the dictatorship."[209] These were the conditions in the plant under which employees had to act – and some paid for their rejection, even if this often only took the form of small asides.

One parameter that may indicate the degree of nazification in the plant is the number of Party members and the role played by other Nazi organizations such as the SA, the SS, and the DAF. But one should be aware that this is only one parameter and one, moreover, that must be used with caution and qualified. In his history of the insurance company Allianz in the Third Reich the historian Gerald Feldman wrote: "how deep nazification went is a more difficult – indeed, impossible – question to answer, and undoubtedly the degrees of commitment to the regime varied from individual to individual and also between 1933 and 1945." Yet altogether, Feldman concluded, "there were enough genuine National Socialists employed at Allianz to ensure that the proper 'spirit' and 'attitude' were maintained by the mass of their colleagues."[210]

In the proceedings of the denazification tribunals after the war it was eagerly pointed out that convinced National Socialists had also acted "decently." Retzinger, for example, was described as a socially minded person who had spoken out against the persecution of priests and Jews. Both a Social Democrat and a factory employee with Communist leanings were able to confirm that Retzinger had interceded on their behalf – and they considered that this showed that he had not been a "proper Nazi." In the case of Hermann Zeh, Retzinger's successor as *Betriebsobmann*, it was again emphasized that he had sometimes behaved "decently," that he had

[208] Cf. Karl Dietrich Bracher, *The German Dictatorship. The Origins, Structure, and Effects of National Socialism*, translated from the German by Jean Steinberg, with an introduction by Peter Gay (Harmondsworth: Penguin, 1973); Dahrendorf, *Gesellschaft und Demokratie*.
[209] Gellately, *Backing Hitler*, p. 262.   [210] Feldman, *Allianz*, pp. 117–18.

even refrained from unduly harassing a neighbor who was married to a "full Jewess." Excuses ranged from having been a "devout Nazi," which strangely enough after the war was considered to be quite positive, to that of having behaved "decently." But in that case who actually was a convinced Nazi?

This question might be answered if we accept membership in the NSDAP as an important indication. People like Retzinger, Zeh, and Wagenheimer, but also Hermann, von Brüning, Lautenschläger, Jähne, or Winnacker, were National Socialists – on the basis of their duties or functions they could even be considered leading or at least prominent Nazis. But in addition to Party membership it is necessary to take note of a person's behavior and way of thinking, even if these are difficult to reconstruct on the basis of the available sources. The Italian feminist and writer Luce d'Eramo, who wrote about her time as a worker in Hoechst during the Second World War in her autobiographical novel *Deviazione* (The Detour), believed that Nazis could primarily be recognized by their behavior. Occasionally using rather drastic language, this point is made in a conversation between the heroine Lucia and a certain Vittorio. Lucia says: "For four months at IG Farben I was convinced that the foreman of Department Ch 89 was a Nazi, and then during the strike of the foreign laborers it turned out that de facto he was on our side. And nevertheless he wore his Party badge in his buttonhole. I thought that another worker without a Party badge was a comrade, because he constantly conspiratorially raised his closed fist when he saw me, winked at me, and didn't maltreat me at work. ... I had interpreted an invitation to fuck as a Communist greeting and believed the most bootlicking whoremonger of the whole department to be a comrade; he wasn't a Party member – on the contrary, he had passed himself off to us as an opponent of the Nazis – but he took advantage of the hunger of the deported women."[211]

During the denazification of the Hoechst plant after the Americans took over in 1945 this distinction was formulated less drastically, but it largely amounted to the same thing. Three lists of names were drawn up: a list of "sympathizing Party members," a list of "not sympathizing Party members," and a list of "sympathizing non-Party members."[212] This categorization seems more true-to-life than a simple record of NSDAP membership, although finding a clear answer remains problematic. It would be naive to assume that people were only either Nazis or uncompromising opponents of the Nazi regime. Very few people are one-dimensional, and

---

[211] Luce d'Eramo, *Deviazione* (Milan: Mondadori, 1979), p. 333; German edition: *Der Umweg* (Reinbek: Rowohlt, 1982), pp. 380–1.

[212] HA, file Entnazifizierung (Vorstandsbüro), Freier Deutscher Gewerkschaftsbund to Military Government, Frankfurt, June 30, 1945, in the appendix the lists dated June 28, 1945 (all photocopied).

only rarely can they be clearly categorized. In this section the complexity of
people in their daily lives, with which the historian is confronted, will be
presented in detail. In the main Nazis tended not to be *only* Nazis, they were
often *also* Nazis. Nor was it necessary to be a member of the NSDAP or
another Nazi organization to serve the Party.[213] This is not hair-splitting
but an attempt to understand a reality, the daily life in the plant in Hoechst
in the years 1933–45.

In the files of the denazification tribunals, so-called decent behavior is not
rarely also found in the person of "convinced" National Socialists. The
example of Retzinger is a case in point, as he was quite prepared to behave
in a humane and "comradely" manner toward old school friends and fellow
sportsmen. Retzinger interceded for them, despite the fact that they were
ideologically on the opposing side, because he knew them well and had
friendly ties with them. He was not just a convinced National Socialist, he
was also a passionate athlete and had strong local ties. And so he – like
many others at the time – attempted to combine his different loyalties.

It is necessary to differentiate when we examine the question of membership in the NSDAP. Prior to 1933 there were already NSDAP members
in the plant in Hoechst, among both the white- and blue-collar employees:
Hirschelmann and Retzinger were blue-collar employees, Schacke and the
Council member Schlichenmaier, who has already been mentioned, were
white-collar employees. The latter is a good example of how hard it is to
define National Socialists simply on the basis of their Party membership,
and we will be looking at even more complex cases. Schlichenmaier, who
allegedly recruited Retzinger for the NSBO, had joined the NSDAP in 1931.
Born in 1896, he started studying chemistry after completing his high school
examinations until his studies were interrupted by the outbreak of the First
World War. He finally completed his doctorate in 1922 and was given a job
in Hoechst, where he worked in the Central Laboratory under Professor
Roser. From 1931 on he was employed in the main Alizarin Laboratory
under Kränzlein. In 1941 he switched to the Synthetics Laboratory.[214] In
1933 he was elected to the Council of Trust in Hoechst as the representative
of the university graduates and remained a member until 1935/36, after
which he was a deputy member until 1939. According to his own statements he had supported Gregor Strasser within the NSDAP and thus
belonged to the "leftist" wing of the NSDAP. The developments after
Hitler's seizure of power then forced him into opposition, particularly
the violence against and persecution of former Jewish colleagues. His
opposition never went so far that he left the NSDAP or had contacts with
resistance circles. But he did demonstrate his solidarity with a neighbor

[213] Cf. Hayes, *Industry and Ideology*, p. 381.
[214] HA, PA Abteilung VI, Schlichenmaier, personal data sheet.

who was married to a Jewess and maintained his ties with the family in question.[215]

Werner Schultheiss also joined the NSDAP prior to its seizure of power. He came to Hoechst in 1928 after studying in Brunswick and completing his doctorate under Professor Karl Fries in 1927. After working in the Central Laboratory, he moved to the main Alizarin Laboratory under Kränzlein in 1931. In April 1932 he applied for Party membership in Hoechst and was admitted to the NSDAP on May 1, 1932.[216] According to his own evidence he joined after moving to Kränzlein's department: "In 1931 I was transferred to another department and made the acquaintance of a number of workers and chemists who were members of the NSDAP. They explained to me that their Party would do away with class struggle and would put the ideals I dreamed of into practice."[217] His former colleagues wrote after the war that he had been "a convinced Nazi" – he had, after all, worked as a political officer of the NSDAP since 1934.[218] Schultheiss transferred to the Azo Dyes Department in 1937 and became its manager at the end of 1938 after the forced retirement of its manager, Hermann Landers.[219]

Quite a number of other important chemists were drawn to the National Socialists in the department of Georg Kränzlein, who was himself one of the most fervent Nazis. Already prior to the seizure of power the main Alizarin Laboratory under Kränzlein had been a hotbed of Nazism, even though Kränzlein himself joined the NSDAP only in 1933. The number of Hoechst employees who had been members of the NSDAP prior to 1933 increased after the seizure of power in many ways. On the one hand, pressure was brought to bear on the plant to hire "old fighters" – and to a certain extent the administration yielded to this pressure.[220] Then there was the recruitment of chemists to the work who had been active National Socialists since their student days; the student associations at German universities had been

---

[215] HHStAW, Abt. 520 F (A-Z), Schlichenmaier, Schlichenmaier to Öffentlicher Kläger bei der Spruchkammer, dated March 13, 1947; statement Hans v. Freyberg, September 5, 1945; statement Albin Hardt (undated); statement Heinz Woelcke, September 2, 1945.

[216] HA, PA Abteilung VI, Schultheiss, Personalblatt and application to appoint Schultheiss a *Prominenter* (this was not an office but a position, which entitled the bearer to many additional privileges) dated February 22, 1943; BA, BDC files on Schultheiss.

[217] HHStAW, Abt. 520 F (A-Z), Schultheiss, Schultheiss to Öffentlicher Kläger Frankfurt, April 17, 1947: Ergänzungen zum Meldebogen.

[218] Ibid., Grosshessisches Staatsministerium, Öffentl. Kläger Frankfurt, working paper on Schultheiss dated November 18, 1946 – testimony Trost (information provided by the works council).

[219] HA, PA Abteilung VI, Schultheiss, Personalblatt.

[220] The most well known were the cases of Zeh, *Ortsgruppenleiter* (local group leader) in Hoechst since 1934 und *Betriebsobmann* from 1940 to 1945, and Sprenger's confidants Wagenheimer and Schilling. For more on the latter, see the chapter on the war.

firmly in the hands of the National Socialists and other ultra-nationalist organizations since the 1920s.[221]

Hoechst also yielded to the pressure by the NSDAP to take on even less promising chemists if they were dedicated National Socialists. In such cases the best one could hope for was a reasonable standard of work. Bernhard Popp is such an example. Popp applied for a job in IG Farben at the beginning of April 1935 "in accordance with my services to the Fatherland and my abilities as stated in the certified copies of my diplomas." In his résumé it became clearer what he meant by services. Born in 1898 in Würzburg, Popp served as a soldier after completing his schooling in 1916 until the end of the First World War. He then began studying chemistry in Würzburg but interrupted his studies in 1919, 1921, and 1923 to fight as a member of a free corps. After completing his dissertation in 1927 he taught in schools and continued his chemical studies. In 1930 he joined the SA, rising to the rank of SA brigade leader.[222] He also joined the NSDAP in 1930.[223] In the testimonial from Otto Hellmuth, *Gauleiter* of the NSDAP for Lower Franconia, his fighting spirit received particular praise: "Pg. Dr. Popp is National Socialist from top to toe. Since being admitted to the SA he has put his whole life behind the cause and has paid the price for this with a prison sentence in November [19]32. We have always been completely satisfied with his work. His leadership qualities are excellent."[224] Based on these qualifications he was taken on in Kränzlein's department, in the main Alizarin Laboratory, in the summer of 1935. His background was duly noted by the Personnel Department with respect to the question of a permanent position: "After five years active work for the SA which he then left last summer after holding the rank of a brigade leader to join us as a chemist, Dr. Bernhard Popp has proved himself during his period of employment in our main Alizarin Laboratory to be an exceptionally diligent and interested chemist; while his knowledge naturally cannot yet measure up to that of a colleague of the same age he has the firm intention and also the capabilities to fill the gaps left by his work for the SA as soon as possible." His performance was rated as "quite satisfactory," and as he was also considered to be "a good comrade and a very pleasant colleague with a faultless character" the suggestion of taking him on permanently was

[221] Cf. Ulrich Wengenroth, "Zwischen Aufruhr und Diktatur: Die Technische Hochschule 1918–1945," in idem (ed.), *Technische Universität München. Annäherungen an ihre Geschichte* (Munich: Technische Universität München, 1993), pp. 215–23.

[222] HA, PA, B. Popp (microfilm), letter from Popp to I.G. Farbenindustrie dated April 11, 1935, with handwritten resumé.

[223] BA, BDC file on Popp, admission dated December 1, 1930, membership number 386,235.

[224] HA, PA, B. Popp (microfilm), NSDAP Gauleitung Unterfranken, passport for B. Popp issued by Gauleiter Dr. Hellmuth dated August 10, 1933 (certified copy).

"warmly" approved.[225] It appears that he gave a very good account of himself professionally, and he became head of the newly founded Technical Department Ch 89, where Luce d'Eramo worked during the war.[226] When *Betriebsobmann* Zeh proposed him as a deputy, the works management demanded that he should give up his position as head of his department – whereupon Popp preferred to remain department manager.[227]

Of course, some of the Nazis newly employed after 1933 were also very competent chemists, such as Ernst Otto Leupold, who came to Hoechst as a pupil of Professor Hermann Staudinger. Leupold, born in 1903 in Waldenburg (Silesia), studied chemistry in Munich and Freiburg im Breisgau, where he completed his doctorate under Staudinger in 1930 and remained as his assistant until 1935. Leupold was a devout National Socialist. He had first made contact with Hoechst while he was still leader of an SA storm troop in 1933. He had sent a request to Kränzlein for used gas mask filter pads to use in paramilitary exercises with his troop.[228] In the summer of 1934 Leupold applied to Kränzlein for a job on the recommendation and with the support of Staudinger.[229] Kränzlein approved of his being taken on, for he knew Leupold to be an excellent chemist and a dedicated National Socialist – in short, to him a true paragon. At the end of 1934 Kränzlein wrote to Staudinger that his own son Paul, currently studying chemistry in Würzburg and also a committed Nazi, was developing "into a little Leupold in Würzburg."[230] Leupold joined Hoechst in April 1935 and initially worked under Kränzlein in the main Alizarin Laboratory, then in the Catalytic Laboratory, and finally in the Department for Acetone and Acetic Acid.[231]

Membership in the NSDAP also increased in Hoechst by persons joining the Party after the seizure of power. The massive wave of admissions into the NSDAP during the first five months of 1933 eventually led to ban on new members. The consequences of this wave of new admissions were also

---

[225] HA, PA B. Popp (microfilm), Kränzlein/Greune to Personalabteilung, May 20, 1936; endorsed by Hermann in May 1936.

[226] d'Eramo, *Deviazione*, pp. 178–9 (*Der Umweg*, pp. 200–2). The heroine Lucia went to the "plant management" and spoke to a "director" called Dr. Lopp, but it is clear that this means the "factory manager," not the plant manager, and is almost certainly a reference to Dr. Popp.

[227] HA, TEA 95f, Niederschrift über die Maingau-Direktionssitzung in Hoechst, December 1, 1941; PA B. Popp (microfilm), personnel and earnings index card.

[228] HA, PA Staudinger, vol. 4, Staudinger to Kränzlein, December 13, 1933; vol. 3, Leupold to Kränzlein, January 29, 1934.

[229] Ibid., vol. 3, Staudinger to Kränzlein, July 28, 1934.

[230] Ibid., vol. 3, Kränzlein to Staudinger, December 3, 1934.

[231] HA, PA Abteilung VI, Leupold, Personalblatt; note by Kränzlein for Personalabteilung dated January 25, 1935; excerpts from letters from Staudinger to Kränzlein dated December 7 and 18, 1934, and letter from Kränzlein's office to Personalabteilung dated January 29, 1935.

visible in Hoechst. Some of the newly joined members clearly wished to prove themselves and were even fiercer in their behavior against opponents or "enemies" of the new regime than the "old fighters." These "turncoats of March" included both established chemists and younger employees with an eye to their career.

Georg Kränzlein, who held an important position in the plant as a director, was one of the most fanatical National Socialists in the plant. Membership in the NSDAP had already been increasing in his department even prior to the seizure of power, and he now took the lead. Kränzlein, born in 1881, had studied chemistry in Würzburg and completed his doctorate there in 1907. At the end of 1908 he joined Hoechst and began working in the Patents Laboratory under Schmidt, where he established himself as an independent researcher. During the First World War he worked in Hoechst as a "war chemist" and collaborated successfully in the "development and introduction of gas and smoke munitions, shells, and tracer and phosphorus ammunition."[232] On the basis of his achievements during the war he was made head of the Alizarin Department in 1920. In 1923 he was made an authorized signatory and appointed director in 1928. In 1930 he became chairman of the newly founded Synthetics Commission of IG Farben.[233] On May 1, 1933, Kränzlein became a member of the NSDAP, after June 1933 he was a candidate for SS membership, and he became a member of the SS in January 1934. The SS praised his zeal and "idealism" but also the opportunities that he provided. As the "foremost gas chemist of IG" Kränzlein possessed important information, and furthermore "SS members received training in the most modern gas and air-raid protection facilities of the IG plant in Frankfurt."[234]

Kränzlein had already been politically active prior to the seizure of power. At the time he had been a member of the DVP and had been politically active in the municipality of Hoechst. In August 1935 he was appointed councilor of the city of Frankfurt upon Main by *Gauleiter* Sprenger, an office held until his death in November 1943, even if this occasionally conflicted with his numerous other duties for chemical trade associations and the Four-Year Plan.[235] As head of research he collaborated closely with important professors, in particular with Hermann Staudinger.

---

[232] HA, PA Abteilung VI, Kränzlein, written confirmation from Hoechst for Kränzlein dated September 29, 1933.

[233] HA, PA Kränzlein, Biographischer Überblick.

[234] BA, BDC file on Kränzlein, Member of the NSDAP since May 1, 1933 (No. 2 399 639); SS identification card No. 124,908; SS-Reitersturm [mounted unit], SS-Oberabschnitt Rhein, December 31, 1934: testimonial of service; 10, SS-Reiterstandarte [mounted regiment] to SS-Oberabschnitt Rhein (Koblenz), August 17, 1935: suggestion to promote Kränzlein.

[235] Institut für Stadtgeschichte Frankfurt am Main, PA 17498 (Kränzlein); Magistrat Nachträge 53, Kränzlein's appointment by *Gauleiter* Sprenger, dated August 5, 1935, and by Mayor Krebs, dated August 9, 1935.

He also worked on projects to reform the the study of chemistry at the universities.[236]

Kränzlein was not only a particularly convinced Nazi, he was first and foremost a fanatical anti-Semite. It was apparently at his insistence that research employees who were Jewish or were regarded as Jews in the Third Reich were sent into early retirement. Even before the outbreak of the Second World War he apparently prophesized "a great cleansing of Jews in Europe including in Russia" with the result "that within five years Europe will have no more Jews."[237] And at the beginning of the Second World War he praised *Gauleiter* Sprenger for his activities in Hesse, above all his "fight against Judaism" in the cities, particularly "in the formerly Jewish metropolis Frankfurt upon Main."[238]

Whether Kränzlein had indeed hoped to become works manager after Hermann's death, as Retzinger stated after the war, a fact that would also explain the choice of the compromise candidate Lautenschläger, cannot be ascertained. It is certainly conceivable, for he was sufficiently ambitious and had the necessary political connections.[239]

Kränzlein was only the most prominent of the so-called "turncoats of March" in the plant. Other "turncoats of March" tried to outdo the "old fighters" in their severity toward putative opponents of the new regime. One example of this is Otto Nicodemus, perhaps one of the worst cases. Otto Nicodemus was born in 1886 and studied chemistry in Heidelberg and Freiburg, where he completed his doctorate in 1907. After working for some time at the University of Gießen he began work in the Central Laboratory of Hoechst in 1911. After the First World War he worked first as an employee, then as the head of the Catalytic Laboratory, where he focused primarily on acetylene chemistry.[240] Nicodemus became a member of the NSDAP and the NSBO at the beginning of 1933.[241] He additionally took on the duties of manager of the cell group for chemistry of the German

---

[236] HA, PA Kränzlein, See, for example, the article "Männer machen Geschichte: Dr. Georg Kränzlein (1881–1943)" from *Die Farben-Post*, June 28, 1962, that does not mention Kränzlein's NSDAP activities; Kränzlein to Rektoren der Universitäten Kiel, Bonn, Marburg, "Stellungnahme zur Neuordnung des chemischen Unterrichts," dated August 16, 1934; Kränzlein: "Lehrplan der Chemiker-Schule des Hauses der Technik, Frankfurt," dated January 23, 1942.

[237] HA, PA Abteilung VI, Greune, with excerpts from a letter from Kränzlein to Greune dated August 24, 1939.

[238] HA, PA Kränzlein, manuscript of Kränzlein, "Aus dem Zeitgeschehen des Gaues Hessen-Nassau," part 1, dated August 20, 1940.

[239] HA, PSW 648, note from the company archive dated April 28, 1978, on an interview with Retzinger of April 26, 1978.

[240] HA, PA Abteilung VI, Nicodemus, Personalblatt, and application to appoint him a *Prominenter* dated March 23, 1938.

[241] HHStAW, Abt. 520 F (A-Z), Nicodemus, Meldebogen dated April 26, 1946.

Association of Technicians. In this function he gave a talk at the end of June 1933 in which he tried to explain the role of the DAF: "The task of the German Labor Front is that of teaching [people] to become a community, and it expresses the fact that all creative people are workers in the best sense of the word. It also means that never again shall employers and employees confront another as enemies on opposing sides. Both have the same wish that the plant should prosper for the greater benefit of the German people, both are indissolubly connected to one another as companions with a common destiny. The German Labor Front wishes to bring all German people together and, if they do not wish to, will force them to come together." In his talk he explained the structure of the DAF and emphasized that membership in the DAF should be mandatory except for farmers and civil servants. He then explained the structure of the chemistry cell group in the plant and appointed the block wardens, who were organized according to the different departments, calling them "representatives of the works groups." While he mentioned that no "old fighters from the movement" had been appointed as block wardens, he excused this on the grounds that they were "already doing so much for us and are still burdened with excessive duties." He considered attendance at meetings of the cell group to be obligatory, and the block wardens were expected to monitor the presence of employees at the meetings, because the "intelligentsia" owed the victory of National Socialism "to the self-sacrifice and willingness to make sacrifices of the simplest, often the poorest men of the people." Now was the time "to honestly overcome all class differences, prejudices, and interests, to suppress all egoism within one's self in the face of vital interests of the German people." Nicodemus continued: "I will not deceive myself: this battle, which must be fought against one's self, against the Jewish spirit in the professions, the economy, and the accumulation of capital, will be terribly difficult."[242]

To be able to carry out "educational work" according to National Socialist principles in his cell group – a group that consisted of around 250 university graduates – he also wished to be allowed to impose sanctions. "Carrying out the training demanded by the leader of the Labor Front will be impossible without punitive measures."[243] He took vigorous action against dissidents. Thus, in a report given at the end of February 1934, he wrote: "Finally I wish to draw attention to the severe fight which I initiated and carried out with the support of the NSBO leadership against two politically unreliable chemists holding senior positions in this plant (one of

[242] HA, Pol 20, Ansprache des Zellenleiters Dr. Nicodemus in der ersten Vollversammlung der Zelle Chemie, dated June 23, 1933.

[243] HHStAW, Abt. 520 F (A-Z), Nicodemus, Nicodemus (Deutscher Technikerverband Ortsgruppe Frankfurt-Hoechst) to Reichsgeschäftstelle Deutscher Technikerverband, October 13, 1933.

them a deputy member of the board of directors of IG Farben, a supporter of the Center Party and a French citizen involved in separatist concerns, the other a leader of the Center Party and a separatist on the official separatist list), a fight which ended after the intervention of the *Gaubetriebszellenleiter* and the *Gauleiter* in the retirement or removal of both gentlemen from IG."[244] The former of the two gentlemen was probably Martin Rohmer, a member of IG Farben's Executive Board; the identity of the latter could not be established. In 1938 Nicodemus together with two colleagues probably informed on Hermann Landers, so that, as he wrote after the war, he was obliged to give up his "position overnight to escape the Gestapo and a concentration camp."[245]

There were also other successful chemists in addition to Nicodemus among the "turncoats of March" who showed their readiness to cooperate with the Nazi state, whether from conviction or opportunism. One of them was Karl Ferdinand Blumrich, born in 1888 in Reichenberg in Bohemia, who had been an authorized signatory since 1928 and was head of the Analytical Laboratory from 1930. He joined the NSDAP after the seizure of power, became block warden for the chemistry cell group, and "political leader" of the local branch of the NSDAP in Frankfurt-Unterliederbach. In 1937 he was coopted to attend the *Reichsparteitag*. On the occasion of the twenty-fifth anniversary of his employment by the plant he celebrated the event in front of a "Hitler altar."[246]

Even for persons affected by the moratorium on new members there were still opportunities to demonstrate their closeness to the new regime. Works manager Hermann, as mentioned before, was allowed to join despite the ban.[247] A number of employees were in the SA but not in the NSDAP; some of them, including Karl Winnacker, became members of the NSDAP in 1937/38. Winnacker wrote in his memoirs that "together with many of my

[244] Ibid., Nicodemus to *Ortsfachobmann* Schörg, February 26, 1934; report dated February 26, 1934, by the spokesman for the local trade group (*Ortsfachgruppenobmann*) in Hoechst and the county trade group spokesman (*Kreisfachgruppenobmann*) for Greater Frankfurt concerning the previous work in the trade groups.

[245] Ibid., affidavit by Hermann Landers dated October 1, 1946 (certified copy).

[246] HA, PA Abteilung VI, Blumrich, Flach on Blumrich, August 5, 1938; NSDAP Gau Hessen-Nassau, August 24, 1937, and August 27, 1937 (copies); Hoechst to Blumrich, October 29, 1945.

[247] BA, BDC file on Hirschel, One of Hermann's closest associates in the Directorate T, the former Stahlhelm member Otto Hirschel, was also apparently admitted to the NSDAP in April 1936 by special permit, in contrast to his own statement made after the war that he became a Party member only in 1937: see his application for admission to the NSDAP dated February 8, 1935, and his admission on April 1, 1936, membership number 3,775,074; see also HHStAW, Abt. 520 F (A-Z), Hirschel: In the questionnaire for the denazification proceedings (*Meldebogen*) from April 24, 1946, he wrote that he had joined only in December 1937, but after payment of his membership dues the date of his joining the Party had been antedated.

colleagues at the institute" of Professor Ernst Berl's at the Technical College in Darmstadt he had "joined the SA storm troopers in Darmstadt in the spring 1933" and continued to belong to the SA while he was employed in Hoechst until about 1936. Winnacker continues: "Fortunately, because of my fading interest, I did not achieve any distinction in the SA, so that eventually I could quietly take my leave of it."[248] However, Winnacker joined the NSDAP in the fall of 1937.[249] Apparently after pressure by the NSDAP, Lautenschläger and Jähne also joined the Party at the same time: they applied for admission at the end of April 1938, although membership in both cases was antedated to May 1, 1937.[250]

Until the ban on new members was lifted there was another form of ingratiating oneself with the new regime: membership in the SS. In Hoechst several employees were members, the most important of them being director Georg Kränzlein, mentioned above. Schacke, *Betriebsobmann* in 1934/35, has also been mentioned, likewise his expulsion from the SS in 1935. During the period of the ban on new members in the NSDAP, the SS offered an opportunity to join the Party, albeit indirectly. A prominent example, but not the only one, was Hans Streeck. Born in 1905 in Bremerhaven as the son of a captain, Streeck studied chemistry in Jena and Munich, where he completed his doctorate in 1931. Subsequently he worked as a university assistant in Erlangen and later in Leipzig. In June 1934 Streeck joined Hoechst and began working in the main Alizarin Laboratory, that is, in Kränzlein's department. In 1936 he moved to the central testing room to work as an assistant; after that he worked in different dyes works, where he was responsible for both operative and scientific problems. This indicated that he had been earmarked for a leading position. He was able to prove himself after 1942 on the staff of Otto Ambros, a member of the IG Farben Board of Directors, during the planning and construction of the company's plant at Auschwitz.[251] Streeck was in the SA from November 1, 1933, until March 25, 1935. On the latter date he officially resigned from the SA and joined the SS; by the beginning of 1938 he had become SS training officer for the 2nd SS *Standarte* in Hoechst. After the ban on new members had ended, he applied for admission to the NSDAP on August 16, 1937, and became a member as of May 1, 1937.[252]

[248] Winnacker, *Challenging Years*, p. 54.
[249] BA, BDC file on Winnacker, application for admission: September 20, 1937, admission dated May 1, 1937.
[250] BA, BDC file on Lautenschläger, application for admission: April 29, 1938, admission dated May 1, 1937, and on Jähne, application for admission: April 30, 1938, admission dated May 1, 1937; on the pressure exerted at the time on businessmen to join the NSDAP, see Hayes, *Industry and Ideology*, p. 169.
[251] HA, PA Abteilung VI, Streeck, Biographisches, Berufliche Tätigkeit.
[252] BA, BDC file on Streeck, SS membership number 276,220, NSDAP membership number 4,703,829.

The membership of employees of the Hoechst plant in the NSDAP or other Nazi organizations, whether from conviction or opportunism, was only one aspect of nazification. The other side was the exclusion, the "cleansing" of the work of "elements" that were no longer tolerated in the so-called plant community, whether for political or "racial" reasons. The "cleansing" affected not only workers and lower-ranking employees. One of the first victims was a member of the IG Farben Board of Directors. Thus, it was not just the new factory code that no longer made any distinctions between employees and workers – both groups were also equally liable to suffer exclusion or persecution.

When employees were sent into early retirement or were dismissed, it is important to look carefully whether they were political or "racial" victims of persecution or whether they were removed from the plant for other reasons. It would be wrong to assume that everybody sent into early retirement between 1933 and 1945 was a victim of the Nazi regime. While this was indeed often the case, some were also dismissed, furloughed, or retired for other reasons. Victims were for the most part Jews, persons regarded as Jews, or persons who were politically objectionable. In addition there were a number of early retirements for operational or personal reasons. An experienced and highly respected chief engineer was forcibly pensioned off in May 1936 when Hermann regarded his continued presence in the works as insupportable and no longer wanted him in the plant. The person in question had been accumulating debts since 1933 and had repeatedly requested advance payments, which, as he was enjoying the privileges of "Department VI," had usually been given to him on generous conditions. After the letter of dismissal was sent out in May 1936, Hermann personally gave instructions in October of the same year that this person should no longer receive any financial assistance.[253] Another case of early retirement was that of the head of the Patents Department, a deputy director, who had an affair with his secretary and was also considered as not quite "reliable" politically. Forged letters to the works management informed on him, and he was obliged to leave the works in August 1936 from one day to the next.[254]

Among the senior employees, early retirement was quite often for operational reasons, not least because, as Lautenschläger wrote in 1940, there were "too many authorized signatories in different departments" within Hoechst.[255] In April 1935 Wilhelm Stellmann was obliged to surrender his commercial power of attorney, which he had held since 1925 for "reasons of internal organization." In the copy for the personnel file the

[253] HA, PA Abteilung VI, Chief engineer Anton E.   [254] HA, PA Abteilung VI, E. Hübner.
[255] HA, PA Lautenschläger, Sekretariat Lautenschläger, Personalangelegenheiten, Lautenschläger to Schnitzler, February 16, 1940 (on Stellmann).

works manager added in his own hand: "to make room for someone else."
Stellmann finally had to leave on July 1, 1937. "For organizational reasons," the letter of the works management to Selck ran, the plant was
prepared to "dispense with his services already today." He was only officially pensioned off on January 1, 1939. He was not a victim of political
persecution: during the Second World War he was keen to work in the
occupied Soviet territories – he had worked for Hoechst in Russia before the
First World War – and Hoechst approached Krauch on his behalf. Moreover Stellmann traveled several times to Paris in 1942 on behalf of Hoechst
to recruit "foreign workers" for the plant.[256] Deputy director Adolf
Steindorff, who held a senior position in the Pesticides Department, was
also sent into temporary retirement on July 1, 1937. A note to the IG
Personnel Department curtly stated: "The reason for this arrangement is
that after carrying out a reorganization we can do very well without the
services of Dr. Steindorff." Steindorff went into regular retirement only on
September 30, 1940. During the Second World War he worked for the
Ministry of Armaments as an industrial commissioner in Prague, where he
held the rank of lieutenant-colonel – it was therefore unlikely that political
unreliability was the reason for his early retirement.[257] Involuntary early
retirement could also affect declared National Socialists such as a chief
engineer in Griesheim, who strongly resisted his retirement and received the
support of the DAF.[258]

It is very difficult to determine the reasons for early retirements or dismissals if one has not examined the personnel files. But these too may often
conceal as much as they reveal, particularly concerning the treatment of
Jewish employees and employees regarded as Jews. Thus, one must be
careful about drawing conclusions.[259] A number of early retirements and

[256] HA, PA Abteilung VI, Stellmann, Hermann to ZA-Büro, Fft-Grüneburg, April 3, 1935;
Hermann to Stellmann, June 30, 1937; Hermann to Selck, July 5, 1937; Lautenschläger to
Stellmann, November 30, 1938; Hoechst to the Mayor of Königstein/Taunus, March 25,
1939; IG to Krauch, March 16, 1942; Hirschel to Flach, April 10, 1942 and June 23, 1942.

[257] HA, PA Abteilung VI, Steindorff, Hermann to Selck, January 5, 1937; Selck to Hermann,
January 9, 1937; Hermann to Steindorff, January 12, 1937 and June 30, 1937;
Lautenschläger to Pensionskasse IG, August 15, 1940; Steindorff to Hoechst (Flach),
December 30, 1943.

[258] HA, PA Abteilung VI, Hebeler, Becker (DAF) to Schnitzler, August 5, 1938 (copy);
response from Schnitzler, August 25, 1938; Jacobi (Werksleiter Griesheim) to Hebeler,
September 7, 1938; note dated September 8, 1938; ter Meer to Becker, DAF, November 2,
1938; retirement as of December 31, 1938, IG Direktion Hoechst (Lautenschläger) to
Hebeler, December 20, 1938.

[259] Cf. on this point also HA, PA Abteilung VI, Rösner, Labor für Patentsachen: Rösner was
notified by a letter dated June 26, 1933, that as an authorized signatory he would be sent
into early retirement; as of 1934 he received waiting pay until the date of his regular
retirement in 1936. He was probably sent into retirement for operational reasons; in
contrast to this compare PA Department VI, Strewinski, Kalkulations-Abteilung

dismissals were clearly politically and "racially" motivated. While the next chapter will be devoted to victims of "racial" persecution such as Jews and persons considered by law to be Jews, this section will treat politically motivated persecutions.

As head of the cell group Chemistry for the German Association of Technicians and a member of the NSBO since 1933, Nicodemus pursued his aim of getting Martin Rohmer, deputy member of the IG Farben Management Board, dismissed. Martin Rohmer was born in 1878 in Berlin; his father came from the Alsace, and after the First World War Martin Rohmer chose to become a French citizen. He had worked as a chemist at Hoechst since October 1902. During the First World War he was substantially involved setting up the nitric acid and explosives production in Hoechst and was made an authorized signatory in 1916. In 1919 he became a deputy member of the Management Board of Hoechst and in 1925 a deputy member of the Management Board of IG Farben. He retired on November 28, 1933, although he was not due for retirement until January 1, 1939. He died during the war in May 1941 – by which time he had applied for and received German citizenship. But his dismissal and the reasons for it remained in effect even after his death, for Lautenschläger, who was works manager at the time, refused to allow an obituary notice by Hoechst "because consideration must be taken of the political tensions at the time which led to the voluntary retirement of Dr. Rohmer."[260] One should not forget that Rohmer's French citizenship must have been a great advantage for Hoechst in the years from 1919 to 1930, the years of French occupation. However, after the so-called national awakening, this was no longer relevant, and Rohmer could easily be sacrificed. Rohmer was not badly off; he had a lot of money at his disposal and after his dismissal whiled away his time with building projects, described by him as "psychological stopgap work," to console himself for the loss of his work in the laboratory, which he missed[261] – the position that he had filled very successfully up until then that had been taken away from him.

Another senior employee sent into early retirement against his will on December 31, 1938, was Hermann Landers, who had been a supporter of

---

(Accounting Department): on March 7, 1938: he applied for early retirement due to ill health and retired a short time later. As he had apparently abstained from setting up a "Hitler altar" on the occasion of the celebration of his forty years' employment by the company, in his case it is quite possible that political reasons played a not unimportant role in his early retirement, particularly as this occurred in 1938.

[260] HA, PA Rohmer, Biographisches; PA Abteilung VI, Rohmer, Meldebogen, Rohmer to W. vom Rath, August 24, 1939, in the appendix letter from Rohmer to *Landrat* Brunnträger, August 24, 1939, Aktenvermerk by Lautenschläger dated June 23, 1941.

[261] Bayer WA, Personalia 271/2.1 Rohmer, Duisberg to Rohmer, December 21, 1933; Rohmer to ter Meer, November 22, 1935.

the Catholic Center Party during the Weimar Republic and had also held political positions. Landers joined Hoechst in 1907 after studying chemistry in Munich and Strasbourg, where he wrote his dissertation under Professor Johannes Thiele. After the First World War he headed a small department and in 1932, as a so-called *Prominenter*, he became head of the Azo works and the attached departments.[262] Even after 1933 Landers was not infected by the anti-Semitic campaign and gave an impressive eulogy for his old friend Franz Henle, who came from a Jewish family, on the occasion of Henle's enforced retirement. Henle was very moved by this gesture of solidarity and thanked Landers in a letter for this "mark of your time-tested friendship," which "at this moment and in these times" was a "doubly agreeable expression of friendship."[263] In the summer/fall of 1938 some chemists keen to advance their own careers took action against Landers. Allegedly Nicodemus, Schultheiß, and another chemist informed against Landers, so that he was obliged to give up his job on one day's notice.[264] In a directors meeting held in Hoechst at the beginning of October 1938 it was curtly noted that the removal of Landers was now known to his co-workers, and a corresponding circular could now be sent out concerning the reorganization of the factory.[265]

In addition to politically motivated early retirements, dismissals – primarily of workers rather than managers – also occurred. Unfortunately the sources for this part of the history of the plant are quite scarce. Two particularly well-documented cases will serve as examples of the pressures that workers, primarily those members of the workforce who had formerly been organized, were subjected to and how much they were mistrusted in the plant. As the case of the unskilled laborer Wilhelm Binnewies shows, dismissal for political reasons was not exceptional – in 1936 Hoechst explicitly declined to reemploy him on the grounds that this would otherwise create a precedent for others who had also been dismissed for political reasons.

Wilhelm Binnewies, born in 1881, worked for Hoechst from June 1912 until October 10, 1935, as a manual worker. On that day he was arrested in Hoechst by the Frankfurt state police. In a note by Hans Pöhn, head of the factory administration and Defense Commissar, the Social Department was

[262] HA, PA Landers, Biographisches; PA Abteilung VI, Landers, letter on his promotion to the rank of *Prominenter*.
[263] *Die Anfänge der Kunststoffwerkstätte in Hoechst*, Dokumente aus Hoechster Archiven 40 (Frankfurt am Main–Hoechst: Farbwerke Hoechst AG, 1969), pp. 61–4: Henle to Landers, April 3, 1936.
[264] HHStAW, Abt. 520 F (A–Z), Nicodemus, affidavit by Hermann Landers, dated October 1, 1946 (certified copy).
[265] HA, TEA 95e, Niederschrift über die technische Direktionssitzung in Hoechst, October 3, 1938.

informed that the state police had arrested a "number of persons" in Hoechst on October 10 and 11, "who had been active on behalf of the former SPD [Social Democratic Party]." The "worker Wilhelm Binnewies employed in the plant in the Acetic Acid Department" had also been among the arrested persons. It was as yet not known, according to Pöhn, to what extent he was incriminated "as the interrogations were not yet over." Schwamborn added a handwritten comment to this, noting that according to the state police "a longer prison sentence was to be expected."[266] Because of his arrest Binneweis was dismissed from Hoechst in October 1935.[267]

According to the indictment that was completed by the middle of December 1935 the charges leveled against Binnewies and seventeen other men from Frankfurt and the Taunus area were that they had attempted in 1934 "to revive the former Social Democratic Party in Frankfurt upon Main and its environment." Binnewies had belonged to the SPD since 1927; "his Party membership booklet had been found during the search of his apartment." Furthermore, "after initially denying the charges" Binnewies had "essentially confessed."[268] Finally, in February 1936 the criminal division of the Higher Regional Court in Kassel sentenced him "to ten months in prison for aiding in the preparation of a treasonable undertaking."[269] The grounds for his conviction were that in the winter of 1934/35 he had several times subscribed to the illegal magazine *Sozialistische Aktion* (Socialist Action).[270] In a letter to the works management in which he requested his reinstatement Binnewies wrote in 1937 that he had lost his position in the factory because of an affair into which "he had gotten involved without thinking and through no fault of his own": "In 4 cases an illegal magazine was brought to me unbidden to read outside the works." He had not been aware of the consequences of his accepting the magazine – nor had the publication contained any Communist ideas. Moreover he paid for his action by spending ten months in custody.[271] Binnewies spent the term of his imprisonment in Frankfurt, Kassel, and the penal camp of

---

[266] HA, PA Binnewies (microfilm), Fabrikverwaltung to Sozialabteilung, October 14, 1935 – "confidential re W. Binnewies," with handwritten note by Schwamborn dated October 15.

[267] Ibid., Weigand, a lawyer, to IG Farben, Grüneburg, on Binnewies dated September 25, 1936: request to be reinstated after release from prison – "the end of October 1935" is given as the date of Binnewies's release. The records do not give a more precise date.

[268] HHStAW, Abt. 518, Nr. 165, Auszug aus der Anklageschrift, Der Generalstaatsanwalt, Kassel, dated December 18, 1935 (O. Js. 147/35).

[269] Ibid., Oberstaatsanwaltschaft beim OLG für Hessen, Zweigstelle Kassel, Confirmation for Binnewies dated March 8, 1950.

[270] HA, PA, Binnewies (microfilm), Weigand (lawyer) to IG Farben, Grüneburg, September 25, 1936, on Binnewies: It was asserted there that he had subscribed to the newspaper twice – in the letters of Binnewies the number of times varies.

[271] Ibid., Binnewies to Direktion IG Hoechst, June 6, 1937.

Börgermoor (in East Frisia), after which he was released at the beginning of August 1936. He had apparently been severely abused during this time because when he returned, his health had been very much impaired.[272]

After his release Binnewies, who was over fifty years old, wanted to return to his former workplace in Hoechst. His lawyer therefore wrote to IG Farben at the end of September 1936 that Binnewies had been sentenced to ten months' imprisonment for "aiding in the preparation of high treason," had now "served his sentence," and had "acknowledged that he had been at fault." Binnewies was asking to be reemployed to help him "return to the national community [*Volksgemeinschaft*]." According to the reference given by his employer, Hoechst, Binnewies had been a "willing and diligent" worker: "His conduct and work were very good." Even the chief public prosecutor in Kassel could see no reason "to continue to exclude him from the *Volksgemeinschaft*" now that he had completed his term of imprisonment. Furthermore he had worked "for almost the whole period of his life" in Hoechst, and at his age "it would not be possible for him to find work elsewhere."[273]

Hoechst reacted cautiously but not completely negatively to the letter. The Social Department wrote to Binnewies's lawyer at the end of October that the matter had been very thoroughly investigated but unfortunately a reemployment of Binnewies was "out of the question for the time being." The decision had not been easy as Binnewies "was appreciated as being a good worker." However, Hoechst did not currently require additional workers. Another reason was also given that makes it clear that Binnewies was not an isolated case. The letter went on to state: "Regrettably, since dismissals for similar reasons as those with which Mr. B. was charged have repeatedly occurred, it was not possible to come to any other decision so as not to establish a precedent." The future would show, the letter concluded, whether there would be any opening for Binnewies one way or another.[274]

Binnewies did not give up and even went to the local Gestapo, which then demanded that the plant should consult it concerning questions of reemployment. Binnewies's lawyer therefore requested the authorities in Hoechst "to get in touch with the secret state police [Gestapo]." However,

---

[272] HHStAW, Abt. 518, Nr. 165, Regierungspräsident Wiesbaden, Feststellungsbescheid, dated November 28, 1951; Regierungspräsident Wiesbaden, Betreff Schaden an Körper und Gesundheit, dated February 11, 1952; The official notification from the district president of Wiesbaden dated June 15, 1953, did not see it in quite that light. However, the notification did not question the fact of his having been maltreated, but merely doubted whether the abuse had been responsible for certain after-effects.

[273] HA, PA Binnewies (microfilm), Weigand (lawyer) to IG Farben, Personalabteilung, Frankfurt-Grüneburg, dated September 25, 1936, on Binnewies.

[274] Ibid., Sozialabteilung Hoechst to Weigand (lawyer), October 23, 1936, on Binnewies.

works manager Hermann personally took it on himself to decline to do so.[275] At the beginning of June 1937, still trying to be reinstated, Binnewies wrote a personal letter to the works management in Hoechst in which he listed his "faults," described himself as "amenable to reason," expressed his worries about his family, included a reference to his time as a soldier at the front in the First World War, and ended by requesting one more chance. He had repeatedly read in the newspapers, so he wrote, that the NSDAP allowed persons "who had stumbled once" to be given work "in order to bring them back into the national community again." Binnewies, as was clear, was trying desperately to be reinstated – and the DAF was even inclined to consider giving him another chance. Once again, his petition failed because of Hermann's intransigence, as he personally requested the Social Department to refuse "as a matter of principle."[276]

Binnewies was employed again in Hoechst only in the fall of 1940. He had once again requested to be reinstated, as his current employment as a laborer in Sossenheim was due to end soon, but he was only taken back after the Defense Commissar in Hoechst, Pöhn, made inquiries of the local Gestapo as to the advisability of doing so.[277] Binnewies continued to work for Hoechst until his retirement in the spring of 1949.[278]

The case of Binnewies clearly demonstrates the persecution of left-wing workers during the "years of peace." But it also shows that in its behavior toward workers the works management allowed itself to be guided primarily by the current labor market. Hoechst showed itself to be even more intransigent against workers with leftist leanings than the Nazi authorities demanded. The works management was prepared to be accommodating only if it required the workers' services. In the spring of 1937 the works management of Hoechst decided not to hire any new workers "since there is a risk that if there is a shortage of materials we shall be obliged to make persons redundant."[279] In a directors' meeting in Hoechst, held in the summer of 1937 when Binnewies was asking to be reemployed, Hermann reported that the number of workers had "unduly increased recently," so that it would be necessary to be "extremely critical of any new appointments."[280] Under such conditions it was easy to dispense with the

[275] Ibid., Weigand (lawyer) to Hoechst, November 4, 1936, below a handwritten note by Schwamborn: "should be turned down according to Dr. H./consequences."

[276] Ibid., Binnewies to Direktion IG Hoechst, June 6, 1936; handwritten note by Schwamborn dated June 15, 1937: "should be turned down – fundamental decision by Dr. Hermann."

[277] Ibid., Binnewies to Hoechst, September 14, 1940; Pöhn to Gestapo Frankfurt, September 18, 1940; Einstellungsblatt dated October 15, 1940.

[278] Ibid., index card on Binnewies.

[279] HA, TEA 95d, Niederschrift über die technische Direktionssitzung in Hoechst, April 5, 1937.

[280] HA, TEA 95d, Niederschrift über die technische Direktionssitzung in Hoechst, June 30, 1937.

services of someone who had been previously convicted for political reasons, even if he had been a good worker – whereas in 1940 the same person was needed and duly taken on again.

A second case is that of the worker Bernhard Hartmann, born in 1899 and employed by Hoechst since 1918. Hartmann was arrested at his place of work on May 12, 1937, and taken away by the Gestapo.[281] On May 18 he was retroactively dismissed as of May 12 "on the orders of the leader of the plant" because "of the impossibility of continuing the employee-employer relationship."[282] In a reference dated the beginning of December Hoechst confirmed that Hartmann's performance during the almost nineteen years he had worked there had been good: "Mr. Hartmann carried out his work circumspectly and diligently. His conduct was very good."[283] In the meantime Hartmann had been sentenced to two years' imprisonment by the Higher Regional Court in Kassel on September 30, 1937, "for making preparations for a treasonable undertaking." He was released after more than two years' imprisonment in the Frankfurt-Preungesheim jail on June 30, 1939.[284] Hartmann was arrested and sentenced because, according to his own statement, he had "donated 3 Marks in aid of the KPD [Communist Party] at the beginning of 1933."[285]

After serving his sentence he wanted to work in the plant again, and unlike in 1936 and in 1937, Hoechst was in great need of labor in 1939. Thus the Hoechst Social Department wrote to Retzinger as the *Betriebsobmann* and member of the Council of Trust that Hartmann had asked to be reemployed: "With due consideration of the conditions also known to you regarding the obtaining of workers we request you to find out from the DAF whether there would be any objections to a reinstatement of H." They added: "his plant leader described him as a circumspect and hard-working man."[286] After consultation with the DAF Retzinger declared that despite his conviction for political reasons he should "definitely" be given the opportunity "to be reintegrated into working life." However, the proviso was added that Hartmann should not be taken back in his old

---

[281] HHStAW, Abt. 518, Nr. 2514, Antrag Hartmann auf Wiedergutmachung Teil E. dated March 30, 1950.

[282] HA, PA Hartmann (microfilm), Abt. f. Arbeiterangelegenheiten to Dr. Eisenmenger, Zwischenpr. Nord, dated May 18, 1937.

[283] Ibid., Reference from Hoechst for Hartmann dated December 2, 1938.

[284] HHStAW, Abt. 518, Nr. 2514, Oberstaatsanwalt Kassel, confirmation on the sentence passed on September 30, 1937 dated January 20, 1951; certificate of release from Frankfurt-Preungesheim dated June 30, 1939: this states that he was in prison from May 15, 1937, until June 30, 1939.

[285] HA, PA Hartmann (microfilm), Hartmann to Direktion der IG (Hoechst), July 10, 1939.

[286] Ibid., Hartmann to Direktion der IG (Hoechst) dated July 10, 1939, with the request to be reinstated; letter from Sozial-Abteilung Hoechst to Retzinger (Council of Trust) dated July 26, 1939.

department as there "is otherwise the danger that old connections will be revived there."[287] The Council of Trust also decided in principle that persons who had been previously sentenced for political reasons could be reemployed, although the Council was of the opinion that "in every case the unqualified permission and consent of the political authorities and the DAF must be given."[288] This was the reason why Hartmann failed to be reemployed. An enquiry by Hoechst, sent to the *Kreisleitung* of the NSDAP in Frankfurt, which did not hesitate to mention the fact that Hoechst was "urgently" in need of manpower, was turned down: "Vg. [*Volksgenosse*, German comrade] Hartmann is politically unreliable because of his previous conviction."[289]

At the beginning of 1941 Hartmann attempted once more to be hired by Hoechst as he had heard that Hoechst was now reemploying persons who had been convicted for political reasons. Once again he failed. In 1942 Hartmann was drafted into the army, where he served until 1944/45, after which he remained in Soviet captivity until 1948. After his return from captivity he was rehired in Hoechst at the end of June 1948 and worked there until his retirement in 1958.[290]

As the shadow of war loomed, conditions for political dissidents worsened during the so-called Sudeten crisis. Denunciations now tended to have very serious consequences for the persons affected. Hugo Gärtner, who had worked for Hoechst as a chemist since 1923, was just one victim of such a denunciation, together with his family. On October 5, 1938, "plant leader" Lautenschläger sent him a registered letter. According to a communication received from the state police in Frankfurt, Gärtner "was being held in custody because of offenses against the *Heimtücke-Gesetz* [Treachery Act] to await trial." Gärtner was summarily dismissed in accordance with the Factory Code. Furthermore he was expected to observe a contractual restriction on working for any commercial competitor to Hoechst "without any claim for compensation" during the waiting period.[291] This was an exceptionally harsh procedure on the part of the works management, since it did not

---

[287] Ibid., *Betriebsobmann* Retzinger to Sozial-Abteilung Hoechst, dated July 28, 1939, on Hartmann.

[288] Ibid., Schwamborn (Sozialabteilung Hoechst) to authorized signatory Flach, August 3, 1939.

[289] Ibid., Hoechst (Sozialabteilung) to NSDAP Kreisleitung Gross-Frankfurt, July 29, 1939; response from NSDAP Kreisleitung dated August 16, 1939.

[290] Ibid., Hartmann to Hoechst, received January 10, 1941; confirmation of Farbwerke Hoechst U.S. Administration for Hartmann dated March 25, 1950; Belegschaftshilfe Farbwerke Hoechst to Regierungspräsident Wiesbaden, February 23, 1960; HHStAW, Abt. 518, Nr. 2514: Antrag auf Wiedergutmachung (petition for compensation) – part E, dated March 30, 1950.

[291] HA, PA Gärtner (microfilm), Werksleitung Hoechst (Lautenschläger) to Gärtner, October 5, 1938.

just mean that Gärtner was dismissed but also virtually disbarred him from practicing his profession without the compensation that would ordinarily have been paid to ensure that a former employee did not work for a competitor.

Gärtner was arrested on September 30, 1938. According to the "facts of the case" as stated by the Labor Court in Frankfurt at the end of March 1939, he had expressed himself in such a way "in his family circle" that "his rejection of the present state" was obvious. In the exact words of the charge: "During the critical days in September of the previous year, when Chamberlain visited the *Führer* in the Obersalzberg, he [Gärtner] was said to have become upset because 'that bigwig Hitler' had not driven out to meet the old man and furthermore had declared to his family that if war broke out, Hitler would bear the blame."[292] The many subjunctives used in the original German transcript are surprising, as they indicate how much of the accusation rested on hearsay. Gärtner had probably made the comment, but, as was also noted, he had uttered it within his family circle. His comment was spread by his wife among her circle of friends and by their six-year-old son at school. A denunciation brought it to the ears of the Party, who then passed it on to the Gestapo. After three weeks' detention in custody awaiting trial Gärtner was released for lack of proof on October 20, 1938. While proof of his wife's participation in the "deed" was considered to exist, she had not been arrested since she was in the final stages of pregnancy. The case against her was also dropped.[293]

The works management in Hoechst insisted on upholding the summary dismissal. Thereupon Gärtner lodged a complaint against Hoechst at the Labor Court. Things could hardly become any worse for him: his wife was highly pregnant, he had two small children and no job, and he was subject to very severe restrictions on his employment elsewhere. On January 14, 1939, the public prosecutor's office called the works to inform the Personnel Department that Gärtner's lawyer had requested a statement from the prosecuting attorney's office that there were no official objections to his reinstatement. However, the public prosecutor's office declared that such certificates were never given "as a matter of principle" and asked the works management to comment on the case. Director Schwamborn replied that "according to special information" any employment in Hoechst or in IG Farben was out of the question.[294] Two days later Lautenschläger and Schwamborn agreed to inform Gärtner "on the part of the plant about the

[292] Bezirksregierung Düsseldorf, Entschädigungsakte (compensation file) Gärtner, certified copy of the sentence of the Arbeitsgericht Frankfurt am Main dated March 31, 1939, in the proceedings Gärtner against the IG Farben plant Hoechst.
[293] Ibid., certified copy of the sentence of the Arbeitsgericht Frankfurt am Main dated March 31, 1939, in the proceedings Gärtner against the IG Farben plant Hoechst; Dr. Hugo Gärtner, "Verfolgungsvorgang" (description of events) dated December 1, 1953.
[294] HA, PA Gärtner (microfilm), "Akten-Notiz betr. Dr. Gärtner" dated January 16, 1939.

hopelessness of his efforts regarding IG." Wagenheimer was also to be informed "as a precaution." In the words of the note: "Wagenheimer was immediately informed and was asked to obtain the letter from the Gestapo, which had already repeatedly been demanded, and to apply once more to the *Gauleitung* in order that proceedings on the part of the plant can continue as indicated." Wagenheimer, who was responsible for direct contacts with the *Gauleitung* and the Gestapo, was expected to obtain from them a firm justification for Gärtner's summary dismissal; apparently Hoechst wanted to be able to emphasize that it was "acting under binding orders." In the Factory Code summary dismissal for "subversive behavior" was a discretionary provision. The employee in question could be dismissed, but it was not absolutely necessary. Gärtner moreover had been released without being sentenced. Wagenheimer did his duty and contacted the *Gauleitung* on the same day. On the evening of January 16, he was able to pass on a verbal report: "The works management, that is, only Messrs. Prof. Lautenschläger or Schwamborn, can and should tell Mr. Gärtner verbally that the *Gauleitung* objects to his reemployment in Hoechst and in the Maingau plants, but has no objection against Gärtner looking for other employment in non-IG factories."[295]

After Gärtner nevertheless took his case to court – and it is not clear whether Hoechst had indeed passed on the information to him – the works management in its turn demanded a written declaration by the *Gauleitung*, once again passing on its request through Wagenheimer either in writing or verbally. Because of the political importance of the case, the works management wrote, "close contact" to the *Gauleitung* had been maintained from the start and the measures against Gärtner had been merely "in accordance with the instructions given to us." His immediate dismissal had been undertaken "in accordance with the *Gauleitung* and at its request," and the refusal to reemploy Gärtner or to find him another job in one of the other IG Farben plants was based on "our assessment of the offences committed and is supported by the Council of Trust." "Together with the *Gauleitung*, the secret state police [Gestapo], and the representatives of the followers [i.e., the workforce] we regarded such a rehabilitation as intolerable." The letter ended with the words "We would ask the *Gauleitung* to confirm briefly the correctness of these details."[296]

In the proceedings before the Labor Court in Frankfurt the political pressure on the works management – which, however, it had never

[295] Ibid.

[296] Ibid., Führer des Betriebs (IG Werk Hoechst) to NSDAP Gauleitung Frankfurt, March 16, 1939: the letter is not signed and was therefore possibly not sent. However, as it has not been crossed out, labeled as a draft, or removed from the file, it can be assumed that it was either sent or its contents passed on by Wagenheimer to the *Gauleitung*. This would be borne out by the handwritten rectification of the date.

attempted to resist – became glaringly obvious. The defendant company, represented by the lawyer Josef Hirsch from Hoechst's Legal and Personnel Department, demanded that the action be dismissed. Gärtner's immediate dismissal was considered justified on the grounds that the plaintiff, Gärtner, had absented himself from work for four successive days because of his arrest, which was prohibited by the Factory Code. Moreover Gärtner's remarks had been of such a "serious nature" that it would have been impossible to continue to employ him. The works management and plant community agreed in this, as was evinced by a vote of the Council of Trust; their opinion was shared by the *Gauleitung* and the Gestapo, indeed, "the *Gauleitung* had demanded the summary dismissal of the plaintiff." The Hoechst plant, again "in agreement with the *Gauleitung* and at its demand, and buttressed by the negative ruling of the Council of Trust" had refused to reemploy Gärtner after his release from custody or to obtain a different position for him "since it was intolerable for the plant to reemploy a man who had insulted the leader of the German Reich at such a critical moment." Reemploying Gärtner would cause "the greatest discontent" among the other employees in the plant and would have "detrimental effects."[297]

The plaintiff Gärtner pleaded that his being remanded in custody did not provide grounds for an immediate dismissal, particularly as the matter was not pursued any further by either the public prosecutor or the Gestapo. Moreover, the Reich Ministry of Justice and the Gestapo had expressed themselves to the effect that any decision to dismiss him or to reemploy him was solely at the discretion of the works management. However, Gärtner's lawyer was not permitted to argue as freely as he had apparently originally planned, since he "had been induced by the Gestapo to withdraw the most essential points of the charge."[298]

The court, as was to be expected by this time, followed the defendant's line of reasoning. The action was dismissed, although the arguments of the court were more of a political than a legal nature: "Let us leave aside the question of which words or sentences the plaintiff used against the *Führer*; at all events it is certain, based on the criminal records, that he commented in a derogatory manner on the *Führer*." After taking all of Gärtner's comments in the file into consideration together with the statement of "the witness *Betriebsobmann* Retzinger" the court was not convinced that the questionable remarks of the plaintiff did not arise from a negative attitude

---

[297] Bezirksregierung Düsseldorf, Entschädigungsakte (compensation file) Dr. Gärtner: certified copy of the sentence of the Arbeitsgericht Frankfurt am Main dated March 31, 1939, in the proceedings Gärtner against the IG Farben plant Hoechst.

[298] Ibid., certified copy of the sentence of the Arbeitsgericht Frankfurt am Main dated March 31, 1939, in the proceedings Gärtner against the IG Farben plant Hoechst; quotation from Gärtner, "Verfolgungsvorgang," December 1, 1953.

toward the Nazi regime. This alone would be sufficient to justify an immediate dismissal as such behavior was "subversive" and it was "irrelevant" whether the subversive behavior had been "carried out in public or not." The Labor Court considered Gärtner's behavior to be "particularly serious" – and therefore implicitly criminal – as he had made the remarks in question during the Sudeten crisis when every German should have rallied "together with the other *Volksgenossen*" to back Hitler and "thereby strengthen his [Hitler's] position against the enemy." The court particularly emphasized that the plaintiff Gärtner was in no way exonerated by the fact that his remarks had been made in private and could therefore not be prosecuted: "Once his remarks had become public, irrespective of how, he must be held responsible for this fact, which additionally aggravates the charge. Nor is behavior which is liable to prosecution a requirement for dismissal without notice." In plain words, even if nothing could be proved against someone and he could therefore not be sentenced, it was still possible to sanction him in industrial law. *In dubio pro reo* certainly no longer applied; political will took precedence over law. At the end of the reasons given for judgment the court stated that "political instances" had also involved themselves in the matter, and they had considered a dismissal of Gärtner without notice as "necessary," as was borne out by the "statement of the witnesses and the confirmation of the Party *Gauleitung* presented by the defendant." For these reasons "the dismissal of the plaintiff without notice was considered to be justified."[299]

There was no longer even an attempt to clothe the affair in an appearance of legality; its political dimension was explicitly emphasized. It also became clear how important informal communications between the works and the *Gauleitung* and the Gestapo were – and how important Wagenheimer's position was – and how willing the works management, Lautenschläger and Schwamborn, were to conform. For they yielded to a demand of the *Gauleitung*, which at first had only been expressed verbally, even if in the course of the proceedings this was followed up by a written confirmation.

Gärtner was not the only case. But because of the publicity of the case in the plant – and this became very obvious during the proceedings – it was enough to silence many people. The paeans of loyalty – of which there are many in the sources – must also be read keeping in mind the employees' justified anxiety concerning their own and their families' livelihoods. It is difficult to distinguish between compulsion and consent here. Even persons who only voiced criticisms in private could – as Gärtner's case had shown – endanger their very existence. And sanctions could be brought to bear

---

[299] Ibid., certified copy of the sentence of the Arbeitsgericht Frankfurt on Main dated March 31, 1939, in the proceedings Gärtner against the IG Farben plant Hoechst.

against their families, which might even include the removal of their children. Such a case clearly demonstrates how intimidating such a threat can be. The few weeks he spent in custody and his dismissal from Hoechst made Gärtner a particularly silent citizen of Nazi Germany, for he was unlikely to endanger his family and his existence again – the events and their emotional strains had made his pregnant wife seriously ill.[300]

It was not just in its treatment of Gärtner that Hoechst showed itself to be particularly unforgiving. Hoechst even provided detailed information about the events to another company where Gärtner had applied for a job, with the injunction to treat this information as strictly confidential and without conceding "any liability for our information in principle."[301] Although Gärtner did succeed in getting a job in Wuppertal in the spring of 1939, he had the threat of his past as an obvious "enemy of the state" hanging over him like a sword of Damocles and was aware that he could come under suspicion again at any time. Gärtner continued to work there until 1952, when he returned to Hoechst.[302]

At the same time as these charges were being brought against Gärtner, a young woman was dismissed who had been hired but who had not been allowed to take up her position. Emilie Thoennissen was the daughter of an active officer, lieutenant-colonel Max Thoennissen. In December 1938, after successfully completing her degree at the interpreter institute of the University of Heidelberg, she applied for a job in Hoechst, where, after being invited for a job interview, she was taken on by the Patents Department and was due to start work in the middle of January 1939. On January 5, 1939, she was sent a letter stating that "her services which would have started on the 10th of the month are not required."[303] After Emilie Thoennissen wrote in mid-January to enquire whether she could apply for a job in another of the IG Farben works, the Personnel Department answered after consultation with Wagenheimer that her records would not be passed on. The letter, as dictated by Wagenheimer, continued: "Our rejection to your application was carried out at the urging of the *Gauleitung* of Frankfurt. We are unable to give you any reasons for this."[304] The solution can be found in the file on Gärtner, for on January 19, 1939, after Wagenheimer had visited the *Gauleitung* to discuss the Gärtner case, he

[300] Ibid., petition of Gärtner to the restitution court of Frankfurt dated May 28, 1952.

[301] HA, PA Gärtner (microfilm), Hoechst (Schwamborn) to Firma Wolff & Co, Walsrode, February 16, 1939.

[302] HA, PA Gärtner (microfilm), correspondence and a number of notes of 1952.

[303] HA, PA Thoennissen (microfilm), Thoennissen to Personalabteilung IG Hoechst, December 19, 1938; Hoechst to Thoennissen, December 21, 1938; Bewerbungs-Fragebogen für kaufmännische Angestellte dated December 28, 1938 with appendices; Hoechst to Thoennissen, December 28, 1938; Hoechst to Thoennissen, January 5, 1939.

[304] Ibid., Thoennissen to Hoechst, January 18, 1939, with handwritten note by Schwamborn on consultation with Wagenheimer; Hoechst to Thoennissen, January 19, 1939.

subsequently verbally informed the works management that in the matter of Emilie Thoennissen the *Gauleitung* had authorized Schwamborn "to explain to the applicant or her father that the *Gauleitung* forbade the employment of Miss Th. in the Hoechst plant. Reasons: contempt and openly expressed negative attitude to the Party."[305] Even a letter that Thoennissen wrote to Hoechst at the end of January and in which she referred to differences between herself and the local NSDAP, which had in the meantime been cleared up, did not change Hoechst's attitude.[306]

These examples clearly show that dismissals were increasingly politically motivated and arbitrary. It was not even necessary to do anything wrong; sometimes a suspicion, a denunciation, or even a remark made in private sufficed. It was not necessary to politically oppose the regime; not being a convinced supporter or even the suspicion of a lack of dedication was enough. As Robert Gellately wrote, the willingness of "ordinary citizens" to collaborate with the regime and to inform on others made life difficult and usually stifled the will to offer resistance.[307] Gärtner's case also confirms Gellately's comments on the Gestapo: it was willing to act on the slightest suspicions, and since the evidence often was not sufficient, many, indeed most, cases did not land in court. Yet, as Gellately correctly remarks, the Gestapo was by no means "inefficient."[308] As the case of Gärtner has shown, being taken into temporary custody was sufficient to tremendously endanger a person's material existence. Although he had been dismissed, Gärtner was still considered an "enemy of the state" – to be sanctioned by forbidding him to work in the area and whose past was passed on to potential new employers.

The historian Michael Schneider wrote that "force and terror" were of "central importance" for the exercise of power by the Nazis. Opponents of the regime were brutally persecuted, and a climate of fear was created by information about the modus operandi of the Gestapo and the existence and the horror of concentration camps: "Furthermore, the fear of spying and informing increased the distrust against neighbors and colleagues, so that people avoided making critical statements in public and only, if at all, said something within the circle of their most intimate friends. Thus the direct terror, to which a minority of the population was exposed, created a climate of repression also for the large majority which aroused feelings of intimidation and fear." Yet Schneider added that it would be wrong to "attribute the readiness of the broad mass of the population to follow to a

---

[305] HA, PA Gärtner (microfilm), "Akten-Notiz betr. Dr. Gärtner" dated January 16, 1939. This also includes comments on the Thoenissen case, which was passed on only by word of mouth but not in writing as was also the case with Gärtner.

[306] HA, PA Thoennissen (microfilm), Thoennissen to Hoechst, January 25, 1939; handwritten notes dated January 30 and February 8, 1939.

[307] Gellately, *Backing Hitler*, pp. 261–2.   [308] Ibid., p. 201, see also pp. 188–91.

fear of reprisals alone." Of course, fear of political persecution and denunciations existed: "But equally in evidence were the testimonies of – more or less – enthusiastic approval about specific areas of National Socialist politics and in particular support of the '*Führer*.' "[309] As we have seen, this picture largely corresponds to the microcosm found in the Hoechst plant.

3.6. NOT A "NATIONAL COMRADE": JEWISH AND PARTLY JEWISH EMPLOYEES

After 1933, employees who were Jewish or regarded as Jewish were subjected to particular ostracism, particularly in the wake of the Nuremberg Laws passed in 1935. The historian Peter Gay, born in Berlin as Peter Joachim Fröhlich and brought up from childhood as a convinced atheist, described the effects of the National Socialist seizure of power on himself and his family in his memoirs: "There are three ways of becoming a Jew: by birth, by conversion, by decree. Brushed by only a breath of the first, I was forcibly enlisted in the third group after January 30, 1933."[310] At the time few would have thought to find themselves as outcasts again, as most would have shared the belief of Gay and his family: "But we *were* Germans; the gangsters who had taken control of the country were not Germany – *we* were."[311] Like Peter Gay and his parents many of the Jewish employees or employees from Jewish families could not have imagined what was in store for them – whether they were atheists, had converted to Christianity, or were professing Jews: "But Hitler's threats were so utterly implausible that we regarded them as unreliable guides to future conduct. They were literally incredible. Germany, after all, was the most civilized of countries; it was the country that, next to the United States, was the haven of choice for Eastern European Jewish emigrants looking for a tolerant society relatively free of anti-Semitism."[312] But the time when the relationship between the Jewish and non-Jewish population in Germany continually improved had disappeared, once and for all, on January 30, 1933.

With Hitler's seizure of power in 1933 began the exclusion, deprivation of rights, and persecution of Jewish citizens or those whom Nazi Germany regarded as Jews.[313] In the first months of 1933 vicious anti-Semitic attacks occurred. In the fall of 1935 Jews and so-called "Jews by decree"

---

[309] Schneider, *Unterm Hakenkreuz*, pp. 480–1.

[310] Peter Gay, *My German Question: Growing Up in Nazi Berlin* (New Haven, CT: Yale University Press, 1998), p. 48.

[311] Ibid., p. 111.   [312] Ibid., p. 112.

[313] Cf. on this point in detail, Saul Friedländer, *Nazi Germany and the Jews, vol. 1: The Years of Persecution, 1933–1939* (New York: HarperCollins, 1997); at a glance, Michael Burleigh, *The Third Reich: A New History* (New York: Hill & Wang, 2000), pp. 279–342.

(*Geltungsjuden*) were excluded from the National Socialist "national community" (*Volksgemeinschaft*) by the Nuremberg Laws. In the first supplementary decree to the Reich Citizenship Law "all persons who had at least three full Jewish grandparents, or who had two Jewish grandparents and were married to a Jewish spouse or belonged to the Jewish religion at the time of the law's publication, or who entered into such commitments at a later date" were regarded as Jews.[314] On the so-called *Reichskristallnacht* or "Night of Broken Glass," the night of November 9 to November 10, 1938, all restraints toward Jewish citizens were finally abandoned. The property of Jews was not just destroyed and their synagogues burned down: Jews were degraded, beaten, some murdered; several thousand Jewish men, referred to as "Action Jews" (*Aktionsjuden*) were arrested and thrown into concentration camps. Those who were still able to emigrate tried to do so. Others chose suicide. Some Jews in so-called "mixed marriages," that is, who were married to an "Aryan" partner, hoped to sit out the events and remained in Germany.[315]

A number of Nazis considered IG Farben to be a "Jewish" company, not least because the Supervisory Board included some prominent Jews or people from Jewish families. In 1931 and in the first half of 1932 the Nazi newspaper *Völkischer Beobachter* carried repeated attacks against IG Farben, describing it as "un-German" and referring to it as "Moloch IG," which was in the hands of the "international capital" and the "Jew Warburg."[316] In the preface to the new edition of his book on IG Farben in the Third Reich, published in 2001, Peter Hayes wrote that he had overlooked an "anomaly" in the first edition of 1987: although IG Farben strove to protect many of its Jewish employees and the Jewish members of its Supervisory Board after 1933, it had elected no Jews to its Management Board in the eight years prior to the National Socialist seizure of power, and their number had fallen from six to zero.[317] If, on the other hand, one looks at the Supervisory Board of IG Farben, then it is apparent that Jews or persons descended from Jewish families were allowed to remain on the Supervisory Board until 1938, and in 1935 it was still possible for Richard Merton, a man with four Jewish grandparents, to become a member of the Supervisory Board. In 1938 there was even a danger of IG Farben becoming classified as a "Jewish" company, because a law passed on January 4, 1938, stated that a company would be considered "Jewish" if it was "dominated by Jews," in other words, if it belonged to Jews or if a quarter of the

[314] Friedländer, *Nazi Germany*, p. 149.
[315] Ibid., pp. 269–305; Burleigh, *Third Reich*, pp. 325–37.
[316] Hayes, *Industry and Ideology*, pp. 65–6.
[317] Ibid., p. ix; cf. also Peter Hayes, "Big Business and 'Aryanization' in Germany, 1933–1939," *Jahrbuch für Antisemitismusforschung* 3 (1994), p. 256, where the case of Curt Hans Meyer is mentioned explicitly.

Supervisory Board were Jewish. In July 1938 a single Jewish member on a Supervisory Board sufficed for a company to be regarded as a "Jewish enterprise."[318] At the beginning of 1938 IG Farben still had five Jewish members on its Supervisory Board: Otto Mendelssohn-Bartholdy, Richard Merton, Ernst von Simson, Arthur von Weinberg, and Carl von Weinberg. All were obliged to resign at the end of May 1938 to avoid the threat of the company's being classified as a "Jewish enterprise." The Central Committee of IG Farben, which was responsible for all important personnel decisions, had already decided on April 25, 1938, to dismiss all "non-Aryan" employees.[319]

In his study of IG Farben in the Nazi period Hayes wrote that Carl Bosch, the chairman of the Management Board, opposed the anti-Semitic policies of the Nazi regime and made this clear in a conversation with Hitler.[320] After his intervention with Hitler had borne no fruit, IG Farben, according to Hayes, began "regretfully" to adapt to the new conditions in Nazi Germany: it began to transfer Jewish employees abroad where this was possible; it left the Jewish members of its Supervisory Board in their positions and in 1935 even elected a Jew to the board; it tried to provide some means of support to Jewish scientists, both in Germany and abroad; and it publicly praised the achievements of Jewish employees such as Arthur von Weinberg or Fritz Haber. At the same time, according to Hayes, it tried to downplay the racism of the Nazis while attempting to offset its reputation as a "Jewish" company at home.[321] In a book on BASF and its position in IG Farben written in 2002, Raymond Stokes noted: "There is little evidence that the IG took advantage of these policies to gain the property of dispossessed Jewish owners or to get rid of Jewish employees. Most of the Jews who had been in positions of responsibility at the IG had been relocated to foreign subsidiaries before 1938."[322] However, according to an article by Peter Löhnert and Manfred Grill, the behavior of IG Farben toward its Jewish employees and the employees married to Jews in the company's plant at Wolfen was more complex and more ambivalent.[323]

---

[318] Martin Fiedler, "Die 'Arisierung' der Wirtschaftselite. Ausmass und Verlauf der Verdrängung der jüdischen Vorstands- und Aufsichtsratsmitglieder in deutschen Aktiengesellschaften (1933–1938)," in *"Arisierung" im Nationalsozialismus. Volksgemeinschaft, Raub und Gedächtnis, herausgegeben im Auftrag des Fritz Bauer Instituts von Irmtrud Wojak und Peter Hayes* (Frankfurt am Main: Campus, 2000), pp. 61–3; Hayes, *Industry and Ideology*, pp. 196–8; Plumpe, *I.G. Farbenindustrie*, p. 695.

[319] Peter Löhnert and Manfred Gill, "The Relationship of I.G. Farben's Agfa Filmfabrik Wolfen to Its Jewish Scientists and to Scientists Married to Jews, 1933–1939," in John E. Lesch (ed.), *The German Chemical Industry in the Twentieth Century* (Dordrecht: Kluwer Academic Publishers, 2000), p. 134; cf. Plumpe, *I.G. Farbenindustrie*, pp. 695–6; Stokes, "IG Farben Fusion," pp. 291–2.

[320] Hayes, *Industry and Ideology*, pp. 92–3.

[321] Ibid., pp. 91–4; cf. also Plumpe, *I.G. Farbenindustrie*, p. 695, also n. 27.

[322] Stokes, "IG Farben Fusion," p. 291.   [323] Löhnert and Gill, "The Relationship."

This chapter will examine the behavior of the Hoechst plant and its executives toward Jewish employees and those considered to be Jews as well as toward so-called "persons of mixed race" and what the measures undertaken against them in Nazi Germany implied. It will become clear that there was no uniform policy toward them in the plant. While some employees who were Jewish or were regarded as Jews had been forced to take early retirement even prior to the Nuremberg Laws, a so-called "full Jew" was allowed to remain at the head of a laboratory until 1938. Expediency as well as personal sympathy characterized the different behavior exhibited toward Jews – but in the end the interests of the company always prevailed.

The first forced early retirement in Hoechst of a senior Jewish employee or one who was considered to be a Jew was that of Louis Benda, born in 1873 in Fürth in Bavaria but of Swiss nationality. As regards his religion the Hoechst personnel files contain only the comment, added at a later date, "Jew."[324] According to the personnel files of the University of Frankfurt, Benda was Jewish and nondenominational.[325] After studying at the Swiss Federal Institute of Technology in Zurich, Benda began working for Cassella in 1899, which formed an alliance with Hoechst in 1904 under the management of Arthur and Carl von Weinberg and became part of IG Farben in 1925. In the course of the reorganization of different lines of production, Benda was transferred to Hoechst in 1931. In Cassella, Benda, a close colleague of the famous chemist Paul Ehrlich, had been head of the Pharmaceutical Department and a director. He expected to hold an equivalent position in Hoechst, also on the basis of his prior achievements. However, in 1932 it was already being debated whether he should be sent into early retirement as he would turn sixty in 1933. In a discussion held in July 1932 between Paul Duden, the outgoing works manager of Hoechst, and Ludwig Hermann, the future works manager, Benda was classified as "dispensable." This was despite the fact that Duden had informed Benda that he would be able to remain, unless illness, an inability to work, or statutory retirement age required his removal. In December 1932 Hermann and Lautenschläger finally agreed to talk to Arthur von Weinberg and Benda about his retirement; according to Lautenschläger Carl von Weinberg had already agreed.[326]

In June 1933 Hermann and Benda discussed the details of his retirement. Hermann and Lautenschläger assured Benda that they found it "painful" to

---

[324] HA, PA Abteilung VI, Benda, [personnel file]; on Benda see also Ritter, Louis Benda 1873–1945, in *Chemische Berichte* 90 (1957), pp. I–XIII.

[325] Renate Heuer and Siegbert Wolf (eds.), *Die Juden der Frankfurter Universität*, unter Mitarbeit von Holger Kiehnel und Barbara Seib, mit einem Vorwort von Notker Hammerstein (Frankfurt am Main: Campus, 1997), pp. 31–2.

[326] HA, PA Abteilung VI, Benda, Benda to Duden, January 23, 1931; several short notes on discussions concerning Benda on a single piece of paper.

have to do without "one of our most capable employees" and that they were obliged to inform him that his work in Hoechst "would have to be terminated earlier than we suspected only a short time ago." This was a heavy blow to Benda. In the past two years he had agreed to take a massive cut in salary as he had been assured that he would be able to continue to work for IG Farben for some years to come.[327] Heinrich Hörlein, the manager of the company's plant at Elberfeld and the leading man in pharmaceutical research in IG Farben, took up Benda's cause. He pressed the management that Benda should be accorded "the most benevolent treatment," a request that Hermann tried to fulfill financially and that was approved by the Central Committee on July 17, 1933.[328]

For Benda retirement was nevertheless hard to bear. In December 1932 he had, as he wrote to the works management, agreed to a cut in salary, as von Weinberg and Hörlein had persuaded him that the most important thing was "to remain in the service of the company for many years to come." Now the management had waited only until his sixtieth birthday to dismiss him. "What this means to me emotionally during the heavy shocks of this year: the loss of my academic position because of my not being Aryan, the almost daily insults which, even if they are not directed personally at me, nevertheless strike me; the complete inner readjustment – (it is not easy if one knows that one is regarded as a member of an inferior race by other people), the dismissal by IG at the same time even if not for political reasons, I don't want to waste any more words on this; I do believe, though, that IG should have taken these circumstances into account when it made its decision."[329] However, Hoechst and IG Farben did not take the new circumstances into account, and Benda was obliged to take his retirement. How far Benda's Jewish origin played a role in his being forced into retirement is difficult to say – it cannot be ruled out.

Benda continued to live in Franfurt until April 1939, after which the "Swiss and Jew" as he was now termed, emigrated to Switzerland although he thereby lost a large part of his income and his fortune. According to calculations made after the war he lost almost 80 percent of his savings by the move.[330] Once again it was only due to Hörlein's efforts that any money

[327] Ibid., several short notes on discussions concerning Benda on a single piece of paper; Hermann and Lautenschläger to Benda, June 26, 1933. Benda to Duden, July 4, 1931.

[328] Ibid., Hermann to Benda, June 26, 1933; Benda to Direktion der IG in Hoechst, July 4, 1933; Hermann to Fritz ter Meer, July 14, 1933; Duisberg to Hermann, July 20, 1933; Hermann to Benda, July 24, 1933.

[329] Ibid., Benda to Hermann, October 21, 1933, with an exposé on his retirement.

[330] HHStAW, Entschädigungsakten (compensation files) Benda, Regierungspräsident, Entschädigungsbehörde: draft of a preliminary notification, dated April 26, 1956, to the petition of Benda's heirs dated May 20, 1951.

at all was transferred to Switzerland.[331] Lautenschläger had already expressed "very considerable doubts" that he voiced "in consideration of the present situation" against the wording of a letter of the Personnel Department written to Benda in August 1938: "Your pension has been fixed at a specific amount and to a large part – almost half – it is in recognition of your merits with regard to the invention of important drugs, which required a number of years until they were fully developed and which have remained sources of income to us and still remain so after your retirement."[332] Apparently Lautenschläger thought that a "Jew" no longer merited such acknowledgments. Benda died in Switzerland in July 1945; in his last years he had fallen seriously ill.[333]

The Hoechst works management, supported by the IG Farben Central Committee, had thus obliged its first senior Jewish employee to go into retirement in the summer of 1933. The next two forced early retirements of senior employees occurred apparently after pressure was brought to bear by Georg Kränzlein.[334] At Kränzlein's urging Hermann sent two very capable employees into retirement. Both were Jews who had converted to Protestantism and who lived in so-called "mixed marriages," that is, their wives were "Aryan." They were sent into compulsory retirement already in the spring/summer of 1935, before the Nuremberg Laws came into force.

Erwin Hoffa, born in Frankfurt upon Main in 1875, had worked for Hoechst since 1908. He was pensioned off on July 1, 1935, although his retirement in March 1935 anticipated his sixtieth birthday. A letter, signed by Hermann, from April 1935 thanked him for his "long-standing and successful services" and wished him "many happy years in full possession of his faculties of mind and body in his well-deserved retirement."[335] Hoffa was reluctant to go into retirement and did so only after pressure had been brought to bear on him by the works management. His reluctance to retire was compounded by a heavy cut in income, and he wanted to be sure that his wife and small daughter would be well cared for. After his retirement Hoffa moved to Munich, where he lived through the "Night of Broken Glass." After this experience he requested an interview with the factory management, and at the beginning of December 1938 he went to Frankfurt to find out from his old company whether he could leave the country and

[331] HA, PA Abteilung VI, Benda, correspondence concerning payments made to Benda at Hörlein's instigation, 1939–41.
[332] Ibid., Hoechst to Benda, August 13, 1938, with a handwritten note on Lautenschläger's "very serious misgivings" "against this version," i.e., against a letter that so clearly paid tribute to Benda's achievements.
[333] Ibid., Ritter (1957), p. I.
[334] Cf. HA, PA Abteilung VI, Hoffa, Vorsitzender des Entschädigungsausschusses to Hoechst, July 27, 1955.
[335] Ibid., Hermann to Selck, March 18, 1935; Hermann to Hoffa, April 25, 1935.

would be able to transfer his pension and the payments due to him during the time he was contractually restricted from working for a competitor. As a "non-Aryan Christian," to use his own words, he felt himself to be in a difficult position – while he was a Protestant, his wife and his eight-year-old "little daughter" were Catholic. After consulting the head office the management of Hoechst denied the possibility of funds being transferred.[336] In April 1939 Hoffa therefore wrote personally to Carl Bosch, requesting him to use his influence on Hoffa's behalf. He described his personal circumstances, stressed his achievements for the company, and emphasized his service in the First World War. Bosch did not answer the letter himself, but left this to Fritz ter Meer, who briefly wrote in reply that the company was not able to help him. Hoffa then wrote to ter Meer directly to enlist his help so that he would be able to emigrate. Ter Meer only repeated "with regret" what he had written in his previous letter – that the company was unable to help him.[337]

IG Farben not only refused to help Hoffa, it also tried to prevent his last attempt to get away from Nazi Germany. On September 4, 1940, the Economic Group for Chemistry wrote to Hoechst in a letter marked "secret" that Hoffa wished to go to Japan to work for a chemical plant in Niihama that was part of the war effort and was therefore answerable to the Japanese War Department. Hoffa had asked to be released from his time of competitive restraint. On the next day, September 5, Kränzlein replied that "at all costs" an objection must be raised against Hoffa's emigration to Japan. "Quite apart from the fact that the aforementioned person is not Aryan – he was obliged to leave our service for this reason in 1935 – he is extremely well informed about the synthesis and production of Azo dyes."[338]

Hoffa was not permitted to leave Germany and was compelled to do forced labor in Munich until the end of the war. He survived, not least

---

[336] Ibid., Hoffa to Hoechst, December 5, 1938; Hoffa to Hoechst, December 10, 1938, with handwritten notes on questions that needed to be discussed (divorce, the company's assistance in emigrating, etc.); correspondence inside IG Farben, then Hoechst to Hoffa, January 3, 1939; see also Bayerisches Hauptstaatsarchiv München, EG 34598/I/8636 Entschädigungsakte (compensation file) Hoffa: Hoffa's petition on the basis of the indemnification law for national socialist injustices (*Entschädigungsgesetz*), dated December 7, 1949, with appendices; Hoffa's petition to be recognized as a victim of Nazi persecution at the Landesentschädigungsamt (regional indemnification office) in Munich, dated February 3, 1953; decision of the Landgericht München I, 1. Entschädigungskammer dated September 2, 1954.

[337] HA, PA Abteilung VI, Hoffa, Hoffa to Bosch, April 29, 1939; ter Meer to Hoffa, June 17, 1939; Hoffa to ter Meer, June 20, 1939; ter Meer to Hoffa, June 26, 1939.

[338] Ibid., Wirtschaftsgruppe Chemie to Hoechst, September 4, 1940; Hoechst to Wirtschaftsgruppe Chemie, September 5, 1940; Wirtschaftsgruppe Chemie to Hoechst, September 10, 1940; as a "non-Aryan" Hoffa would not have met the criteria of the Japanese Ministry of War either.

thanks to his wife who stood by him and did not leave him. What he had gone through is hinted at in a letter that he sent to the new factory management of Hoechst on July 30, 1945. He asked about being employed in Hoechst or abroad. He considered himself still physically capable of such employment for, as he wrote: "I don't seem to be in too bad a way because I was able to endure the hard physical work which the late government expected of me up until the end; this was combined with the severest strain on my nerves, because without wishing to look back or to go too much into humiliating details, I would like to add that it takes a lot to cope with daily new threats against one's own person and family."[339] And he considered the possibility of going abroad and leaving Munich behind: "It would not be any great sacrifice for me to live abroad, and for my family it would be very desirable to find a place containing fewer memories of the starkness of the last years." His desire to be employed by Hoechst and to move away from Munich was not fulfilled. He died at a very advanced age in Munich in July 1967.

Franz Henle, born in Munich, was forced into retirement at almost the same time as Hoffa. Henle, born in 1876, came from a prominent Jewish family. His father, who had converted to Protestantism, had been a lieutenant-colonel in Bavaria.[340] After studying in Munich and Strasbourg where he had completed his *Habilitation* under Professor Johannes Thiele, Henle joined Hoechst in 1907. In December 1906 Gustav von Brüning wrote to Thiele in Strasbourg that Henle had been given a job in Hoechst: "he makes a pleasant, astute impression, and I think that we have made a good acquisition with him."[341]

Shortly after Henle began working in Hoechst, he married Helene Vogt, the daughter of one of the most important civil servants in the Alsace. In Hoechst he switched to the Free Church while his wife and two sons remained in the established regional Protestant church. One of his sons was an officer in the *Reichswehr:* he was dismissed in 1934, even before his father, because of being a "half Jew." Henle obviously was very popular in the plant; the works management and Hermann himself attempted to protect him and wished to continue employing him. But in 1935 there was neither the will nor, perhaps, the wish to withstand pressure from Kränzlein, the NSDAP, or its representatives in the plant.[342]

To ensure that Henle still had some means of subsistence, he – like Benda and Hoffa – was not dismissed but sent into early retirement. On June 27,

---

[339] Ibid., Hoffa to Hoechst (Bockmühl), July 30, 1945.
[340] On Henle, see *Neue Deutsche Biographie* 8 (1969), pp. 530–1.
[341] HA, PA Abteilung VI, Landers, letter from G. v. Brüning to Thiele, December 5, 1906.
[342] See Waltraud Beck, Josef Fenzl, and Helga Krohn, *Juden in Hoechst. Die vergessenen Nachbarn* (Frankfurt am Main: Jüdisches Museum, 1990), pp. 47–8; oral communication by Oskar Henle, May 10, 2002.

1935, Hermann informed Henle, apparently in person, that he would be retired sometime during the first quarter of 1936 and that already prior to this date he would no longer be required to come to work but would continue to receive his salary.[343] At the end of March 1936 Hermann Landers, his direct supervisor with whom he had studied in Strasbourg, held a dignified farewell celebration for Henle. Landers stressed Henle's achievements as a chemist and referred to his service as a soldier in the First World War. In Landers's own words: "How he was still able to look after his employees in addition to the abundance of management tasks described previously and also find time to work personally on a large number of problems must remain his secret. I only can say: I view this achievement with honest admiration." Landers did not hesitate to add that Henle's work showed "that in him was to be found an excellent synthesis of the worker of the intellect and the worker of the hands."[344] Henle was very much moved by these words, which by then were far from customary when sending off "non-Aryan" employees. In a letter to Landers Henle cordially thanked him for this "testimony" to their friendship and regretted that the time of their collaboration had ended. However, he looked forward to the fact that "our personal relations – hopefully – need not end completely."[345]

At the beginning Henle was allowed to stay in his company apartment in Frankfurt-Hoechst. After 1940 he was forced to sweep the streets and work in the cemetery.[346] During the Second World War a decree of the *Führer* of August 30, 1941, ordered that his son Karl be treated as the equivalent of a "person of German blood with all rights and duties arising therefrom" and reinstated him in the army as an officer on September 1, 1941. However, Karl, by then a captain, was killed in August 1942 on the Eastern Front.[347] Despite the fact that Henle's son had been an officer who had been killed fighting for *Führer* and Fatherland, the harassments against Henle continued to increase, and the family was obliged to move, as Henle's former supervisor Kränzlein had long demanded that Jews should no longer be allowed to occupy company apartments. By permission of the local Gestapo Henle and his wife were allowed to eat at the kitchens of the NSDAP Public Welfare Organization. However, Henle was arrested for this on April 1, 1944. The Gestapo officer who had given the permission could not be reached, and Henle was to be held in custody until he could be found. After the war his widow reported: "After receiving this news my husband went to the toilet and there took the cyanide capsule he had always carried with him

[343] HA, PA Henle, handwritten note by Hermann, dated June 27, 1935, on a conversation of the same day with Henle.
[344] Ibid., speech of Landers, March 27, 1936 (copy, signed by Landers).
[345] *Die Anfänge der Kunststoffwerkstätte*, pp. 61–64, Henle to Landers, April 3, 1936.
[346] Beck et al., *Juden in Hoechst*, p. 48.
[347] Private files Henle, Urkunden; oral communication by Oskar Henle, May 10, 2002.

during the past months. I saw my husband return from the toilet, fall down and immediately lose consciousness. After one hour the doctor who had been summoned gave the cause of death as poisoning."[348] On the burial certificate the cause of death was recorded as "probably poisoning," and it was noted: "Police case, no funeral ceremony!"[349]

While these cases of compulsory retirement occurred prior to the Nuremberg Laws of 1935, the year 1938 signified the end of professional life in IG Farben and thus in Hoechst for Jewish employees or employees regarded as Jews. The protocol of the meeting of the Technical Directorate in Frankfurt-Hoechst on July 13, 1938, records: "The dismissal of Jews from the plant was discussed."[350]

The situation for Jews in Germany did not just worsen, their chances of emigrating also decreased. In the protocol of the meeting of the Technical Directorate in Frankfurt-Hoechst on January 9, 1939, it is recorded that works manager Lautenschläger reported on the meeting of the IG Farben Management Board held on December 16, 1938, during which the "Jewish question" had also been discussed. It would be necessary "to determine in which cases of dismissed Jews it would be possible to dispense with the prohibition on emigration."[351] This does not sound like IG Farben wished to offer support to its Jewish employees, and Hoffa's case has already shown that he did not receive the help from the senior management of IG Farben he had hoped for. Yet after the war, there was in some cases even a refusal to acknowledge that certain employees had been dismissed in 1938 solely on the grounds of their Jewish origin.

This was particularly blatant in the case of the engineer Franz Michalson (or Michalsohn). Born in 1880 as the son of the merchant and factory-owner Max Michalson in Graudenz (West Prussia), Franz Michalson studied at the Technical College Mittweida and the Technical College of Karlsruhe. After completing his studies he worked as an engineer in various companies and served in the army for four years during the First World War. In December 1920 Michalson became a technical officer in Hoechst. He was married, had one child, and was a Lutheran.[352]

---

[348] HHStAW, Abt. 518, Nr. 14826, Amtsgericht Frankfurt am Main, June 26, 1951, affidavit by Helene Henle concerning her husband's death on April 1, 1944; Abt. 520 F (A-Z), Hirschel, statement Helene Henle, dated December 30, 1945, and statement Friedrich Meissner, dated February 14, 1946; Beck et al., *Juden in Hoechst*, p. 48.

[349] Ibid., Stadt Frankfurt an Main, Bestattungsamt, Feuerbestattung Henle, Franz Israel, dated April 12, 1944 (copy).

[350] HA, TEA 95e, Niederschrift über die technische Direktionssitzung in Ffm.-Hoechst, July 13, 1938.

[351] HA, TEA 95e, Niederschrift über die technische Maingau-Direktionssitzung in Ffm.-Hoechst, January 9, 1939.

[352] HA, PA Michalson (microfilm), dates according to the resumé from his job application; employment per December 1920.

In the summer of 1936 the works management resolved to replace him "at the earliest possible moment by a younger and more efficient man."[353] In the spring of 1937 it was decided to pension him off at the end of the vacation season.[354] However, his services were apparently still required, because it was only in March 1938 that the Personnel Department finally informed Michalson that he was given leave "with immediate effect until the regular period of termination of his employment."[355] In the application for pension payments the fifty-seven-year-old was described thus: "Michalson must be considered as worn out and no longer equal to the increased requirements."[356] Dr. Adolf Baldus, the medical examiner for Hoechst, certified Michalson as having "arteriosclerosis of the brain and mental disturbances, intestinal sluggishness." Michalson was therefore ready for retirement as "this impairment of his reason poses a risk for the plants." With Jähne's consent, director Schwamborn gave instructions that Michalson be pensioned off and be given leave as soon as possible.[357]

A note on Michalson's early retirement shows that this was politically motivated. Siegfried Kiesskalt, Michalson's supervisor, left it to chief engineer Berger to inform Michalson of his dismissal. After the conversation Landmann from the Personnel Department had a conversation with Michalson and informed him that Hoechst had decided that he should be retired. Michalson replied that Berger had informed him only that he would be given leave as of June 1; he had not been informed whether he would be pensioned off or be given redundancy pay. "However, he had assumed straight away that his retirement was planned. This assumption had been confirmed by the fact that he had been examined by the medical examiner of the plant with the purpose of sending him into retirement." Thereafter Landmann informed Michalson that he had been instructed "to inform him that on the basis of the medical findings an application had been filed for his retirement as of 6/1/38." Michalson, according to Landmann's notes, replied that "he must resign himself to it, but requested the works management to leave him his salary during a transition period." Michalson added: "The company normally is very willing to oblige in similar cases."[358]

Michalson was sent into early retirement on September 30, 1938. An entry can still be found in the wartime personnel files: In mid-November

[353] HA, TEA 95d, Niederschrift über die Vorstands- und technische Direktionssitzung, August 17, 1936.

[354] HA, TEA 95d, Niederschrift über die technische Direktionssitzung, May 3, 1937.

[355] HA, PA Michalson (microfilm), Personalabteilung to Michalson, dated March 4, 1938.

[356] Ibid., IG Hoechst, application for maintenance dated March 4, 1938, on behalf of Franz Michalson.

[357] Ibid., note by Personalabteilung dated March 11, 1938, with a report of the findings of Dr. Baldus and a handwritten note by Schwamborn.

[358] Ibid., note Landmann (Personalabteilung) dated May 11, 1938.

1941 Director Otto Hirschel forbade "Mr. Franz Isr. Michalson" to be "admitted to our works department store."[359] The "retired engineer Franz Israel Michalsohn," "Protestant, formerly Mosaic" died from "cardiac insufficiency, veronal poisoning" on April 28, 1942. Michalson, like many other victims of racial persecution, had finally chosen suicide.[360] As Franz Henle's widow confirmed after the war, Michalson had "been persecuted by the Gestapo because of his Jewish descent" until he killed himself at the end of April 1942.[361] However, after his retirement he did not just suffer from persecution by the Gestapo but also under the compulsory labor imposed on him and his exclusion from German society and from the plant. His widow wrote in 1957: "after his early retirement in 1938 only wrongful detentions, compulsory menial work, and these methods of persecution finally drove him to suicide in 1942. His family suffered irreplaceable damage from this."[362]

However, no compensation was given for Michalson's persecution and death until 1957, because on enquiry from the compensation agency Hoechst insisted that Michalson's early retirement had been exclusively for physical reasons: "The circumstance that Mr. Michalson was of half Jewish descent certainly played no role in his retirement and in the granting of a pension from the salaried employees' insurance fund on 10/1/1938." There was nothing in the personnel files that would "indicate that the deceased had suffered from persecution due to National Socialist measures." The reason for his retirement was "because both in the opinion of the medical examiner of the pension fund, Dr. Baldus, and that of the private medical expert Dr. Jonas, MD from Griesheim, he was pensioned off because he had increasingly suffered from arteriosclerosis of the brain, mental disturbances, and intestinal sluggishness since 1936."[363] If one recalls that there had been repeated plans after 1936 to send Michalson into retirement and that as a "Jew by decree" he had increasingly suffered from persecutions both at his place of residence and at his place of work in Hoechst, it is not at all improbable that he suffered from "mental disturbances." In the mid-1930s arteriosclerosis of the brain could only be diagnosed on the basis of symptoms, that is, when the person in question acted extremely confused in his daily life, and at the age of fifty-seven Michalson was quite young for

---

[359] Ibid., Hirschel to Michalson dated October 16, 1941 – not sent, as Mrs. Michalson was informed by word of mouth on October 17, 1941.

[360] HHStAW, Abt. 518, Nr. 36537, Sterbeurkunde Michalsohn (dated February 9, 1957); cf. Konrad Kwiet, "The Ultimate Refuge. Suicide in the Jewish Community under the Nazis," *Leo Baeck Institute Yearbook* 29 (1984), pp. 135–67.

[361] HHStAW, Abt. 518, Nr. 36537, Affidavit by Helene Henle dated December 28, 1951.

[362] Ibid., Statement Ella Michalson dated February 6, 1957 regarding Franz Michalson's "Verfolgungsvorgang."

[363] Ibid., Pensionskasse der Angestellten der IG Farben, Frankfurt-Hoechst, to Regierungspräsident Wiesbaden, dated March 21, 1952.

such an illness. So this "diagnosis" gives the impression that an organic illness was invented in addition to the mental problems to be able to force him into retirement without difficulty.

Michalson's conversation with Landmann appears to bear this out. In the spring of 1956, Michalson's widow wrote that her husband had been "fit as a fiddle" at the time and the reasons given for his retirement "were the diplomatic excuses used at the time in those special circumstances."[364] In the fall of 1956, after her claim for compensation had been turned down, she conceded that "it could not be doubted" that "only" the state of her husband's health had "played a role" in his retirement. But she did ask "when reviewing the case to also take the conditions at the time against a certain 'race' into consideration." She added: "my husband's early retirement was followed by years of persecution and humiliation, which were such that in 1942 he chose suicide."[365] On the basis of existing material and the personnel files it is quite clear that Michalson was harassed and bullied at his workplace and fell ill in consequence. The organic illness that was subsequently diagnosed seems rather far fetched. And the "mental disturbances" were almost certainly a result of his persecution as a "Jew according to state decree."[366] Michalson was accorded the status of a victim of National Socialism only fifteen years after he had committed suicide, and the compensation to his family was given in the form of a settlement. His widow died only a few months after the settlement and was unable to derive much benefit from it.[367]

The case of the chemist Ernst Ludwig Sander is not less tragic. His father was the "in Darmstadt well-known and respected Jewish court banker and director of Deutsche Bank, *Geheimrat* [Privy Councilor] Paul Richard Sander," his mother "the full Jewess Bettina Weil."[368] According to the personnel files, when he joined Hoechst in May 1926 Sander was a Protestant. He had studied at the Technical College of Darmstadt, where he completed his dissertation at the end of 1923 when he was about thirty. Afterwards he worked as a scientist in Darmstadt and at the University of Lausanne until the beginning of 1926.[369] In Hoechst he began working in the main Azo Laboratory under Henle, who thought very highly of him and

---

[364] Ibid., Ella Michalsohn to Regierungspräsident Wiesbaden, Entschädigungsbehörde, April 28, 1956.

[365] Ibid., Ella Michalsohn to Regierungspräsident Wiesbaden, Entschädigungsbehörde, October 31, 1956.

[366] I am grateful to Dr. med. Michael Lindner, Universität zu Köln, for his valuable explanations.

[367] HHStAW Abt. 518, Nr. 36537, Regierungspräsident Wiesbaden, Auszahlungsanordnung (payment authorization) dated May 7, 1957; Ella Michalson died on September 14, 1957.

[368] Ibid., Fritz Frey to Hessisches Staatsministerium, Minister des Inneren, Abt.-VI Wiedergutmachung, dated March 27, 1950.

[369] HA, PA Ludwig Sander (microfilm), Sander's application dated May 14, 1926.

praised his independent work: "His diligence is untiring. His interest in the work arises not just from his pronounced sense of duty but also from his intellectual alertness and good inclination for chemistry." For this reason Henle suggested that Hoechst should soon grant Sander "a substantial bonus."[370] During the following years Sander had a brilliant career as a researcher; his annual income increased between 1932 and 1938 from the already substantial sum of 7,455 RM to 11,000 RM.[371] Thus Sander had an annual income that was far above average. However, after 1933 the organics research in which Sander worked until 1936 was part of Kränzlein's department. As the company wished to continue employing Sander because of his excellent work, he was not dismissed but was initially moved to the Literature Department, where he was responsible for evaluating foreign journals. At the end of 1936 he switched to the Laboratory for Intermediate Products headed by director Staib and after Staib's death by Valentin Hilcken – which meant that he was removed from Kränzlein's direct sphere of influence. Here again Sander showed his worth, as is also demonstrated by the amount of his salary.

At the beginning of July 1938 Hoechst wished to extend Sander's time of competitive restriction, as he had worked in a number of new fields. On July 8, 1938, he had a conversation with his supervisor Hilcken, probably already on his impending dismissal. Following their talk Sander wrote to Hilcken: "I feel impelled to confirm to you in writing that in future I will continue to the best of my ability to serve the interests of IG as I do today and have done for the last 12 years."[372] His expression of loyalty was not reciprocated. On July 11 he had a meeting with works manager Lautenschläger, who informed him that "due to the new regulations" he, "being a full Jew," would be dismissed on September 30, 1938, effective from April 1, 1939, and after that time he would be subjected to a two-year time of competitive restriction for the areas stated in his employment contract. Sander replied to Lautenschläger "that this notice of termination hit him hard as it made it impossible for him to find any job in Germany." He requested Lautenschläger to dispense with the time of competitive restriction and promised to look for a job abroad that would not touch on his previous fields of work. If this were not possible, he

---

[370] Ibid., Henle on Sander, dated March 17, 1927.

[371] HHStAW, Abt. 518, Nr. 27372, Sander's petition on the basis of the indemnification law for national socialist injustices (*Gesetz zur Wiedergutmachung nationalsozialistischen Unrechts, Entschädigungsgesetz*), dated March 24, 1950, Section E. "prejudice to economic advancement" with a certified copy of the letter from the IG plant Hoechst dated February 17, 1939, to Mertens (lawyer and legal representative of Sander at the time).

[372] HA, PA Sander (microfilm), Hoechst (Personalabteilung) to Sander (Zwischenprodukte Abt. Nord), July 5, 1938, on period of restriction; Sander to Hilcken dated July 8, 1938.

requested that he be sent into retirement. Lautenschläger turned down both suggestions but did offer to go into the matter again after his vacation. However, during the course of the conversation Lautenschläger gave Sander his notice, and Sander accepted it.[373]

While Lautenschläger went on vacation, the life for which Sander had worked so hard collapsed completely – for the sole reason that he was a Jew according to state decree. An attempt by *Geheimrat* Sander to ensure that his son was given "the possibility of starting his life anew" abroad by curtailing the waiting period was unsuccessful. *Geheimrat* Sander traveled especially to Hoechst on July 14, 1938, to request that the time be shortened, as reported by Sander's supervisor Hilcken. "*Geheimrat* Sander offered the company his private fortune as a security for the activities of his son during the time of competitive restriction abroad." Hoechst was prepared under certain circumstances to shorten the waiting period but not to waive it. On the same day Ernst Ludwig Sander was given leave of absence.[374]

Sander went on vacation, during which he suffered a nervous breakdown – he seemed to be incapable of getting over his dismissal and almost appeared not to want to realize it. There is no other way of explaining Sander's letter to Hilcken on August 12, 1938, from the Mont Riant Clinic near Montreux, Switzerland. There he wrote that after a "lengthy indisposition" he had "fallen seriously ill" during his vacation and therefore it would "unfortunately not be possible" for him to fulfill his duties in the coming week. Nevertheless he hoped "to be able to return to Hoechst soon." On the same day, however, he wrote to Hoechst confirming that he had sent all papers concerning his work to the departmental managers and was returning his works identity card.[375] In the return letter, Sander's letter to Hilcken was referred to and Sander was informed that he had been given a leave of absence since July "and therefore a resumption of your activities is out of the question during the period of notice." Furthermore Hoechst explained that he was dismissed as of March 1939 and that it would not be possible to waive his time of competitive restriction, although Hoechst was prepared to limit the period to the end of September 1940.[376]

Sander confirmed the receipt of the letter with its repeated notice of dismissal, which he now seemed prepared to accept. Once again he asked

[373] Ibid., Lautenschläger, Niederschrift "on the discussion with Dr. Sander on July 11, 1938" dated July 12, 1938.

[374] Ibid., notes by Hilcken, dated July 14, 1938, on a discussion with *Hofrat* Paul Sander, Darmstadt, on the same day; Hoechst (Personalabteilung) to Sander, August 5, 1938, requesting him to return his factory pass as he had been suspended since July 14.

[375] Ibid., Sander to Hilcken (Hoechst), August 12, 1938; Sander to IG Hoechst, August 12, 1938, with a handwritten note that his factory pass arrived on August 16, 1938.

[376] Ibid., Hoechst (Lautenschläger/Schwamborn) to Sander, August 24, 1938.

for the period of competitive restriction to be shortened "in consideration of the extremely difficult situation into which I have been brought by my dismissal through no fault of my own." He combined this request with a clearly stated avowal of his attachment to the plant in which he had successfully worked for so long. "As I have worked with pride and joy in IG for 12 years, I will continue to feel the same connection to the Hoechst plant and IG as before." He also requested a reference for his work in the plant that should "at least briefly touch upon the reason for my dismissal," so that no one might think that he had neglected his duties toward IG Farben. He requested that the reference be sent to the clinic in Switzerland "since I will be obliged to stay here for some time."[377]

The Personnel Department drew up a reference on the basis of short reports by his former supervisors, in which he was given a very positive assessment. It included the statement that "he carried out the tasks assigned to him, which were essentially to create new pigments or the necessary intermediate products, diligently, with perseverance and great skill and in many cases successfully solved these problems independently on his own." Sander's behavior was confirmed to have been "impeccable," and the reference continued: "Dr. Sander is leaving us because he is not of Aryan origin." However, Schwamborn added in his own hand that "no reasons [for the dismissal] can be given in the reference." Lautenschläger even decided not to reply to Sander's letters as long as Sander remained abroad.[378]

However, these instructions by Lautenschläger met with resistance in the Personnel Department, as this was felt to threaten Hoechst's position toward Sander. For if Hoechst did not answer Sander's letters and refused to meet its obligations to continue to pay his salary and to write him a reference, Sander would be able to take the line that he was not obliged to comply with the period of competitive restraint either. Lautenschläger therefore gave instructions to pay Sander's salary into a blocked account. Furthermore Sander's case was discussed with Wagenheimer, the plant's contact man with the *Gauleitung* and local Gestapo, "to prevent false interpretations and any resultant difficulties."[379]

The works management was clearly more worried about the reactions of the *Gauleitung*, as the consultation with Wagenheimer indicates, than

[377] Ibid., Sander to Hoechst (Personalabteilung) dated August 31, 1938.

[378] Ibid., internal memo from Hilcken to Personalabteilung Hoechst, September 7, 1938; internal memo by Literatur-Abteilung to Personalabteilung, September 8, 1938; draft of a reference from the Personalabteilung dated September 9, 1938, with handwritten comments by Schwamborn made on the same day. The reference and the correspondingly drafted letter were not sent due to Lautenschläger's intervention.

[379] Ibid., note by Landmann for Schwamborn dated November 3, 1938, with handwritten annotation by Schwamborn; Aktennotiz (Personalabteilung) dated November 7, 1938, on consultation with Wagenheimer on Sander.

about the future of loyal former employees. In the interests of the plant a solution to the problem was pursued dilatorily; the motto appeared to be "wait and see." Sander officially gave notice that he would remain in Switzerland, where he attempted to find work. He was desirous, however, as his lawyer explained to the works management, "to remain within the legal framework toward IG at all costs." Furthermore Sander would not be able to find a job in Switzerland "either in an official capacity or at the university" as long as his relationship to IG Farben was not "completely clarified." His lawyer additionally requested that the plant "waive the time of competitive restriction retroactively," adding that "legally in view of the fact that the payment and transfer of the compensation for the period of competitive restriction was rendered impossible" Sander was "free of his obligations" toward Hoechst. After that Hoechst agreed to dispense with the period of competitive restriction.[380] However, Sander's lawyer asked for a written confirmation of the waiver and for a reference. Both were granted, although the final reference was less positive than the original draft. At Lautenschläger's instigation the grounds for his dismissal were not included, and positive adjectives like "great" and "active" and the phrase "independently on his own" were also removed.[381]

However, neither the end of the period of competitive restriction nor the reference were of any help to Sander, as he was given only a temporary residence permit and no work permit in Switzerland. However, he succeeded in obtaining a visa for England. At the end of March 1939 he arrived in England, but once again was not given a work permit and was obliged to live at the expense of his relatives. He subsequently emigrated to Palestine in May 1939, where he found a job as a chemist in the Daniel Sieff Research Institute (Weizmann Institute) in Rehovoth. He continued to work there until the beginning of 1948. "In 1948 I decided because of the unbearable conditions in Palestine at the time, which brought about a return of my former nervous illness, to leave the country and return to England." There he found a low-paid position at the Imperial Chemical Industries, where he worked until he had reached retirement age in 1956. He was not entitled to a pension since he had been too old for admission to the pension fund at the commencement of his work there. None of his family survived the Nazi period: his mother "had already died in 1936 out of grief about the insults and abuses suffered by her relatives after 1933," his father had died in

---

[380] Ibid., note on Sander's moving away dated November 8, 1938; note by Schwamborn for Hoechst on a visit by Dr. Mertens, Sander's lawyer, dated January 31, 1939; additional note dated February 2, 1939, on a waiver of the period of restriction.

[381] Ibid., Mertens (lawyer) to Personalabteilung Hoechst (Landmann), February 13, 1939; draft of a reference for Sander dated February 10, 1939, with proposals for amendments by Lautenschläger; reference for Sander dated February 15, 1939.

1940, and his sister had disappeared in 1942 during an attempt to flee from the Gestapo to Denmark.[382]

Max Sorkin was also dismissed in 1938, aged barely thirty. Like all the other cases described before, Sorkin was another "Jew according to the letter of the law" – and like Benda he was nondenominational. He was the son of the master locksmith Nikolaus Sorkin, who had been born a German national in Vilnius, which belonged to Russia at the time. Sorkin's mother, also born in Vilnius, was a Russian who received German citizenship after her marriage. Sorkin studied chemistry in Basel, then in Freiburg im Breisgau, where he completed a doctorate on cellulose under Professor Hermann Staudinger and continued his research in this field for a further year as Staudinger's assistant.[383] During the cooperation between Kränzlein and Staudinger, which will be discussed in more detail later, Sorkin began working for Hoechst. Sorkin and Kränzlein signed the employment contract in April 1937, and Sorkin began working for Hoechst in June of the same year.[384] During the following fifteen months Sorkin successfully worked as a research chemist on synthetic materials, and the success of his work was mirrored by his share in various patents.

However, in September 1938 his life changed dramatically. On September 12, 1938, he called on works manager Lautenschläger to report "that during the examination of the papers of his brother living in Berlin it had been determined that he was not Aryan." In this case, Sorkin said, "this would also apply to himself; he was unable to explain it as until then he had been firmly convinced that he was an Aryan." Sorkin wished to know from Lautenschläger what "should be done." Sorkin was requested to return on the following day when, according to director Schwamborn, Sorkin informed him that he wished "to be dismissed and stop work." The question of his time of competitive restriction should be decided after consultation with the heads of the departments.[385] What followed was a controversial discussion in Hoechst concerning his period of competitive restriction. While his former supervisor Kränzlein in particular insisted on the time of competitive restriction being maintained, others regarded it as "not necessary." Lautenschläger finally decided that

---

[382] HHStAW, Abt. 518, Nr. 27372, affidavit by Sanders dated February 11, 1960; Frey (lawyer) to Hessisches Staatsministerium, Minister des Inneren, Abt. VI Wiedergutmachung dated March 27, 1950.

[383] Schweizerisches Bundesarchiv, Einbürgerungsdossier K 55346 Max Sorkin, Kantonspolizei Zürich, Rapport dated December 14, 1973, Bürgerrechtsdienst Winterthur.

[384] HA, PA Max Sorkin (microfilm), Hoechst (Kränzlein) to Sorkin, April 13, 1937; contract of employment dated April 13, 1937; Staudinger to Kränzlein, May 21, 1937 (copy); confirmation by Director Schwamborn to Sorkin on his term of service dated October 3, 1938.

[385] Ibid., Akten-Notiz (Schwamborn) on Sorkin dated September 14, 1938 – two copies, one with handwritten notes.

Sorkin should sign an obligation to comply with a period of competitive restriction, after which it would be possible to reconsider whether to insist on the period or not. Schwamborn and the head of Hoechst's Legal Department, Josef Hirsch, finally came to an agreement with Sorkin that the latter should hand in his notice himself. Sorkin was given leave from September 19. Hoechst wished to retain the option to demand a waiting period of one and a half years and make the corresponding payments or to waive the right partly or completely. "The question of Dr. S., whether he is also bound by the period of competitive restriction if he is abroad and the company is unable to transfer the compensation payments for this period of waiting, was answered to the effect that if the contract cannot be fulfilled by one side, the other party is then automatically released from his obligations." This interpretation was confirmed in a subsequent telephone call between Hirsch and Lautenschläger [386] – which was then the basis for the treatment of all emigrated Jewish employees. Initially, however, Lautenschläger decided that Hoechst should insist on the period of competitive restriction.[387]

Sorkin, who had returned to his family in Rheinfelden in the meantime, answered the letter from Hoechst in such a manner that made clear that he had not given notice voluntarily: "I can confirm that after my communication concerning my descent, my continued employment in the plant was seriously called into question so that I handed in my notice or rather it was suggested that I hand in my notice." Sorkin informed Hoechst that his family had "lodged an appeal against the ruling of the Reich Kinship Bureau [*Reichssippenamt*]" and asked whether Hoechst would be willing to reemploy him "if the ruling is revoked." He also asked about his share in certain patents and requested a reference. In the reply to his letter Hoechst took note of Sorkin's appeal against the Reich Kinship Bureau. Moreover it was suggested that "under the circumstances" Sorkin should forego his being mentioned as an inventor in those patent registrations that concerned him. This would not affect his rights to a share of the profits as inventor. At the same time he was placed under an obligation to maintain secrecy concerning the object of the inventions. He was not given a reference but instead was issued a certificate confirming that he had been employed in Hoechst from June 7, 1937, until September 19, 1938.[388]

[386] Ibid., Note Schwamborn and Hirsch dated September 17, 1938, on Sorkin's period of restriction.

[387] Ibid., draft of a letter from Hoechst (Schwamborn) to Sorkin, September 19, 1938, on his dismissal, leave of absence, and period of restriction; below, handwritten annotation by Schwamborn on Lautenschläger's decision; Hoechst (Lautenschläger/Schwamborn) to Sorkin, September 19, 1938.

[388] Ibid., Sorkin to Hoechst, September 23, 1938; draft by Dr. Spiess (Patentabteilung) on Sorkin's inquiry into patents, dated September 27, and handwritten draft (undated);

Unlike Sander, Sorkin was apparently self-confident enough not to be intimidated by Hoechst and the increasing number of measures being undertaken against German Jews. He thanked Hoechst for the certificate of employment but firmly requested the reference promised him by Schwamborn. He would, of course, maintain secrecy concerning the objects of his inventions; however, he was unable to see why he should forego the mentioning of his name "on the 3 main patents and the 2 ancillary patents": "Nobody has any reason to be ashamed of me. I always have done my duty, and no man can do more."[389]

This view was perhaps shared by Schwamborn, but not by Sorkin's former supervisor, Kränzlein. Schwamborn was clearly aware of Kränzlein's attitude and pointed out to him that according to the legal regulations and following Lautenschläger's decision Sorkin should be given a reference. Kränzlein was told to request the necessary documents for a reference from his co-worker Arthur Voss and send them to the Personnel Department: "We assume that you do not wish to sign the reference yourself and therefore request your comments." It was also pointed out to Kränzlein in the matter of Sorkin's share in the patents and the inclusion of his name in the registration that the Patents Department of Hoechst was of the opinion that an objection by Sorkin might invalidate the patent.[390]

However, Kränzlein not only declined to sign the reference, he objected as a matter of principle to any reference being written. In a detailed letter to the Personnel Department Kränzlein wrote that he considered any mention of Sorkin as inventor as wrong because "I do not take the view that Sorkin had independently invented something important." The carrying out of copolymerization was not indicative of "any particular inventive step" since the procedure was known and the mixing ratio did not constitute an inventive activity. Kränzlein wrote that with the upcoming Patents Act "a case worker like Dr. Sorkin, if he had not done any special intellectual work or had carried out any particular feats," could no longer expected to be "referred to as an inventor." He added: "In my opinion it has gone too far if such concessions which exceed the obligations of the plant are made, as they were made in this case, for a case worker who had to be dismissed without notice." And finally Kränzlein addressed the point of providing a reference for Sorkin and his employment in Hoechst. "The writing of a reference for Dr. S. is unnecessary in my opinion since, as a Jew, Dr. S. will not find employment in any German company. If the reference is written it will only allow him to join a foreign company afterwards. However, we are not interested in declaring abroad in

Hoechst (Schwamborn/Spiess) to Sorkin, October 3, 1938; confirmation by Schwamborn on Sorkin's term of service dated October 3, 1938: granted leave of absence as of September until the expiration of his employment contract on March 31, 1939.
[389] Ibid., Sorkin to Hoechst (Schwamborn), October 7, 1938.
[390] Ibid., Personalabteilung (Landmann) to Kränzlein, October 12, 1938.

writing that Dr. S. has worked in synthetics." Kränzlein went even further in his rejection of Sorkin: "In my opinion the employment of Dr. S. has been declared null and void, just as certain marriages are dissolved by a court ruling declaring them void, i.e. as if they had not existed at all. I similarly take the view that Sorkin has therefore not existed, neither as an employee for the plant nor as a subordinate to myself."[391]

The events during the following weeks in the wake of the "Night of Broken Glass" as the violent climax of the measures taken against Jews convinced Sorkin and his family that they would no longer be able to stay in Germany. In December 1938 Sorkin moved "from Germany to Basel with a border permit."[392] He informed Hoechst about his departure at the end of January 1939 and repeated his request once again in February that he be given a reference and furthermore that the payments for his time of competitive restriction be paid in Switzerland.[393] While Schwamborn was prepared to send Sorkin a reference, he declined his request to have the payments transferred to Switzerland in Swiss francs. But the letter drafted by Schwamborn was not sent, as Lautenschläger decided "for the time being" to pass over the question of the reference. Moreover the Gestapo had been informed about Sorkin by Wagenheimer on February 15, whereupon the decision was taken in Hoechst not to send the reference to Sorkin – he could collect it himself from Hoechst. As this was "surely impossible" and Kränzlein refused to write a reference, Schwamborn was of the opinion that Wagenheimer should "obtain written instructions from the Gestapo that we should decline to write a reference (reason: crossing the border without permission)." Correspondingly the reply to Sorkin stated only that it was not possible to transfer the money to him abroad and that he should state where the money should be transferred within Germany.[394]

Sorkin responded only in April to the letter, after he had searched for a means of receiving the payments for his period of competitive restriction. He needed the money, not least because he was unable to find any employment due to this contractual obligation. As Hoechst could not transfer the money to Switzerland and was not interested in continuing to pay the money for Sorkin into a blocked account, an agreement with Sorkin was sought. Hoechst offered to waive the period of competitive restraint,

[391] Ibid., Kränzlein to Personalabteilung (Landmann), October 17, 1938; the letter from the Personalabteilung to Kränzlein dated October 12, 1938, already bears a handwritten comment: "Dir. Dr. Kränzlein has refused to give a reference."

[392] Schweizerisches Bundesarchiv Bern, E 4264 1988/2/1186, petition of Max Sorkin for an identity card or a so-called Nansen passport, dated June 15, 1953.

[393] HA, PA Max Sorkin (microfilm), Sorkin to Hoechst (Schwamborn), January 26, 1939; Sorkin to Hoechst (Schwamborn), February 10, 1939.

[394] Ibid., draft of a letter from Hoechst (Personalabteilung) to Sorkin dated February 14, 1939, below handwritten annotation by Schwamborn on Lautenschläger's decision, dated February 16, 1939; Hoechst (Schwamborn) to Sorkin, dated February 17, 1939.

which Sorkin agreed to immediately – alluding once more to the reference, which was still not forthcoming. The Personnel Department decided in July 1939 to wait "a bit longer" before asking Kränzlein for a reference.[395]

Sorkin received a reference from Hoechst only in the middle of October 1939, that is, after the beginning of the Second World War. The reference indicates that he had been much more than a case worker: "You should be apprised that your absence is particularly noticeable in the Department of Synthetic Materials. But as you know yourself – there is nobody who cannot be replaced – particularly in these difficult times. You may be assured, however, that I much valued your work, indeed I valued it very highly. You came to us not just with very extensive knowledge of and exceptional talent for chemical work, but you also adjusted with remarkable speed to our line of business and accomplished things which few will be able to equal. I also particularly valued your friendly relations to your colleagues and employees. You will certainly go far in life."[396] Both the tone and the contents of this reference differ greatly from the previous declarations of Hoechst, primarily those made by Kränzlein, concerning Sorkin's role and show that Sorkin had been considerably more than a "case worker." The decree by the state declaring him to be a Jew had not just made it impossible for him to continue his profession in Germany but made his continued existence in Germany impossible; Sorkin had to emigrate.

At the beginning he found it difficult to find employment and therefore decided to extend his knowledge of chemistry through self-study in Basel. After labor service in two work camps he found a job as a scientist at the pharmaceutical institute of the University of Basel in 1941, where he remained until 1946. Afterwards he was head of the Research and Development Department of Carl Weber AG in Winterthur until 1967, subsequently switching to the Swiss Materials Science and Research Institute for Industry, Building, and Construction in St. Gall.[397]

---

[395] Ibid., Sorkin to Hoechst (Schwamborn), April 3, 1929; Schwamborn to IG Bankabteilung, Frankfurt-Grüneburg, dated April 12, 1939, concerning the payment of a salary to a former Jewish employee now residing abroad; IG Zentral–Finanzverwaltung Berlin to Oberfinanzpräsidenten (Devisenstelle) Karlsruhe, April 26, 1939 (Durchschlag); Sorkin to Hoechst (Schwamborn), April 27, 1939; Hirsch (lawyer) to Schwamborn, May 9, 1939, re "Sorkin"; Hoechst (Schwamborn) to Sorkin, May 10, 1939; Sorkin to Hoechst (Schwamborn), June 11, 1939; IG Frankfurt to Hoechst, June 28, 1939, below, handwritten note by Landmann (Personalabteilung) dated July 17, 1939.

[396] Schweizerisches Bundesarchiv Bern, Einbürgerungsdossier K 55346 Max Sorkin, Kantonspolizei Zürich, Rapport dated December 14, 1973, Bürgerrechtsdienst Winterthur, the reference from Hoechst dated October 15, 1939 is extensively quoted in addition to other job references – unfortunately without naming the author; the author of the reference was a superior: Voss or Starck or Möller.

[397] Ibid., Kantonspolizei Zürich, Rapport dated December 14, 1973, Bürgerrechtsdienst Winterthur, includes a detailed list describing his scholastic, university, and professional history.

Questioned about his feelings toward Germany and his Jewish origin in the 1970s Sorkin told the Zurich cantonal police that he "disliked" totalitarian regimes, which was why he had always rejected National Socialism "for humanitarian and political reasons." In the file on Sorkin it was stated: "he has no more contacts with his former country of origin. He was expatriated on 2/22/1939 on the grounds of the National Socialist legislation because of his alleged non-Aryan descent which went back on his mother's side to the third or fourth generation. After his country of origin was no longer prepared to accept him as a citizen on these flimsy grounds, after the end of the war he did not opt for Germany either and has preferred to remain stateless." In reply to the question on religion Sorkin answered that he had been nondenominational since his earliest youth, as his parents had been. Sorkin declared further that he would remain in this "state" since "he cannot understand that there should be different religions in human life while there was only one God."[398]

The differences in the treatment of Jewish employees are particularly evident in the case of Robert Julius Schnitzer, who was perhaps not very religious, but was nevertheless a professed Jew.[399] He was born in 1894 in Berlin the son of the writer Manuel Schnitzer and his wife Johanna, née Krämer, a journalist. After studying medicine in Berlin he began work at the Department of Chemotherapy of the Robert Koch Institute, where from 1919 to the end of 1924 he was the closest collaborator of the renowned Professor Julius Morgenroth, a former colleague of Paul Ehrlich and cofounder of chemotherapy. In a reference by the Prussian Institute for Infectious Diseases "Robert Koch," dated the beginning of 1926, the following description is given: "He is the well-known colleague of Julius Morgenroth. During the long illness of his teacher he independently continued to carry out the work with much success."[400] A reference from March 1927 goes into greater detail, stating that as a collaborator of Morgenroth he had "carried out a number of important scientific works, some of them independently," particularly in the fields of chemotherapy, immunoresearch, and the infections caused by streptococci and pneumococci. His work had led to important new results in some fields. Altogether Schnitzer was summed up as having "exceptional scientific talent coupled with an unusually wide-ranging education" and

---

[398] Ibid., Kantonspolizei Zürich, Rapport dated December 14, 1973, Bürgerrechtsdienst Winterthur, detailed information on his political views and in the addendum on his religion.

[399] Private files Schnitzer/Mota, interview with Robert Julius and Peggy (Eva) Schnitzer by their granddaughter Gail Mota dated February 11, 1984.

[400] Ibid., reference by Preuss. Institut für Infektionskrankheiten "Robert Koch," Chemotherapeutische Abteilung, dated January 9, 1926.

that due to his personal manner he was held in the greatest esteem by his colleagues.[401]

On the basis of such references he joined Hoechst in October 1928 as a "chemotherapist." While still in Berlin he married Eva Rosenberg in 1926, whom he had met at Morgenroth's Institute, where she worked as a technical assistant and published her results together with Morgenroth and Schnitzer.[402]

After his move to Frankfurt Schnitzer was head of the Chemotherapeutical Laboratory in the Pharmaceutical Department of Hoechst until July 1938. To be allowed to keep him in this important and senior position – among other things he had collaborated in the development of a highly effective remedy for malaria and was working on a chemotherapeutical cure for tuberculosis – Hoechst apparently made large payments between 1933 and 1938 to the NSDAP Winter Relief Organization.[403] In 1938, however, Schnitzer too was dismissed, although it was arranged that he should receive his salary until February 1939. In a talk on July 5, 1938, Lautenschläger told Schnitzer that "according to the new decree Jews must leave our plant." Lautenschläger dismissed Schnitzer at the beginning of October 1938, effective March 1939, and demanded a period of two years' competitive restraint. Lautenschläger suggested that Schnitzer should try to be taken on at the Jewish hospital in Frankfurt, a move Schnitzer had apparently also considered. If Schnitzer could find a job there, he should train his successor part time until September.[404]

Schnitzer received his written notice on August 24, in which he was thanked for his "valuable collaboration in the completion and development of new compounds." The letter concluded with the words "respectfully" and bore the signatures of Lautenschläger and Schwamborn, as the director of the Personnel Department; however, according to a handwritten annotation by Schwamborn, it was suggested that the complimentary ending "respectfully" should in the future no longer be used when addressing Schnitzer. Schnitzer was given leave on August 27, 1938.[405] On September 30, 1938, he was informed in a letter from the Reich State Medical Board/ State Medical Board of Registration of Hesse-Nassau that his approbation to practice as a doctor had been withdrawn: "To clear away any possible doubts and to spare you from trouble we wish to point out to you that

[401] HA, PA Schnitzer (microfilm), reference by Preuss. Institut für Infektionskrankheiten "Robert Koch," dated March 7, 1927 (copy).

[402] Private files Schnitzer/Mota, Reference for Frl. Eva Rosenberg by Prof. Morgenroth, dated November 22, 1921.

[403] Cf. article "Robert Schnitzer – 70 Jahre," in *Der Aufbau*, May 15, 1964, p. 10.

[404] HA, PA Schnitzer (microfilm), note Lautenschläger, dated July 5, 1938, on conversation with Schnitzer of the same day.

[405] Ibid., Hoechst (Lautenschläger/Schwamborn) to Schnitzer dated August 24, 1938.

from this date onwards all medical work is forbidden to you and you must take down your doctor's nameplate immediately."[406] After that, Schnitzer worked in the Jewish hospital "cleaning the dishes," as he commented in an interview.[407]

At what point Schnitzer began to think of emigrating is not clear. At all events, in the summer of 1938 he probably did not yet have any emigration plans. In reply to the question of an American journalist why he had not left Germany prior to his arrest in November 1938 he said: "I was too uninterested. I didn't talk to people."[408] As part of the general abuses following the "Night of Broken Glass" Schnitzer was arrested by the Gestapo as a so-called "Action Jew" on November 11/12 like thousands of other Jewish men and sent to the concentration camp at Buchenwald, where he was given the prisoner's number 26,687, and where he was held until released at the beginning of January 1939.[409] On his release from the camp he was obliged by the Gestapo to promise in writing to leave Germany within the next four weeks.[410]

During the time of his arrest his wife requested Hoechst not to transfer his salary to his bank account anymore but to bring it to her at their apartment. For their two children, aged eleven and nine years old at the time, the couple was able to find places in one of the Children's Transports to Belgium, where they were taken in by a host family in January 1939. In February 1939 the couple fled to France; they were completely destitute because they were only allowed to take 10 RM over the border. In their flight from Germany they were helped by a former young laboratory employee of Schnitzer's in Hoechst, who had connections in France and used them to help them get a visa for that country. In Paris the Schnitzers were able to reunite their family; in the meantime they received financial support from Eva Schnitzer's brother, who worked as a doctor in Cairo.[411]

---

[406] Private files Schnitzer/Mota, Reichsärztekammer, Ärztekammer Hessen-Nassau, September 30, 1938, to Schnitzer (signed Dr. Zöckler, Deputy Director of the Medical Board [*Ärztekammer*]).

[407] Ibid., Schnitzer in interview dated February 11, 1984.

[408] Ibid., Article "Master of Art and Science" (by Michael C. Pollak) on Schnitzer, from *The Record*, June 4, 1984, rubric "North Jersey Originals."

[409] HHStAW, Abt. 518, Nr. 20455, Allied High Commission of Germany. International Tracing Service, Certification of arrest No. 44352 dated October 9, 1954; on his arrest and deportation to Buchenwald, cf. also Wolf-Arno Kropat, *Kristallnacht in Hessen. Der Judenpogrom vom November 1938* (Wiesbaden: Kommission für die Geschichte der Juden in Hessen, 1988), pp. 167–70 and 225–8: document 88, "Transport in das KZ Buchenwald," report by an unknown author.

[410] Article "Robert Schnitzer – 70 Jahre"; HHStAW, Abt. 518, Nr. 20455, Max L. Cahn to Landgericht Wiesbaden. 2. Entschädigungskammer, November 5, 1959, in the appendix affidavit by Schnitzer dated October 30, on his arrest and emigration.

[411] HA, PA Schnitzer (microfilm), Eva Schnitzer to Personalabteilung IG Farben (Hoechst), November 20, 1938; HHStAW, Abt. 518, Nr. 20455, , Max L. Cahn to Landgericht

After Schnitzer's flight Hoechst immediately stopped its payments to Schnitzer on Lautenschläger's instructions. According to Schwamborn, Lautenschläger had been "very astonished" that Schnitzer moved abroad before his contract had ended at the end of March and "without informing the works management about this fact or about his intentions to exercise his profession in the future."[412]

On March 6, 1939, Lautenschläger received a letter from Emil Barell, the chairman of Hoffmann La Roche, where Schnitzer had applied for a job, asking Lautenschläger for a reference. Lautenschläger wrote back on March 8 that "Dr. Schnitzer (– full Jew –)" had previously carried out research at the Robert Koch Institute on chemotherapeutical compounds, primarily preparations from Hoechst, and had been taken on after Professor Morgenroth's death and given a department of his own. "Dr. Schnitzer carried out independent work in the different bacteriological fields and was primarily concerned with the evaluation of chemotherapeutic agents for different infectious diseases and for use in tropical medicine for us. He has also carried experiments in the field of cancer in different areas. Dr. Schnitzer is a very skillful experimenter with good ideas of his own and a comprehensive knowledge of German and foreign literature. He was a valued employee for us, both with respect to his scientific and technical achievements and to his character." Lautenschläger wrote a similarly positive reference for Schnitzer for the School of Tropical Medicine at the University of Liverpool.[413]

In France Schnitzer worked for Rhône Poulenc for several months. At the same time he wrote numerous letters in an attempt to get a permanent position. On the basis of Schnitzer's reputation as a scientist as well as various contacts – among others through his wife Eva's brother – he received several attractive offers. Both Barell from Hoffmann La Roche, mentioned above, and Chaim Weizmann, the future president of Israel, were interested in employing him, Weizmann for his research institute in Rehovoth, Palestine. Neither was able to obtain a visa for him, or, at all events, not immediately. Schnitzer therefore decided take up the offer conveyed to him by Professor Hermann Fischer of a two-year contract at the Connaught Laboratories of the University of Toronto, where he wished to collaborate with Charles Best, the co-inventor of insulin. In July 1939 he and his family boarded a ship to take them to

---

Wiesbaden. 2. Entschädigungskammer, November 5, 1959, in the appendix affidavit by Schnitzer dated October 30, on his arrest and emigration; private files Schnitzer/Mota, interview of February 11, 1984.

[412] HA, PA Schnitzer (microfilm), Aktenvermerk Hoechst (Landmann), dated March 3, 1939.

[413] Ibid., Barell to Lautenschläger, March 6, 1939; Lautenschläger to Barell, March 8, 1939; Barell to Lautenschläger, March 10, 1939; Lautenschläger to Prof. Warrington Yorke, School of Tropical Medicine, University of Liverpool, March 15, 1939.

Canada; their meager financial resources obliged them to travel in the cheapest class.[414]

At the end of 1941 Schnitzer moved to join Hoffman La Roche in the United States in Nutley, New Jersey, where he became head of the Chemotherapeutical Laboratory. His greatest and most spectacular success in this position was his substantial contribution to the development of the active agent for the treatment of tuberculosis, isoniazide, which is still known and used today, and for which he and his laboratory were awarded the prestigious Lasker Prize.[415] When already in "retirement" he continued to work at the Mount Sinai School of Medicine as a "professorial lecturer" in New York. He died at an advanced age in April 1987.[416]

In addition to the Jewish employees of Hoechst mentioned on these pages there may have been others who were also considered to be Jews according to the definition of the Nazi state; yet even most of the persons referred to above were Protestant or nondenominational, and thus difficult to trace.[417] According to the investigations by a team of local historians in Frankfurt-Hoechst there were three Jewish workers in Hoechst; however, no documents concerning them were found in the Hoechst archives.[418]

The actions of the Nazi state against Jews and "Jews by decree" also increasingly began to affect so-called "half Jews" or "individuals of mixed race" as well as persons in "mixed marriages," who began to feel more

[414] Private files Schnitzer/Mota, interview of February 11, 1984, register of Schnitzer's correspondence from February 28, 1939, to November 19, 1939, and telegram from Fischer with the offer from Toronto dated April 5, 1939; article "Robert Schnitzer – 70 Jahre"; HHStAW, Abt. 518, Nr. 20455, Max L. Cahn to Landgericht Wiesbaden. 2. Entschädigungskammer, November 5, 1959, in the appendix affidavit by Schnitzer dated October 30, on his arrest and emigration.

[415] See Stephan H. Lindner and Michael Lindner, "Dr. Robert Julius Schnitzer – ein führender Forscher auf dem Gebiet der Chemotherapie," in Albrecht Scholz and Caris-Petra Heidel (eds.), *Emigrantenschicksale. Einfluss der jüdischen Emigranten auf Sozialpolitik und Wissenschaft in den Aufnahmeländern* (Frankfurt am Main: Mabuse, 2004).

[416] Private files Schnitzer/Mota, interview dated February 11, 1984; documents on the life of Schnitzer; oral communication with the Schnitzer family dated January 8, 2003.

[417] Cf. Beck et al., *Juden in Hoechst*, pp. 47–9, who give detailed information only on Henle; the only reference to Michalso(h)n states that no information on him was found in the Hoechst Archives, otherwise only the names of a few workers and so-called "half Jews" and employees married to Jewish women are mentioned. Benda, Hoffa, Sander, and Schnitzer are not mentioned. See also HA, PA Abteilung VI, Paul Heisel, Werksleitung Gersthofen to Gestapo Augsburg, September 9, 1938 (Confidential): report that in the subsidiary plant Gersthofen near Augsburg the "full Jew" Siegfried Wiesenthal, commercial clerk, was dismissed at the beginning of August 1938 on the instructions of the Hoechst works management.

[418] Cf. Beck et al., *Juden in Hoechst*, p. 49. No information on the three persons was found in the Hoechst Archives.

and more ostracized.[419] In the summer and fall of 1938 the works management of Hoechst went even beyond the demands of the Nazi regime and wanted to dismiss not only those of its employees who were Jewish or were regarded as Jews but also to proceed against the so-called "individuals of mixed race" and against employees married to Jewish wives. In the latter group were Walter Herrmann, head of the Salvarsan operations, and Albin Hardt, an employee in the Coloristic Department. Their dismissal was planned – unless they were prepared to leave their wives. Just how the works management intended to deal with them, as always in consultation with Wagenheimer, would depend on the outcome of a legal action brought against the plant by a brother and a sister who had been dismissed as "half Jews" and were suing the plant for rein-statement.[420]

The brother and sister in question were Karl and Erna Grossmann, who had both been working for Hoechst only since the middle of the 1930s; they had a Jewish mother and an "Aryan" father who had fallen in the First World War. In the case of the commercial clerk Karl Grossmann, the Labor Court of Frankfurt upon Main ruled on September 22, 1938 – only a few weeks before the "Night of Broken Glass" – that Hoechst was obliged to continue to employ him further or to indemnify him with a single payment of 1000 RM. Like Hardt, Grossmann had been employed in the Coloristic Department, and he had been dismissed in August 1938 on the grounds that he was a "full Jew." Although he furnished proof on the same day that he was a "half Jew," Hoechst refused to take back its notice of dismissal. The court gave as the reasons for its ruling that, as a letter from the Reich Ministry of Economics to the plaintiff stated, "Jewish individuals of mixed race who like the plaintiff fulfilled the conditions for temporary citizenship of the Reich were to be treated economically as equivalent to persons of German blood" and that companies "who employ such individuals of mixed race would suffer no disadvantages on the part of the authorities." Hoechst's objection that as a "company important to the war effort" it should no longer employ Grossmann was rejected as "not con-clusive": "considering the size of the defendant company it would have been possible without any problem to employ the plaintiff in another area, if the company had had well-founded doubts about leaving him in his previous department." What also spoke in favor of the plaintiff was his

[419] Cf. Jeremy Noakes, "The Development of Nazi Policy towards the German-Jewish 'Mischlinge' 1933–1945," *Leo Baeck Institute Yearbook* 34 (1989), pp. 291–354; Beate Meyer, "*Jüdische Mischlinge*," *Rassenpolitik, Verfolgungserfahrung 1933–1945* (Hamburg: Dölling und Galitz, 1999); Ursula Büttner, "The Persecution of Christian-Jewish Families in the Third Reich," *Leo Baeck Institute Yearbook* 34 (1989), pp. 267–89.
[420] HA, PA Hardt (microfilm) and Herrmann (microfilm); Beck et al., *Juden in Hoechst*, pp. 48–9.

involvement in the NSDAP since 1930: "Furthermore, with the submission of a declaration in which three old fighters of the movement have testified to his enthusiastic involvement in the movement from 1930 until it came to power, the plaintiff has proved that, despite his nature as an individual of mixed race, he must be regarded as having his roots in the present state."[421]

In the case of his sister, Erna Grossmann, the Labor Court of Frankfurt upon Main also ruled on September 27, 1938, that Hoechst must either continue to employ her or pay her a compensation of 750 RM. Reading the opinion of the court makes it very clear that her dismissal was not demanded by the political leadership but by the works management itself, which, to quote the historian Ian Kershaw, was "working toward the *Führer*"[422] – and to a far greater extent than demanded by Hitler or his henchmen. The court stated that Erna Grossmann's origins had already been known to Hoechst at the time when she was first taken on by the company in 1936. It was also indisputable that the dismissal had been given "on the defendant's [i.e., Hoechst's] own initiative" and "without the defendant being prompted to do so by the pressure of a state or Party office or because of ill feeling within the workforce." Hoechst had even emphasized during the proceedings that it had "taken the decision itself to demand the highest standards concerning the purity of the blood of its followers; that from fundamental considerations, without the prompting of any state or Party office it had given notice to all its employees of mixed race." However, the court countered by pointing out that the Reich Ministry of Economics had particularly ordered that "in their economic activities individuals of mixed race are to be treated as equivalent to persons of German blood and they should not be subjected to any particular restrictions." Therefore, unless "special circumstances existed," it would not be possible to agree with Hoechst "if, in defiance of those guidelines, it had only dismissed Jewish individuals of mixed race because they were Jewish half-breeds." In Erna Grossmann's case the court was of the opinion that special mitigating circumstances existed in her favor. Her father had fallen in the First World War, and, as her brother's lawsuit had shown, her brother had "since 1930 unselfishly placed his whole person at the service of the local branch of the NSDAP-Hoechst, the NSBO operation cell of IG Farben, and other organizations of the NSDAP." The court therefore came to the conclusion "that the element of German blood predominates so strongly in the family of the plaintiff that no fears of any type can be

---

[421] HA, Pol 20, Arbeitsgericht Frankfurt am Main, judgment in the lawsuit Karl Grossmann against IG Farbenindustrie AG, represented by the Management Board in Frankfurt-Hoechst dated September 22, 1938, Chief Judge: *Amtsgerichtsrat* Dr. Kalb (certified copy).

[422] Ian Kershaw, *Hitler* (London: Allan Lane, 1998–2000), vol. 1, pp. 527–31; cf. also Feldman, *Allianz*, p. 76.

associated with the further employment of the plaintiff," and her dismissal had therefore "been wrongful."[423]

Despite the ruling Hoechst upheld both dismissals – and paid the compensations. Although Erna Grossmann was not employed by Hoechst again, at a later date Karl Grossmann was again taken into Hoechst's employ. In 1941 he was called up for service after having been declared an "Aryan," apparently because of his involvement in the NSDAP, for as an "individual of mixed race" he would have been "unfit for military service." He was killed on the Eastern Front at the beginning of March 1944.[424]

After the court ruling Hardt and Herrmann, who were both married to Jewish wives, were allowed to continue working at Hoechst, although Herrmann's children were obliged to leave school because they were "individuals of mixed race." Both found jobs in Hoechst, apparently on the intervention of Alfred Fehrle, the head of the Pharmaceutical Department and Herrmann's supervisor.[425]

The "half Jew" Michael Erlenbach was also allowed to retain his position despite the works management's demands concerning the "purity of blood" of its employees referred to in front of the Labor Court. A note in his personnel file made at the beginning of August 1938, that is, prior to the ruling of the Court, states: "according to verbal communication of Wagenheimer on 8/1/38 Erlenbach is *Aryan*. Confirmation by letter shall follow."[426] If the works management had not decided on its own initiative to also dismiss "half Jews," such an entry would never have been necessary. But keeping Erlenbach in the plant was important for the works management. For Erlenbach was one of the most talented researchers in IG Farben in the field of pesticides. Moreover he was a nephew of the former works manager in Wolfen, Arnold Erlenbach, and family loyalties did exist within the company.[427] But primarily it was his abilities that made him "indispensable" and thus gave rise to Wagenheimer's communication.

[423] HA, Pol 20, Arbeitsgericht Frankfurt am Main, judgment in the lawsuit Erna Grossman against IG Farbenindustrie AG, represented by the Management Board in Frankfurt-Hoechst dated September 27, 1938, Chief Judge: *Amtsgerichtsrat* Dr. Kraft (certified copy).

[424] HA, Pol 20, transfers to Erna and Karl Grossmann; PA Erna Grossmann (microfilm) and Karl Grossmann (microfilm): he had originally joined in January 1937; in the entry in his file, dated April 7, 1944, the date of his joining is given as December 9, 1937, which points to a new membership and a new date of admission. His file also mentions that he was called up for military service on March 10, 1941, and killed on March 1, 1944, "in the East."

[425] HA, PA Hardt (microfilm) and PA Hermann (microfilm); Beck et al., *Juden in Hoechst*, pp. 48–9; HHStAW, Abt. 520 F (A-Z), Alfred Fehrle, expert opinion by Dr. Walter Herrmann, August 8, 1946; Abt. 520 F, Nr. 4815 (Zeh), Protokoll der öffentlichen Sitzung, January 12, 1949, there, Dr. Herrmann's testimony.

[426] HA, PA Abteilung VI, Erlenbach, personal data sheet, June 21, 1938, handwritten note by Schwamborn, dated August 6, 1938 (underlining in original).

[427] Cf. Heine, *Verstand*, pp. 83–4 on Arnold Erlenbach; see also pp. 167–70 on Max Ilgner, whose uncle was Hermann Schmitz, chairman of the Board of Directors from 1935 to 1945.

Erlenbach, born in 1902 in Nuremberg, was a Protestant. His grand-parents on his father's side were both Jews but had already left the Jewish community in 1888.[428] Erlenbach studied chemistry in Freiburg, Erlangen, Leipzig, and finally in Munich and completed his dissertation there in 1927 under Professor Heinrich Wieland. After working for some time as Wieland's assistant he wished to move into industry, whereupon Wieland recommended him to Hoechst. Wieland described Erlenbach as "a skillful and absolutely reliable experimenter" who had found certain questions "a tough nut to crack and had the necessary tenacity." He believed him capable of doing "valuable work" in the field of pigments synthesis and praised his ability of "independently going his own way."[429] Erlenbach began work in the Patents Laboratory under Schmidt, then moved to the Pesticides Laboratory, where he focused on finding new synthetic fungicides and insecticides to replace the poisonous, metalliferous combinations commonly used. His greatest success was his role in the development of Nirosan, the first insecticide that did not contain arsenic and when intro-duced for use in viniculture in 1940 quickly replaced other, more poisonous agents. Even after the war Nirosan continued to be one of Hoechst's most successful products.[430]

Because of his extraordinary achievements Erlenbach was temporarily entrusted with the management of the laboratory in the fall of 1939, when the head of the Pesticides Laboratory, Kaspar Pfaff, was called up. Even the findings of the police headquarters in 1940 that he was a "1st degree Jewish, mixed race individual" were not a barrier. Only in 1944, when the punitive measures undertaken against "mixed race individuals" increased and they began to be subjected to forced labor or deported, was Erlenbach sent to the West Wall in Lorraine as an "excavations laborer" to build fortification trenches. He survived this very dangerous deployment and returned to Hoechst at the end of 1944, where he continued to work in the Pesticides Laboratory until the end of the war. He was appointed trustee of the works under the American occupation in 1946. He did not owe his return to the efforts of the works management, despite the fact that in another case, that of the "half Jew" Paul Heisel from the IG Farben plant at Gersthofen, the management had exerted its best efforts to prevent his being sent to work on a construction site of the Ministry of Armament's *Organization Todt* or being deported to a work camp or concentration camp.[431]

---

[428] HA, PA Abteilung VI, Erlenbach, personal data sheet; Beck et al., *Juden in Hoechst*, p. 48.

[429] HA, PA Abteilung VI, Erlenbach, Wieland to Hoechst, July 6, 1928; response from Director Schmidt, July 21, 1928; the date of joining is given as September 10, 1928.

[430] Bäumler, *Century*, pp. 314–15.

[431] HA, PA Abteilung VI, Erlenbach, personal data sheet; Military Government of Germany Fragebogen; cf. also HA, Hirschel, "Aus meiner I.G. Zeit," pp. 35–6. On Paul Heisel see

The extent to which cost-benefit calculations affected the management's considerations – Heisel was closely involved in the war production in Gersthofen – can be established for one enterprise, the Wacker works in Burghausen, 50 percent of which belonged to IG Farben. At the beginning of November 1944 a confidential note stated that five "followers, who were either Jews according to the Nuremberg Laws or married to Jewesses," had received a summons from the labor exchange that they would have to report for an examination. One of them had received a "summons to present himself to the secret state police [Gestapo]," and Wacker feared that the others would receive similar summons. The loss of these employees was considered a "severe burden"; however, only in two cases was their potential removal considered a "danger for the continuation of operations." Accordingly, the note continued: "We therefore regard it as right that in our attempts to gain their release we shall concentrate on these cases." The other three cases could be "accepted at a pinch," especially if they could be used to "make it easier" to gain the release of the other two.[432] It is not possible to formulate the company's own interests any more clearly. In Hoechst this was almost certainly the motivation behind any efforts the works management were prepared to expend for an "individual of mixed race" or an employee married to a Jew.

What is also important about the fates of the individuals briefly given here is not just how different their treatment could be, but also that Hoechst was prepared to offer some support to Jewish employees or to those considered to be Jews as long as this was not to the detriment of the company or if the company even benefited from such support – Schnitzer and Erlenbach are good examples of the fact that Jewish employees or employees of Jewish origin were allowed to remain in senior positions if their work was valuable.

Altogether, it is possible to say for all of IG Farben that support and/or protection were given only if this assistance was not in any way injurious to the company. To first ensure the protection of the company and then that of former employees was the motivation behind the behavior of the "Aryan" senior management in IG Farben toward its employees who were Jewish or regarded as Jewish. This was particularly evident in the case of the Jewish director of the IG Farben plant at Wolfen, Richard May; none less than Fritz ter Meer attempted to prevent him from emigrating in the second half of the 1930s, as he did not wish May's knowledge and abilities to be placed at the service of a foreign company.[433]

PA Abteilung VI, Heisel; because of the focus on the Hoechst plant the issue of Paul Heisel will not be addressed further.

[432] Wacker UA, 4 A 5, Notiz C/Dir. Ra/Fi 2480 vom 2. Nov. 1944: A 5911 (Confidential).

[433] Cf. HA, TEA 2011 (previously 832), includes extensive correspondence on the case of Director Richard May of the IG works at Wolfen.

3.7. THE SURRENDER OF THE HOECHST MANAGEMENT TO THE NSDAP DURING THE WAR

Since the 1930s the factory management had maintained direct contact with the *Gauleitung* and the local Gestapo through Hans Wagenheimer – with whose help the executive attempted to avoid or anticipate any problems with the Nazi regime. The good reputation Hoechst enjoyed at the *Gauleitung* is demonstrated by the fact that Anneliese Sprenger, daughter of the *Gauleiter*, accepted a position in the plant in the summer of 1940. Anneliese Sprenger had attended a prestigious boarding school in Erfurt, where she completed secondary school, after which she went to a school in Freiburg to receive some training in housekeeping and finally spent half a year at a foreign language school in Hamburg.[434] Her father apparently chose that city because her uncle was living there: "Sprenger thus knew that the young lady would be well taken care of in Hamburg." After completing her exams "Anneliese returned to her parents, so that her father once again had her under his eye." *Gauleiter* Jakob Sprenger, as his biographer wrote, was a "person with a conservative worldview, whose ideals included the preservation of marriage and the family."[435] Only two weeks after her return from Hamburg, Anneliese Sprenger, now nineteen years old, applied for a position in Hoechst. She was taken on as a commercial clerk in the Patents Department on July 18, 1940, the same department that had dismissed Emilie Thoennissen on the instructions of the *Gauleitung* one year earlier. However, from the start her job was envisaged as only temporary. Hoechst therefore refrained from giving her an examination and from the usual probationary period as she was employed only "for the purposes of continuing her training" and was not going to become a permanent employee.[436] At the beginning of November 1941, Anneliese Sprenger left her position quite suddenly "in accordance with an agreement made between Prof. Dr. Lautenschläger and the *Reichsstatthalter Gauleiter* Sprenger" to work as a secretary for her father in the *Gauleitung*. Her positive employer's reference bore Lautenschläger's signature.[437]

---

[434] HA, PA Anneliese Sprenger (microfilm), letter of application from A. Sprenger dated July 15, 1940, with attachments; Sprenger had thus not concluded an apprenticeship as a commercial clerk, as Zibell had stated – Stephanie Zibell, *Jakob Sprenger (1884–1945). NS-Gauleiter und Reichsstatthalter in Hessen* (Darmstadt: Hessische Historische Kommission, 1999), p. 378.

[435] Zibell, *Jakob Sprenger*, p. 378.

[436] HA, PA A. Sprenger (microfilm), "Betr.: Anstellung, 16.7.[1940]," handwritten notes in Sprenger's personal data sheet, obviously from Schwamborn; Hoechst to A. Sprenger, July 17, 1940.

[437] HA, PA A. Sprenger (microfilm), memo dated November 7, 1941, noting that A. Sprenger would leave the company on November 8 to begin working for her father on November 10; reference dated November 8, 1941; see also Zibell, *Jakob Sprenger*, p. 379.

Considering the care Sprenger took of his daughter and the number of companies in Frankfurt where she could have worked, the choice of Hoechst must be considered a sign of the *Gauleiter's* confidence in the plant and its management – this favor was clearly owed to Wagenheimer.[438]

Shortly before Sprenger's daughter was employed in Hoechst, Ludwig Retzinger was removed from his position as *Betriebsobmann;* Wagenheimer allegedly had a hand in the matter. *Betriebsobmann* Retzinger had never been an uncontroversial figure and, as already mentioned, had been removed from office once before. After the war Retzinger explained his renewed dismissal from office in February 1940 as being ostensibly due to charges of incompetence, while in fact he had been dismissed because he had "uncovered the machinations of an SD man (Pg. Wagenheimer)" and had spoken out "against him and other older Party comrades."[439] These reasons for his dismissal are by no means implausible, as we shall see later.

The office of *Betriebsobmann* in the plant was then given to the head of the local group of the NSDAP in Frankfurt-Hoechst, the graduate engineer Hermann Zeh. Born in 1890, Zeh had studied at the Technical College of Darmstadt; his studies were interrupted by the First World War, during which served four years as a soldier. Zeh had worked for Hoechst as production engineer in the water supply works from 1923 to 1930; he was dismissed at the end of 1930 when there was no more work for him. Difficult times followed, for in the next few years Zeh was unable to find a job. An attempt by Hoechst to find him a job in Moscow in 1931 failed.[440]

Only two weeks prior to Hitler's seizure of power Zeh applied once again to Hoechst, asking whether it would not be possible to do something for him; as Hoechst was aware, he was living "in great penury." In the meantime, on February 1, 1932, Zeh had become a member of the NSDAP

---

[438] HA, PA A. Sprenger (microfilm), Personal-Abteilung to Wagenheimer, August 7, 1940, with the request to complete the documents of A. Sprenger. Anneliese Sprenger's move to the offices of the *Gauleitung* at the end of 1941 may therefore have been connected to Wagenheimer's leaving Hoechst at the beginning of 1941. That Wagenheimer as a trusted confidant of Sprenger was expected to keep a watchful eye on his daughter is not implausible, for when his daughter went to Oslo in 1943, according to Zibell, *Jakob Sprenger*, pp. 379–80, Sprenger engaged "two trustworthy men known to him" to keep an eye on her.

[439] HHStAW, Abt. 520 F (A-Z), Retzinger, Retzinger's petition on denazification dated June 16, 1946.

[440] HA, PA Zeh (microfilm), Zeh to Hoechst, January 8, 1923; Hoechst to Zeh, June 27, 1930; reference for Zeh dated July 1, 1930; Ernst May (Moscow) to Orth (Hoechst), January 16, 1931 (copy); Orth (Hoechst) to May, January 22, 1931; Orth to Zeh, April 13, 1931; Hessisch-Nassauische Lebensversicherungsansalt to Hoechst, December 23, 1931, and response from Hoechst, December 28, 1931; on the role of the *Ortsgruppenleiter* (local NSDAP leaders) cf. in detail Carl-Wilhelm Reibel, *Das Fundament der Diktatur: Die NSDAP-Ortsgruppen 1932–1945* (Paderborn: Schöningh, 2002).

and soon afterwards a leading executive of the local chapter in Hoechst.[441] During his visit to Hoechst on January 17, 1933, Zeh stated, according to the relevant memorandum, "that he was close to despair since all his efforts to somehow find paid employment had failed." In response to the question put to him by the Hoechst employee Zeh was visiting, a Mr. Landmann from the Social Department, as to whether as an official in the Nazi Party Zeh did not receive some form of remuneration, Zeh answered "that his activity was completely voluntary and he did not receive a singe penny for it." Zeh added that he had taken on this work "just to have some intellectual activity again." Landmann promised to find out whether there might be some position available for Zeh. Yet works manager Hermann personally decided, although "the matter was regrettable," that because of numerous similar cases nothing should be done, and that Zeh should neither be given a temporary job nor any financial assistance.[442]

Two weeks later, after the appointment of Hitler as German Chancellor and the seizure of power, the situation in Germany changed and with it that of Hoechst. In mid-July 1933 Zeh was given a temporary position, at first limited to a period of six months, primarily on the basis of his position in the NSDAP. The written confirmation from Hoechst stated: "it will not be possible to admit you to our pension fund for employees since your employment with us is only of a temporary nature."[443] Zeh was also employed at a considerably lower salary. His direct supervisor, chief engineer Jähne, finally decided in July 1934 that "for the time being, Zeh shall continue to be employed." The time limit of six months, always previously mentioned, now was omitted. Although at the time it was still strongly suggested to Zeh that he should endeavor to find himself "another, permanent employment," Hoechst hired him on a permanent basis in November 1934 – the same year that he became head of the local group of the NSDAP (*Ortsgruppenleiter*) in Frankfurt-Hoechst.[444]

Just how much Zeh's employment reflected preferential treatment over other candidates is borne out by comparison with another biography. In the Nuremberg trial Jähne's defense lawyer attempted to show that Jähne was not a National Socialist and referred in this context to a "half Jew" Popp whom Jähne had supported.[445] Indeed, Walter Popp was employed from August 1934 till April 1935 as an engineer hired "as a temporary help for a limited period," but afterwards his contract was not renewed. Popp had applied for reemployment in Hoechst in July 1933 after he had read in the

[441] BA, BDC file on Zeh, NSDAP membership number 953,897.
[442] HA, PA Zeh (microfilm), Aktenvermerk by Landmann dated January 17, 1933.
[443] Ibid., Hoechst to Zeh dated July 21, 1933.
[444] Ibid., Ing. Gelbert (Bauabteilung Hoechst) to Landmann, July 4, 1934; Jähne and Landmann to Zeh, July 10, 1934; internal note for Landmann, November 12, 1934.
[445] HA, IG/50, IG-Prozessakten, Trial letter for Jähne from defense lawyer Hans Pribilla.

newspapers that IG Farben was hiring again and that the engineer Zeh, who had been dismissed at the same time, had found employment in Hoechst. Popp received a negative reply to his letter, in which it was pointed out that these jobs were merely temporary. Moreover, the letter continued: "In the matter of persons being taken on, in the first instance we were obliged to consider the fighters of the national movement, especially those who had particularly suffered great privations because of their social circumstances. Although today we still have a surplus of technical employees, after long deliberations we came to the opinion that we could take on the responsibility of temporarily hiring the engineer Mr. Zeh, as he is the oldest of the engineers who were dismissed at same time together with yourself and, due to his marriage and his lack of financial resources, he appears to be affected most by unfavorable economic circumstances."[446]

In the first elections to the Council of Trust held in 1934 Zeh became a deputy member of the Council. He took over from Retzinger as *Betriebsobmann* for the Hoechst plant in 1940; in 1941 he finally became senior *Betriebsobmann* for the Maingau works. Zeh was also able to strongly improve his financial position in the years between 1934 and 1945 – his income increased from 7,200 to 15,000 RM.[447] As *Betriebsobmann* he apparently approved promotions of employees and foremen only if these persons were members of the NSDAP or Nazi organizations and could be considered "politically irreproachable." Thus he opposed the employment of a young woman, despite her membership in the League of German Girls, because, according to Zeh, she spent more time and was more involved in the Catholic Church.[448] He was willing to proceed against "fellow workers" on the slightest suspicions or denunciations, even demanding that the Gestapo arrest employees. At his instigation the Gestapo arrested the employee Ernst Krell in the summer of 1944 after someone had informed against him. Krell was released again only in the spring of 1945 when at a hearing in front of a "special court" he "was acquitted for lack of proof."[449] Zeh also alleged that at the end of 1944 a certain Georg Hissenauer had stated that the Nazi government was using women and children as cannon fodder and that Hissenauer had referred to the government in a not very roundabout fashion as "blackguards." Zeh therefore demanded

[446] HA, PA Walter Popp (microfilm), Popp to Hermann, July 25, 1933, Hoechst to Popp, July 28, 1933 (from where the quotation is taken); Hoechst (Jähne/Landmann) to Popp, July 25, 1934; reference from Jähne for Popp dated April 30, 1935.

[447] HHStAW, Abt. 520 F, Nr. 4815 (Zeh), Meldebogen and Military Government of Germany, Questionnaire of Zeh dated December 7, 1946; testimony of Zeh according to Protokoll der öffentlichen Sitzung der Spruchkammer, January 12, 1949.

[448] Ibid., Zeh to Sozial-Abteilung, July 30, 1941, re. "Aenne Kopp."

[449] Ibid., testimonies of Pöhn, Zeh, and Krell according to Protokoll der öffentlichen Sitzung der Spruchkammer, January 12, 1949.

that he be arrested by the Gestapo: "Hissenauer is employed in our main workshop, the members of which do not always show a clearly positive attitude. Since I regard Hissenauer as one of the men who has an unfortunate influence on the mood of fellow workers, I request that vigorous action be taken immediately and that you arrange the arrest of Hissenauer by the Gestapo." Ironically Hissenauer never learned about this denunciation – luckily for him the letter, sent shortly before the end of the war in the Rhine-Main area, yielded no results.[450]

Despite the value Zeh placed on "political reliability," from time to time he did exhibit some humane traits, and at his appearance in front of the denazification tribunal after the war he tried to present himself as "decent" and "humane." Thus a senior employee of Hoechst, who was married to a Jewish woman, apparently suffered no disadvantages from his marriage – as the employee himself testified after the war. Nor had Zeh prevented this man's son, who had had to leave school because of being a "half Jew," from being employed in the plant. He emphasized, however, that despite being neighbors they had avoided each other and did not greet each other.[451] But as the denazification tribunal stated, the occasional exhibition of humane traits did not lessen his harsh stance toward opponents of the regime. With his willingness to back informers Zeh contributed substantially to the conformist climate in the works.

As discussed in the last two chapters, the works management of Hoechst was already highly compliant before the war; indeed, the Party had substantially intervened in the works. Not just the daughter of the *Gauleiter* but also a number of "old Party members" had found work in Hoechst in the 1930s, and they now sought to consolidate their positions. During the war in particular they managed to occupy important posts, primarily but not only at the expense of senior employees who were not members of the NSDAP. The latter persons felt increasingly pressured to join the NSDAP or to otherwise run the risk of losing their position or even being dismissed – no matter how well they cooperated with the Nazi representatives in the works.

In the hope of keeping their positions or even improving them, several senior employees who had previously maintained a certain distance to the regime joined the NSDAP during the war.[452] Director Josef Nüsslein, who had formerly worked in the IG Farben plant at Ludwigshafen and switched to Hoechst in 1934 to take over the management of the Coloristic

[450] Ibid., Zeh to DAF, Kreisobmann Frankfurt, December 5, 1944; see also testimony of Hoechst Works Council (Trost) dated October 31, 1947, and Protokoll der öffentlichen Sitzung der Spruchkammer, January 12, 1949.

[451] Ibid., written testimony of Dr. Herrmann and Protokoll der öffentlichen Sitzung, January 12, 1949. Cf. on this topic also n. 425 above.

[452] Cf. Hayes, *Industry and Ideology*, pp. 199–203.

Department, applied and was admitted to the NSDAP in 1941. As a devout Roman Catholic deeply involved in the Church, he had for a long time apparently taken a skeptical view of the Party and the new regime. However, in 1940 he decided to join, not least, as he said after the war, because of his children. The situation in the plant also played a decisive role because, as he also stated after the war, he was "being increasingly bothered by complaints from National Socialist circles." Finally, while his application for membership was still being processed, someone denounced him to the Gestapo. However, the proceedings against him were quashed after he had been subjected to several interrogations. In 1941 he was accepted into the NSDAP and subsequently felt safe from such threats. As he himself wrote, he was "unable to report any heroic deeds" for that period, but he had apparently never greatly exposed himself politically.[453] The authorized signatory and head of the Central Laboratory, Heinrich Greune, also joined the NSDAP in 1941 – clearly after strong pressure was brought to bear on him by Kränzlein, to whom he was not just subordinate but with whom he was also obliged to share an office during the war. Greune apparently made no secret of his opposition to the NSDAP and the Nazi regime, yet he nevertheless finally chose to adjust by applying for Party membership when it was a question of keeping his position in the plant.[454] Ludwig Orthner, head of the Textile Additives Laboratory, also became a member of the NSDAP only during the war. In the case of this severely disabled veteran of the First World War and outstanding chemist, who had studied under the "Jewish" professors Kurt Hans Meyer, Richard Willstätter, and Stefan Goldschmidt and had written his *Habilitation* under the last, still another motive was involved. Despite having moved into industry, Orthner was an enthusiastic lecturer and held the position of an adjunct professor at the University of Frankfurt beginning in 1939. After he heard that only Party members would be allowed to lecture at the university in the future, he applied for membership in the NSDAP early in the summer of 1941 and became a Party member in the summer of 1942. After the war Orthner was not only able to claim that he had not been politically active, he was even able to point to certain positive actions on his part toward "foreign workers."[455] Julius Weber, the head of the pharmaceuticals

---

[453] HA, PA Abteilung VI, Nüsslein; HHStAW Abt. 520 F (A-Z), Nüsslein, Meldebogen dated May 4, 1946, with attachments; cf. also testimonies and Urteil der Spruchkammer dated May 28, 1947.

[454] HA, PA Abteilung VI, Greune; HHStAW Abt. 520 F (A-Z), Greune, Meldebogen and testimonies; Urteil der Spruchkammer dated June 25, 1947: classified as "fellow traveler" (*Mitläufer*).

[455] HA, PA Abteilung VI, Orthner; HHStAW Abt. 520 F (A-Z), Orthner, see his own statement in Orthner to Kammer dated October 17, 1946; testimonies of the French chemists Cumet (October 11, 1946) and Deremenesnil (October 16, 1946), together with testimonies of a number of employeees – in this case the declarations do appear credible. In

research office, also joined the NSDAP at the same time, after problems with the Gestapo because of his contacts with the Catholic opposition. At the time Wagenheimer helped him to get his Party membership and settled his problems with the Gestapo, for which Weber continued to be grateful after the war.[456]

Thus, adjustment to the Nazi regime now encompassed people who had managed to keep their distance during the 1930s. Worry about keeping their positions, whether in industry or in education, led them to become "fellow travelers" and to join the NSDAP. With this step they demonstrated, at least externally, that they supported the regime and the Nazi state, whatever their individual motives might have been. Just how all encompassing the nazification of Hoechst's senior executives was became apparent after the war. A transcript on the Technical Directorate, written around the middle of May 1945, states that of the "prominent employees" (*Prominente*), that is, the senior executives in the so-called Department VI, "25% were not Party members" – in other words, three out of four senior employees belonged to the NSDAP.[457]

If this figure is compared with the figures for the staff employed in the plant, categorized according to occupational group, it becomes obvious that in Hoechst the quota of Party members was disproportionately high among the senior executives. In August 1945, 32.8 percent of the chemists, dyes specialists, and pharmacists, 34.4 percent of the graduate engineers, 34.4 percent of the engineers and technicians, 22 percent of the commercial clerks, 18.8 percent of the master craftsmen and cost accountants, 15.4 percent of the laboratory workers, and 13.5 percent of the other employees were classed as "offenders" (*politisch belastet*), that is, active members of the NSDAP. Among the foremen the number of Party members was considerably lower: 11.5 percent of the workmen, boilermen, and engineers, 7 percent of the unskilled foremen, and a mere 3.6 percent of the women were Party members.[458]

Despite the fact that such a high number of executives in Hoechst joined the NSDAP, not everyone followed suit – and not all of those who were not

the denazification proceedings Orthner was classified as a "fellow traveler"(*Mitläufer*), and the proceedings were then dropped following the Christmas amnesty – he was therefore in the same group as the former *Betriebsobmann* Retzinger.

[456] HHStAW, Abt. 520 F (A-Z), Julius Weber, Weber to Oberbürgermeister, December 20, 1944; according to BA, BDC index card on Weber, accession dated May 4, 1940, application for accession dated July 1, 1940.

[457] HA, TEA 95f, Niederschrift über die technische Direktionssitzung in Hoechst, May 12, 1945.

[458] HA, file Pensel-Entnazifizierung, Allgemeiner Schriftwechsel, Personalabteilung, statistical table of white-collar employees employed in the plant grouped according to professions dated January 27, 1949: the percentage of politically incriminated employees was added by hand in August 1945; list of the foremen employed in the plant grouped according to non-Party members and Party members, also dated January 27, 1949.

members lost their jobs. The most prominent examples of executives in senior positions who were not NSDAP members were Director Max Bockmühl, manager of pharmaceutical research, and Director Paul Roth, head of acetone–acetic acid operations. However, both men were also outstanding experts in their field and would have been difficult to replace. At least one of them was politically "safeguarded" by his relatives.[459] In the nontechnical areas, executives who did not join the NSDAP could apparently be replaced more easily, and during the war some of them found themselves facing the end of their professional career – even though they had made more concessions to the regime in their professional life than had been necessary.

The most prominent case is that of Director Wilhelm Schwamborn, who, as head of the Social Department, was responsible for questions concerning the staff in the plant as well as the company health insurance fund, department store, work savings bank, and welfare services and works housing. Since 1938 he had been extremely cooperative, almost obsequious, toward Wagenheimer. His close cooperation with the plant's contact man to the *Gauleitung* and the Gestapo not only contributed to the harsh treatment of Gärtner by the plant, but Schwamborn had also tendered his services to the regime in the matter of Jewish employees or employees considered to be Jews. In fact, he had even gone beyond what was required in his treatment of *Mischlinge*, those of partly Jewish ancestry. This conformism finally led by 1939/40 to all new appointments in Hoechst and in the other Maingau plants being checked by the *Gauleitung*.[460] After the war a member of the works council said about the Hoechst plant: "If someone was promoted in the Third Reich, not his ability but his political reliability was decisive."[461] However, even Schwamborn's wide-ranging concessions toward the demands of the regime and his close cooperation with Wagenheimer were of increasingly little help for Schwamborn himself.[462]

The decision to replace Schwamborn was taken even prior to his sixtieth birthday. After consultation with IG Farben's Management Board

[459] HA, PA Bockmühl und PA Roth; Entnazifizierungsakten (Vorstandssafe), Brisbois et al. to Lt. Col. Percival, July 23, 1945 (copy).

[460] HA, PA Lautenschläger, Sekretariat Lautenschläger, Personalangelegenheiten, Schlick to Lautenschläger, December 20, 1940, Schlick emphasized that as a "mark of confidence" in him as the new head of the Social Department in Hoechst, *Gauleiter* Sprenger had at the end of 1940 decided to limit "the till then customary checks of new appointments in Maingau to the new appointments of leading members of the workforce."

[461] HHStAW, Abt. 520 F, Nr. 4815 (Zeh), testimony of Trost in the trial dated January 12, 1949.

[462] HA, PA Abteilung VI, Schwamborn, personal data sheet, including an article from the "Main-Taunus-Zeitung" dated October 28/29, 1939, on the occasion of Schwamborn's sixtieth birthday.

Lautenschläger chose Heinrich Schlick, a Party member, as Schwamborn's successor. At the beginning of October Lautenschläger and Schlick held initial talks, about which Schwamborn apparently was not informed. Thus, at the beginning of October 1939 Schlick wrote to Lautenschläger that he inferred from Schwamborn's remarks during a conversation that he regarded Schlick's transfer to Hoechst as a "probationary period of employment" and not as his training period prior to taking over the succession: "This incorrect opinion would not affect me much, if I had not at the same time gained the fixed impression during the conversation that on my transfer to Hoechst I will doubtlessly have to contend with extreme difficulties."[463]

At the end of 1939 Lautenschläger and Schwamborn agreed that Schwamborn should retire at the end of 1940, and his departure was sweetened by very generous benefits.[464] Schwamborn, who had apparently not expected his removal and about whom rumors began to circulate, including even "doubts concerning his proper behavior" when in office, fell ill in the fall of 1940. After his recovery he did not return to his job, but wanted to discuss the rumors buzzing around with Lautenschläger "on neutral ground," preferably with Jähne present. While Lautenschläger declined to include Jähne in their talk, he offered to hold the conversation with Schwamborn in his office at the IG Farben headquarters in Frankfurt, since Schwamborn "no longer had a car at his disposal" and Lautenschläger wished "to save him the long journey."[465] In view of this correspondence and the rumors that continued to circulate, most of which referred to a slightly dubious purchase of flashlight batteries, the letter sent to Schwamborn on his enforced retirement must have seemed like sheer mockery, despite his princely severance pay: "As is generally known, it was considered that your contract of employment for an indefinite period would be allowed to run for a certain period beyond your 60th year of life. But we are able to release you from your responsible activities and have therefore come to an agreement with you that you already can retire on December 31, 1940."[466]

Originally Schlick had feared that he would have problems with his predecessor, Schwamborn. However, his concerns proved to be unfounded.

[463] HA, PA Lautenschläger, Sekretariat Lautenschläger, Personalangelegenheiten, Schlick to Lautenschläger, October 7, 1939.

[464] HA, PA Abteilung VI, Schwamborn, Aktennotiz by Lautenschläger dated November 23, 1939, with addition dated December 10, 1939.

[465] HA, PA Lautenschläger, Sekretariat Lautenschläger, Personalangelegenheiten, Schwamborn to Lautenschläger, October 28, 1940; Schwamborn to Lautenschläger, December 29, 1940; Lautenschläger to Schwamborn, January 3, 1941.

[466] HA, PA Abteilung VI, Schwamborn, Lautenschläger und Jähne to Schwamborn, December 31, 1940. It must be added that Schwamborn was not satisfied with the very generous compensation awarded him, and this led to disputes with Lautenschläger.

After Schlick wrote to Lautenschläger about possible problems at the beginning of October 1939 and informed his previous supervisor, Carl Wurster, the IG Farben member of the Board of Directors, about them, Lautenschläger attempted to clarify possible problems. After his discussion with Lautenschläger Schlick expressed himself as "very much satisfied," although the same could not be said about Schwamborn. Moreover, it was agreed with Fritz ter Meer at the beginning of December 1939 that Schlick should become an authorized signatory as soon as possible, preferably around the middle of 1940.[467]

As everything had run smoothly for Schlick so far and according to his ideas, he rapidly started to overestimate his own position and soon faced serious problems in the plant and with the *Gauleitung*. He had been in Hoechst only a few months when as a Party member he was given the opportunity by Adalbert Gimbel, a district leader in the NSDAP and a member of the Reichstag, to meet the *Gauleiter*. Schlick was happy to take up the offer, and the meeting took place, apparently around the end of October 1940.[468] However, in the conversation, Schlick did not content himself with an exchange of compliments, but also appealed directly to Jakob Sprenger about problems with two employees in the work: the "old fighter" Walter Schilling, who worked in the company health insurance fund, and Wagenheimer, above all.

In another conversation with Sprenger at the beginning of November 1940, now in the presence of *Betriebsobmann* Zeh, Schlick drew Sprenger's attention to the fact that "rumors were circulating about Pg. [Party comrade] Wagenheimer in Hoechst, which did not exactly speak for Pg. Wagenheimer." He also mentioned "the obtaining of flashlight batteries" for the plant's department store. Schlick suggested talking the matter over with Wagenheimer "as Party comrade to Party comrade" to "clear up the affair." Sprenger appeared to agree with Schlick; he even commented to Schlick "so to speak as a guideline" that "he had turned over a district leader with whom he had cooperated for a long time to the public prosecutor." Schlick understood this statement as an invitation to proceed against Wagenheimer, who had in the meantime had gotten wind of the matter, and preempted Schlick by himself requesting an examination of the matter by the *Gauleitung*.[469]

[467] HA, PA Lautenschläger, Sekretariat Lautenschläger, Personalangelegenheiten, Wurster to Lautenschläger (persönlich), October 9, 1939; Wurster to Lautenschläger und ter Meer, December 8, 1939, in Anlage Aktenvermerk by Wurster dated December 6, 1939, on discussion of ter Meer, Lautenschläger, and Wurster on December 5, 1939.

[468] Ibid., Gimbel to Schlick, December 14, 1940; note by Zeh dated November 29, 1940, on "Pg. [=Party comrade] Dr. Schlick"; Schlick to Lautenschläger, December 20, 1940.

[469] Ibid., note by Zeh dated November 29, 1940 on "Pg. [=Party comrade] Dr. Schlick"; Schlick to Lautenschläger, December 20, 1940.

Even if at the beginning of November Sprenger had initially expressed himself "in a favorable manner" about Wagenheimer toward Schlick, Schlick apparently now saw an opportunity to considerably improve his own position in the plant and to oust Wagenheimer, even if he expected to encounter problems with the works management. Thus Schlick emphasized to Lautenschläger, who was outraged that the matter had not been settled internally, that the *Gauleitung* placed great confidence in himself. Sprenger had given instructions to the head of the *Gau* Labor Office, Holländer, in Schlick's presence that "the customary checks carried out until then of all new appointments in Maingau were to be restricted to new appointments of leading members of the workforce." Schlick self-confidently reported that Sprenger had indicated that "I could consider this order as a mark of his confidence in my person." Furthermore, Schlick wanted to clearly demonstrate that he had behaved correctly and had been right to report Wagenheimer. In his position "as a Party member and in the interests of the Party's reputation" Schlick had wished to obtain "the *Gauleiter*'s authorization" in order "to talk about these questionable events openly with Wagenheimer from Party comrade to Party comrade." He made it clear just how seriously he took the matter of the dubious purchase of batteries by adding "Should the matter be of a more serious nature, the *Gauleiter* will expect a further report anyway, whereby the *Gauleiter* gave me to understand that in future his door would be open to me in all important matters." Schlick also emphasized that he had chosen this means of proceeding to forestall "an otherwise inevitable official examination of the matter" and that he had at first only considered his activities as a "purely Party affair." However, Wagenheimer had anticipated a talk Schlick had planned to hold with him, for he had "received information about the matter by unknown means" and had presented "the controversial action in a form which was favorable to himself." Schlick still wanted to proceed against Wagenheimer and felt himself to be supported in this matter by *Betriebsobmann* Zeh, who procured a legal opinion concerning the sale of the batteries from the assessor at the *Gau* court and president of the professional association of lawyers for Hesse-Nassau, Counselor of Justice Leuchs-Mack. According to this opinion Schlick felt that it was "quite obvious" that in the matter of the purchase of the batteries Wagenheimer, in his capacity as go-between, had violated the ordinance prohibiting price increases and was answerable for attempted and actual fraud according to the German Criminal Code. "This proves that the rumors concerning Wagenheimer were by no means unfounded."[470]

Thus Schlick and Zeh formed an alliance with the intention of exposing and possibly even criminalizing Wagenheimer and used the story of the purchase of the batteries to remove him from his preeminent position with

[470] Ibid.

respect to the factory management and the *Gauleitung*. Schlick and Zeh were accusing Wagenheimer of nothing less than profiteering, and part of their accusation was based on the legal opinion. However, Wagenheimer had gotten wind of the matter in due time, recognized its explosiveness, and taken the offensive. In the previous few years Wagenheimer had not just helped the plant by clearing up tricky questions concerning staff members in advance; on the basis of his connections to the *Gauleitung* and to other important persons he also had managed to procure various items for the plant that were otherwise very difficult to obtain – and that even included horses from the SA for the Behring works.[471]

The purchase of 2,000 flashlight batteries for the works' department store, on which the whole affair rested, had been originally ordered by Schwamborn in August 1940, and had constituted a basis for rumors against him at the time. Schwamborn had informed Wagenheimer that he wished to obtain flashlight batteries for the works' department store, and Wagenheimer was soon able to report to him that such batteries could be delivered. At Schwamborn's request Wagenheimer procured the batteries via an "agent" who, however, wanted a "commission." Wagenheimer informed Schwamborn, who agreed to pay and reward the agent's efforts. According to Schwamborn, he had even been prepared to dispense with "a profit margin in view of the well-known difficulties of procurement" and in view of the fact that "it had been possible to meet the pressing demand for batteries created by the dangers resulting from the required blackout in the plants."[472]

The head of the Legal Department for the *Gau* of Hesse-Nassau judged the matter accordingly. He did not consider "a one-off procurement of a batch of flashlight batteries without a trading license" as a punishable offence on the part of the agent nor did he regard the transaction as questionable with respect to the "legality of the price." Nor was he of the opinion that "Pg. Wagenheimer was in breach of duty as an employee of IG," since Wagenheimer's intentions had been sincere. According to the head of the *Gau's* Legal Department Wagenheimer's intention "had only been to render a special service to hard-working national comrades, namely the workers of IG."[473]

[471] Ibid., Lautenschläger to Wagenheimer, April 21, 1941. This appears to have been one of Wagenheimer's specialties. According to HA, Hirschel, Aus meiner I.G. Zeit, p. 35, during the Nuremberg trial he supplied Lautenschläger and Jähne with food and cigarettes.

[472] HA, PA Lautenschläger, Sekretariat Lautenschläger, Personalangelegenheiten, declaration of Schwamborn dated January 10, 1941, concerning the purchase of 2,000 flashlight batteries in August 1940, Schwamborn commented initially that he was making this declaration "solely at the express request of Mr. Wagenheimer."

[473] Ibid., NSDAP Gauleitung Hessen-Nassau, Gaurechtsamtsleiter (Dr. Vogt) to Gauinspekteur Reisse, December 20, 1940 (copy).

With this rehabilitation Schlick's report against Wagenheimer within the Party was summarily rejected. But even prior to this Schlick's relations with Lautenschläger had already cooled, as Lautenschläger was incensed that Schlick had gone over his head on an internal matter. Lautenschläger wrote indignantly to Christian Schneider, the member of IG Farben's Management Board responsible for personnel matters, that his confidence in Schlick had been "very much shaken" since Schlick "had gone to the *Gauleiter* without my knowledge and reported matters which should at the very least have been discussed with the plant leader beforehand." He felt himself to be in an awkward position because Schlick's conversation with Sprenger had created "a very unpleasant situation" and "the *Gauleiter* has withdrawn his confidence from him and expects me to fill the vacancy of Dr. Schlick at your earliest convenience." In order to "prevent the *Gauleiter* from installing someone in the Social Department," Lautenschläger had immediately decided to place that department under Otto Hirschel. At the same time Lautenschläger also demanded the immediate removal of Schlick.[474] In the report of the Technical Directorate of Hoechst at the end of January 1941 this was presented somewhat less dramatically: "Lautenschläger announces that Dr. Schlick will be leaving the Hoechst plant at his own request on 1/28/41 to join the IG plant at Heydebreck near Breslau. Dr. Hirschel will take over the management of the Social Department."[475]

With his brash manner and circumvention of Lautenschläger Schlick had not just "committed the grossest breach against the bond of trust to his plant leader." He had also showed Lautenschläger up in front of the *Gauleitung* and within IG Farben. Sprenger had even approached Carl Krauch in the matter.[476] Moreover Sprenger informed Lautenschläger "in a personal letter" about the result of the internal Party inquiry into Wagenheimer's behavior: "It turned out that the accusations against Wagenheimer were unfounded and W. has emerged from the investigations as irreproachable, both personally and in his behavior. As far as the Party is concerned there will be no consequences against Wagenheimer."[477] This

[474] Ibid., Lautenschläger to Schneider, December 21, 1940; Schneider had written to Lautenschläger on December 16, 1940, that Schlick himself had already requested "a position elsewhere within the company."

[475] HA, TEA 95f, Niederschrift über die technische Direktionssitzung in Hoechst, January 27, 1941. After the war Schlick rose to become a high-ranking representative in various associations of the chemical industry and the federation of employers. He was well known for his work on company pension schemes and as a judge of the Labor Court of the *Land* and of the Federal Labor Court (documents on Schlick from the BASF archives).

[476] HA, PA Lautenschläger, Sekretariat Lautenschläger, Personalangelegenheiten, Lautenschläger to Schneider, December 21, 1940, and note on "Fall Dr. Schl" and "Fall W." dated January 12, 1941 (no author).

[477] Ibid., Lautenschläger to Schneider, January 6, 1941.

written communication of the Party's findings made an independent examination by the factory management more difficult. But even after an investigation was carried out within the plant itself no other conclusion was reached: Wagenheimer had procured the batteries on the orders of the plant; not he, but Schwamborn, had been responsible for consenting to pay the commission, and even the higher price resulting from that payment was legally defensible. Moreover, according to Wagenheimer the agent had intended to use the commission "to go drinking with his soldiers" so that in the end the money benefited soldiers. And finally, by procuring the batteries Wagenheimer had intended to help the workforce. In the end, it was concluded "it would be wrong not to reject the accusations made against W."[478] So it appeared that Wagenheimer was the victor in the affair – but he too was soon to leave the plant despite his excellent connections to the *Gauleiter*. Only a short time later he was called up.

Not only was Lautenschläger's reputation in tatters, but he also lost his closest confidant and helper against encroachments by the *Gauleitung* and various Party members within the plant. It now became more difficult for the plant to obtain certain items, so it comes as no surprise that at the end of April 1941 Lautenschläger still turned to Wagenheimer, then already serving in the army, asking him to procure certain articles.[479] Above all, it now became far more difficult for Lautenschläger to ward off the ambitions of influential NSDAP members.

After the difficulties with the *Gauleitung* triggered by Schlick, Lautenschläger recognized the danger of a close friend of the *Gauleiter*'s taking over the Social Department. In his discussion with Sprenger Schlick had also mentioned a certain Walter Schilling and referred to a rumor currently circulating in the plant that the relationship between the *Gauleiter* and Schilling was "under a cloud." Sprenger had thereupon not only spoken "very positively" of Schilling but added: "The man who can disturb this bond of trust has yet to be born."[480] Sprenger even "expressly" instructed Schlick to inform Lautenschläger of his "attitude toward Pg. Schilling." Walter Schilling, born in 1907, was an "old fighter" who had already become involved with the NSDAP in 1923 and joined the Party in 1925. He was regarded as a close friend of Sprenger's, who was godfather to his first child. Schilling had trained as a commercial clerk and worked for the same company until 1930. From 1930 until the seizure of power in 1933 he was unemployed, during which time he took on a number of Party functions. Between 1933 and 1935 he had primarily been occupied with setting up the Leather Plant Community (*Gaubetriebsgemeinschaft Leder*) for the districts

---

[478] Ibid., note on "Fall Dr. Schl" and "Fall W." dated January 12, 1941 (no author).

[479] Ibid., Lautenschläger to Wagenheimer, April 21, 1941, and response from Wagenheimer, May 2, 1941.

[480] Ibid., note by Zeh dated November 29, 1940, on "Pg. Dr. Schlick."

of Hesse and Hesse-Nassau.[481] In the summer of 1935 he applied for a job in Hoechst and received the reply that although at present there was no "suitable vacant position" for him, he should nevertheless come for an interview. At the urging of Sprenger and of Becker, the *Obmann* of the DAF, he was nevertheless taken on in summer 1935 at a disproportionately high salary.[482] In the following years he worked in the Social Department, where he rose to be deputy head of the company health insurance fund and apparently gave a good account of himself.[483]

Whether Sprenger's remarks about Schilling made Lautenschläger fear that Schilling wanted to take over Schlick's position is not clear – but it is certainly plausible. For at almost the same time Schilling attacked Georg Müller, the head of the company's health insurance fund, with the intention of having him removed from office and taking over his position. Here too, Lautenschläger cut a bad figure. Once again it was demonstrated both to him and to a broader audience that he wielded no authority as works manager. For contrary to the express prohibition of Lautenschläger, who wanted to settle the matter within the plant, Schilling called on the Social Disciplinary Court to adjudicate the affair with Müller; according to the AOG and the Factory Code the function of the Social Disciplinary Court was to deal with problems that could not be solved within a plant. The works management was unable to prevent Schilling's action. As a senior employee of the Legal Department wrote, Schilling had "waited precisely until the day on which Mr. W. was called up, who would otherwise have protected the company and the fund, and would probably have indicated to Mr. Sch. on the orders of the *Gauleitung* that it was time for him to pack his bags." Lautenschläger now no longer had a direct line to the *Gauleitung* through Wagenheimer, who probably would have been able to forestall this renewed humiliation of Lautenschläger by Schilling. But Wagenheimer had left the plant.[484] Müller, who apparently had an alcohol problem, was relieved of his position at the end of March 1941 and was sent into

[481] HA, PA Schilling (microfilm), Schilling to Kränzlein with resumé and reference (obviously from June 1935); HHStAW, Abt. 520 F (A-Z), Schilling, Meldebogen dated April 23, 1946, there NSDAP membership number 17,395.

[482] HA, PA Schilling (microfilm), Schilling to Kränzlein with resumé and reference; Schwamborn to Schilling, July 2, 1935, with handwritten note "M 500 ab 1/8.35"; HHStAW, Abt. 520 F (A-Z), Schilling, Hoechst Betriebsrat (Trost) to Spruchkammer, September 24, 1947, "Stellungnahme."

[483] HA, PA Schilling (microfilm), see draft of a reference for Schilling dated November 29, 1945; HHStAW, Abt. 520 (A-Z), Schilling, IG Hoechst reference from Dr. Roth for Schilling dated December 4, 1945.

[484] HA, PA Lautenschläger, Sekretariat Lautenschläger, Personalangelegenheiten, memo from Dr. Hirsch for Dr. Hirschel dated March 6, 1941, and March 9, 1941 (quotation from there); Hirsch to Lautenschläger dated March 25, 1941; see also Wagenheimer to Lautenschläger, May 2, 1941, who expressed similar views to those of Hirsch.

retirement at the end of the year by the Social Disciplinary Court. After that Schilling took over Müller's position as head of the company's health insurance fund.[485]

For Lautenschläger the loss of face in front of the *Gauleitung* and within IG Farben was probably very great. Altogether the result of the intrigues swirling around Wagenheimer, Schlick, and Schilling was a disaster for the works management. It can safely be assumed, as it is suggested by the chronology, that the grooming of Karl Winnacker as Lautenschläger's successor, which occurred almost simultaneously in the summer of 1941, was connected to the events surrounding Wagenheimer, Schlick, and Schilling.

The capitulation to the regime did not just concern questions of personnel and of membership in the NSDAP, but went much further. The works management had not just handed over a large part of its authority and allowed the NSDAP to interfere with the running of the plant. It had sacrificed its values and awareness of what was right and legal. This was made clear in the case of Gärtner and in the treatment of Jewish employees or employees regarded as Jews. This was particularly illustrated at the end of March 1941 in a plea written by Josef Hirsch, the head of the Legal Department who had represented Hoechst in the Gärtner case, for certain NSDAP members now had him in their sights. Hirsch, who was ousted by Schilling at the same time as Müller, made a "political statement of belief" at the time, whose frankness drastically demonstrates the full extent of the adjustments and moral corruption that had occurred in the years since 1933.

Josef Hirsch, born in 1880, had come to Hoechst from Kalle in 1924. On January 1, 1942, he was pensioned off at Schilling's instigation.[486] In a personal letter to Lautenschläger Hirsch sharply rebutted Schilling's allegations of "an attitude opposed to the Reich, the Fatherland and the Party." His letter is primarily an admission of having seen and recognized the crimes of the regime and then rationalized them out of existence; it is a devastating document, highly discrediting to its author, of adjustment and moral decline. For it shows that someone saw what was happening, thought about it, but drew no other consequence than to assure a regime recognized as criminal of continued loyalty. The letter therefore deserves quotation at length as an indication of the moral surrender and corruption of the works management.[487]

[485] Ibid., Hirsch to Hirschel, March 9, 1941, and note on "Angelegenheit Müller/Frl. K." dated March 7, 1941; HHStAW Abt. 520 F (A-Z), Schilling, IG Farben Hoechst reference from Dr. Roth for Schilling dated December 4, 1945; statements by Trost (Hoechst Works Council) dated June 18, 1947, and by Embs on Schilling; Protokoll der öffentlichen Sitzung der Spruchkammer, April 5, 1948; HA, PA Georg Müller (microfilm).

[486] HA, PA Abteilung VI, Josef Hirsch.

[487] HA, PA Lautenschläger, Sekretariat Lautenschläger, Personalangelegenheiten, Hirsch to Lautenschläger, March 25, 1941: the *Glaubensbekenntnis* dates from March 24, 1941.

Hirsch was, as he wrote, a member of the nationalist ex-servicemen's organization *Stahlhelm* (Steel Helmet). Very early on he believed the party system of the Weimar Republic to be inadequate, and at the beginning of the 1930s he placed his hopes – "which soon proved to be illusory" – in Franz von Papen. Thereafter he had supported Hitler "without reservations." Together with Papen and Franz Seldte (the head of the *Stahlhelm*) and the NSDAP, "we, the *Stahlhelm* veterans," had worked for the "rebuilding" of the German Reich. Nor had he given up hope, even when in the spring of 1933 "certain things occurred which painfully affected every friend of the Fatherland and even more so every German who upholds the law." Hirsch quite specifically stated what these "things" were: "I am thinking of the destruction of non-Aryan businesses and the refusal of the relevant authorities to deploy any police for their security. At the time, we lived under the protection of established laws of the Reich and did not yet recognize that in order to effect a complete change, perhaps other means than mere compliance with norms were necessary." At all events this had not shaken his trust in the *Führer:* "Nor was it affected by the events of 1934 which – once again without due legal procedure – cost many innocent people their lives in addition to the guilty (Röhm and his comrades). At the time every German man, who took the welfare of the Fatherland seriously, criticized these things clearly and sharply. That, seen today, they were perhaps unavoidable does not change these historical reflections." The successes of the Nazi regime in the following years, which included the military reoccupation of the Rhineland and the annexation of Austria, had turned all "worried patriots" into "fighters for the *Führer* and the movement." If one did not now join the NSDAP, then it was only because one "did not wish to be classed among the latecomers and opportunists." Hirsch conceded, however, that even in those years there was a problem: "We experienced a brief setback in November 1938 when the synagogues of the Jews were set alight, an action which probably only few thinking persons did not disapprove of, as fears were felt concerning the very serious damage which might result from this measure for the German people ... I too expressed myself on the subject of this specific event as well as on other matters which followed (the removal of numerous non-Aryans into detention camps) unequivocally with disapproval and certainly used words which left no room for doubt." Since Hirsch had to defend himself against a serious accusation, he added: "I spoke out with particular severity against the lootings and thus only expressed the opinion of the government; for looters were severely punished." And to prove his Nazi views during the war, he continued: "It was only at the beginning of the war, for which Jews carry a fair bit of blame, that I learned to appreciate the purposes of these undoubtedly unpleasant activities and belatedly approved these purposes; for one cannot desire success and discuss the means afterwards." Thus he had promoted the war and loyally supported

Hitler. He did not consider himself "answerable" to any "informer"; this letter commenting on the accusations leveled against him was intended solely for the works leader.

The letter clearly shows that Hirsch had been quite aware of the undermining of the rule of law but nevertheless contributed to the stabilization of the Nazi regime and the end of the constitutional state. In his position he had not only been one of the persons responsible for the wording of the factory codes, he had also represented Hoechst in court against Gärtner, who had been the victim of vicious denunciations – and together with the *Gauleitung* had robbed him of his rights and endangered his livelihood. Hirsch had actively contributed to the system of oppression in Hoechst, although as an "upholder of the law" he had recognized how the law was being treated. When he finally became the victim of an intrigue himself and was obliged to justify himself to the Social Disciplinary Court there was no longer anyone around who could or wanted to defend him – and he himself only considered the possibility of fighting a duel with Schilling. However, unlike Gärtner and his family, Hirsch was not forced to leave his home nor was his livelihood threatened; he was only sent into early retirement. It is impossible to demonstrate the surrender of the works management and its senior employees in front of the NSDAP and the Nazi regime any more plainly than this letter does. Reading the letter in the spring of 1941, Lautenschläger, to whom Hirsch's letter was personally addressed, must have become conscious of the role he played for his employees as a works manager in a state which no longer followed the rule of law.

The threat of forced retirement confronted not just senior employees who had not joined the NSDAP – even if they had previously done their utmost to ingratiate themselves with the Nazi regime. In the climate of intrigue and denunciation "old fighters" could also meet this fate. Ernst Schmidt, born in 1885, had worked as a commercial clerk in Hoechst since 1904. In 1929 he joined the NSDAP, and for a time was district treasurer for his home town of Wiesbaden. In the spring of 1932 he received a sharp rebuke from the works management for political agitation on the premises – he had distributed NSDAP handbills during working hours at the plant. A letter expressing "most severe disapprobation" on the part of the works management was signed by Hermann and Schwamborn, and Schmidt was threatened with "the most serious consequences" in the event of a repeat offence.[488] After the seizure of power, however, he went on to make a political career for himself. In 1934 he became a member of the Council of Trust and worked in its office; after 1935 he was a city councilor for the

---

[488] HA, PA Schmidt (microfilm), Hoechst to Schmidt, April 21, 1932; HHStAW, Abt. 520/BW Nr. 4007 (E. Schmidt), Meldebogen and letter by Hoechst (Frank, Trost).

NSDAP in Wiesbaden. Statements concerning Schmidt and his activities are very contradictory. Schwamborn spoke very positively, noting that Schmidt "alone in the whole Council of Trust had exhibited a faultless and exemplary behavior during his whole period in office." Schmidt's "courageous frankness and his sense of justice" had driven him to combat "lies, slander and denunciations all the way up to the *Gauleitung*" without respect of persons and without taking account of the prestige of the NSDAP. With this behavior he had incurred "the deep hatred" of various NSDAP members and their bosses.[489] Other witnesses, however, spoke very negatively about Schmidt. Thus other tenants living in the same apartment block talked about the political pressure that had been brought to bear on them, even of threats that they would be reported.[490] In December 1942 Schmidt resigned from the DAF and stepped down from office as member of the Council of Trust. At the same time steps were taken for him to go into early retirement, and he finally retired in 1943. However, the works management granted him a generous transitional allowance.[491] The reason for the sudden loss of all his political offices – he was also obliged to stand down from the city council – was apparently his continued membership in the Protestant church. It appears that he had been told to leave the church and had declined to do so since, as he wrote after the war, he did not intend "to tolerate the Party's intervention in religious matters."[492] This case once again shows how difficult it is to talk about someone being a "Nazi" since Schmidt was an active Party member and had actively supported the Nazi regime during his time in the plant as member of the Council of Trust, but subsequently lost his position because of his religious beliefs. Schmidt suffered no persecution. While he lost his position, he was financially secure and experienced no political sanctions.

The case of the worker Elisabeth L. is as depressing as that of Gärtner prior to the war and far worse than the downfall of Hirsch or Schmidt. It is another graphic example of the effect denunciations had at the level of the workers and how the plant reacted to them. Hans Pöhn, the works Defense Commissar, declared after the war that he himself had always favored solutions within the plant. He had allegedly always tried to keep the Gestapo outside the works. In one case, which Zeh passed on to him to hand over to the Gestapo, he had allegedly managed to ensure that

[489] HHStAW, Abt. 520/BW Nr. 4007 (E. Schmidt), testimony by Schwamborn dated September 17, 1946.
[490] Ibid., report and testimonies.
[491] HA, PA Schmidt (microfilm), Schmidt to Lautenschläger, December 7, 1942; Aktennotiz by Hirschel on a discussion with Schmidt dared December 7, 1942; Hoechst to Schmidt, June 24, 1943.
[492] HHStAW, Abt. 520/BW Nr. 4007 (E. Schmidt), Meldebogen dated September 3, 1946, and resumé dated March 27, 1947.

the matter was "settled within the plant." However, this account was contradicted by the person in question, who spent several months in custody.[493]

Contrary to Hans Pöhn's claims, the case of a twenty-year-old woman who had been employed as a worker in the plant since the beginning of 1940 shows how readily the works management of Hoechst opened the doors of the plant to the Gestapo.[494] As Defense Commissar Pöhn wrote to Hirschel, then the new head of the Social Department, on July 2, 1941, Werner Schirmacher, the head of the department for the bottling of acetic acid, had observed at the end of June that the worker Elisabeth L. had been loitering "in a conspicuous manner in the immediate proximity of French prisoners of war" who worked in the plant. He had thereupon "most strongly warned her and referred to the regulations about contacts with prisoners of war." At the same time he had requested the foreman to watch her. L., who had fallen in love with one of the prisoners, wrote him a letter warning him not to speak to a certain Anna anymore since she had betrayed her to the foreman. This letter was found by another worker, who gave it to the foreman, who in turn passed it on to Schirmacher. Schirmacher had a comparison made of the handwriting and passed the matter on to Pöhn with the additional comment that L. had not attempted to come into contact with prisoners of war to betray company secrets, although she had committed a punishable offence. Either because Pöhn was absent or because the matter was the responsibility of the head of the Social Department, Hirschel reported the charge against L. in a letter to the chief of police written on the following day; he gave a detailed report of the matter ending with the words: "We leave it up to you to initiate legal proceedings against the accused."[495]

In September 1941 L. was sentenced to four months' custody on the basis of the report. She accepted the sentence but pleaded for clemency in the form of transmuting the prison term into a fine. In her plea she expressed her deep regret about her "recklessness" as she had not been aware of the consequences of her behavior. The sentence would not just affect herself, her mother had been "so upset by the incident that she had become ill and bedridden." L. had written her plea for clemency with the help of the works advisory service, which she consulted after being sentenced. The works management was at first prepared to wait and see whether her plea for

[493] HHStAW, Abt. 520 F, Nr. 4815 (Zeh), Trial dated January 12, 1949, testimonies by Zeh, Pöhn, and Krell.

[494] HA, PA Elisabeth L. (microfilm), personal data sheet dated March 5, 1940.

[495] HA, Pol 15, Pöhn to Hirschel, July 2, 1941, with a copy (photograph) of the letter from L.: at the end of his letter Pöhn requested that "I shall be informed of what arrangements are made"; Hirschel to Polizeipräsident c/o 17. Polizei-Revier Ffm-Hoechst, July 3, 1941, note at the top: "Mr. Pöhn is taking a break."

clemency would be granted; however, it was turned down.[496] Schirmacher informed the Social Department on October 1, 1941, that L. would now "be absent as she was serving her prison sentence." L. was thereupon summarily dismissed on October 1 for being "unable to work in our plant." However, she was given the option to apply for a job in the plant after having completed her sentence.[497] With the help of the Labor Office she did so in February 1941 on her release from jail in Frankfurt-Preungesheim, but was not taken on. Her application was reconsidered only in October 1942 because of the severe shortage of workers, although Defense Commissar Pöhn demanded that in the event of her being reemployed she was under no circumstances to be permitted to work in an area in which she would come into contact with prisoners of war. She was rehired and continued to work in Hoechst until the end of the war.[498]

The conclusion to be drawn from these cases is that not only did the works management under Lautenschläger largely give in to the NSDAP, its organizations, and representatives, but its behavior was often more conformist than necessary. The works management and the senior employees acted with the greatest severity against employees who did not conform absolutely. In the case of Elisabeth L. an internal caution by the works management and the threat to notify the police of any recurrence would have been entirely sufficient to discipline her. Under Hermann far worse offences had been punished with small fines. However, Otto Hirschel, the new head of the Social Department, had virtually requested the Gestapo to procede against L. – and he had done so immediately and without hesitation. This was one way of enforcing discipline within the plant, but at the expense of the plant's independence. It was a form of surrender to the Gestapo and the *Gauleitung*, and Hoechst became an extremely conformist workplace where rash words or deeds could have devastating consequences. Word got around fast and resulted in fear and conformist behavior – and it gave free rein to informers.

In his book *Die Rotfabriker* Bäumler largely ignored these unsavory parts of the "family history" and painted a far better picture of Nazi followers in Hoechst such as Retzinger than they deserved. Nor is it possible

[496] Ibid., L. to Amtsgericht Frankfurt, September 4, 1941, handwritten addendum below: "has been refused"; HA, PA L. (microfilm), note by Sozialabteilung dated September 5, 1941.
[497] HA, PA Elisabeth L. (microfilm), memo from Schirmacher to Sozialabteilung, October 1, 1941, and IG Hoechst to L., October 7, 1941.
[498] Ibid., Arbeiterannahme to Abwehrbeauftragter, October 2, 1942; response from Pöhn dated October 5, 1942; during the war L. once again faced difficulties in 1943: she was arrested again and subsequently placed under the supervision of the Central Office for the People's Social Welfare of the NSDAP in Bad Soden. She worked for Hoechst until the end of the war and toward the end of the war was assigned to work in other plants near Hoechst.

to concur with the comment made by Karl Winnacker, at the time the "crown prince" in Hoechst, on the relationship between the works management and the Nazi regime. Winnacker wrote in his memoirs: "Even when National Socialism was at its zenith, the management had largely succeeded in keeping party politics out of the company's affairs. Many people who were politically or racially at risk were able to work in Hoechst until the end of the war. Those in power did not succeed in laying their hands on them."[499] As shown both in this chapter and in the two previous chapters, not even the most generous interpretation of events supports this claim.

3.8. THE RISE OF THE "CROWN PRINCE": KARL WINNACKER

Just how much Lautenschläger was and remained a compromise candidate as works manager of Hoechst became clear during the war in 1941, when Karl Winnacker became the "crown prince" of Hoechst. It is impossible to say with any certainty whether Lautenschläger's role in the conflicts with Zeh and Schlick concerning Wagenheimer and in those with Schilling concerning Müller and Hirsch induced the management of IG Farben to build up a crown prince. However, the chronology of events makes this very plausible. Moreover the situation within Hoechst was not at all to the liking of the IG Farben management, not least due to the very great number of workers calling in sick.

In Winnacker the company management had a candidate who had risen very quickly through the ranks in the previous few years. Karl Winnacker was born on September 21, 1903, in Barmen as the son of a high school teacher who died in 1914, and studied at the Technical Colleges of Brunswick and Darmstadt. He completed his doctorate under Professor Ernst Berl in Darmstadt on oxidation processes in hydrocarbons and then worked as Berl's private assistant, which included making contributions to Berl's compendium *Chemical Engineering Technology*. As a Jew, Berl lost his chair at the Technical College in the spring of 1933 and subsequently emigrated to the United States. Before he finally left, in addition to winding up his own affairs, he also took pains to secure the future of his employees and colleagues. He recommended that Winnacker should move into industry and procured him a job in Hoechst.[500]

Berl's letter of recommendation to Hoechst praised Winnacker's technical and didactic talents. Because of his work Winnacker knew "the field of chemical engineering technology as well as it is possible to know it by working at a university." In Berl's opinion Winnacker would "do an excellent job both in the research laboratory and in the company but also,

[499] Winnacker, *Challenging Years*, p. 114.   [500] Ibid., pp. 54–56.

should the occasion arise, in dealings with buyers." Berl added that he considered Winnacker to have "an irreproachable character" and that he "particularly valued him for his personal qualities."[501] Not least due to Berl's intervention Winnacker's application was successful. He began working in Hoechst in September 1933, although the financial terms of his contract were relatively poor.[502] After joining Kränzlein's department, Winnacker promptly came into conflict with Georg Kränzlein, who had problems with Winnacker's self-confidence. In a letter to his subordinate Heinrich Greune, who wondered where to deploy Winnacker at the beginning of September 1933, Kränzlein wrote: "As I already informed Dr. Tampke, Dr. Winnacker will be under his authority and work together with him and Kiesskalt, Funke, [and] Boedecker on fine pastes and fine powders. Winnacker will have the opportunity to make full use of his technical abilities of which he is so convinced. He should also immediately begin working on pad prints and sodium chloride pastes and introduce a strong scientific note into our empiricism."[503]

From the beginning Winnacker appears to have been very successful in his work in Hans Tampke's department for alizarin and vat dyes, because Tampke gave him an exceptionally favorable assessment. Above all, Tampke was impressed by Winnacker's knowledge of physical chemistry.[504] From 1933 to 1936 Winnacker worked in Tampke's department "on all manner of questions concerning colloid chemistry." He was responsible for the finishing of dyestuffs: "New methods for the production of brands for vat dyeing and printing pastes for textile dyeing with vat dyestuffs were developed."[505] At first Winnacker's work was mainly concerned with improving the manufacturing processes for dyes, but he increasingly turned to other problems such as the dehydration and deashing of lignite. This led to friction with the Dyes Department. Hermann finally decided that Winnacker should not just continue working in these fields, but he even

[501] Hoechst Human Resources, PA Abteilung VI, Winnacker, Berl to Direktion IG Farben, Werk Hoechst, May 28, 1933.
[502] Ibid., Winnacker's application dated May 29, 1933; memo from Kränzlein to Jähne dated June 22, 1933, on Berl's support for Winnacker; memo from Kränzlein to Personalabteilung, August 18, 1933, Hermann and Kiesskalt approved of hiring Winnacker; Winnacker to Personalabteilung Hoechst, August 28, 1933; Hoechst Personalabteilung to Winnacker, August 29, 1933; Winnacker's application dated September 1, 1933; Paritätische Kommission gemäss §16 des Reichstarifvertrages für die akademisch gebildeten Angestellten der chemischen Industrie to IG Farben, Frankfurt-Hoechst, dated October 18, 1933, on Winnacker's below-tariff salary; cf. also Winnacker, *Challenging Years*, pp. 56, 64–5.
[503] HA, PA Abteilung VI, Greune, Greune to Kränzlein, September 1, 1933, and response from Kränzlein, September 4, 1933.
[504] Hoechst Human Resources, PA Abteilung VI, Winnacker, Tampke to Personalabteilung, July 19, 1934.
[505] Private files Winnacker, resumé dated June 26, 1946.

gave him his own independent department for "process engineering" under chief engineer Jähne at the end of 1936.[506]

This newly created department under Winnacker consisted initially of around ten researchers and combined basic research on physical chemistry and a physical laboratory. Its tasks were summarized by Winnacker as follows: based on his work in Tampke's department further colloid chemical observations had been made "which in 1936 led to the foundation of an independent physical chemistry department which was placed under my authority and which had the task of carrying out general colloid chemistry and physical chemistry research for all departments in Hoechst. This work led to new knowledge of the processes of polymerization, dyeing methods, and textile printing as well as colloid chemical methods for the deashing and refining of carbons and more. Furthermore the department was responsible for carrying out general physical and physical chemistry research for all parts of the plant." Winnacker now independently headed a laboratory that worked for all departments of the plant.[507]

Ernst Struss, managing director of IG Farben's Technical Committee, emphasized after the war that it was Winnacker's professional achievements that underpinned his fast-track career. Under Tampke Winnacker had worked in colloid chemistry in the Alizarin Department, which had resulted in "externally visible successes." Winnacker focused on reducing alizarin dyes to "as fine a powder as possible," which allowed improved processing of the pigments in the dye and print works. In 1936 he left the department and began working together with the engineer Siegfried Kiesskalt in the Process Engineering Laboratory, where he remained until the summer of 1938: "Here polymerization methods were examined and an interesting method for coal conversion was developed until it was ready for production. These successes made Dr. Winnacker stand out among the young chemists in IG and, as Hoechst had lost three leading chemists (Dr. Hermann, Dr. Staib, Dr. von Brüning) by their deaths, in the summer of 1938 he [Winnacker] was entrusted with the management of the two large departments 'Inorganics' and 'Nitrogen.'" Ter Meer had also become aware of Winnacker at the time. Struss wrote, "as particularly efficient persons among the new generation of IG, who were suited for leading positions, were very rare, Dr. ter Meer supported Dr. Winnacker in all respects and by giving him special tasks ensured that he received a general idea of IG."[508]

After the war, Winnacker's achievements as head of the Process Engineering Laboratory were also acknowledged by Franz Patat, Winnacker's

---

[506] Winnacker, *Challenging Years*, pp. 73–80.
[507] Private files Winnacker, resumé dated June 26, 1946.
[508] Ibid., Struss to Winnacker dated August 15, 1946, in the appendix a testimonial from Struss on Winnacker dated the same day, apparently for the denazification tribunal – quotes from the appendix.

successor in office and later professor for process engineering at the Technical University of Munich: "Dr. Winnacker was by far the most qualified of the younger generation in the plant, and this was ungrudgingly recognized by all objective colleagues." Patat emphasized that Winnacker always had a "strictly matter-of-fact attitude" and that he had "not been a 'pleasant' boss in the usual sense of the word": "He expected the same full commitment from his employees as he did from himself, but in return he energetically stood up for their interests regardless of any difficulties until success was finally achieved." After leaving process engineering Winnacker therefore had a career that was "fully and completely" commensurate with "his personal and professional capabilities" and not politically motivated.[509]

On August 1, 1938, Winnacker was transferred to the Inorganics and Nitrogen Department "to support Dr. von Brüning." After Brüning's suicide Winnacker became departmental head of the Inorganics Department at Hoechst.[510] But to his disappointment this appointment did not include his being made an authorized signatory. The works manager Lautenschläger was thereupon treated to a taste of Winnacker's self-confidence. On the occasion of the impending appointment of Schlick and two others to become authorized signatories Lautenschläger informed Winnacker at the beginning of 1940 that he had been removed from the list of appointments, "since Hoechst could not be allowed a greater number" of signatories. Lautenschläger also informed him that ter Meer had given "his fundamental consent" to Winnacker's appointment as authorized signatory, and it was being considered for the coming year. However, Winnacker had no intention of accepting this slight. According to Winnacker, the new appointees were in charge of "considerably smaller areas of responsibility" than he was; moreover one of them had been employed in the plant for only three months. In consideration of the fact that the management of the Inorganics Department had been entrusted to him one and a half years ago and no one had ever voiced any criticism of his administration, "this preferential treatment" constituted "an undeserved slight which apart from the personal aspect places me at a disadvantage when carrying out my duties, particularly as otherwise the conferment of such an area of responsibility always includes being made authorized signatory." To make sure that Lautenschläger would not ignore him a second time, Winnacker ended his letter with the statement that Lautenschläger had "recognized"

---

[509] Ibid., Patat to Winnacker, March 21, 1947, in the appendix "Zeugnis über Dr. Ing. Karl Winnacker über Tätigkeit in Hoechst" dated March 22, 1947 – quotes from this reference letter.

[510] HA, PA Winnacker, Allgemein, Rundschreiben Nr. 71 der Direktions-Abteilung T (Hirschel) to plant leaders and heads of laboratory dated July 29, 1938; Winnacker, *Challenging Years*, p. 91.

the "fundamental legitimacy" of his claim but had felt himself to be "committed by older promises." Winnacker added: "You have promised me that these circumstances will be altered at the end of the year."[511]

His unmistakable protest bore fruit; in November 1940 Lautenschläger requested that Winnacker also become an authorized signatory, and the company's Central Committee granted the request in December 1940.[512] Backed by ter Meer, Winnacker's career progressed by leaps and bounds. At the beginning of September 1941, only a few months after being made authorized signatory Winnacker was transferred to the company's plant at Uerdingen so that as a future director of the IG Farben group he could become acquainted with other plants. His work in Uerdingen brought him into contact with Ulrich Haberland, the crown prince of the Lower Rhine Works Group, with whom, according to Winnacker, he was to cultivate a close relationship. In addition to his work in Uerdingen, Winnacker was still expected to manage his department in Hoechst, and he also regularly attended meetings of the Board of Directors.[513] In 1943 Winnacker was sent to the company's plant at Schkopau, an ultra-modern factory, to gain more experience. In the fall of 1943 he finally returned to Hoechst.[514]

After his return Winnacker became the second most important man in Hoechst. In September 1943, ter Meer wrote a letter to Carl Krauch, chairman of the Supervisory Board of IG Farben, and to the members of the Central Committee requesting their consent to the promised promotion of Winnacker to director, which had already been announced in June 1943: "I explained in the C.C. [Central Committee] at the time and also on earlier occasions that with the 2-year sojourn in the Uerdingen and Schkopau works I was training Dr. Winnacker in many different fields and for an extended area of responsibility with a view to giving Dr. Winnacker the management of the Department for Intermediate Products and Solvents in Hoechst in addition to the management of the Inorganics and Nitrogen Department, which he has already been in charge of since several years." Ter Meer added: "Dr. Winnacker will of course be subordinate to Professor Lautenschläger, the head of the plant at Hoechst. However, I would imagine that his relations to the other directors of the Hoechst plant will be that of a primus inter pares." This elevated position was emphasized even more by Winnacker's appointment to the Technical Committee.[515] On

---

[511] HA, PA Lautenschläger, Sekretariat Lautenschläger, Personalangelegenheiten, Winnacker to Lautenschläger, April 13, 1940 (handwritten); Winnacker, *Challenging Years*, p. 91–2.

[512] Hoechst Human Resources, PA Abteilung VI, Winnacker, application for his appointment to become authorized signatory dated November 6, 1940, approved by the Central Committee (*Zentralausschuss*) on December 11, 1940.

[513] Winnacker, *Challenging Years*, pp. 98–101.   [514] Ibid., pp. 103–4.

[515] Hoechst Human Resources, PA Abteilung VI, Winnacker, Fritz ter Meer to Krauch and members of the Central Committee, September 14, 1943.

October 1, 1943 Winnacker became a director – "on the application of Dr. ter Meer" according to the letter of the Central Committee office to Lautenschläger.[516]

Thus the choice of Lautenschläger's successor in Hoechst by the IG Farben Executive Board had already fallen on Winnacker, at ter Meer's suggestion, during the war. Winnacker also was given greater authority than ordinary directors of other plants. For example, he had to carry out negotiations with the Nazi authorities and government ministries on behalf of the indecisive and unassertive Lautenschläger concerning the allocation of forced labor and raw materials.

Winnacker's membership in the NSDAP may have helped him; at the end of the war some of the employees certainly considered him to be a particularly staunch Nazi. Winnacker had joined the SA only in the spring of 1933, that is, after the seizure of power, and subsequently he became a corporal. He joined the NSDAP in the summer of 1937 and apparently tried without success to end his SA membership. To what extent he actually was a Nazi is difficult to say; the statements are too contradictory, and the reasons behind the respective statements of his opponents and friends are too obvious. He was apparently not anti-Semitic, for he kept in touch with his academic teacher Berl until the late 1930s.[517]

What is evident in all of this is his take-charge personality as well as his lack of scruples in his work. He ensured that the plant continued to function by requesting forced laborers and raw materials,[518] and his process engineering work in the field of carbons brought him into contact with Auschwitz in 1943. In a talk with an employee of the IG Farben plant at Leuna in July 1943 the employee had mentioned to Winnacker "that Auschwitz offered an exceptionally fortunate raw materials basis for the problems Hoechst was working on." The work being carried out in Hoechst concerned the problem of electrode coke production, for which the coal in Auschwitz was particularly suitable. Winnacker therefore was advised to cooperate with Auschwitz, and a "local meeting in Auschwitz to discuss

[516] Ibid., ZA-Büro to Lautenschläger, October 2, 1943; however, Lautenschläger had already drafted an application for Winnacker to be promoted to director in 1942 but with a handwritten addendum: "not in 1942."

[517] Cf. BA, BDC file on Winnacker, member of the NSDAP since May 1, 1937, application for admission dated September 20, 1937; Nordrhein-Westfälisches Hauptstaatsarchiv Düsseldorf, NW 1004 G 6.3 208, questionnaire on Winnacker's denazification, dated December 23, 1946; HA, Entnazifizierungsakten (Vorstandssafe), letter from Freier Deutscher Gewerksschaftsbund to Military Government, Frankfurt, June 30, 1945 (copy).

[518] Cf. HA, PA Lautenschläger, Sekretariat Lautenschläger, Vertrauliche Korrespondenz, draft from Winnacker for Lautenschläger dated October 25, 1943, re. "Schreiben Krauchs vom 7.10.1943"; private files Winnacker, Winnacker to Lautenschläger dated June 4, 1945 (copy).

further joint work" was agreed on in July.[519] Whether this meeting actually took place is not recorded, but as part of his work Winnacker did come to Auschwitz – and even if he saw nothing, what was happening in Auschwitz could still be smelled, as a former inmate of the concentration camp declared in the Wollheim trial in 1953.[520]

With his qualities as a "doer" Winnacker managed to prevent the Reich authorities from shutting down Hoechst and to maintain production until the end. He was awarded the War Distinguished Service Cross (*Kriegsver-dienstkreuz*) for his achievements. He was so proud of what he had achieved for Hoechst that in 1945, after the occupation of the plant by the Americans, he wrote to Lautenschläger, who apparently wanted to dismiss him, that in contrast "to numerous colleagues in IG and also in the Hoechst works" he had not held any office in any of the organizations of the Ministry for Armaments, in any other of the Ministries, or in any of the local organs of the NSDAP. However, he had had "fierce disputes, particularly with many departments of the *Gebechem* [*Generalbevollmächtigter für Sonderfragen der chemischen Erzeugung im Vierjahresplan*, Plenipotentiary General for Special Questions of Chemical Production in the Four-Year Plan] and the Office for Raw Materials because of the irrational decisions made there to the disadvantage of our plant." He had focused all his energies on "improving the already difficult situation of the Hoechst plant again and again," even in the teeth of the authorities who objected to the plant being kept running. "The accusation that I tried with all my strength to maintain production up until the end can only rebound to my honor, unless a soldier is to be reproached for carrying out his duties until receiving the order to surrender."[521]

Winnacker thus considered himself to be a "soldier." In this capacity it was incumbent on him to "do his duty" without asking questions – and as he saw it, this meant ensuring the continued functioning of the Hoechst works during the war and obtaining raw materials and workers for Hoechst. During the war, however, it was only possible to get hold of "foreign workers."

---

[519] HA, PA Winnacker, Allgemein, Aktennotiz by Winnacker, July 16, 1943 re. "Besprechung mit Dr. Braus, Leuna, am 9.7.1943 in Schkopau betr. Elektrodenkoks-Fabrikation."

[520] Cf. Winnacker, *Challenging Years*, pp. 104–6; oral communication of Albrecht Winnacker on Karl Winnacker's sojourn in Auschwitz, in response to Winnacker's question what was going on there, Ter Meer had replied: "You don't want to know what is going on in Auschwitz." Winnacker had accepted this answer; oral communication by Herbert Spahn: after a visit to Auschwitz Winnacker had subsequently asked ter Meer what was going on there; thereupon ter Meer had answered that terrible things were happening there and that Winnacker should keep out of it; IfZ, ED 422, vol. 1, statement of Benedikt Kautsky in the Wollheim trial during the public hearing on November 30, 1953: "I am certain that a person who kept his eyes open when he walked around the grounds in Buna must have seen those things if he wanted to see them, and if he didn't want to see them, then he must have smelled them."

[521] Private files Winnacker, Winnacker to Lautenschläger dated June 4, 1945 (copy).

FIGURE 1. The Hoechst Works in August 1930 (Hoechst Archives)

FIGURE 2. Ludwig Hermann (Hoechst Archives)

FIGURE 3. Celebration of the 75th Anniversary of Hoechst in January 1938
(Hoechst Archives)

FIGURE 4. Carl Ludwig Lautenschläger (Hoechst Archives)

FIGURE 5. Ludwig Retzinger (Bundesarchiv Berlin)

FIGURE 6. Hans Wagenheimer (Bundesarchiv Berlin)

FIGURE 7. Anniversary of Karl Ferdinand Blumrich in front of a "Hitler altar" (Hoechst Archives)

FIGURE 8. Franz Henle (Hoechst Archives)

FIGURE 9. Franz Henle's son Karl Henle (Hoechst Archives)

FIGURE 10. Hoechst's Pharmaceutical Research Laboratory in 1931
(Hoechst Archives)

FIGURE 11. Karl Winnacker (Hoechst Archives)

FIGURE 12. The Arrival of Female Eastern Workers (*Ostarbeiterinnen*) at Hoechst during the War (Hoechst Archives)

FIGURE 13. The Barracks of the Eastern Workers – the so-called *Russenhof*
(Hoechst Archives)

FIGURE 14. Georg Kränzlein (Hoechst Archives)

FIGURE 15. Hoechst's Pharmaceutical Bureau, headed by Julius Weber
(Hoechst Archives)

FIGURE 16. Carl Ludwig Lautenschläger at the IG Farben Trial before
the Nuremberg Military Tribunals (Hoechst Archives)

3.9. NOT PART OF THE "FOLLOWERS" IN THE WAR:
THE "FOREIGN WORKERS"

The workforce of Hoechst had increased in the 1930s. During the war the number of workers stagnated, although it remained high until 1944 despite numerous call-ups for military service. Because of the increasing labor shortages Hoechst tried to hire women. The chemical pharmaceutical industry had generally employed higher numbers of women than other industries, and during the war the percentage of women employed in Hoechst increased – despite the fact that in the early 1930s many women had been pushed out of work at the instigation of the Nazis.[522] Before the war the ratio of men to women in the offices of Hoechst was five to one, but by the end of the war the ratio stood at two to one. The age distribution in Hoechst also changed during the war, since it was mainly the younger men who were called up, leaving the women and older men behind. In some departments the age pyramid had been virtually stood on its head. By the end of the war, for example, 41 percent of the employees of the Factory Accounts Department were between fifty and sixty years old, 32 percent were between forty and fifty, 16 percent between thirty and forty years of age, and only 11 percent were under thirty years of age.[523]

Lautenschläger wrote in his memoirs toward the end of 1942 that more and more employees were conscripted in the course of the war, even employees who had initially been considered indispensable. By the middle of 1942, out of some 11,000 employees 2,200 had been called up. By the end of 1942, 143 of the conscripted employees had died, and some had returned seriously wounded and disabled. Many of those who returned disabled were often unable to perform their former jobs and therefore had to be retrained; nevertheless, they were taken on in increasing numbers, not merely out of humanitarian impulses but because of the shortage of manpower. At the beginning of the war conscripted employees deployed in locations where fighting was suspended were given "economic leave" during the winter months, which allowed them to return to work in the companies from which they had been conscripted, including Hoechst. However, since they could be recalled at any time, they could not be considered as fully available. Lautenschläger was of the opinion that women were likewise of only limited use, in part because they were physically incapable of performing all the tasks carried out by men,

[522] HA, PSW 648, note from the company archive dated April 28, 1978, on an interview with Retzinger of April 26, 1978.
[523] HA, file Pensel-Entnazifizierung, allgemeiner Schriftwechsel, note from Personalabteilung (Frank) for Dr. Bockmühl, October 12, 1945.

TABLE 3.5. *Number of Persons Employed in Hoechst from 1938 to 1945 According to the Company History of Hoechst AG Published in 1990, Not Including "Foreign Workers"*

Year	Blue-Collar Employees	White-Collar Employees	Total
1938	7,287	2,293	9,580
1939	7,933	2,339	10,272
1940	6,693	2,116	8,809
1941	6,551	2,209	8,760
1942	6,293	2,234	8,527
1943	6,512	2,209	8,721
1944	6,388	2,414	8,802
1945	3,939	1,612	5,551

Source: Chronik der Hoechst AG, p. 357.

but because in addition to their work in the plant they were also needed in the home.[524]

Table 3.5 is based on the company history of Hoechst AG published in 1990 and gives the numbers of persons employed in Hoechst between 1938 and 1945. Interestingly, these statistics do not include "foreign workers" – even in 1990 they were still not regarded as forming part of the workforce. According to these figures, the number of persons working in Hoechst decreased during the war from about 10,000 to between 8,500 and 8,800. Lautenschläger mentioned around 11,000 persons working in Hoechst in mid-1942. In his memoirs written in 1952 he refers to 12,000 persons, consisting of around 9,000 men, 2,300 women, and 700 adolescents, working in Hoechst during the war.[525] These numbers considerably exceed the figures given in the company history because of the inclusion of "foreign workers"; their number in Hoechst increased substantially in the course of the war.

The numbers given in this company history also tally with the figures that Otto Hirschel lists in his history of the plant. However, Hirschel not only fails to supply any figures concerning the foreign workers, he does not mention them at all – as if they had not existed.[526] This is even more astonishing in view of the fact that Hirschel, who was head of the Social Department or "Followers Department" as it was called during the war, was also responsible for the foreign workers.

[524] HA, PA Lautenschläger, Lautenschläger, Erinnerungen, vol. 3 [pp. 123–7]. They were not supposed to replace the men; cf. circular on women's areas of work.

[525] Ibid., vol. 8, [pp. 173–4].

[526] HA, 12, Hirschel, "Das Werk Hoechst im Verbande der I.G.," p. 38, "Belegschaftsentwicklung," foreigners are mentioned only on p. 33 on March 24, 1945, where they are referred to as "looters."

Hirschel, born in Frankfurt in 1872, studied chemistry in Jena and Gießen. His studies were interrupted by the First World War, in which he served as a soldier. He completed his doctorate in Gießen after the war, joined Hoechst in 1921, and began working in the Patents Laboratory under Professor Schmidt. In 1926 he transferred to the administration and was primarily responsible for cost accounting and for suggestions on how to reduce costs. He apparently gave a good account of himself because Hermann appointed him to be a member of the "Directorate T" when it was founded in 1933, and there he prepared IG Farben meetings, edited cost estimates, and carried out cost controls. As Hermann was completely satisfied with his work, he was promoted in 1937. After Hermann's death and Brüning's suicide, Hirschel, as head of the Directorate T, became one of Lautenschläger's closest colleagues. After Schlick's enforced departure from the Social Department, Lautenschläger appointed Hirschel to be Schlick's successor at the end of January 1941, and he was given commercial power of attorney at the end of April 1941.[527]

In the Weimar Republic Hirschel had been a member of the DNVP (German National People's Party) and the *Stahlhelm*. He officially joined the NSDAP in April 1936, during the official ban on new members. After the war he claimed to have joined in 1937 and to have become a Party member only under pressure; the date of his joining had subsequently been brought forward. However, he had already applied for membership in February 1935, and his application was approved in July 1936 by the local branch leader. Whether he was then admitted to the NSDAP or only in 1937 and the date of his admission was then antedated cannot be ascertained. But contrary to his statement after the war, he was not forced to join the NSDAP.[528]

As head of the Social Department, Hirschel had to cooperate closely with *Betriebsobmann* Zeh. According to his own statement, his task as the successor to Schwamborn and Schlick consisted of requesting and hiring employees, settling questions respecting wages and salaries, and making arrangements for the accommodation and food for the foreign workers.[529] Yet his functions went further than that, as his role in the case of the worker L. shows. In the event of any disciplinary transgressions it was the Social

[527] HA, PA Abteilung VI, Hirschel, Hirschel's job application as a chemist; application to appoint Hirschel an authorized signatory dated March 11, 1941, approved by the Central Committee (*Zentalausschuss*) on April 24, 1941; Hirschel, "Erinnerungen aus meiner IG-Zeit," pp. 14–15.

[528] BA, BDC file on Hirschel; HHStAW, Abt. 520 F (A-Z), Hirschel, Hirschel to Öffentlicher Kläger der Spruchkammer Main-Taunus, August 22, 1947, and Hirschel's declaration according to Protokoll der öffentlichen Sitzung dated March 1, 1948 – he described his joining the NSDAP as coerced and attempted to portray this as a "sacrifice" made for his plant.

[529] NI-2973, affidavit by Hirschel dated January 16, 1947; HHStAW, Abt. 520 F (A-Z), Hirschel, Hirschel's statement according to Protokoll der öffentlichen Sitzung dated March 1, 1948.

Department's job to punish employees.[530] And what this could mean was shown by the case of L., who was handed over to the Gestapo.

In general, the morale of the German employees during the war does not appear to have been very high, and this was reflected in the extremely high number of persons calling in sick, the only remaining form of protest in the Nazi state. In the winter of 1941–42 the situation in the plant was dramatic. The number of sick persons increased first to 7 percent, then to over 9 percent – normally these figures were around 1 percent and not more than 3 percent. In 1942 Hoechst had the highest rate of employees absent from work in the whole of IG Farben.[531] The number of sick persons was particularly high for the conscripted German employees: the figure stood at 7.2 percent for men and 7.1 percent for women – and among the women it was particularly high among the so-called labor maids. To reduce absences from work, Hoechst arranged for a medical examiner in the employ of the *Gauleitung*, a Dr. Gutermuth, to examine those who were not bedridden. The number of sick persons in Hoechst promptly dropped to 3.41 percent.[532] However, Gutermuth had to intervene a second time at the request of the factory management of Hoechst, and he volunteered to send a doctor to Hoechst to examine those who claimed illness "respecting their fitness for work." The "examination" resulted in a miracle cure – only a short time later 68 percent of the patients had returned to work in the plant.[533]

With such measures, which were not enforced by the factory management's own authority but were due to fear of the regime and its representatives, Hoechst was able to achieve some short-term successes with its employees. But as the war dragged on and became a "total war," it was evident that the problem of the underlying refusal to work, of which this was only a manifestation, could not be properly dealt with, since the workers were well aware of how indispensable they were. According to Lautenschläger, discipline continued to deteriorate – which in its turn led to more punishments being meted out by the factory management.[534] What this meant specifically can be gathered from the records of a meeting of the Directorate held in summer 1942. The management noted that the punishment of employees for "idling" had recently "extraordinarily increased" and that even fines amounting to a full days' earnings were "ineffective."

[530] HA, TEA 95f, Niederschrift über die technische Direktionssitzung in Hoechst, June 7, 1943.

[531] HA, TEA 95f, Niederschrift über die technische Direktionssitzung in Hoechst, January 26, 1942; Niederschrift über die technische Maingau-Direktionssitzung in Hoechst, February 2, 1942, and May 5, 1942.

[532] HA, TEA 95f, Niederschrift über die technische Maingau-Direktionssitzung in Hoechst, June 22, 1942.

[533] HA, TEA 95f, Niederschrift über die technische Maingau-Direktionssitzung in Hoechst, January 24, 1944 and February 7, 1944.

[534] HA, PA Lautenschläger, Lautenschläger, Erinnerungen, vol. 3, [pp. 128–30].

TABLE 3.6. *Number of Hoechst Employees Who Fell in the War or Were Reported Missing, Not Including "Foreign Workers"*

Group	Fallen	Died in Air Raids	Missing	Did Not Report Back
White-collar employees	97	9	30	7
Blue-collar employees	578	19	223	52
TOTAL	675	28	253	59

Source: HA, WIN 653, Büro Winnacker/Personalia Allgemein 1952/55, "Die gefallenen und vermissten Werksangehörigen des 2. Weltkrieges, zusammengestellt von der Sozialabteilung Hoechst" (undated).

The management drew the following conclusion from this: "The leader of a factory should be given an even greater authority to mete out punishments; this will be discussed with the Reich Trustee."[535] The management of Hoechst apparently considered the already severe punishments that could be meted out to employees as inadequate. And the management correspondingly cooperated with the Gestapo and the *Gauleitung* to achieve the objective of guaranteeing production.

By the end of the war, almost one thousand of Hoechst's German employees had fallen, been reported missing, or died in Allied air raids. The figures, given in Table 3.6, do not include the numbers of foreign workers who died. If a large number of German employees suffered under the increasingly tough measures implemented by the factory management and the climate of fear and denunciations in Hoechst, the foreign workers suffered even more.

In his seminal study on the use of foreign labor in Germany during the war, written in the 1980s, the historian Ulrich Herbert wrote that the most important archival materials on this topic would "surely be found in the factory archives," but at the time these remained closed to him: "The special role which 'slave labor' played in the Nuremberg industrial trial has led to a defensive attitude on the part of large West German companies, which – by restricting access to archives – allows legends and speculations to proliferate and makes any differentiated approach impossible."[536]

---

[535] HA, TEA 95f, Niederschrift über die technische Maingau-Direktionssitzung in Hoechst, July 6, 1942.

[536] Ulrich Herbert, *Fremdarbeiter. Politik und Praxis des 'Ausländer-Einsatzes' in der Kriegswirtschaft des Dritten Reiches*, 2nd ed. (Berlin: Dietz, 1986), p. 21; the abridged English edition: *Hitler's Foreign Workers: Enforced Foreign Labor in Germany under the Third Reich*, trans. William Templer (Cambridge: Cambridge University Press, 1997).

However, this attitude of many major companies changed, particularly in the 1990s, and a number of them commissioned studies on the history of their companies in the Third Reich. These investigations were not carried in the spirit of apologetic commemorative volumes, as was still the case in the 1970s and even the 1980s, but dealt critically with this important aspect of the companies' past.[537]

To illustrate the not unimportant contribution of voluntary foreign workers and the differences in the treatment of foreign workers, the term "slave laborers" will not be used here as a general term to describe foreign workers. When examining the fate of foreigners working in the "German Reich," it is important to distinguish between more or less voluntary civilian workers, forced labor, and prisoners of war: their legal status and their living conditions were different.

Prisoners of war were subject to the Geneva Convention of 1929 according to which they were entitled to humane treatment as well as adequate accommodation and food. The Convention stated that enlisted men could be compelled to work. However, the international law of war stated that it was not permitted to force them to do work for the armaments industry. Their working and living conditions were monitored by representatives of a "protecting power," that is, a country that "represented" the enemy state, and by the International Red Cross. The employment of prisoners of war was not governed by civil law but was considered to be "a particular kind of legal relationship under public law." Prisoners of war from a detention camp (*Stalag*) who were considered fit to work were sent by representatives of the armed forces to a company, and the company was informed of the conditions under which these persons would be allowed to work. Such treatment, which followed the international rules of war, was refused outright to Polish and Soviet prisoners of war and to the so-called Italian military internees, while French and Yugoslav prisoners of war were accorded at least some limited rights due to their status; on the other hand, British and American prisoners of war were largely treated in accordance with international law.[538]

The contractual relationship between foreign civilian workers and companies was subject to private law and was similar to that of German workers. This legal construct remained effective even when labor conscription by a German labor office prevented foreign workers from returning to their own countries after their original contracts had terminated. After October 1942 foreign workers could be compulsorily

---

[537] Cf., e.g., the studies on Volkswagen (Mommsen and Grieger, *Volkswagenwerk*) or on Degussa (Hayes, *Cooperation*).

[538] Mark Spoerer, *Zwangsarbeit unter dem Hakenkreuz. Ausländische Zivilarbeiter, Kriegsgefangene und Häftlinge im Deutschen Reich und im besetzten Europa 1939–1945* (Stuttgart: DVA, 2001), pp. 99–105.

conscripted for labor, unless they came from Bulgaria, Denmark, Italy, Croatia, Romania, Slovakia, Spain, or Hungary. By means of this legal act, many "foreign civilian workers who had originally come to Germany voluntarily" from countries that were not regarded as "friendly" like most of those listed above were turned into conscript "forced laborers."[539]

According to the definition given by the historian Mark Spoerer, forced labor was characterized by two main features: by the legally institutionalized indissoluble nature of the employee-employer relationship, which was "for an indefinite period of time," as well as by the limited possibility of having "any considerable influence on the circumstances of the deployment of labor." Spoerer differentiates between workers from Western Europe, who were able to effect some improvements of their living conditions even when these were only insignificant, and workers from Eastern Europe, who "in effect hardly had rights" and had considerably lower chances of survival.[540] The employment relationships of "Eastern workers" were not governed by civil law but were considered as legal transactions governed by the law of obligations, which meant that work was performed in return for remuneration without the employer having any further obligations under social law.[541]

In this study foreign civilian workers, including "Eastern workers" and prisoners of war, will be referred to as "foreign workers," a term that Ulrich Herbert also uses in his study. The foreign workers in Hoechst were never part of the "followers" but were always second-class employees or even lower. It is astonishing that, as evidenced by the figures above, in the chronicle of Hoechst published in 1990 they were still not counted as part of the workforce.

Ernst Bäumler's book *Die Rotfabriker*, published in 1988, devotes a few lines to them. According to Bäumler no blame could be attached to Hoechst for using foreign workers: "When [the number of] German workers no longer sufficed because of the many called up for military service the labor offices assigned foreigners to Hoechst." Bäumler added: "Altogether they numbered 5,077, of which one-quarter were women."[542] Bäumler stated that the first foreigners to arrive were Polish prisoners, followed by French, Belgians, Danes, Dutch, Italians, Croatians, Lithuanians, and Russians. "Their treatment was very different, depending on whether they were citizens of states allied with Germany like Italy until 1943 or citizens from 'enemy states.' Those of the 'Eastern peoples' had the hardest time." According to Bäumler, the factory management saw to it that sufficient food and clothing was provided. In the matter of any "abuses" that occurred, particularly in the treatment of Soviet prisoners, Bäumler quotes Lautenschläger, according to whom "the state authorities must alone be

---

[539] Ibid., pp. 96–7.   [540] Ibid., pp. 15–16.   [541] Ibid., pp. 96–7.
[542] Bäumler, *Rotfabriker*, p. 313.

held responsible." Bäumler finally mentions that "numerous examples of spontaneous help on the part of German workers" occurred.[543]

This section will examine whether and to what extent these statements are correct. The available source material will be presented together with the number of foreign workers who worked for Hoechst during the war. The question as to whether the factory management actively requested the allocation of foreign workers or whether the workers were assigned by the labor offices will be addressed. And finally the treatment these foreign workers received in Hoechst and the differences in treatment will be investigated.

Ulrich Herbert had suspected that important source material was to be found in the archives of factories. However, he also quotes Heinrich Vesper in the Nuremberg trial on the question of slave labor. Vesper, who had been responsible for foreigners in the IG Farben plant Autogen between 1941 and 1945, stated: "Shortly before the arrival of the Americans the order went out from the management of Hoechst that all papers and documents concerning the employment of foreigners must be destroyed." The order was passed on verbally by the works manager, and the documents were destroyed despite Vesper's protests.[544] When he was questioned by the Americans, Vesper surmised that all documents in Hoechst had also been destroyed since the order had come from there.[545]

The Nuremberg trial showed that this was not the case. As part of an attempt to discredit Johannes de Bruyn, a Belgian and a former employee of Hoechst who spoke extremely critically about the working conditions of foreign workers in Hoechst,[546] it became clear that de Bruyn's file, his pay sheets, and even his medical file were still in Hoechst.[547] However, according to information provided by the director of the Hoechst archives, the files of foreign workers were destroyed in the 1970s together with other files of the Department for Human Resources and Social Affairs to make room for a table-tennis facility.[548]

Thus, at the beginning of the investigations for this study, the archives in Hoechst contained only a few pertinent documents. These consisted primarily of circulars from the Social Department or the Department for Followers, the files of an employee of the Social Department, a list of outstanding wages due to "foreign workers," and a list compiled in Hoechst

---

[543] Ibid., pp. 313–14.

[544] NI-2995, affidavit by Heinrich Vesper, December 9, 1946; Herbert, *Fremdarbeiter*, p. 364, n. 57.

[545] IfZ, ZS 1556, Vesper Heinrich, interrogation of Vesper by Cooper, December 5, 1946.

[546] NI-11613, affidavit by Johannes de Bruyn dated September 26, 1947.

[547] HA, IG/50, IG-Prozessakten, Jähne document 57, affidavit by Josef Ems dated January 8, 1948; Jähne document 59, affidavit by Adolf Baldus dated March 9, 1948, with attachments.

[548] This at least was the story which passed down orally in the plant according to a communication by the head of the Hoechst Archives, Dr. Wolfgang Metternich.

after the war of "members of the United Nations and other foreigners" put together on the orders of the Ministry for Work and Welfare of Greater Hesse. This list constituted the basis for Bäumler's reported number of 5,077 foreign workers.[549] However, the records concerning foreign workers were considerably augmented by the discovery of the "employment books" of the Department for Personnel and Social Welfare (PSW). These did not just permit the number of foreign workers working in Hoechst during the war to be determined more precisely, but brief comments on individual persons added in the margins sometimes also give an indication of their subsequent fate.[550] In addition to these documents in Hoechst the documents of the Nuremberg IG Farben trial were also consulted. Then there is also the book *Deviazione* by the Italian feminist Luce d'Eramo, although – as this mixes novel and autobiography – its usefulness is limited. In her book d'Eramo describes her time in Hoechst, which she came to under the name Luce Mangione as a voluntary foreign worker on February 14, 1944.

According to statements of witnesses made during the Nuremberg IG Farben trial, an average of 3,000 foreign workers worked in the plant. Otto Hirschel, who as head of the Social Department during the war could be expected to know most about these matters, declared that the total number of employees during the war without counting those called up for military service had amounted to around 10,000 to 11,000 employees, "of which approximately 3,000 were foreign workers, prisoners of war and civilian workers."[551] The figures provided by Karl Trost, a member of the Hoechst Works Council, are similar. According to his statement, during the war the workforce of Hoechst amounted to around 12,000 workers, of which approximately 3,000 were foreign workers. Of these, 400 were French prisoners of war, whose employment relationships were subsequently transformed into civilian employment relationships; around 1,500 were "forced labor convicts," in the main "Russians, Poles, and Yugoslavians"; and the rest were "so-called voluntary foreign workers": "Belgians, Dutch, Danes, Norwegians, French, Italians, etc."[552] These figures are largely in agreement with those of Ferdinand Pensel, who was head of the Social Department in Hoechst after the war and was responsible for the denazification of the plant. According to his statement the number of "foreign workers" increased from about 2,800 in 1942 to over 3,200 in 1944 (Table 3.7).[553]

[549] HA, "Liste der Angehörigen der Vereinten Nationen und anderer Ausländer in Hoechst," signed by Roth and Pensel, compiled for the Grosshessisches Staatsministerium für Arbeit und Wohlfahrt, probably from the end of 1945/early 1946.

[550] HA, PSW, Betriebseintrittsbücher, 1939–1945.

[551] NI-2973, affidavit by Otto Hirschel dated January 16, 1947.

[552] NI-2993, affidavit by Karl Trost, January 15, 1947.

[553] HA IG/50, IG-Prozessakten, Jähne document 49, affidavit by Ferdinand Pensel on expenses for foreign workers dated January 13, 1948.

TABLE 3.7. *Number of Employees of the Hoechst Plant, 1942–1944, Including "Foreign Workers"*

Year	Blue-Collar Employees	White-Collar Employees	Foreign Workers	Total
1942	6,293	2,234	2,814	11,341
1943	6,512	2,209	3,130	11,851
1944	6,388	2,414	3,219	12,021

Sources: HA IG/50, IG Prozessakten, Jähne document 49, affidavit by Ferdinand Pensel on expenses for foreign workers dated January 13, 1948; *Chronik der Hoechst AG*, p. 357.

It was not possible to ascertain whether all employees in Hoechst were recorded in the books of the Personnel Department; moreover many workers from outside companies also worked in Hoechst.[554] Nevertheless, the figures gleaned from these employment books are the best figures currently available and most nearly approximate the actual number of foreign workers. The books give the name, date of birth, date of commencing employment, and status (prisoner of war, sent by the Labor Office, or the like). After subtracting 827 who were repeatedly mentioned, a total number of 8,095 foreign workers who worked for shorter or longer periods in Hoechst during the war years could be established. This figure includes all foreign workers listed in the employment books. However, this figure may also include some "ethnic Germans" (*Volksdeutsche*), while there were a few persons who could not be precisely classified.[555]

If the total number of 8,095 foreign workers working in Hoechst during the war is compared to the average number of about 3,000, then there must

---

[554] In Hoechst there appear to have been many other workers from other companies and subcontractors, as well as prison inmates, etc. See Wagner, *IG Auschwitz*, pp. 263 and 332, on the contents of the document NI-11412. Wagner reports the figure as referring to the number of "concentration camp inmates" in Hoechst – this will be discussed in more detail later on.

[555] HA, PSW, Betriebseintrittsbücher 1939–1945, 827 names that were included several times over were subtracted as well as twenty-two so-called SS-resettlers who were not included among the foreign workers. Another category that was not counted consisted of fifty-six men with German names with the qualifier "from Auschwitz" next to their names. It is not clear whether these were workers from the plant or from the concentration camp. It appears as though at least some of them were workers who had been sent to work at Auschwitz and returned to Hoechst. The French prisoners of war and the so-called Pardini workers (i.e. the workers recruited via the French company Pardini) were subtracted from category 1 "TA-forced laborers" and were included in the group of French workers; the same applies to category 2 "unknown." Multiple entries sometimes occurred after small spelling mistakes, but in the main repeated entries tended to occur when a worker left or was jailed and later returned to the plant.

have been considerable fluctuations in the plant. Thus, either many of the foreign workers were reassigned by the labor offices or Hoechst returned them, or they themselves terminated the employment relationship or did not return from vacation – in the event that the latter two options were even open to them.

The term "Russians" in the employment books was apparently used to cover all "Eastern workers." It has not been possible to ascertain whether Soviet prisoners of war also worked in Hoechst, who then would not have been regarded as Eastern workers. According to a definition from 1942 Eastern workers were workers of non-German origin from the Reich Commissariat for the Ukraine, the General Commissariat White Ruthenia, or other areas to the East that bordered on these areas or adjoined the former countries of Latvia and Estonia – in other words, predominantly Russians, Ukrainians, and Byelorussians. The territorial principle was crucial, so that a Polish person from one these areas was regarded as an Eastern worker, while Ukrainians living in the *Generalgouvernement* or in annexed areas were not.[556]

The "Russian legionaries" also listed in Table 3.8 were disabled members of the Wlassow army no longer able to fight because of their injuries.[557] With their "undisciplined behavior" they occasioned no small trouble to the Hoechst factory management. Some legionaries who proved themselves to be "completely useless" in the plant were even sent back.[558]

Algerian and Moroccan workers who had apparently been recruited in France were listed as "French" in the employment books, although some Algerians were also listed as "Algerian." For greater clarity all Algerians have been subsumed in the table as French.

Finally, the last column details the number of arrests made – without differentiating whether the persons arrested were simply placed in "custody," incarcerated in the "Frankfurt Preungesheim" prison, or sent to an *Arbeitserziehungslager* (i.e., Labor Education Camp, a euphemistic description of what was actually a concentration camp for disobedient workers).

As Table 3.8 shows, the total percentage of female foreign workers was around 15.5 percent – not a very high percentage in all. However, the percentage of female workers differed greatly depending on the country of origin. Thus, around 73 percent of the Croatian workers were women. The

---

[556] Spoerer, *Zwangsarbeit*, pp. 94–5.

[557] HA, IG/50, IG-Prozessakten, Jähne document 19, affidavit by Alex Snessarow dated January 9, 1948.

[558] HA, TEA 95f, Niederschrift über die technische Direktionssitzung in Hoechst, August 28, 1944, and Niederschrift über die technische Maingau-Direktionssitzung in Hoechst, September 4, 1944.

TABLE 3.8. *"Foreign workers" at the Hoechst Plant, 1940–1945 as Stated in the Employment Books*

"Foreign Workers"	Total	Female	Younger than 14 Years of Age	Male	Prisoners of War	Younger than 14 Years of Age	Arrests
TA forced laborers	14	1		13			
Unknown	40	3		37			
Belgium	1,800	67		1,733	2		16
Bulgaria	1			1			
Denmark	301	123		178			6
France (including Algeria)	1,962	29		1,933	787 (partial "exchange")		133
Greece	1			1			
Italy	1,192	22		1,170	127		24
Yugoslavia	2			2			
Disabled Russian legionaries	256			256			
Croatia	393	287		106			9
Lithuania	146	10		136			19
Luxembourg	1			1			1
Netherlands	274	5		269			6
Poland	242	34		208			4
Romania	1			1			
Russia	1,469	678	9	791	(?)	13	46
TOTAL	8,095	1,259		6,836	916		264

Source: HA, PSW, Betriebseintrittsbücher 1940–44/45.

percentage of women workers among the Russian or Eastern workers was more than 46 percent. Contrary to Lautenschläger's comment on the limited utility of women as workers, these women carried out heaviest types of men's work – for minimal wages.

According to the figures in the employment books, 13.5 percent of foreign workers were prisoners of war. Yet here again numbers varied very much according to the country of origin. Almost 41 percent of the French were – at least originally – prisoners of war. In contrast to this, only around 11 percent of the Italian male workers were so-called Italian military internees, or IMIs. After Marshal Pietro Badoglio capitulated to the Allies and the subsequent capture of Italian soldiers by German troops Hitler refused to accord these soldiers the status of prisoners of war commensurate with the international law of war.[559] The records in the employment books do not provide any information as to whether Soviet prisoners of war were included in the "Russian" contingent. Hoechst repeatedly applied for Soviet prisoners of war, so it cannot be excluded that some of the "Russians" were prisoners of war, even if according to a witness for the defense "Russian prisoners of war never worked" for Hoechst.[560]

This brings us to the question concerning Hoechst's responsibility for the allocation of foreign workers, which also formed part of the charges respecting "slave labor" in the IG Farben trial in Nuremberg. At the time Friedrich Jähne said that Hoechst had "not needed to make use" of the "recruitment of foreign workers" that IG Farben had carried out in countries like Italy and Belgium. "An office in Berlin" had been responsible for foreign workers, although the name of the office now escaped him.[561] Jähne's defense lawyer, Hans Pribilla, went so far as to claim that the company had been "forced by official orders" to employ "foreign workers." IG Farben and thus Hoechst as well had only employed foreign workers "out of necessity" in order to fulfill its production obligations. As no German workers were assigned to them, Hoechst had no choice but to take on the foreign workers: "This was absolutely compulsory."[562]

On the other hand, Lautenschläger testified in Nuremberg that when the question concerning the employment of foreign workers first arose, at the time both he and Jähne, as well as the other members of the Management Board, "regarded the suggestion entirely positively": "as there were not sufficient German workers available, we welcomed the employment of foreign workers."[563] Pribilla, who was also Lautenschläger's lawyer, tried to play down this statement in the trial.[564] However, it was confirmed by Otto Hirschel, the head of the Social Department. Using almost the same words as Lautenschläger, Hirschel said that at the time when the idea of using foreign workers was first mooted, "the management regarded this suggestion entirely positively."[565]

At the beginning of the war foreign workers still were being offered to Hoechst, and the works management did not have to take any steps itself. In

[559] For the essential study on "IMI" cf. Gabriele Hammermann, *Zwangsarbeit für den 'Verbündeten.' Die Arbeits- und Lebensbedingungen der italienischen Militärinternierten in Deutschland 1943–1945* (Tübingen: Niemeyer, 2002).

[560] HA, IG/50, IG-Prozessakten, Jähne document 19, affidavit by Alex Snessarow dated January 9, 1948.

[561] NI-5168, affidavit by Friedrich Jähne dated May 29, 1947.

[562] HA, IG/50, IG-Prozessakten, trial letter for Jähne from defense lawyer Hans Pribilla, pp. 37 und 40.

[563] NI-6415, affidavit by Lautenschläger dated March 26, 1947 (handwritten correction).

[564] HA, IG/54, IG-Prozessakten, trial letter for Lautenschläger from defense lawyer Hans Pribilla, pp. 7–8.

[565] NI-2973, affidavit by Otto Hirschel dated January 16, 1947.

March 1940 at a meeting of the management of the IG Farben Maingau Works Group a great shortage in the number of available workers was ascertained for Hoechst. The decision was therefore taken to increase the working hours of the entire workforce from 52 to 56 hours per week, at first for a limited period of time.[566] The labor shortage was again "discussed in detail" in April; the advent of 200 Slovaks of German origin "was promised" for April. Furthermore it was planned to remedy the labor shortage by conscripting labor and shutting down smaller workshops.[567] At the end of May 1940 it was reported during a meeting of the Technical Directorate that "several workers" had been transferred to Hoechst from companies that had been closed down in the Taunus area. Furthermore 250 Polish civilian workers were going to be allocated, and there was a promise of a further 200 Italians.[568]

In mid-June 1940, a report given during a meeting of the Technical Directorate stated that Hoechst had been assigned 200 workers from the West Wall by *Organization Todt* and promised a further 380 Polish workers – of which Hoechst "wanted only to take on 200."[569] By the middle of June, a total of 120 Poles, fifty-three of which were trained workmen, had finally arrived in the plant, but it still was short some 300 workmen and workers.[570] By July 1940 the labor situation had eased, and hopes were voiced that the war would soon be over. In summary, it was stated that Hoechst had been allocated 361 workers in the past few weeks: 34 Upper Silesians, 119 civilian Poles, and 208 "workers of German origin by means of conscription." Furthermore, another 200 German workers and possibly fifty Belgians were expected. Accordingly, the management of Hoechst considered that the need for labor had been "satisfied for the time being."[571]

The first French prisoners of war finally arrived in the Hoechst plant in the fall of 1940.[572] And the inflow of foreign workers continued. In

---

[566] HA, TEA 95e, Niederschrift über die technische Maingau-Direktionssitzung in Hoechst, March 4, 1940.

[567] HA, TEA 95e, Niederschrift über die technische Direktionssitzung in Hoechst, April 8, 1940.

[568] HA, TEA 95e, Niederschrift über die technische Direktionssitzung in Hoechst, May 27, 1940.

[569] HA, TEA 95e, Niederschrift über die technische Maingau-Direktionssitzung in Hoechst, June 10, 1940.

[570] HA, TEA 95e, Niederschrift über die technische Direktionssitzung in Hoechst, June 17, 1940; Niederschrift über die technische Maingau-Direktionssitzung in Hoechst, June 24, 1940.

[571] HA, TEA 95e, Niederschrift über die technische Direktionssitzung in Hoechst, July 2, 1940.

[572] HA, TEA 95e, Niederschrift über die technische Maingau-Direktionssitzung in Hoechst, September 16, 1940.

December 1940 the factory management, which anticipated "the allocation of additional foreign workers," felt it necessary to extend the hut camp to hold 800 to 900 workers. In January 1941 plans were already being made to accommodate 1,000 persons, and in February this rose to 1,200 – it was planned to accommodate them in two hut camps, in the municipal water-works, in the main hall of three inns, and in the company's own residence hall for unmarried workers.[573]

By the end of June 1941 after Germany had attacked the Soviet Union, the situation on the labor market once again became difficult because many young men were now being called up. Even in IG Farben the individual plants primarily looked after their own interests. Thus, 215 Italians Ludwigshafen had promised to Hoechst did not arrive, whereupon the situation in Hoechst became "rather difficult." After that, Kränzlein wished to attempt to keep 200 to 250 students for a period of ten weeks. As Defense Commissar for the district, he additionally received information from the Reich Minister for Armaments and Munitions that between 200 and 300 German workers or workers "of foreign origin" from the Reich autobahn administration would be made available to Hoechst.[574]

Subsequently more and more foreign workers were recruited. In the summer of 1941 Hoechst recruited around 120 Belgian workmen.[575] But as a whole the labor shortage worsened, and in Hoechst the situation was "catastrophic"; Lautenschläger intended to report as much to Carl Krauch, who was then head of the Supervisory Board of IG Farben and Plenipotentiary General for Special Questions of Chemical Production in the Four-Year Plan (*Gebechem*).[576] The situation in Hoechst was aggravated by the fact that Italian and Dutch workers left the plant "in large numbers without permission" or did not return from their vacations.[577] Hoechst therefore increasingly attempted to recruit workers in France.[578] However, the efforts made to obtain French, Croatian, Italian, Belgian, Dutch, and Polish workers were not enough to remedy the shortage of labor in Hoechst.

---

[573] HA, TEA 95e, Niederschrift über die technische Maingau-Direktionssitzung in Hoechst, December 9, 1940; TEA 95f, Niederschrift über die technische Maingau-Direktionssitzung, January 6, 1941, and Niederschrift über die technische Direktionssitzung in Hoechst, February 24, 1941.

[574] HA, TEA 95f, Niederschrift über die technische Maingau-Direktionssitzung in Hoechst, June 30, 1941.

[575] HA, TEA 95f, Niederschrift über die technische Maingau-Direktionssitzung in Hoechst, August 25, 1941.

[576] HA, TEA 95f, Niederschrift über die technische Direktionssitzung in Hoechst, September 1, 1941.

[577] HA, TEA 95f, Niederschrift über die technische Maingau-Direktionssitzung in Hoechst, September 22, 1941.

[578] HA, TEA 95f, Niederschrift über die technische Direktionssitzung in Hoechst, October 13, 1941, and October 20, 1941.

By the middle of November 1941 the factory management had agreed that it would be necessary to fall back on another group which was now available: "The labor situation in the plant at Hoechst makes it necessary to use Russian prisoners of war. In the Construction Department there is employment for about 40, in the Department for the Shipping of Nitrogen for about 50 prisoners. The building of a camp for 400 Russian prisoners is planned."[579] However, the management did not intend to leave the matter at that, since the labor shortage in the factories was becoming "increasingly appreciable." The works management planned to recruit Poles: "In Poland Hoechst intends to set up a recruitment office similar to what Leverkusen has already successfully carried out, which will recruit Polish civilian workers. The Regional Labor Office will be applied to for permission to recruit 1,000 Poles."[580]

At first, however, these efforts met with only limited success. In mid-December 1941 it was noted that "the efforts of different offices to obtain workers of foreign origin" had been "unsuccessful" up to then.[581] One week later only 18 German and Dutch workers from the Reich autobahn administration had been sent on loan to Hoechst; even worse, "the application of Hoechst for permission to recruit 500 workers in Poland was approved by the Reich Ministry for Labor but turned down by the government of the *Generalgouvernement*, since Polish workers who were still in Poland were required for other urgent tasks." The works management decided therefore with the help of the Hessian representative in Berlin to "obtain permission to recruit Serbians." Moreover workers "capable of Germanization" from Croatia should come to the plant to work.[582]

At the beginning of January 1942 Hoechst noted that "of the Italians given leave over Christmas, more than 100 have not yet returned by now," and in addition the management was expecting numerous workers to be called up into the army. In addition to Serbians it planned increasingly to recruit workers from Eastern Europe, as well as to request the assignment of Russian civilian workers. However, according to the factory management, a prerequisite for the assignment of workers from Eastern and Southeast Europe was "that a delousing facility and a hut for sick persons should be built in Hoechst."[583]

[579] HA, TEA 95f, Niederschrift über die technische Maingau-Direktionssitzung in Hoechst, November 17, 1941.

[580] HA, TEA 95f, Niederschrift über die technische Direktionssitzung in Hoechst, November 24, 1941.

[581] HA, TEA 95f, Niederschrift über die technische Maingau-Direktionssitzung in Hoechst, December 15, 1941.

[582] HA, TEA 95f, Niederschrift über die technische Direktionssitzung in Hoechst, December 22, 1941.

[583] HA, TEA 95f, Niederschrift über die technische Maingau-Direktionssitzung in Hoechst, January 5, 1942.

Soon afterwards the situation in Hoechst once more improved considerably. In mid-January 1942 it was recorded that the "first transportation of Lithuanian Poles" was expected. These persons would first be accommodated in the "quarantine camp" in Kelsterbach and then distributed to the various industrial plants. Between 100 and 200 Poles from this transport were earmarked for Hoechst. Furthermore, "100 women from *Ostland* [occupied territories in Eastern Europe] had been offered, and they should be accepted." Because of the "want of coal" in France and Belgium, the result of enforced shipments of coal to Germany that increased unemployment in those two countries, there was also a possibility of getting workers from French and Belgian companies: "300 Frenchmen or Belgians have been requested for Hoechst."[584]

In the winter of 1941/42 the labor situation in the plant became very difficult as the number of sick persons increased from the already very high percentage of 7 to 9 percent. The Reich Ministry for Labor had instructed the recruitment office in Paris to assign 500 French civilian workers to Hoechst.[585] Nevertheless, in order to find a solution for the difficult situation Hoechst decided it needed to become active itself – for as Lautenschläger and Hirschel said, Hoechst regarded the utilization of "foreign workers" very positively. In February 1942 Hirschel, head of the Social Department, was able to report: "An agreement has been concluded with a Parisian businessman, which will provide us with 300 manual workers as well as 30 skilled workers and craftsmen. The first transport will arrive in Hoechst at the end of this week."[586]

According to Hirschel, the procedure to obtain workers during the war was generally as follows: The *Gebechem* office in Berlin or its branch office in Wiesbaden informed Hoechst as the seat of the works management of the IG Farben Maingau Works Group when foreign workers were available. After the number of workers required had been determined within the plant by Hirschel asking the different departments or by the departments independently requesting more workers, Hoechst then declared how many workers were required. The labor office, the proper authority with the responsibility of providing workers, allocated too few workers to Hoechst, and Hoechst therefore appealed directly to the *Gebechem* in Berlin, from whom they received their allocation of Russian workers.[587] In contrast to this, according to Hirschel, the assignment of French workers had

[584] HA, TEA 95f, Niederschrift über die technische Maingau-Direktionssitzung in Hoechst, January 19, 1942; on France and Belgium cf. Spoerer, *Zwangsarbeit*, pp. 60–3.
[585] HA, TEA 95f, Niederschrift über die technische Direktionssitzung in Hoechst, January 26, 1942, and Niederschrift über die technische Maingau-Direktionssitzung in Hoechst, February 2, 1942.
[586] HA, TEA 95f, Niederschrift über die technische Direktionssitzung in Hoechst, February 9, 1942.
[587] NI-2973, affidavit by Otto Hirschel dated January 16, 1947.

progressed "only very slowly and sluggishly." Hirschel therefore suggested driving to Paris and recruiting workers there himself. The Directorate approved his proposal, and in consequence he traveled three times to Paris, where he received a promise for French workers. On February 7, 1942, Hirschel signed a contract with the company Entreprise J. Pardini in Paris for a total of 332 workers to come to Hoechst. As Hirschel still believed that the labor problem was not yet solved, he prevailed on the factory management to send Wilhelm Stellmann, an authorized signatory who had gone into early retirement, to Paris to recruit further employees.[588]

By the end of March, only 249 of the approved 800 French civilians and the 200 "persons from *Ostland*" had arrived in the plant. The factory management therefore felt obliged to attempt "to obtain Russian prisoners of war."[589] At the end of March 1942 the works management discussed the labor situation and recorded: "The shortage of workers becomes greater from week to week. Only very few of the 500 French civilians promised by the Reich Department of Labor or of the 300 approved French civilians recruited by the company's own efforts have arrived. Last week a total of only 13 French persons were referred."[590]

Stellmann said after the war that Hirschel had asked him in February 1942 whether he was prepared to travel to Paris "to accelerate" the recruitment of French workers. When after a few weeks almost no workers from France had come, he had gone to Paris again "at the request" of Lautenschläger. However, Stellmann considered his mission not very successful: "As a result of the sharp competition of German companies endeavoring to obtain French workers, my efforts were more or less unsuccessful."[591] Indeed, from the middle of February until the beginning of April and again from the beginning of May until the middle of June 1942 Stellmann was in Paris working for Hoechst "on questions related to the deployment of labor," for which he was well remunerated.[592] It should perhaps be mentioned that, based on the records of the employment books, over 300 workers arrived from Pardini.

This was not the only time that Hoechst was directly involved and concluded independent contracts with companies that arranged to send workers to Hoechst. In 1944 Hirschel sent another employee, Johann Simon, to Italy to recruit workers. However, as other IG Farben plants were also looking for workers there and the *Gebechem* had correspondingly

[588] Ibid.; HA, Wk 61, Treaty IG Hoechst (Hirschel) with Pardini dated February 7, 1942.
[589] HA, TEA 95f, Niederschrift über die technische Direktionssitzung in Hoechst, March 23, 1942.
[590] HA, TEA 95f, Niederschrift über die technische Maingau-Direktionssitzung in Hoechst, March 30, 1942.
[591] NI-2991, affidavit by Wilhelm Stellmann dated January 16, 1947.
[592] HA, PA Abteilung VI, Stellmann, Hoechst to Krauch, March 16, 1942; notes by Hirschel for Flach dated April 10, 1942 and June 23, 1942.

drawn up quotas based on urgency, Simon was apparently only moderately successful. In a meeting of the management held in May 1944 the minutes even recorded that the result had been "very bad."[593]

Hoechst was therefore by no means merely a passive recipient of allocated foreign workers. On the contrary, Hirschel personally exerted himself to obtain greater numbers of foreign workers; after 1943 he was supported in this by Winnacker.[594] "Compulsion" existed only insofar as Hoechst, in its attempts to prove that it was a profitable company within IG Farben, was driven by entrepreneurial logic to be involved in production for the war at all costs and accordingly needed workers. However, this was not the sort of "compulsion" Friedrich Jähne and his lawyer were referring to. In the pursuit of its goals the works management acted independently, and in the question of the allocation of foreign workers it was an active agent.

This brings us to the question of the treatment of foreign workers, and here it is important to make careful differentiations. In the summer of 1942 Hoechst employed about 2,600 foreigners, who consisted, according to Lautenschläger, of "Italians, Poles, Lithuanians, Slovaks, Croatians, French prisoners of war, French civilians, Dutch, Belgians, Serbians, Danes, and Russians." There also appeared to be clear differences with regard to performance and willingness to work of the foreign workers: "Among the workers of foreign origin, the French, Belgians, Danes, and Russians performed best, while other foreigners, for example the Dutch and Italians, barely achieved the average output of a German unskilled laborer worker; they were also very unreliable in complying with working hours, while on the other hand they made the greatest demands respecting treatment and food." Lautenschläger commented that the attitude of the foreign workers toward Germany was "very changeable" and took its cue "primarily from events at the front and news which reached them from their homeland."[595]

But it was not only the attitude of the foreign workers that varied according to the course of the war. The attitude of the German political leadership and of the companies also changed over time. The difficult labor situation led to the appointment of Fritz Sauckel, the *Gauleiter* of Thuringia, as Plenipotentiary for the Deployment of Labor in March 1942, and he attempted to recruit workers for Germany all over Europe, both by force and coercion and by offering incentives.[596] Likewise German

[593] NI-2974, affidavit by Johann Simon dated January 17, 1947; NI-2973, affidavit by Otto Hirschel dated January 16, 1947; HA, TEA 95f, Niederschrift über die technische Maingau-Direktionssitzung in Hoechst, May 8, 1944.

[594] HA, PA Lautenschläger, Sekretariat Lautenschläger, Vertrauliche Korrespondenz, draft by Winnacker for Lautenschläger dated October 25, 1943 re "Schreiben Krauchs vom 7.10.1943."

[595] HA, PA Lautenschläger, Lautenschläger, "Erinnerungen," vol. 3, [pp. 128–30].

[596] See Spoerer, *Zwangsarbeit*, pp. 35–40.

companies also tried "to spur on" their "foreign workers," most of which were now indeed forced laborers, and increase their efficiency.

In 1943 a propaganda leaflet from Sauckel even used the term "guest workers" (*Gastarbeiter*) in Germany in addition to the term "foreign worker" (*Fremdarbeiter*). The accompanying text to the illustrations showing how "pleasant" working in and for Germany was: "The millions of busy guest workers in Germany have every reason to be satisfied with their present lot. They have work and bread, receive their productivity bonuses, work in clean, bright, modern companies."[597]

Corresponding photographs of workers, their accommodation, sanitary facilities, and "leisure-time activities" were also made in Hoechst, but these propaganda photos did not show the real conditions but rather how they were supposed to appear. Even if the accommodation for foreign workers in Hoechst was not as far away from the plant as Luce d'Eramo describes, conditions in the huts, in the rented inns, and in the company's residence for unmarried workers were in some cases unhygienic and often very cramped. At all events the "foreign workers" were certainly not living in the beautiful and spacious conditions shown in the photographs.[598] The minutes of a management meeting held at the end of October 1942 recorded: "The danger of contamination of the factories and laboratories by vermin from the camps of the foreign laborers was discussed. The vermin in the huts must be combated by all possible means."[599] And again in spring 1943: "Following a circular from the Department of Followers the work clothing of foreign workers will be deloused. To ensure complete success, the simultaneous delousing of the foreigners themselves as well as their accommodation is regarded as expedient."[600] This indicates that within the space of half a year problems with vermin such as lice once again had emerged, and the spread of vermin was fostered by the cramped quarters and the existing overcrowding of the huts. Such problems did not accord well with the idyllic conditions that the photographs, particularly of the "Italian village," pretended to mirror.

Let us now look at the treatment of Italians and other foreign laborers, referred to by Karl Trost as "so-called voluntary foreign workers." To create an incentive for workers from these countries, they were given bonuses in addition to their regular wages, so that some of them cost Hoechst more than German workers would have – and these figures do not include the costs for the newly built hut camps. However, it is not certain

---

[597] Friedrich Didier, *Europa arbeitet in Deutschland. Sauckel mobilisiert die Leistungsreserven* (Berlin: Zentralverlag der NSDAP Franz Eher Nachf., 1943), [p. 40].

[598] See d'Eramo, *Deviazione*; the German edition *Der Umweg* has a map on p. 219.

[599] HA, TEA 95f, Niederschrift über die technische Direktionssitzung in Hoechst, October 26, 1942.

[600] HA, TEA 95f, Niederschrift über die technische Maingau-Direktionssitzung in Hoechst, March 1, 1943.

that the Italian workers finally did end up with more money in their pay packets.

In the matter of the Italian workers Hoechst recorded with displeasure in March 1941: "Up to now Hoechst has not paid any transfer compensation (separation allowance and expenses for lodging) to workers who are forced to maintain separate households. Since the authorities in Berlin demand the payment of transfer compensation to Italian workers, this question must generally be settled for the Maingau plants to ensure that German followers are not placed at a disadvantage compared to foreign workers. This is still being negotiated with the Reich Trustee."[601] Director Paul Roth remarked in September 1941 that with the payment of the transfer compensation Italians received higher wages than German workers, who felt themselves to be "disadvantaged" by this.[602] However, Hoechst could not change this. Thus Hoechst had recruited Italians as workmen at an hourly wage of 82 Pfg. (*Pfennige*); the recruited Italians apparently proved to be unsuitable for the tasks assigned them and were subsequently employed and paid as unskilled laborers. In a meeting held with representatives of the German Labor Front, some higher officials of the *Gau*, and an Italian delegate, Hirschel was obliged to pay the Italians a compensation of an additional 14 Pfg. per hour, otherwise "the Italians would have had the right to leave the plant." As some of the German workers had also been transferred to other departments in which they earned less than they had earned before, Hoechst determined that for the duration of the war German employees of many years standing would also receive compensation that would "not to put them at a disadvantage compared to foreign workers."[603]

However, as the number of workers lessened, they became more expensive – for Hoechst needed them so urgently that the management was prepared to keep them on despite the additional payments. This applied both to German workers and to foreign workers in general, not just to the Italians. How far Hoechst was prepared to go was demonstrated in the fall of 1942. At the end of September, it was noted that "time and again it was difficult" to obtain "foreign workers for the rates of pay paid in Hoechst." Thus, the Belgian company Leclercq was supposed to provide transport workers at an hourly wage of 68 Pfg.; but because companies in Hamburg offered wages of 90 Pfg., "not a single worker could be got hold of for Hoechst." As the needed workers remained employees of the Belgian company, Hoechst finally agreed to pay them the rate for transport workers

[601] HA, TEA 95f, Niederschrift über die technische Maingau-Direktionssitzung in Hoechst, March 17, 1941.
[602] HA, TEA 95f, Niederschrift über die technische Maingau-Direktionssitzung in Hoechst, September 8, 1941.
[603] HA, TEA 95f, Niederschrift über die technische Direktionssitzung in Hoechst, September 16, 1941.

of 81 Pfg. per hour. This had a drawback, namely, "the unpleasant situation that German followers performing the same work receive lower wages."[604]

In the contract with the company Pardini, Hirschel, acting for Hoechst, offered French workers the same standard wages as German workers together with the standard bonuses. In addition, family allowance, separation allowance, and bonuses for performance were also paid as well as contributions toward traveling expenses and journeys home to visit families. All in all, the remuneration was quite good. Moreover, Hirschel added another request, apparently for the particular benefit of the French civilians: "In order to make the stay of the foreign workers in Hoechst as pleasant as possible, regular movie screenings and other events should be held."[605]

Hoechst apparently went to great lengths to be a relatively attractive employer for workers from "friendly" states or Western European countries – because workers were desperately needed and the competition from many other companies in Germany, even from other IG Farben plants, was fierce.

Food was an important factor that strongly affected the satisfaction of the workers, and Hoechst was well aware of this. According to the statement of the manager responsible for food and catering in Hoechst, Albert de Vries, the food was very good, and attempts were even made to take account of the traditions of respective countries. Thus there was even "Italian cuisine." As part of his defense in the Nuremberg trial, de Vries even managed "to find some bills of fare" as evidence of how good the meals had been – based on this testimony the meals would indeed have been quite good.[606] Yet according to what the historian Valentina Maria Stefanski was able to ascertain for the IG Farben plant at Leverkusen, the statements concerning the alleged diet of the foreign employees of IG Farben made in the course of the trial were "diametrically opposed" to the memories of former Polish forced laborers and slave laborers questioned by Stefanski. Stefanski commented that it was difficult to conceive of more contradictory statements: "on the one hand more than 2,000 calories as normal rations and over 4,000 for heavy laborers – on the other hand hunger day in, day out, and the overpowering wish to eat one's fill at least once more in this life."[607]

---

[604] HA, TEA 95f, Niederschrift über die technische Direktionssitzung in Hoechst, September 28, 1942.

[605] HA, Wk 61, Treaty IG Hoechst (Hirschel) with Pardini dated February 7, 1942; TEA 95f, Niederschrift über die technische Direktionssitzung in Hoechst, February 9, 1942.

[606] HA, IG/50, IG-Prozessakten, Jähne documents 17 and 51, affidavits by Albert de Vries dated January 6, 1948, and March 10, 1948, the latter with the alleged "bills of fare"; cf. TEA 95f, Niederschrift über die technische Maingau-Direktionssitzung in Hoechst, June 23, 1941.

[607] Valentina Maria Stefanski, *Zwangsarbeit in Leverkusen. Polnische Jugendliche im I.G. Farbenwerk* (Osnabrück: fibre, 2000), pp. 172–5.

The time of people's arrival, the duration of their stay, their gender, and their national category all determined what sort of food and how much a worker received. At the end of May 1941, for example, at a meeting of the Maingau Technical Directorate, it was noted: "Lately, there have been repeated complaints on the part of the workers housed in the camps about insufficient food. Talks will be held with the factory inspectorate and the nutrition office about whether it will be possible to increase the allocation of food ration cards."[608] The French prisoners of war were apparently also among those who received insufficient food. This became obvious after Hirschel negotiated with the factory inspectorate in the summer of 1941 about increasing the amount of food allocated to the French prisoners of war: "The bread allocation, which had been reduced to 2/3rds for heavy laborers, should be increased to 100%."[609]

While such prisoners were able only to voice their complaints, volunteer workers were able to take action – even if only by absenting themselves. Thus, it appears that in the spring of 1942 a number of volunteer civilian workers returned to their native countries after their contracts had terminated, for the most part on the grounds "that the food is insufficient." However, Jähne reported to the management meeting that after a careful examination of the catering facilities he had come to the conclusion "that nothing is being neglected with respect to the feeding of the foreigners." Jähne even said that the food provided to the foreigners was "more abundant than that of the German workers" and the complaints that were being made were "unjustified."[610] However, this is unlikely to have been accurate.

Poles did not even have the possibility of complaining, as they had no rights whatsoever and were obliged to eat whatever was given them. They were regarded as *Untermenschen* or subhuman creatures by the Nazis, who would have preferred that no Poles lived in the Reich. However, since they were urgently needed for German agriculture and industry, a compromise was hammered out between Nazi ideologues and the spokesmen for economic interests with the enacting of the so-called *Polenerlasse* (Edicts concerning Poles), which strongly discriminated against Poles; Ulrich Herbert describes these measures as "terror as a compromise of rule." Poles were obliged to wear a special badge consisting of a "P" worn on their clothes and to endure many daily restrictions and harassments. Any contact with Germans was strictly forbidden. On the orders of Lautenschläger,

---

[608] HA, TEA 95f, Niederschrift über die technische Maingau-Direktionssitzung in Hoechst, May 26, 1941.

[609] HA, TEA 95f, Niederschrift über die technische Maingau-Direktionssitzung in Hoechst, June 23, 1941.

[610] HA, TEA 95f, Niederschrift über die technische Maingau-Direktionssitzung in Hoechst, April 20, 1942.

the terror against Polish workers embodied in the *Polenerlasse* also affected the Polish workers in Hoechst – and their experiences with the allocation of food probably corresponded to those quoted above regarding Leverkusen.[611]

In general, for reasons of pure necessity, Hoechst showed itself increasingly willing to make concessions toward foreign workers – and simultaneously it exploited them even more brutally. Thus, in order to compensate for the pressing shortage of properly trained workmen the works management requested that foreign workers be retrained in the apprentice workshops of the German Labor Front[612] – which ultimately improved the worker's financial standing. And although all foreign workers could be constrained to work on Sundays, work that consisted of such tasks as the loading and unloading of railroad freight cars, the management deliberated whether "beer or other luxuries could be offered as recognition to those foreigners called on to perform special tasks." On the other hand, the works management put pressure on the foreign workers. To force them to return from their vacations "a certain sum out of their wages was to be retained every month as security money, which is only to be paid out after they have returned."[613] However, for many this was apparently not enough of an incentive, and French and Belgian workers in particular often did not return from their vacations. As many as 210 foreign workers did not return from vacation in the months of June and July 1943 alone.[614] This in turn meant that in summer 1943 there was "an immediate need of manpower in the Hoechst plant" with a shortage of no fewer than 1,000 workers.[615]

To offer an additional incentive to those who worked hard in Hoechst and possibly also to protect them from foreign workers opposed to the Nazi regime, Jähne suggested in fall 1943 "to accommodate the foreign workers who distinguished themselves by their diligence separately within the camps."[616] The works management additionally ruled that the foreigners' huts should generally be heated "in cooler weather regardless of the allocated amounts of coal": "Any additional requirement of coal is to be taken

---

[611] NI-4683, circular from Lautenschläger dated June 18, 1940, re. "Verhalten gegenüber Zivilarbeitern polnischen Volkstums"; on this topic in general, see Herbert, *Fremdarbeiter*, pp. 70–82.

[612] HA, TEA 95f, Niederschrift über die technische Maingau-Direktionssitzung in Hoechst, February 15, 1943.

[613] HA, TEA 95f, Niederschrift über die technische Maingau-Direktionssitzung in Hoechst, March 15, 1943.

[614] HA, TEA 95f, Niederschrift über die technische Maingau-Direktionssitzung in Hoechst, July 19, 1943, and August 16, 1943.

[615] HA, TEA 95f, Niederschrift über die technische Maingau-Direktionssitzung in Hoechst, July 12, 1943.

[616] HA, TEA 95f, Niederschrift über die technische Direktionssitzung in Hoechst, October 4, 1943.

from the coal stores."[617] And the management tried to ensure that sufficient food was available for the foreigners, which was becoming increasingly difficult by the end of 1943: "Matters concerning food for the foreign workers were discussed. The reduction of the potato rations has created a difficult situation. At all costs the attempt must be made to ensure that the foreigners are fed sufficiently in order to preserve their capacity to work."[618]

While Hoechst tried by these means to keep up the strength and spirits of its foreign workers, it also put them under heavy pressure. To cut down the numbers of foreign workers not returning from vacation, members of the same ethnic group were retained in the plant as hostages. At the end of 1943 Sauckel, the plenipotentiary for the deployment of labor, introduced "a new regulation concerning vacations" for foreign workers: "This provides that foreigners of the same nationality will travel home in groups. If not all members of the first group return, then only half of the second group or even fewer will be allowed to travel home."[619]

The minutes of a management meeting held in Hoechst at the beginning of 1944 stated: "The SS shall take action against those foreigners who do not return from vacation at our instigation."[620] Indeed, contrary to the assertion made by one of the witnesses for the defense after the war, Hoechst did report workers who did not return to the plant after their vacation.[621] And the works management decided in May 1944: "For education purposes those workers who are unwilling to work (Germans and foreigners) shall be transferred to a punishment crew which shall be formed immediately and used for heavy labor."[622] The management attempted by these means to keep unwilling workers in the plant, creating special punishment crews for them. "Idling" and theft also apparently increased in the course of the year, and the Gestapo primarily took action against the latter – probably at the request of the Social Department or the Defense Commissar. Yet Hoechst urgently needed its workers, and to prevent them being sent away to an *Arbeitserziehungslager* (Labor Education Camp) the

[617] HA, TEA 95f, Niederschrift über die technische Direktionssitzung in Hoechst, October 18, 1943.

[618] HA, TEA 95f, Niederschrift über die technische Maingau-Direktionssitzung in Hoechst, November 29, 1943.

[619] HA, TEA 95f, Niederschrift über die technische Direktionssitzung in Hoechst, November 1, 1943.

[620] HA, TEA 95f, Niederschrift über die technische Maingau-Direktionssitzung in Hoechst, January 24, 1944.

[621] HA, IG/50, IG-Prozessakten, Jähne document 54, affidavit by Karl Gebhardt dated March 4, 1948; dagegen NI-2974, affidavit by Johann Simon dated January 17, 1947; NI-14824, letter from the French Ministry of Veterans' Affairs to the Chief Counsel for War Crimes, dated February 16, 1948, with index cards of French workers who were searched for after Hoechst had reported them.

[622] HA, TEA 95f, Niederschrift über die technische Maingau-Direktionssitzung in Hoechst, May 8, 1944.

works management wished to establish a satellite camp within Hoechst itself: "The Gestapo takes a tough line against Eastern workers who steal gasoline or methanol and transfers them to an *Arbeitserziehungslager*. During this period the workers are lost to the plant. In order to retain them the attempt should be made to transfer a satellite camp of the *Arbeitserziehungslager* to Hoechst."[623]

According to the records of the employment books a total of 264 foreign workers were sent to an *Arbeitserziehungslager* or to prison. It is impossible to say whether this number gives a full account of all arrests, but certainly the figure is not overstated. All of these foreign workers were turned over to the Gestapo by Hoechst. This number of 264 arrests is considerably higher than the figure quoted after the war by Franz Spiess, an employee of the Social Department who was responsible for foreign workers. According to his statement foreign workers were primarily fined. Sauckel had decreed that "German workers were to be reported to the Trustee of Labor, foreign workers to the *Geheime Staatspolizei*," but this only occurred in cases of "repeated indiscipline." Spiess stated that from the end of 1943 until March 1945 "in my recollection this was only done in 3 or 4 cases."[624] Such a statement simply cannot be right and cannot be explained by a poor memory. Out of the Belgian contingent alone a total of 16 persons were arrested, three of whom were sent to the prison in Frankfurt-Preungesheim and four to an *Arbeitserziehungslager*. Also, 133 French workers were arrested, 49 of whom were sent to Preungesheim and 10 to an *Arbeitserziehungslager*. Eight of the 24 Italians who were arrested were sent to an *Arbeitserziehungslager*; 19 Lithuanians were arrested, two of whom were transferred to an *Arbeitserziehungslager*. Six Dutch were also arrested, and one was sent to an *Arbeitserziehungslager*. Out of 46 arrested Russians, 10 suffered in an *Arbeitserziehungslager*. The French workers suffered most from the disciplinary measures in Hoechst; with a total of 133 arrests listed in the employment books, whether spent in prison or in an *Arbeitserziehungslager*, the numbers were far higher than those of either the Russians or the Italians.

In December 1942, for example, a French civilian worker was arrested by the criminal police on "the urgent suspicion" of having "helped French prisoners of war to escape." The works management therefore decided to take measures to ensure "that French prisoners of war and French civilians are not working next to one another in any of the factories."[625] This was a

[623] HA, TEA 95f, Niederschrift über die technische Maingau-Direktionssitzung in Hoechst, July 10, 1944.

[624] HA, IG/50, IG-Prozessakten, Jähne document 21, affidavit by Franz Spiess dated January 16, 1948.

[625] HA, TEA 95f, Niederschrift über die technische Maingau-Direktionssitzung in Hoechst, December 7, 1942.

problem, however: "The defense bureau of the armaments command has emphasized that no opportunity should be given to French prisoners of war to meet with French civilian workers. Since a separation in the Hoechst plant is impossible, an attempt will be made to exchange the prisoners of war for civilian workers, particularly since it is to be expected that the French prisoners of war will be gradually discharged home."[626] In general, however, as the statement of Lautenschläger quoted above shows, Hoechst was satisfied with the French civilian workers and the prisoners of war, since the works management took the decision that "A French prisoner of war will be discharged for every civilian Frenchman who begins work in Germany. An attempt will be made to influence the prisoners so that they will remain here to work for us after being discharged and having had a spell of home leave." Moreover a proposal was put forward to make the prisoners of war work night shifts, and permission for this was sought from the *Stalag* in Bad Orb.[627] In March 1943, after the previous positive experience with the Pardini company, Hirschel concluded "another service agreement concerning the provision of 50 construction workers" with Pardini.[628]

On the orders of the Nazi authorities Hoechst was expected to be as obliging as possible toward French civilian workers: "The OKW [*Oberkommando der Wehrmacht* – High Command of the Armed Forces] and the plenipotentiary for the deployment of labor have ordered that the wives of French prisoners of war should be employed in the same factories and should be permitted to live with their husbands sharing a single room together. The companies are urged to provide accommodation for this purpose with single rooms." However, Hoechst foresaw a problem: "As it is not expected that any construction material will be allocated, it is doubtful whether the plan can be realized."[629]

However, in their treatment of French prisoners of war Hoechst doubly violated the existing international law of war. First, it was forbidden for prisoners of war to work in armaments production. Hoechst did not abide by this, although, as Lautenschläger testified after the war, he was aware "that the employment of prisoners of war in the armaments industry represented a violation of the International Laws and Customs of War on Land as determined by the Hague and Geneva Conventions."[630] Second, in 1943 the works management apparently decided to transmute the status of the prisoners of war working in Hoechst into a civilian status. During a

[626] HA, TEA 95f, Niederschrift über die technische Direktionssitzung in Hoechst, January 11, 1943.
[627] HA, TEA 95f, Niederschrift über die technische Maingau-Direktionssitzung in Hoechst, July 6, 1942.
[628] HA, TEA 95f, Niederschrift über die technische Direktionssitzung in Hoechst, March 8, 1943.
[629] Ibid.   [630] NI-6415, Affidavit by Carl Ludwig Lautenschläger dated March 26, 1947.

management meeting held in June 1943 it was reported: "On 6/19 [the status of] the French prisoners of war in the work has been transmuted into a civilian employment relationship; from now on they are to be treated in all respects like French civilian workers."[631] With the alteration of their status to civilians the French soldiers lost all their privileges as prisoners of war and were additionally tainted at home by the suspicion of collaboration, which by 1943 most were anxious to avoid. Such a transmutation of soldiers into civilians was usually carried out against the will of the majority of the persons concerned.[632] There are no indications of any voluntary changes to a civilian status in Hoechst. However, Hoechst attempted to sweeten its violation of international law for the affected French soldiers. On June 19, 1943, the works management announced that "we have taken the wishes of the French workforce into account insofar as from Monday 6/21/43 we will be preparing special food for French persons in one of our kitchens Ch 116." In July 1943 a circular sent out by Hirsch gave the information that "it must be assumed" that certain groups of "former French prisoners of war will be granted leave" to go home. The criteria for permission would be their performance and "attitude toward the German Reich."[633] Whether the food did indeed improve, in view of the increasing scarcity of foodstuffs, and whether the former prisoners of war were allowed to go on leave is not clear. It seems highly improbable that they were allowed to travel to France at the time.

Altogether, the situation of most French workers seems to have been fairly acceptable. In February 1944 a French commission from the Vichy government came to Germany at the invitation of the *Gauleiter* and paid a visit to Hoechst: "The camps of the French workers, the kitchen for the foreigners and some factories were inspected. The commission expressed its satisfaction."[634] Quite obviously, some of what the commission was shown in Hoechst were "Potemkin villages," because after the war French workers were less positive in their description of their treatment, particularly because of the former heavy political pressure in the plant – after all, French workers were those most frequently arrested.[635]

[631] HA, TEA 95f, Niederschrift über die technische Maingau-Direktionssitzung in Hoechst, June 21, 1943.

[632] Cf. Spoerer, *Zwangsarbeit*, pp. 64–5.

[633] HA, Rundschreiben 00002, Hirschel to Abteilungsleiter und Betriebsführer, June 19, 1943; circular from Hirschel dated July 29, 1943, re. "Beurlaubung der ehemaligen französischen Kriegsgefangenen."

[634] HA, TEA 95f, Niederschrift über die technische Direktionssitzung in Hoechst, February 14, 1944, and Niederschrift über die technische Maingau-Direktionssitzung in Hoechst, February 21, 1944 – refer to a visit that took place on February 23, 1944; this must be a mistake as the meeting was held on February 21, 1944.

[635] Cf. HHStAW, Abt. 520 F (A-Z), Orthner, statements of the conscripted French workers A. Dunet (September 3, 1946), L. F. Cumet (October 11, 1946), and R. Derenemesnil

While the treatment of most workers from Western Europe appears to have been fairly good during the first years, in 1943 this changed. Statements exist that the food was considerably worse than was claimed after the war. A witness for the prosecution in Nuremberg said that the quality of the food resembled that of food given to prisoners. And according to the statement made after the war by a German who had had some contacts with Belgian workers, in 1943/44 they were only given "turnips cooked in water and a slice of bread 3 x daily, and this for weeks on end."[636]

Even Lautenschläger had to admit that in the end the so-called Western workers were also mainly forced laborers. He declared that they too "worked under compulsion, since at a certain point during the war all contracts with French workers which had run out had to be extended by decree for an indefinite period of time without consideration of the wishes of the worker in question."[637]

The "Russians" in Hoechst suffered the most. According to Luce d'Eramo, the food of the Italians and other "Westerners" was relatively good compared with the food of the Eastern workers. In her book the heroine Lucia suddenly decides not to eat in one of the canteens of the Westerners but to join the Russians instead: "She fought her way through to a table and already with the first spoonful of soup she had the taste of rottenness in her mouth: no comparison to the soup which the westerners were given to eat. This one consisted only of overcooked beets without a trace of potatoes. And the brown rye bread was so sticky that it stuck to the fingers."[638]

Because of the shortage of labor Hoechst did its best in 1942 to obtain an allocation of workers from occupied areas in the Soviet Union and to facilitate their deployment in the plant. The works management ordered that "various simplifications in camp accommodation and supervision of the Russians" should be made, and so 350 Russian civilians were to be accommodated in the camp at the water tower, and a camp for Russian prisoners of war was to be completed by the middle of June. Moreover a camp was prepared for 100 Eastern women.[639] Finally a Russian woman doctor was sent to Hoechst for "the medical treatment of the Eastern workers" in the summer of 1942.[640]

(October 16, 1946); cf. also d'Eramo, *Deviazione*, pp. 193 et seq. (*Der Umweg*, pp. 217 et seq.), according to which the French workers had allegedly been responsible for organizing a strike in the plant.

[636] NI-11613, affidavit by Joannes de Bruyn dated September 26, 1947; letter from Ms. L. to Dr. Metternich (Hoechst Archives), February 25, 2003.

[637] NI-6415, affidavit by Carl Ludwig Lautenschläger dated March 26, 1947.

[638] d'Eramo, *Deviazione*, pp. 173–4, 177–9 (*Der Umweg*, pp. 194–5, 199–201).

[639] HA, TEA 95f, Niederschrift über die technische Maingau-Direktionssitzung in Hoechst, May 18, 1942.

[640] HA, TEA 95f, Niederschrift über die technische Maingau-Direktionssitzung in Hoechst, August 10, 1942.

Lautenschläger was soon quite aware, as he testified after the war, that "the majority of the Russian workers had not been voluntarily recruited but had simply been registered and deported to Germany to work." He also knew that Russian workers who had reported voluntarily "had been given great promises with respect to their work and conditions of life, which were not complied with in Germany and that it was of course not possible for these so-called Eastern workers to return to their homes." However, Lautenschläger emphasized that the plant had attempted to compensate for the different food rations of Eastern workers and Western workers. But the considerably lower payment of Eastern workers, their visible discrimination by having to wear the East badge (*Ost*) on their clothes, and the fencing off of their accommodations were normal in Hoechst – even according to Lautenschläger.[641] The kidnapping and deportment of the Eastern workers to Germany, which even included children, does not appear to have disturbed any of the Hoechst works management. A cheerful memorandum written at the end of June 1942 simply noted: "The plentiful allocation of Russian civilian workers has alleviated the labor situation."[642]

Hoechst apparently wanted to make concessions toward the Russians, for the payment of Russian workers was placed on a new footing so that they would be left with "more money than previously."[643] In real terms this often meant only that the wages for hard labor were not a complete mockery – and what finally remained of these low wages after the compulsory *Ostarbeitersparen* (forced savings by Eastern workers) contributions had been deducted was next to nothing. This particularly affected the Russian girls in Hoechst, who were expected to perform the roughest and heaviest men's work, despite being underage. Hirschel granted them a pay increase as an additional motivation: "It has turned out that the Russian workers, women in particular, who carry out men's work in our companies, are in a bad position with respect to wages. The relatively low cash payments paid out to these still underage workers has an adverse effect on performance. In the future their wages will be adjusted and their income raised to that of grown workers by granting bonuses when underage Russian workers are employed to carry out a full man's work and achieve the same output as grown men."[644] Even these small wages still had to be earned by what were predominantly young girls working very hard, and despite the payment of the additional allowance Hoechst was still able to save a lot of money.

---

[641] NI-6415, affidavit by Carl Ludwig Lautenschläger dated March 26, 1947.
[642] HA, TEA 95f, Niederschrift über die technische Direktionssitzung in Hoechst, June 29, 1942.
[643] HA, TEA 95f, Niederschrift über die technische Maingau-Direktionssitzung in Hoechst, July 6, 1942.
[644] HA, Rundschreiben 00002, circular from Hirschel dated September 25, 1942, re. "Lohnerhöhung für minderjährige russische Arbeitskräfte, welche Männerarbeit leisten."

The pittance paid out in return for very hard work was one side of the coin. There was also the fact that the living conditions of the Russians for which Hoechst must be held responsible were very bad. At a meeting of the works management in August 1942 it was stated: "A falling off of the willingness to work on the part of the Russian workers has been noted. This has been attributed to the food, some of which is prepared very uncaringly. It has been decided that meals shall be tasted daily by the works chemists."[645] The meals must have been very bad if even the works management was willing to use such words. But Lautenschläger felt it necessary to make a display of severity toward the Eastern workers. At the following meeting he decided that no compromises in the matter of food should be made toward the Eastern workers, despite the decision to the contrary taken at the previous management meeting. He stated that he had personally established – at this point the minutes of the meeting were amended to "it was established during regular rounds of inspection" – that Russians "with a few exceptions are satisfied with the meals." Lautenschläger went even further. He requested that people should "refrain from asking the Russians too much about their wishes, as this leads to more and more wishes being voiced and creates a certain amount of discontent." With these words he was indicating that the discontent of the Russians with the food was merely the result of too many questions, although only a short time before the works management had seen the matter in a different light. The full extent of Lautenschläger's harshness toward Eastern workers is evident from an addendum to the minutes of the meeting: "It is once again emphasized that collections for Russians are forbidden in the factories."[646]

The situation of the Russians appears to have been so bad that other foreign workers as well as German workers had taken up collections to help them. The misery of the Eastern workers aroused even the sympathy of the Polish forced laborers who lived wretchedly enough themselves.[647] And although all contacts with foreign workers and prisoners of war were strictly proscribed for Germans and it was decreed that "followers should at all costs keep their distance from foreigners of other origins,"[648] nevertheless collections for the "Russians" were taken up. In the IG Farben plant at Gersthofen, which reported to Hoechst, the works manager Karl Weber felt impelled to criticize "the behavior of German national comrades" toward Russian prisoners of war and Polish workers at a meeting of the

---

[645] HA, TEA 95f, Niederschrift über die technische Direktionssitzung in Hoechst, August 17, 1942.

[646] HA, TEA 95f, Niederschrift über die technische Maingau-Direktionssitzung in Hoechst am 24.8.1942 dated August 25, and dated August 26, 1942 (the first was a version for internal use that was subsequently shortened and amended).

[647] Stefanski, *Zwangsarbeit*, p. 453.

[648] HA, TEA 95f, Niederschrift über die technische Maingau-Direktionssitzung in Hoechst, March 3, 1941.

Council of Trust: "It must be stressed that these people must be treated severely but fairly and that it cannot be tolerated that foreigners are spared and granted privileges at the expense of Germans." The Russians and Poles there were apparently in such a sad way that German workers did some of their work, gave cigarettes to Russians, and allowed Poles to "push to the front" at mealtimes. That similar scenes also occurred in Hoechst is probable, as indicated by a comment in the minutes of the Maingau directors meeting, but the facts cannot be verified – not least because, unlike in Gersthofen, no further documents of the Council of Trust in Hoechst are extant.[649]

The treatment of Eastern workers appears to have become more and more differentiated, the more they were needed and the more work they carried out in the plant – giving the lie to all racial stereotypes – such as the underage girls referred to above who performed the work of grown men. At the beginning of November 1942 some Russians "were transferred for retraining as locksmiths for the Acetic Acid Department."[650] This indicated a possible rise from laborer to workman with corresponding privileges. At the end of November 1942 the works management made a decision: "To redress the shortage of workmen, suitable Russian workers from the factories shall be trained as semi-skilled laborers. The semiskilled workers shall later be employed in the department which sent them."[651]

Occasionally Eastern workers were also offered a little entertainment: sporadically movie shows were held in the canteen hut.[652] And apparently other entertainment evenings were also held for them.[653] Yet Hoechst refrained from implementing many of the so-called privileges for hardworking and efficient Russian workers thought up by Sauckel or in the other plants. Thus the works management recorded in November 1943: "following a decree by the plenipotentiary for the deployment of labor those Eastern workers who have distinguished themselves by their impeccable conduct are allowed to wear the Eastern workers badge on their left sleeve. The Eastern workers have declared that it is all the same to them where they have to wear the badge; they consider the fact that they are obliged to wear a badge as an insult. The execution of the decree has

[649] Clariant WA, file 542, Aktennotiz Vertrauensratssitzung dated December 14, 1942; cf. also Bäumler, *Rotfabriker*, p. 314.

[650] HA, TEA 95f, Niederschrift über die technische Maingau-Direktionssitzung in Hoechst, November 9, 1942.

[651] HA, TEA 95f, Niederschrift über die technische Maingau-Direktionssitzung in Hoechst, November 23, 1942.

[652] See HA, TEA 95f, Niederschrift über die technische Maingau-Direktionssitzung in Hoechst, January 4, 1943.

[653] Cf. HA, TEA 95f, Niederschrift über die technische Maingau-Direktionssitzung in Hoechst, October 25, 1943 – here, however, refering to the Offenbach plant.

consequently been postponed for the time being."[654] Similarly, when the plant in Leverkusen, following the example of the plant in Ludwigshafen, gave "the Eastern female workers jewelry pendants made of plastic in recognition for special achievements," Hoechst simply noted briefly "we do not expect to achieve any improvement in the performance of the Eastern female workers by such a measure."[655] In other instances, however, Hoechst was willing to follow the example of other IG Farben plants. The company's plant at Merseburg applied for passports for foreign nationals to be granted to so-called intelligentsia Russians, such as chemists and engineers, if they had given a good account of themselves for at least one year in the plant. Such passports entailed certain privileges: "The general police regulations for foreigners apply instead of the police regulations for Eastern workers. They are no longer obliged to wear the Eastern workers badge, and the wages tables for Eastern workers should no longer be used when calculating the monthly wages." Hoechst decided to "proceed in the same manner."[656]

Hard-working employees were generally to be given preferential treatment. Thus Jähne applied to the *Gauleitung* on behalf of Hoechst and expressed the wish that "the Russians who have proved to be good and willing workers should receive the same food as the other foreigners." *Gauleiter* Sprenger was also willing to give his support.[657] At the end of 1944, measures were once again put forward to increase the willingness to work of the Eastern workers; however, Hoechst saw no occasion for this and feared that this might instead lead to conflicts. It was preferred that everyone should have just as much, or rather as little, as everybody else.[658] In the last years of the war, most Eastern workers in the plant were in a bad way, particularly the families, who were hit hardest by the decrease in rations in the course of the war.[659] At the management meeting held at the end of December 1943 it was noted: "In the camp for Russian families almost all the children have fallen ill with measles and pneumonia, nine have already died."[660] Such a high susceptibility to disease was created by

[654] HA, TEA 95f, Niederschrift über die technische Maingau-Direktionssitzung in Hoechst, November 15, 1943.
[655] HA, TEA 95f, Niederschrift über die technische Direktionssitzung in Hoechst, February 28, 1944.
[656] HA, TEA 95f, Niederschrift über die technische Direktionssitzung in Hoechst, December 6, 1943.
[657] HA, TEA 95f, Niederschrift über die technische Direktionsitzung in Hoechst, January 17, 1944.
[658] HA, TEA 95f, Niederschrift über die technische Direktionssitzung in Hoechst, November 13, 1944.
[659] HA, TEA 95f, Niederschrift über die technische Maingau-Direktionssitzung in Hoechst, November 29, 1943.
[660] HA, TEA 95f, Niederschrift über die technische Maingau-Direktionssitzung in Hoechst, December 27, 1943.

the cramped conditions in the huts, the poor diet, and the inadequate clothing: even in the propaganda photographs the children are usually barefoot.

A circular from Hirsch indicates that the children of Eastern workers were also recruited to work – although this was nothing out of the ordinary in the Third Reich during the war.[661] In a circular concerning the "working conditions for Eastern workers" sent to all heads of departments and factory managers in Hoechst on May 11, 1944, Hirschel decreed that the Eastern workers would be given the same wages and salaries as the other foreign workers. This would entitle them to receive productivity bonuses and extra pay for rough working conditions as well an additional allowance for overtime and night work. For all the vaunted equality of treatment, the subsequent paragraph is significant. It is important to be aware that at the time the employable age for children was fourteen years of age: "Insofar as Eastern workers below the age of 14 years are called on to work they shall receive 40–90% of the standard wage for 14-year-olds depending on their performance (40–90% of 30 Pfg. per hour)." They were not eligible for social welfare payments, and after 12 months' employment they were entitled to six days holiday, which could only be spent in camp. All Eastern workers were placed in the tax bracket 1, which meant they had to pay the highest taxes. Moreover they additionally had to pay a "social equalization levy of 15%." And while accommodation of the Eastern workers in the communal residences was free "as it was for the other foreign workers," the food usually had to be paid for, unless they were allowed extra money for food. As a rule, children up to the age of five were free, children below the age of ten cost 50 Pfg., and children up to fourteen years of age cost 75 Pfg. Those receiving normal rations had to pay 1 RM, while heavy laborers paid 1.25 RM, the maximum amount. The incentive to allow their children to work was increased by the fact that for Eastern workers the cost of food for children below the age of fourteen years that "are not in an employment relationship and therefore receive no wages" was borne by the head of the family.[662]

According to the employment books, twenty-two Russian children between the ages of ten and thirteen years worked in Hoechst. Nine girls and thirteen boys had to work in order to survive, despite the fact that they had usually been unwillingly deported from their homes by the Germans. It is difficult to say whether these were the only cases. At all events the fact that Russian children under twelve years of age were working contradicted a circular note sent around by Hirschel in February 1944, probably based on instructions given by Sauckel. In the note it was stated: "Children below

---

[661] Spoerer, *Zwangsarbeit*, pp. 149–50.
[662] NI-4685, circular from Hirschel dated May 11, 1944, re. "Einsatzbedingungen für Ostarbeiter."

the age of 12 years may not be employed to work." Yet children younger than twelve were demonstrably employed in the plant; even a witness to the defense mentioned such a case.[663]

After the war, Lautenschläger claimed that he had interpreted the working of children below the age of fourteen who, according to the guidelines quoted above, were obliged to work for a pittance, as a means of preventing them "from doing anything stupid."[664] But the children were forced to work so as not to be a burden on their parents or on the parent with whom they had been deported,[665] for they also had to pay for their mostly substandard food with the starvation wages they were given. Even if a Russian school and a Russian nursery were later established in Hoechst and the food of the Russians was supposed to be "sufficient," as a Russian witness for the defense said after the war,[666] their lot was the hardest, not least because of Lautenschläger's harshness. And a number of the children who had been deported to Germany alongside their parents fell ill and died in the winter of 1943/44, after they had been severely weakened by the poor diet.

After the fall of 1943 the conditions of the so-called Italian military internees were as bad as those of the Eastern workers, as Hitler refused to accord them the status of prisoners of war. Germans referred to them as "Badoglians" and quite a few Western workers called them "fascist sheep": they were beset from all sides.[667] Ulrich Herbert has already pointed out how unpopular the Italians had been with the Germans as civilian workers. Their work ethic and discipline were criticized, not least because until 1943, as previously mentioned, they had asserted themselves and demanded their rights and their wages.[668] Lautenschläger wrote in his memoirs after the war that the alliance with Italy had been "forced on" the Germans. Lautenschläger believed that it was based "exclusively on the friendship between Hitler and Mussolini, which was not echoed by the German people."[669] After the successes of the Allies in Italy and the fall of Mussolini, the German army intervened in Italy and set up the Republic of

[663] HA, PSW Betriebseintrittsbücher 1940–1945; IG/50, Jähne document 29, circular from Hirschels dated February 23, 1944, re. "Arbeitsschutz für ausländische Arbeitskräfte und Ostarbeiter"; Jähne document 19, affidavit by Alex Snessarew dated January 9, 1948: even when he appeared as a witness for the defense he stated that a "girl of around 11 years" worked in his kitchen; Spoerer, *Zwangsarbeit*, pp. 149–50.

[664] BA, Film 44839, Lautenschläger's testimony to Cooper dated March 20, 1947.

[665] Spoerer, *Zwangsarbeit*, pp. 73–5.

[666] HA, IG/50, IG-Prozessakten, Jähne document 19, affidavit by Alex Snessarew dated January 9, 1948.

[667] Cf. d'Eramo, *Deviazione*, p. 187 (*Der Umweg*, p. 210); Hammermann, *Zwangsarbeit*, pp. 314, 455–6.

[668] Herbert, *Fremdarbeiter*, pp. 100–3.

[669] HA, PA Lautenschläger, Lautenschläger, "Erinnerungen," vol. 6, [p. 37].

Salo under Mussolini, who had been freed by a unit of the Waffen SS. The military failures of the Italians before and after 1943 were commented on scornfully in Germany and in Hoechst.[670] The insults to the Italians occasionally escalated so much that the works management was obliged to take action. In Gersthofen the works manager announced in July 1944: "It has repeatedly happened that prisoners of war, particularly Italian military internees, have been tormented on their way from their huts to our plant and back by children of our factory employees in an insulting manner. Such behavior is uncalled for. Our followers are therefore asked to act on the members of their family so that these excesses will cease in future."[671] Hoechst was allocated ninety-eight Italian military internees – but this was not sufficient to satisfy the demand for labor.[672] According to the employment books a total of 127 IMIs worked in Hoechst.

This discussion of "followers" cannot be concluded without touching on the subject of the utilization of concentration camp inmates by Hoechst. In his study on the IG Farben plant at Auschwitz, the historian Bernd Wagner writes: "With 26.6 percent the share of concentration camp prisoners amounted to more than the triple the numbers of those in Hoechst, which reported the second highest percentage."[673] According to Wagner, who based these figures on a document from the Nuremberg trial, Hoechst had employed a great number of concentration camp inmates: on October 1, 1944, 8.5 percent of the total number of employees were supposedly concentration camp inmates. However, the figure of 8.5 percent for Hoechst given in the document quoted by Wagner referred not only to concentration camp inmates but also to subcontracted workers, forced laborers, and prisoners – as Wagner himself states correctly in his appendix.[674]

While we can therefore assume that the number of concentration camp inmates in Hoechst must at all events have been lower because many subcontracted laborers and prisoners worked in Hoechst, it is impossible to completely exclude the possibility that concentration camp inmates were employed in Hoechst. Indeed, Hoechst was ready to employ Jewish concentration camp inmates. In the summer of 1944 Hoechst received a corresponding offer: "the *Gebechem* Wiesbaden has offered 300 Hungarian Jews to the plant to be employed in closed groups. Their deployment in the individual departments was discussed. The 300 Jews were requested by

---

[670] Ibid., [p. 38].

[671] Clariant WA, file 586, Bekanntmachung by Weber dated July 8, 1944, re. "Verhalten gegenüber Kriegsgefangenen."

[672] HA, TEA 95f, Niederschrift über die technische Maingau-Direktionssitzung in Hoechst, September 27, 1943.

[673] Wagner, *IG Auschwitz*, p. 263.

[674] NI-11412 dated September 24, 1947; Wagner, *IG Auschwitz*, p. 332, Table 3 – Hoechst is not mentioned here.

telephone."[675] Whether these Jewish concentration camp inmates ever arrived in Hoechst is not clear, however. There is no record of them in the documents examined for this study. However, this does not mean that they did not work in Hoechst.

Yet there are some indications that no concentration camp inmates ever worked in Hoechst. It is not just that Luce d'Eramo wrote in her book on Hoechst that "no concentration camps were attached to this IG Farben plant."[676] An internal memorandum by the Executive Board of Hoechst from the summer of 1953 also points in this direction. At the time the Social Department wrote that it was "in the agreeable position" of being able to answer the question posed by IG Farben in liquidation in the negative as to whether Hoechst had employed concentration camp inmates – there had been no concentration camp inmates in Hoechst, that is, in the Hoechst plant during the IG Farben period.[677] So, while it cannot be completely excluded that concentration camp internees worked in Hoechst, it is at all events rather improbable. For this would surely have been handed down at least orally, if not in writing, and would have featured in the IG Farben trial as well.

But even if there were no concentration camp inmates working in Hoechst, the treatment of the workforce, in particular of foreign workers, is anything but a glorious chapter in Hoechst's history. Far from the view, still held in 1988, that the miserable treatment of Eastern workers could be laid at the door of the state alone, examination of the facts clearly demonstrates that these workers were exploited, particularly the young girls forced to perform hard labor appropriate for grown men in return for starvation wages and minimal bonuses. Nor had there been any shrinking back from employing children between the ages of ten and thirteen, even if the work allotted them may not have been heavy work.[678]

According to the information provided by the works doctor Adolf Baldus, between 1940 and 1945 only fifty-seven of the foreign workers died. Baldus declared that this list demonstrated that "the number of the deaths was lower than the mortality rate for the German Reich in the years from 1931 to 1936."[679] However, his list is incomplete, for Baldus gives a figure of six deceased Russians for the year 1943, but nine Russian children alone died in December 1943. It is obvious that these children do not appear in his statistics, but at the same time he compares them with the mortality

[675] HA, TEA 95f, Niederschrift über die technische Direktionssitzung in Hoechst, June 26, 1944.

[676] d'Eramo, *Deviazione*, pp. 196–7 (*Der Umweg*, p. 221).

[677] Clariant WA, file 586, note for Winnacker, Erlenbach, Heisel, Kaufmann, Weil, and Schulz, July 21, 1953, signed Mü/Sch.

[678] HA, IG/50, IG-Prozessakten, Jähne document 19, affidavit by Alex Snessarew dated January 9, 1948; Jähne document 4, affidavit by Alois Brisbois dated January 7, 1948.

[679] Ibid., Jähne document 60, affidavit by Adolf Baldus dated March 9, 1948.

rates for Germany in the early 1930s that included children. The figure of fifty-seven foreign workers who died while working for Hoechst is surely too low, even without counting the children.

This section has discussed the number of foreign workers employed in Hoechst, as well as the committed and independent efforts Hoechst made to obtain foreign workers. The falsity of some of the statements given during the Nuremberg trial concerning the treatment of Eastern workers and their children and the extent of Hoechst's cooperation with the Gestapo and the *Gauleitung* was also demonstrated. While it has been possible to show the most important developments, the numbers, and the differences between individual groups, the lack of source material did not permit the lives of the foreign workers to be presented in any greater detail.

In summary one can state that, faced with a shortage of labor, Hoechst, like the other IG Farben plants, was happy to take on foreign workers and even forced laborers, and expended a lot of time and effort to obtain them. At first glance the costs for these workers appear to have been higher than for German workers because of the allowances paid to Western workers and the creation of a special infrastructure that included huts, fences, guards, kitchens, and canteens. Nevertheless the employment of foreign workers was a profitable undertaking for IG Farben, as Peter Hayes has stated: "Despite the influx of putatively inferior foreign workers, Farben's sales income per worker in the core corporation was 580 marks higher in 1942 and 806 marks higher in 1943 than in 1939; expenditures for wages, insurance, and social services had fallen by 407 and 194 marks per head, respectively." And in Hoechst the hard-working and poorly paid Eastern workers must have contributed to the positive cash flow. Thus the statement made by Hayes for IG Farben as a whole also applies to the company's plant at Hoechst: "For all its defects, the Nazi labor system paid, both for Farben and the Third Reich in general."[680]

If the works management of Hoechst had shown few scruples in employing foreign workers, it showed a similar lack of empathy or solicitude concerning their fate at the end of the war. When at the beginning of 1945 the plant largely discontinued production, Lautenschläger ordered that the foreign workers should be "offloaded," they should be "got rid of."[681] Accordingly, Hoechst came to an agreement with the German National Railways to provide 1,500 of the company's foreign workers, and Hoechst sent another 1,000 of them to the Rhine "to dig trenches."[682]

---

[680] Hayes, *Industry and Ideology*, p. 344.
[681] Clariant WA, file 125, Lautenschläger to Weber, March 16, 1945.
[682] HA, TEA 95f, Niederschrift über die technische Maingau-Direktionssitzung, March 5, 1945, and March 19, 1945, Niederschrift über die technische Direktionssitzung, March 12, 1945.

4

From Self-Sufficiency to War Production, Drugs, and Experiments on Human Beings

4.1. THE REORGANIZATION OF THE PLANT UNDER LUDWIG HERMANN AND FRIEDRICH JÄHNE

One of Ludwig Hermann's first measures as plant manager, as early as February 1933, was to set up a *Direktionsabteilung T(echnisch)* or "Directorate T(echnical)" under his direct control. Its functions were to cultivate contacts between the plant and the "technical IG Farben offices" and to serve as head office for the Middle Rhine (Maingau) Works Group and the Hoechst plant. In the latter capacity it was supposed to ensure that issues of a similar nature arising in different operational areas would be dealt with together and that the statistics for all areas of activity would be recorded in a uniform and standard manner; the Directorate T was also responsible for drawing up and controlling Hoechst's budget. Hermann's closest collaborators in the Directorate were Gustav von Brüning, head of the Operational and Scientific Department, and Otto Hirschel, head of the Cost Accounting Department.[1] One of the first tasks of the new Directorate T, the forerunner of the *Zentrale Direktionsabteilung* (ZDA) or Central Directorate of Hoechst,[2] was the drafting of an inventory in February 1934, which has already been discussed in detail above.

Almost simultaneously Hermann began to tackle the organizational shortcomings of the plant, which were particularly obvious in its production of paints and dyes. On January 31, 1934, a reorganization of the production plants was announced in a circular letter, as it was considered that a reorganization had become "necessary" due to the many changes in staff: "In order to create a uniform structure and in consultation with the heads of the divisions, those departments of the plant which are of an

---

[1] Clariant WA, file 1546, Hermann to department heads and heads of laboratory, February 14, 1933, on the founding of the "Directorate T."

[2] *Chronik der Hoechst AG*, p. 157.

identical nature have been combined to form five groups, each of which will be headed by a specialist in the field." The tasks of the specialists would be "to unite the different points of view while taking the specific and individual characteristics of the different departments into account and to serve as representatives to the works management and outside the company." Their functions within the IG Farben committees would "not be affected," and the "management tasks of the previous departmental heads" would remain in place. According to the new arrangement Hermann would have overall control; von Brüning was his deputy. Wilhelm Plato was specialist for inorganic compounds and nitrogen after Martin Rohmer was forced to retire. Valentin Hilcken, who had transferred from Gersthofen to Hoechst only in 1933, took over intermediate products. Wilhelm Pfaffendorf was the specialist with responsibility for the entire production of paints and dyes, thus terminating the independence of the individual dye departments for azo, alizarin, indigo, and triphenyl sulfur. Paul Roth was responsible for the Solvents Department. And finally Alfred Fehrle became the specialist for pharmaceuticals, although he was subordinated to Lautenschläger.[3] As Otto Hirschel wrote later, this meant that "the Hoechst organization was now in line with the organizational changes previously carried out in IG."[4]

The changes in Hoechst, which Hermann had implemented within the space of one year, are a good example of "organizational learning," that is, the willingness to try out new forms of organization and organizational structures and to improve them as the occasion warrants.[5] The leading IG Farben plants provided the models for the reorganization and modernization of the Hoechst plant – after the "old Hoechst hierarchy" had retired, their organizational structures were also replaced under Hermann's direction. The reorganization was not confined to a restructuring of existing plants and research facilities; new structures were also created, such as the Central Laboratory, which was clearly planned after the model of the main laboratories in Ludwigshafen and Leverkusen.[6] Hermann was also attempting to put his own professional experience gained in Hoechst into practice.

But Hermann did not confine his activities to merely adopting ideas and structures from IG Farben. He also acquired know-how from the company in the form of new senior staff from other IG Farben plants. The most

---

[3] HA, Pol 18, circular from Hermann for the plant leaders dated January 31, 1934, re. "Neu-Organisation der Fabrikations-Betriebe des Werkes Hoechst."

[4] HA, 12, Hirschel, "Das Werk Hoechst im Verbande der I.G.," p. 28.

[5] Cf. Alfred D. Chandler, *Scale and Scope. The Dynamics of Industrial Capitalism* (Cambridge, MA: Belknap Press of Harvard University Press, 1990), which focuses on "organizational capabilities."

[6] Cf. Ulrich Marsch, *Zwischen Wissenschaft und Wirtschaft. Industrieforschung in Deutschland und Grossbritannien 1880–1936* (Paderborn: Schöningh, 2000), pp. 75–81 – on p. 76 there is a reference to a "Central Laboratory" in Hoechst at the end of the 1920s, but this was not identical to Hermann's plans.

prominent of these was without doubt Friedrich Jähne from Leverkusen, the new chief engineer. But Jähne was by no means the only one: Karl Staib, head of the Inorganic Department and potential successor to Hermann, came from Rheinfelden in 1934. For the Coloristic Department there was Joseph Nüsslein from Ludwigshafen, successor to the head of the department Albert Beil, who retired. Adolf Steindorff, head of the Textile Additives Laboratory, was classified as "dispensible" very early on, sent into retirement, and replaced by Ludwig Orthner from Leverkusen, from whom the plant management clearly expected better things. Heinrich Schlick, who succeeded Wilhelm Schwamborn as head of the Social Department, came from another IG Farben plant – however, due to the attacks that he instigated against Wagenheimer and the ensuing conflict with Lautenschläger his contribution was only brief.

The reorganization of the plant and the filling of important senior positions with staff from other IG Farben plants, particularly key departments of Hoechst, meant that the corporate culture in Hoechst increasingly became an "IG culture." In his memoirs Karl Winnacker wrote that the individual works had their own individual *esprit de corps:* "one belonged to the I.G. and was proud of it. But one was also, and sometimes even more so, a man from Hoechst, from Leverkusen or from Ludwigshafen."[7] The scandal that led to the retirement of Weidlich and Ref and that had brought the works management of Hoechst into disrepute, the subsequent reorganization according to the model of Ludwigshafen and Leverkusen, and the filling of senior positions with "IG men" led to the Hoechst corporate culture being largely replaced by an "IG Farben corporate culture" – and this to an extent that was otherwise only achieved by the new plants in central Germany. By comparison, Raymond Stokes notes that the "change in the corporate culture of BASF" during the time it was owned by IG Farben was "not particularly pronounced." This may have been due to the fact that BASF was quite capable of exerting its influence on IG Farben – from asserting its interests when filling leading management positions to implementing its ideas with respect to its product lines.[8]

However, the reorganization of the factories and research facilities by Hermann was only one aspect of modernization; the other, consisting of technical innovations and construction work, was no less important. This had already begun before Hermann took over the works management in January 1933 and was mentioned in the inventory of February 1934. In his memoirs looking at IG Farben in the years prior to the Third Reich

[7] Winnacker, *Challenging Years*, p. 67.
[8] Cf. Stokes, "IG Farben Fusion," p. 211, who considered the extent to which BASF set the tone in IG Farben and whether this had an influence on maintaining separate corporate cultures; see also Winnacker, *Challenging Years*, p. 67, and Haber, *Chemical Industry*, p. 340.

Winnacker wrote that the outlook for Hoechst was particularly bleak during the rationalization phases "when the reckoning began," as the technical conditions were "much less favorable than at Bayer or at BASF." Among other things, the transportation and energy costs were much higher in Hoechst than in the other plants. According to Winnacker, this was due to "the fragmentation of the antiquated production." The IG Farben management had therefore sent Friedrich Jähne from the company's plant at Leverkusen to Hoechst as chief engineer "with instructions to do something about the technical equipment at Hoechst which had been so greatly neglected."[9]

In the fall of 1931 Jähne had already moved from Leverkusen to Hoechst and took over the management of the technical engineering operations.[10] He was very well qualified for this position. Born on October 24, 1879, in Neuss, he completed his schooling and subsequently studied mechanical engineering at the Technical College Berlin-Charlottenburg, graduating as a qualified engineer. Before he joined Bayer in 1921, he worked for almost fifteen years as a production engineer in five different companies, four of them chemical companies in Germany. He was already appointed an authorized signatory in Leverkusen in 1923, and became a director in 1928. During his close cooperation with Carl Duisberg he learned how to build up a chemical plant systematically and logically. When he was transferred to Hoechst to become chief engineer of the plant and of the Middle Rhine Works Group on October 6, 1931, this was with the promise that he would be appointed to the IG Farben Management Board as soon as possible, particularly in view of his function as chairman of the Technical Commission. However, to Jähne's annoyance, this promise was not kept. In the fall of 1932 he complained to Carl Duisberg, whom Jähne reminded that he had taken on the tasks assigned to him in Hoechst only on the assurance of becoming a member of the Management Board. In January 1934 he finally became a deputy member of the Management Board; he had to wait until June 1938 to become a full member.[11] Jähne was not alone in thanking Duisberg for his appointment to the board in 1934, which apparently was only achieved in the teeth of opposition. Hermann, who had become a full member of the Management Board at the same time, expressed his gratitude to Duisberg that he had "not rested" until "my dear colleague Jaehne also joined the Management Board." This had brought Hermann "great

[9] Winnacker, *Challenging Years*, pp. 69–70.

[10] HA, 12, Hirschel, "Das Werk Hoechst im Verbande der I.G.," p. 24.

[11] HA, PA Abteilung VI, Jähne, Personalblatt; questionnaire for Human Resources; questionnaire for the personnel file of members of the Management Board; Jähne to IG Vorstand, Werk Hoechst, October 6, 1931; Jähne to Duisberg, October 13, 1932; article "Friedrich Jähne. In Hoechst wird sein Name weiterleben" from *Die Farben-Post*, issue 1, 1966.

pleasure" as Jähne's achievements for Hoechst and for IG Farben had thereby been duly recognized.[12] The cooperation between Hermann and Jähne was by all accounts very good, and it formed an important basis for the successful technical and operative reorganization within Hoechst.[13]

Jähne joined the NSDAP in 1938 together with Lautenschläger and became a retroactive Party member from May 1, 1937. However, he was never considered to be a keen Nazi. In a report prepared for the *Gauleitung* at the beginning of 1942 by his local branch leader concerning the political reliability of leading businessmen, the author wrote that he was unable to "testify to Jähne being a National Socialist," as he very rarely showed up and maintained "no close contact" with the local branch. The branch leader concluded his report with the words: "His world view is not rooted in Nat. Soz. [National Socialism]. He is also too lukewarm toward our movement."[14] On the other hand, Jähne apparently never had any problems with his company's involvement in Auschwitz, as he not only stayed there several times himself, but his son worked there as an engineer. Lautenschläger later claimed to have learned about Auschwitz and the mass murder of Jews there from Jähne.[15]

Under Lautenschläger Jähne became deputy works manager of Hoechst. He apparently never aspired to become manager of the works or of the Maingau Works Group, although it is not certain if this was ever discussed, as he did not stem from the Works Group, and such a position would have run contrary to Jähne's image of himself as an engineer in a chemical company.[16] In 1951 Jähne wrote about the role of an engineer in a chemical company: "In a chemical company the chemist decides what should be produced and how it will be produced. He knows best which suitable raw materials will be required and which are available and about the laws of their transformation. The engineer is his assistant in achieving the 'how' of the transformation, and if he fills his position properly, he is his most important and most indispensable assistant." Therefore someone who "cannot accept a position in second place within a company" should not

[12] Bayer WA, Autographensammlung Carl Duisberg, Jähne to Duisberg, May 2, 1934; Hermann to Duisberg, May 3, 1934. The conflict surrounding the appointment of Jähne may be linked to his contribution to the expert opinion on whether benzene hydrogenation should be continued. In 1932 Jähne had supported the closing of the facilities – in opposition to Bosch, who finally prevailed. See Hayes, *Industry and Ideology*, pp. 39–40.

[13] Cf. Winnacker, *Challenging Years*, p. 75.

[14] BA, BDC file on Jähne, application for admission, April 30, 1938, admission dated May 1, 1937; HHStAW, Abt. 483, Nr. 10531, information on Jähne for *Gauleitung* dated December 4, 1941, and January 14, 1942.

[15] BA, Film 44839, Interrogation of Lautenschläger by Benvenuto von Halle on April 30, 1947; HA, IG/50, IG-Prozessakten, trial brief for Jähne from defense lawyer Hans Pribilla, pp. 84–9.

[16] Winnacker, *Challenging Years*, p. 89.

work in a chemical company.[17] The fundamental task of an engineer in a chemical company was to ensure that the chemists were able to concentrate on research, development, and production. An engineer must "provide everything" the chemists needed for their tasks. He must first "transport the substances required by the chemist to the proper location at the right time (transportation tasks)," second "provide the equipment necessary to change these substances into the required states under the necessary physical conditions (process engineering tasks)," and third "supply the physical and mechanical conditions to which the chemist wishes to expose his substances for the procedure to be carried out" (energy management tasks).[18]

Jähne set about tackling all three areas in Hoechst. In January 1932 he already presented his plans for alterations and extensions to be made in Hoechst. He planned the modernization and centralization of the energy plants, a solution to the problems of traffic, and a reorganization of the departments in the plant. Among other things this included the removal of annexes that stood in the way of traffic flow and a redistribution of premises for the different departments that would be carried out with due allowance for potential expansions in order to prevent the renewed proliferation of unplanned annexes. The total costs for this expansion plan amounted to between 8 and 9 million RM. It was decided "when carrying out new investments to base them on this expansion plan, even if this would result in slightly higher costs in individual cases, as the implementation of the plan combines considerable advantages for the plant as a whole."[19]

One of Jähne's closest collaborators during these years, the engineer Siegfried Kiesskalt, who was to become professor for process engineering in Aachen after the war, wrote a detailed article in 1960 on the expansion and rebuilding of the Hoechst plant in the 1930s.[20] Kiesskalt referred to the example of Hoechst as "a particularly instructive record" from a "chronological and a critical" point of view. The plant was set up in 1863 on the banks of the river Main, which formed the boundary of the plant in the south. A busy road between Frankfurt and Wiesbaden went through the middle of the plant. There was a single-track train connection to two main railroad lines. According to Kiesskalt, the rapid growth of various organic divisions alongside several older, but bigger, products still being produced forced the plant to expand primarily westwards "crammed in between the road and the river." Because of these special circumstances the layout of the streets in the Hoechst plant was rather narrow: "For this reason a

---

[17] Friedrich Jähne, *Der Ingenieur im Chemiebetrieb* (Weinheim: Chemie, 1951), pp. 11–12.
[18] Ibid., p. 12.
[19] HA, 12, Hirschel, "Das Werk Hoechst im Verbande der I.G.," p. 25 refers to meeting of the TEA Sparte on January 28, 1932.
[20] HA, PA Kiesskalt, article by Kiesskalt: "Zum Fortschritt im Aufbau der modernen chemischen Fabrik" from *Die Farben-Post*, December 21, 1960.

narrow-gauge 750-mm railroad was constructed through the plant with numerous track switches, turntables, and intersections, with stations for reloading onto the main-line train. The only advantage provided by this expensive train system in the plant was that, given the turnover of material at the time, certain classes of goods or even trains could be driven over the narrow bending track right into the factory buildings. This was only an advantage as long as the manufacturing shipments were substantially smaller than they were in our century; later this turned out to be a severe disadvantage." The narrow factory streets, intersected by narrow-gauge rail tracks with barriers, and the small decentralized power stations scattered all over the plant and supplied with coal, transported by road and a small-gauge railroad, demanded a fundamental reorganization.

Step by step Jähne's plans for rebuilding and expansion in Hoechst were realized. A reorganization of energy supplies was crucial: "It was decided to shut down a number of the scattered power stations at once and to build a large, high pressure-counterpressure power plant with Löffler boilers, which had the disadvantage that, because of its design, 90% of the necessary feed-water had to be taken from the polluted Main River; therefore an extensive water-treatment facility was also required in addition to the building of the power plant." However, such an industrial power plant could only be "fully effective if an electricity linkage with public power stations" could be created – in Hoechst's case, this was the Rheinisch-Westfälische Elektrizitätsgesellschaft (RWE). After lengthy negotiations combined operations were agreed on, and the modern power plant became something of a tourist attraction for German and foreign visitors.[21]

Kiesskalt considered the structural and architectural development of the Hoechst plant as an example of "an exceptionally difficult development." The long-term planning goals included the relocation of public roads to outside the plant and the creation of "straight streets within the plant without railway tracks, which could be used by other means of transportation together with a general reallocation and consolidation of buildings." Under Jähne's management numerous suggestions and plans were drawn up. Yet in the 1930s it appeared very improbable that "such a radical realignment could be financed in the foreseeable future: only new plants and ongoing rectifications were consistently carried out in accordance with the new framework." But by 1939 a general layout of the plant had been created that formed a "serviceable basis for the reorganization of the

---

[21] Among other sources HA, ZSLA 169, Letters from MAN and Mitchell Engineering, January 1938; Siemens-Schuckert-Werke to Jähne, July 8, 1938, requesting a visit to the facilities by the vice president of Consolidated Edison, which was not permitted by the Liaison Office W on July 12, 1938; ZLSA 170, Oberkommando der Kriegsmarine to IG Farben, February 16, 1939; idem to IG Vermittlungsstelle W, March 15, 1939, the permission for Swedish engineer Per Parén to view the steam boiler facilities was approved.

plant," although this was only finally implemented at the beginning of the 1950s. By 1939 Jähne had succeeded in "normalizing the layout of the entire plant," incorporating a standard-gauge railway line and creating a clearly arranged network of streets between the factory buildings. Moreover, premises on the south bank of the river Main, purchased in the thirties with an eye to the future, were available for future expansion.

On the factory premises north of the Main numerous new buildings were erected, and existing factory and laboratory buildings were converted before and during the war. The basic concept was to leave ample room when building. In principle, all floor plans had to include a potential for expansion by at least 25 percent of production. The rows of machines and apparatus had to be easily accessible for assembly, repairs, and possible rearrangements. Existing bottlenecks, such as those created by transmission drives, were to disappear and be replaced by independent drives developed and tested in Hoechst: "The ideal to which everything aspired was that of a continual, ongoing flow chart and much of the work of the scientific engineering group was directed toward this goal."

In some departments these suggestions met with stiff resistance because, according to Kiesskalt, they "often appeared revolutionary at the time." But with Hermann's backing and authority they were implemented. Jähne and Hermann were a successful team; Hermann even boasted of the structural changes and improvements to the new rulers. On the occasion of a mass rally held at the end of November 1935 to mark the visit of a high-ranking representative of the German Labor Front, Karwahne, who wished to learn more about the Hoechst works, Hermann stated: "if someone who has not visited us for 3–4 years walks through our plant today, whether he enters by gate 2, gate 3, or gate 5, he will notice quite considerable changes and innovations. The ugly annexes, referred to as swallows' nests, have disappeared. Broad paved streets with carefully laid tracks have taken the place of tortuous lanes. Open green spaces bear witness to the good air in Hoechst. Spacious, well-lit factories have taken the place of outdated buildings. Igepon, acetoacetic ester, patent blue, intermediate products building F 50, sodium chloride, test rooms, dyes storehouses, washroom Ch 131, power plants, dyes works. Every year, during the last few years, 10 million marks have been spent to improve the layout of our plant."[22] On the occasion of the celebration of May 1, 1937, Hermann was even more enthusiastic. Much had been achieved within the last few years since Hitler's seizure of power, much had been altered, some new departments and research establishments had been created: "We need only to look around the factory; many holes and corners of the old factory have been cleaned out and removed, so that many parts of our plant have been given a

---

[22] HA, PA Hermann, typescript of Hermann's talk at the mass rally on the occasion of the visit of the head of RBG Chemie of DAF, Karwahne, November 26, 1935.

new face. Many an old building has been clothed in a new outfit; many factory streets and – even more importantly – what is moved and propelled along these streets have been replaced or improved; large new buildings have arisen which rise majestically from among the old ones and contribute to the embellishment of the whole plant."[23]

This was not just publicity to convince the Nazi representatives of the "beauty of work" in Hoechst. The plant had changed much in the years 1932/33 to 1938/39. An American visitor from the leading chemical company DuPont reported that he had visited a state-of-the-art azo-dye factory in Hoechst. He also reported on new pharmaceuticals buildings: "The next plant visited was the latest type of pharmaceutical plant with tiled floors and walls for the manufacture of Pyramidon and anti-pyrene. It was a beautiful construction in every respect and is used for propaganda purposes and cost RM 4,500,000, or $1,800,000, at the present official rate of exchange. At this plant the aceto acetic acid for all of the I.G. requirements is made by the action of sodium on ethyl acetate. This plant represents one of the finest installations for chemical manufacture that I have ever seen." He concluded his report on Hoechst saying "At the Hoechst plant there is a very large volume of new construction under way."[24]

Indeed, much was being built at the time, and sometimes a problem or an accident became a starting point for promising investments with an eye to Hoechst's future. Thus, after the fire in the Aldol workshop at the end of July 1935 that killed several workers, the Solvents Commission and subsequently the Technical Commission of the corresponding IG Farben division approved an application for the construction of a new building for the Acetic Acid and Solvents Department. The credit required by Hoechst for the construction of the new building was finally approved in February 1936; according to Otto Hirschel after the war, this heralded "a new era for the solvents department in Hoechst": "A modern manufacturing workshop for the production of acetic acid and solvents was built on the Kalkfeld to the west of the factory, which incorporated the most up-to-date knowledge in IG. An open-plan construction style was used as far as possible, which had the advantage on the one hand of costing less and on the other hand of avoiding the danger of room explosions. The operation of the equipment, the towers, and the large catalytic reactors, which stood outside protected only by a roof, was transferred to a control center which ran along the length of the factory. Recording instruments with automatic recorders were installed which continually indicated the temperature, pressure, and amounts of liquid, so that anomalies would be noticed immediately."[25] The

---

[23] HA, PA Hermann, typescript of Hermann's speech on Labor Day (May 1) 1937.
[24] Hagley Museum and Library, DuPont Company Records, Jasper Crane Papers, Series II, Part 2, Box 1038, E. K. Bolton to Jasper Crane, June 11, 1936.
[25] HA, 12, Hirschel, "Das Werk Hoechst im Verbande der I.G.," pp. 91–4.

model for this type of open-plan construction was provided by the IG Farben plants at Leuna and Schkopau, where it had been very successful. Before this, in Hoechst operations of this type were set up in already available buildings that had previously been used for the production of dyes or intermediate products. Now the technical and structural innovations in design and construction from the new IG Farben plants in central Germany were being transferred to Hoechst – here too, a learning process was initiated that affected the entire corporate group and that introduced innovations to the older plants.[26]

All these buildings and new technologies were financed by investment money provided by IG Farben, and the sum was by no means as insignificant as was later alleged. Karl Winnacker's observation is therefore incorrect when he states that with the exception of weapons programs the Nazi regime "had little understanding of the need of scientific and technical endeavour": "Throughout the Hitler era, including the war, this government attitude had one particular result. Hardly any building was possible in Hoechst. The site looked as antiquated as it did at the time of the I.G. fusion and during the economic crisis."[27] This statement is not only contradicted by the speeches Hermann gave at various official celebrations but also by the modernization of the plant and the construction of numerous new buildings and alterations to old buildings, as borne out by both Kiesskalt's article and the report of the DuPont manager.

Of course, investment in Hoechst was not as large as the sums allocated to the new works in central Germany or at Auschwitz. But the infrastructure in Hoechst was already there, and new buildings could be constructed within an existing system – which made them considerably cheaper, as Hoechst was quick to point out in its application for a new building for the Acetic Acid and Solvents Department. But overall, the sums provided for investments in Hoechst were quite considerable. After listing the amounts of investment money given to the plants in central Germany, Ludwigshafen and Oppau, Raymond Stokes writes that the "Hoechst Group," which had received only 258 million RM, had been "relatively neglected" in the years between 1925 and 1944. However, he added: "The main plant at Hoechst, of course, fared best within the group, obtaining RM 220 million in investment capital. In other words, the main Hoechst plant received more capital investment than did the Bayer plant; the Hoechst group as a whole received less than the Bayer group as a whole."[28]

Stokes's statement largely corresponds to the figures available from the Hoechst Archives. According to a list drawn up for the period from 1927 to

[26] HA, Steinberger, "Die Acetylen-Chemie in Hoechst" (undated), pp. 114–17.
[27] Winnacker, *Challenging Years*, pp. 96–7.
[28] Stokes, *Divide and Prosper*, p. 22: all numbers for IG Farben on pp. 20–1, table 2.

TABLE 4.1. *Investments and Repairs in the Large Plants of IG Farben from 1925/27 to 1942/43*

Category	Hoechst	Leverkusen	Ludwigshafen
Repairs	184.2	256.2	259.0
Investments	204.9	179.4	294.8
TOTAL	389.1	435.6	553.8

Source: HA, RFL 78, Statistisches der Farbwerke 1927–1939/48, investments and repairs to the large plants from 1925 to 1943 and from 1927 to 1942.

1942 Hoechst received the sum of 204.9 million RM for investments and 184.2 million RM for repairs – out of a total of 253.1 million RM for investments and 238.6 million RM for repairs given to the entire Maingau group in those years. Leverkusen received only 179.4 million RM for investments but was given 256.2 million RM for repairs, while Ludwigshafen received 294.8 million RM for investments and 259 million RM for repairs.[29]

Gottfried Plumpe estimates that the Hoechst plant received some 142.8 million RM for investments in the period from 1933 to 1945; like the figure arrived at by Stokes, who quotes another source, this sum was larger than of the Leverkusen plant, which was given 130.8 million RM. However, like Stokes, Plumpe compares these figures with the considerably larger investments in other works: these include investments amounting to 425.2 million RM in Ludwigshafen/Oppau, 425.4 million RM in Schkopau, 373.9 million RM in Heydebreck, and a total of 569 million RM in Auschwitz (Table 4.1).[30]

Investments in Hoechst were therefore quite high, even if they were not comparable to the sums expended on Ludwigshafen and the new IG Farben plants. Hoechst could not complain, for the sum of money granted for investments and repairs permitted Hoechst to substantially modernize the plant. It should be added that not only was money liberally available from IG Farben, but through the services of Hans Wagenheimer the plant also had a direct line to the *Gauleitung*, which, in view of the increasing shortage of raw materials and construction materials during the 1930s, must also have helped. Thus, at the beginning of 1938 works manager Hermann received a report that *Gauleiter* Sprenger would be traveling to Berlin to confer with Göring, General Löb, and Colonel Hannecken: "The *Gauleiter*'s goal is to make it clear to the gentlemen that it is not enough to

[29] HA, RFL 78, Statistisches der Farbwerke 1927–1939/48.
[30] Plumpe, *I.G. Farbenindustrie*, pp. 594–6, especially table "Verteilung der I.G. – Investitionen auf alte und neue Grosswerke," p. 595.

grant someone a quota to build industrial buildings without at the same time providing the iron for the apparatus to be installed in these buildings. As Mr. Wagenheimer informed me, the *Gauleitung* is confident that after the consultation with Göring, concessions will be made to the *Gauleiter*, as the area he is in charge of is a depressed area. IG could only welcome this, because it would also benefit."[31]

Ludwig Hermann is said to have once told his close colleague Otto Hirschel after a meeting of the Technical Committee that Hoechst was "the stepchild of IG." This was surely an exaggeration, Hermann expressing his frustration about the meeting he had just returned from.[32] However, despite the reorganization, the investments, and the modernizations, Hoechst did not become the equal of its two "big sisters" at Leverkusen and Ludwigshafen. While the alleged words of Hermann about Hoechst's being the "stepchild of IG" were doubtlessly an exaggeration, the comments by his successor Carl Ludwig Lautenschläger that under Hermann Hoechst had "once again become a strong pillar within the entire IG group" and "one of the foremost of the large plants of IG Farben" were equally exaggerated.[33] Hoechst developed into a relatively modern plant with numerous products, old and new, according to Hermann's plans for research, development, and production in Hoechst. But Hoechst did not become a plant comparable to its two "big sisters" at Ludwigshafen and Leverkusen. Hoechst did not expend comparable sums on investments and repairs nor was it capable of establishing "colonial plants" as Ludwigshafen did with Auschwitz or Gendorf, where Otto Ambros played the leading role. Nor did Hoechst acquire leading positions on the IG Farben Management Board as the two larger plants succeeded in doing.

4.2. RESEARCH AND DEVELOPMENT

One of the most important authors on IG Farben's research and development policies is the English historian Peter Morris. He found it "a little ironic" that the profits from dyes and pharmaceuticals permitted IG Farben to compensate for the almost complete commercial failure of high-pressure chemistry during the Great Depression: "The rapid development of the embryonic heavy organic chemicals and polymers sector became all the more urgent. Ter Meer maintained the existing strategy of diversifying away from dyes by sponsoring a number of new areas, namely, synthetic rubber, other polymers, acetylene chemistry, and pharmaceuticals. In a very real sense, ter Meer and his colleagues saw Buna rubber as standing in the

[31] HA, PA Hermann, Bormann (Sekretariat Hermann) to Hermann, February 1, 1938.
[32] HA, Hirschel, "Aus meiner I.G. Zeit," p. 19.
[33] HA, PA Lautenschläger, Lautenschläger, "Erinnerungen," vol. 2, [unpaginated].

same tradition of industrial synthesis as alizarin and indigo."[34] Hitler's seizure of power and his policies directed toward self-sufficiency and a war economy forced ter Meer to change his plans. According to Morris, IG Farben now directed its research and development toward the creation of strategically important materials – even at the expense of what were probably more profitable fields of research. The profitable special product Buna N, for example, was sacrificed to the mass product Buna S, which was more suitable for the production of tires. As there was a certain time lag between planning and production, the new policies had few consequences for Division 2 of IG Farben during the first years of the Nazi regime – with the important exception that considerably more money was spent on research. The research budget of Division 2, the division to which most of Hoechst belonged, doubled between 1934 and 1937.[35]

An important document summarizing the research and production targets of IG Farben in the Third Reich exists; it has already been discussed by Peter Hayes in his study on IG Farben in the Third Reich.[36] Gottfried Plumpe seems to underestimate the importance of this document; at all events he does not mention it. Plumpe consistently emphasizes the dependence of IG Farben on world markets as well as the economic and technical dimensions of its corporate planning. Although elsewhere he talks of the significance of politics for the company, he immediately takes this back again.[37] Yet the National Socialist economic control, the quest for self-sufficiency, and the armaments economy created its own market laws and market forms in Germany, which bore little resemblance to the economy prior to the seizure of power – the market had become deeply politicized.

The initiative for the aforementioned document came from Director Wilhelm Gaus, head of the IG Farben plant at Ludwigshafen. At a meeting of the Technical Directorate of Ludwigshafen on May 4, 1934, he developed "his fundamental thoughts concerning the expected future development of IG in the long term." According to Gaus, it was to be expected that the industrialization of countries that had previously produced only raw materials would continue slowly but "unstoppably," and this would be accompanied by a fall in exports. This would inevitably lead to IG Farben's export and output volumes suffering heavy losses. Previously the company had been able to compensate for its losses by its foreign contracts and

---

[34] Peter J. T. Morris, "Ambros, Reppe, and the Emergence of Heavy Organic Chemicals in Germany, 1925–1945," in Anthony S. Travis et al. (eds.), *Determinants in the Evolution of the European Chemical Industry, 1900–1939. New Technologies, Political Frameworks, Markets and Companies* (Dordrecht: Kluwer Academic Publishers, 1998), pp. 89–123, here pp. 102–4.

[35] Ibid., p. 104.

[36] Hayes, *Industry and Ideology*, pp. 129–30.

[37] Plumpe, *I.G. Farbenindustrie*, contradictory statements in the introduction on pp. 15 and 17; cf. Wagner, *IG Auschwitz*, p. 11, who already touched on inconsistencies in Plumpe.

rationalization measures, but if decreases in output and export volumes continued, this might no longer be possible. Nor was there much expectation of growth in hitherto existing areas: "In this context, the problem of autarky appears in a different light. We are convinced that autarky is inexpedient and that a maximum of exports is vital for Germany. However, IG should not underestimate the facts which great upheavals create in the existence of nations, and with this in mind consider the problem of how in later decades Germany can be fed and employed. It is worth reflecting whether it [IG] should not already now begin to shift its research toward production for the domestic market, which would then compensate for IG's reduced volume of output." Numerous imported goods could be replaced without difficulty by "substitute chemical products," for example, textile fibers, rubber, tanning agents, or gum: "IG should not refrain from carrying out research in these areas even if the processes are at present still unprofitable."[38]

Only a few weeks later on June 30, 1934, Arthur Reithinger from IG Farben's *Volkswirtschaftliches Archiv* (Economic Department) presented a study undertaken on the initiative of the IG Farben directors Gaus and Max Ilgner in Berlin: "New Goals and Tasks for the Research and Production Policy of IG Arising from the Developments in Germany and on the World Market." The document presents the case for a reorientation of IG Farben's production and research, giving a detailed justification of such a reorientation.[39] The exposé consisted of two parts, a first part on "Basic Considerations" in which the structural developments and changes in Germany and the world were presented, and a second part in which conclusions were drawn for IG Farben.

Reithinger first noted that about two-fifths of IG Farben's production in the past years had been sold abroad and around three-fifths had been sold within Germany. Since the beginning of the slump export volumes had declined by around half, while domestic sales had fared better. The reduced utilization of industrial capacities and the fall in sales revenues had largely been absorbed by rationalization measures and cost reductions, but now "a limit has been reached which in the opinion of technicians cannot be appreciably reduced." The developments both in Germany and on the world market had created two fundamental problems for IG Farben's research and production policy: "1. How can chemical research and production take account of conditions on the German *domestic market* which on the one hand will make it necessary to set up new capacities for raw material substitutes, while on the other hand production in previously

---

[38] BASF UA, IG Archiv C 121/11, Bericht über die 44. Sitzung der Technischen Direktion am 4. Mai 1934, Ludwigshafen, dated May 4, 1934 (Schoenemann).

[39] HA, 3/0 IG, Exposé "Neue Ziele und Aufgaben für die Forschungs- und Produktionspolitik der I.G." dated June 30, 1934, by Volkswirtschaftliches Archiv der IG, Dr. R.[eithinger].

important areas is likely to fall?" And: "2. How can German *chemical exports* be increased to keep pace with the recovery of the world economy such that the expected losses of important areas in the domestic market are compensated for?"

The answer to these two questions depended, according to the study, largely on whether the current problems in Germany and on the world market were considered to be the result of economic cycles or whether they were in fact structural problems, in other words, whether Germany's need for foreign currency was structural or temporary, and whether the halving of chemical exports had economic reasons or resulted from changes in the world market and world trade structures: "Today the economist can answer both questions with overall certainty to the effect that the general German problem of exports and foreign currency as well as the particular difficulties facing our chemical exports are occurrences of longer duration, which can be expected to persist until the middle of this century. Thus the question has already been answered in the affirmative that the responsible authorities within the group must increasingly take account of the changed conditions in Germany and on the world market when setting the goals for chemical research and for the entire production and sales policy." They would be forced to do so by: "1. the necessity as a private-sector company of adjusting to changes in foreign and domestic demand which have already occurred and which are likely to be even more extensive, as well as by our interest in an adequate utilization of available capacities." Furthermore, they had: "2. the economic task of promoting as far as possible the necessary expansion of our domestic raw material base as a qualified producer on the one hand and on the other hand, as a principal export industry, of increasing the meager flow of foreign currency of the German economy."

As a result of the high imports, reduced exports, and the outcome of the First World War, the foreign currency difficulties remained unchanged, so that Germany, previously a creditor nation, became a debtor country. The end of Germany's technical lead and the creation of numerous industries in former raw-material-producing countries made the situation more difficult for the chemical industry, and political developments were an impediment to increasing exports: "The growing uncertainty in Europe and in the Pacific and the open resumption of rearmament in the world also had a significant impact on the developments of the chemical industry. Since chemistry has been recognized as one of the most important armament industries, all politically important countries have attempted to set up their own chemical industry and will continue to do so in the future without any consideration of private-sector costs or competitive conditions."

The conclusions that the chemical industry must draw from the fact that the German foreign currency situation was of a structural nature and that the fall in exports was only partly caused by economic developments but was also the result of long-term and worldwide technical, economic, and political developments varied, depending on whether the main focus was on the domestic or the world market. There certainly appeared to be opportunities in the domestic market. Prospects were expected to be favorable "for all divisions dealing with the production of raw materials or raw material substitutes, or which focused on changing over to cheaper foreign raw materials, increasing the yield of materials from raw materials, the conservation of material and reuse of scrap or which help processing industries with the necessary changeover to substitute production." As possible new areas of interest the study suggested focusing on the use of mineral or vegetable raw materials that were freely available domestically, such as gypsum, potatoes, or scrap materials; a switch to cheaper foreign raw materials in place of those currently being used; an increase in the yield of material derived from processing foreign raw materials such as ores and tanning substances; better conservation of foreign raw materials; and, finally, reuse of scrap material. This would have an important impact on the flow of foreign exchange, as in 1933 alone the sum of 660 million RM was spent merely on raw materials for textiles. However, it would be necessary to verify which "substitute production would be cheapest" – that carried out by the chemical industry, by the farming industry, or by the mining industry.

This statement was followed by a paragraph that expressly dealt with armaments and that is worth being quoted in full: "All of the bigger items are essential raw materials which may also become of importance militarily. Research in these fields therefore remains important, even if the foreign currency situation should make domestically produced substitutes super- fluous. In addition, a military interest exists for a number of key products which have only an insignificant impact on the foreign currency situation (e.g. nickel, hydrogenating catalysts). The question of qualitative com- petitiveness or cheaper production compared to foreign products may have to take a back seat to other considerations." However, this statement was qualified by the comment "If they are only of military interest, then in most cases it might suffice if development is carried out until products are ready to go into production, if domestic production is considerably more expensive than foreign production."

In addition to the opportunities offered by the domestic market, the prospects of the German chemical industry on the world market were assessed as not bad. However, the sales policy would have to take account of the great changes in the world economy, which, although they primarily affected nitrogen production and the dyes and chemical divisions, also had an impact on the Pharmaceutical Department. The prospects for exports were considered especially promising in countries that had no links to large economic blocs, while "with respect to the rest of the world market, the focus should be on products which are not immediately essential to the war effort and which are not mass products but rather the result of high-quality research and industrial work as well as the interplay of highly ramified branches of manufacturing."

The study concluded with the comment that the consequences for IG Farben's research, production, and sales policy were still incalculable and should be studied intensively. The study was intended only as a basis for further discussion. But even so, it clearly indicated the future path to be taken. Even if, as Gaus had put it at the beginning of May, little could be said in support of self-sufficiency, nevertheless both self-sufficiency and armaments offered real and possibly the only opportunities for growth. Exports should mainly be limited to high-quality products, an important and lucrative market, although it remained doubtful whether the turnover and profits obtained from such exports would be able to compensate for losses in other export areas.

The important fact was that a watershed had been reached. The management of IG Farben was prepared to carry out research and production in fields that were unprofitable in the short and medium terms but were regarded as lucrative in the longer run, and these included war goods. In his study on IG Farben Plumpe has argued that while the share of "autarky areas," that is, areas that received particular support by the state, in IG Farben's turnover increased, the consequences of the state industrial policy "were not very big," if one takes the share of turnover as an indicator. He does concede, however, that other areas of IG Farben's production "were affected and influenced directly or indirectly by the state industrial policy." He finally comes to the conclusion: "While economic policy had an influence and in some areas certainly made an impact, it should not be overestimated. The assumption that IG had adapted its entire business policy to bring it in line with the state economic policy certainly misses the mark."[40] Yet Plumpe's conclusion is not tenable in this form. Dyes and textile additives had to be adapted to take account of the demand for self-sufficiency, and more attention was given to the production of staple fibers and rubber. Furthermore, the aim was now to create detergents that would not contain any fat, as far as this was possible, in order to reduce the "fat scarcity" – that is, the shortage of fats in Nazi Germany.

Not only this memorandum, but the company's plans for mobilization as well, which existed since 1935, and the establishment of the "Liaison Office W" (with W standing for *Wehrmacht*) in September 1935 prove how early IG Farben prepared for a potential conflict – one year prior to the official Second Four-Year Plan of 1936 in which Hitler demanded that the preparations for war be completed within four years. Georg Kränzlein, head of organic research in Hoechst, proudly stated in 1943: "By the time the *Führer* passed the Four-Year Plan on to Reich Minister Hermann Göring in 1936, IG with its farsighted research and development plans

[40] Plumpe, *I.G. Farbenindustrie*, pp. 555–6.

had already considered about 80% of the demands submitted there, for example, in the area of new, high-performance explosives, newly developed types of Buna rubber, synthetic materials, textile additives, detergents, tanning substances, camouflage materials, pharmaceutical products such as serums, vitamins, hormones, agents for the control of epidemics, pesticides, and at the same time solved many other new problems."[41] While this report may be somewhat exaggerated, nevertheless after 1933/34 research and development had begun to focus more on autarky and on products that were of particular interest for a war economy. It is therefore certainly not wrong to speak of a self-mobilization on the part of IG Farben.

But to assume that the company was now focusing entirely on autarky and a war economy falls short of the mark. The Management Board of IG Farben, in particular the management of Division 2 under ter Meer, continued to pursue a strategy of diversification and modernization in the Third Reich. The management wanted to move away both from dyes production and from "ruinously expensive" high-pressure chemistry: "During the Third Reich, ter Meer and Ambros also sought to diversify away from 'political' products such as Buna S toward potentially more profitable consumer goods, such as oil-resistant Buna N, nylon, and polyurethanes."[42] While demand for strategic materials and substitute substances was very big in the Third Reich, modernization and diversification implied the introduction of new products and new production methods, not just the adoption of autarky and substitute products. Growing demand by the automobile industry was an important driving force behind the search of the chemical industry for solvents and varnishes. Peter Morris has correctly pointed out that focusing exclusively on demand gives us an incomplete picture of technical innovations in the organic chemical industry.[43]

This leads to the question as to what type of research and development Hoechst focused on and how it implemented the above-mentioned targets of IG Farben. In his personal memoirs Carl Ludwig Lautenschläger wrote almost euphorically about the atmosphere of optimism that prevailed at Hoechst after Ludwig Hermann had taken over management of the plant. By dint of "arduous work" the scientists and works managers had managed to discover "new and valuable products" in different dyes groups, intermediate products, and synthetic materials and "to document the profitability and key role of the Hoechst plant for the new IG concern."[44]

[41] HA, PA Kränzlein, "Kunststoffe für den Flugzeugbau," manuscript by Kränzlein dated April 19, 1943.
[42] Morris, "Ambros," p. 118.   [43] Ibid., p. 119.
[44] HA, PA Lautenschläger, Lautenschläger, "Erinnerungen," vol. 2, [unpaginated].

Hermann himself does not seem to have been entirely satisfied by what was achieved under his management. After a meeting of the Technical Committee in 1937 he dejectedly told his colleague Otto Hirschel that ter Meer had spoken out vehemently against his plans to begin carbide production with a view in the longer term to setting up through-flow production for solvents: "Dr. ter Meer had interrupted him and declared that Hoechst should banish the thought. He acknowledged that Hoechst could take credit for the technical implementation of acetylene chemistry. However, Hoechst had subsequently rested on its laurels and not delivered anything new. The leadership in the field of acetylene had devolved to Ludwigshafen. If one day carbide production will need to be enlarged, then the furnaces will not be in Hoechst but in Ludwigshafen."[45] Indeed, Hoechst, which under Paul Duden had once led the field of acetylene chemistry, had surrendered its preeminence to Ludwigshafen, where research and development into acetylene chemistry was intensively pursued under Walter Reppe.[46]

Thus Hoechst was obliged to submit and had to relinquish its leading position in an area that it had pioneered to a competitor inside IG Farben. But in the 1930s there were other possibilities for chemical companies, in both inorganic and organic chemistry, and Hermann tried to grasp as many opportunities as possible. In Gersthofen he had made the plant profitable by constantly introducing new product lines, and he intended to do the same in Hoechst. He wrote to Karl Weber, his successor in Gersthofen in 1936 about the goals he had pursued: "So you should always be going upwards and onwards, keeping an eye open to see if you can spot something new for Gersthofen, whether there or from the outside under license."[47]

He also furnished Weber with a description of his method of proceeding in scientific work, which he divided into two parts "with great success." First, there were "those tasks concerned with examining every stage of individual operating procedures in detail and completely without bias, no matter how self-evident they appear to be." Second, "New work." This division meant, on the one hand, being innovative and increasing the yield and/or reducing production costs for already existing products and production processes, while simultaneously developing completely new products and production processes. Hermann's program focused on what is referred to in innovation research as "incremental innovations," that is, improvements of existing processes, and on "radical innovations," or completely new developments.[48]

---

[45] HA, Hirschel, "Aus meiner I.G.-Zeit," p. 19.   [46] Cf. Morris, "Ambros."
[47] HA, PA Abteilung VI, Karl Weber, Hermann to Weber (Persönlich), April 21, 1936.
[48] Cf. Christopher Freeman and Luc Soete, *The Economics of Industrial Innovation*, 3rd rev. ed. (Cambridge, MA: MIT Press, 1999); Joel Mokyr, *The Lever of Riches. Technological Creativity and Economic Progress* (Oxford: Oxford University Press, 1990).

Hermann emphasized that it was important not to try to economize by reducing the number of scientists: "One should accordingly not be stingy with the number of chemists, particularly as there are many of them around, in order to achieve a maximum yield of materials in the factories."[49] This was entirely in the tradition of how IG Farben carried out its research, as Jasper Crane, a senior manager of DuPont, described it in 1930. Crane had not just noted the large sums that IG Farben spent on research and development but also the optimism of the company's management, which was convinced that more and more natural substances could be substituted by producing them artificially, and their belief that with the help of research and development it would be possible to develop something new, meaningful, and profitable every couple of years: "Their faith has indeed been justified by the impressive list of achievements to their credit, and examination of their affairs shows that without the developments of their own research they would be very badly off financially. We cannot help but be impressed, therefore, by the I.G.'s vision of the possibilities of research." But he also noted the drawbacks: "They do not hesitate to spend money extensively and sometimes lavishly on these experimental projects. Materials, equipment, men, are massed in the attack on the problem in hand." This dependence on a "mass attack" as propagated by Hermann appeared to Crane to be a weakness of the IG Farben system since research was not carried out in any great depth and did not tackle fundamental problems, which resulted in a lack of imaginative and really original work, but instead proceeded without regard to money and time to produce as much of the new material as possible while attempting to incorporate it into as many items as possible.[50]

But this systematic approach favored by the company also had big advantages. The scientists in IG Farben worked to develop and improve their products and product lines still further as well as to imitate those of their competitors' patents and publications that appeared interesting and subsequently systematically circumventing those patents – a task wherein they were often very successful. Although Carl Duisberg, the "founding father" of IG Farben, was unable to see any "spark of genius" in such "scientific mass production," he nevertheless promoted it.[51] According to the historian William Kingston, the company's "scientific mass attack" and the systematic approach involved were transferred from chemistry to pharmaceuticals. In the United States this resulted in enormous successes, for example, the discovery of streptomycin as an active agent against tuberculosis. Selman Waksman, the "father" of this active substance, declared later that

---

[49] HA, PA Abteilung VI, Heisel, Hermann to Weber, March 16, 1931.

[50] Hagley Museum and Library, Accession 1416, box 7, speech of Jasper Crane, May 26, 1930.

[51] Marsch, *Zwischen Wissenschaft und Wirtschaft*, pp. 84–5.

streptomycin was found because it had been methodically and purposefully searched for. However, after switching to "systematic series investigations" it was difficult to state the share of individual researchers involved in a discovery: "This change was adoption of the methods of mass screening that had been so successful in the German dyestuffs industry as it diversified into pharmaceuticals. For example, Salvarsan was number 606 in the sequence of compounds containing arsenic, which Ehrlich tested, and its improvement, Neosalvarsan, was number 914."[52] The historian John Lesch also recognized this "industrial" approach in sulfonamides, in which the role of Gerhard Domagk was overrated and the methodical, systematic cooperation of the other chemists involved was underplayed.[53]

Research and development enjoyed the highest priority in IG Farben even in times of war and preparation for war. Sometime in 1942 Lautenschläger wrote about his work during the years 1936 to 1942: "In my sphere of activity at Hoechst research stood at the forefront of my activity – despite many other important tasks – because only with this pioneering work can the high standard of our German chemical industry be maintained for any length of time."[54] Its importance could be gauged by the fact that IG Farben spent immense sums on research and development. In the years from 1925 to 1939 the company spent an average of 7 percent of turnover on research and development, more than any other company in the world, far more than most chemical companies after 1945 when the share was between 3 and 5 percent. Between 1925 and 1931 it was even between 7 and 10 percent, but this suffered a massive slump during the Great Depression – in 1933 it was only 4.9 percent. But between 1934 and 1939 the share of research and development costs in relation to turnover crept up to between 5 and 6 percent again.[55] In the years after 1933 this was helped by the fact that IG Farben froze the distribution of dividends at the level of 1932, which stood at 7 percent. This was because all profits above this amount had to be invested in government bonds in which the company had no interest. IG Farben therefore invested large sums in the expansion of its companies, laboratories, or testing facilities.[56]

The heads of the leading research departments in the IG Farben plants played a key role in research and development. Carl Ludwig Lautenschläger as the head of the Pharmaceutical Department was the most important man

[52] William Kingston, "Streptomycin, Schatz v. Waksman, and the Balance of Credit for Discovery," *Journal of the History of Medicine and Allied Sciences* 59 (2004), pp. 441–62, here p. 454.
[53] John E. Lesch, "Chemistry and Biomedicine in an Industrial Setting: The Invention of the Sulfa Drugs," in: Seymour H. Mauskopf (ed.), *Chemical Sciences in the Modern World* (Philadelphia: University of Pennsylvania Press, 1993), pp. 158–215.
[54] HA, PA Lautenschläger, Lautenschläger, "Erinnerungen," vol. 3, [p. 4].
[55] Freeman and Soete, *Economics*, p. 111.
[56] Marsch, *Zwischen Wissenschaft und Wirtschaft*, p. 83.

in this core line of business for Hoechst. Georg Kränzlein was appointed manager for the entire organics research on the strength of his success as a researcher – Kränzlein had originally worked in the Patents Laboratory, which focused on getting around patents of other companies. After his appointment he reorganized the Alizarin Department. Both Kränzlein and Lautenschläger headed a Central Laboratory; in Kränzlein's case this was the Alizarin and Vat Dyeing Laboratory, although the range of products which the laboratory worked on was considerably larger.[57] Both men were also in charge of the other laboratories and development departments in their respective divisions. Apart from the central or main laboratories, the application engineering departments, technical rooms, and testing stations were immensely important for IG Farben as research was no means over after laboratory testing was concluded: "The development work, whereby a synthesis first tested in laboratories was then carried out on a grander scale, then in pilot plants and finally put to the test in a trial operation, was an integral part of the research process for both the chemists involved and the management."[58]

Ludwig Hermann was responsible for setting up a Central Laboratory that was obviously modeled on the examples of Ludwigshafen and Leverkusen.[59] The original idea had been that the chemists working there would include not only organics chemists but also inorganics chemists, as both groups worked in completely separate laboratories up until then: "The Central Laboratory will consist of 40 to 45 chemists; it will freely carry out research in all fields of chemistry and additionally has the task of training young chemists for the laboratory and workshop." The "basic principle" behind work carried out in the Central Laboratory was that "once something has become technically interesting, products developed there [in the Central Laboratory] will be passed on as quickly as possible to the production departments for technical elaboration as they are more experienced."[60] This corresponded exactly to the division of labor in research and development described above for the main laboratories, special laboratories, and technical testing rooms.

One of the reasons why a Central Laboratory was specially set up and kept well supplied with materials and chemists to work on basic innovations and carry out fundamental research was the lack of success of the existing laboratories and their heads, in whom Hermann had originally placed so much faith. Certainly in 1934 Hermann was repeatedly

---

[57] Cf. HA, 12, Hirschel, "Das Werk Hoechst im Verbande der I.G.," pp. 26, 157, 160–7.
[58] Marsch, *Zwischen Wissenschaft und Wirtschaft*, p. 82.
[59] Cf. ibid., pp. 78–82.
[60] HA, Pol 18, Aktennotiz by Hirschel dated June 25, 1938, on "Besprechung betr. Laboratoriumsfragen am 24.6.1938 zwischen v. Brüning, Kränzlein, Pfaffendorf und Hirschel."

heard to complain that the costs for the laboratories had increased very much and would have to be pared back again.[61] In 1937 he again complained about the high costs – and noted at the same time that the "creative achievements of the Hoechst plant are beginning to fall off again," which was also borne out by the decrease in the number of patents filed by Hoechst.[62]

The duties of the Central Laboratory were finally limited by a decision of Lautenschläger's, who clearly did not want his Pharmaceutical Department to be merged with Kränzlein's department, as the Central Laboratory had been put under Kränzlein's control. Accordingly, in mid-July 1938 Lautenschläger ordered that the duties of the Central Laboratory under Heinrich Greune should be to carry out "free research in all chemical areas with the exception of pharmaceutics" while also training "all young chemists with the exception of the pharmacists and inorganics chemists."[63] In the second half of the 1930s fundamental research in Hoechst was being propagated comparable to that which Kurt Hans Meyer and Hermann Mark had carried out in their main laboratory in Ludwigshafen up until 1932. Since the Inorganic Department had by now pulled out of the newly created Central Laboratory, in Hoechst this basic research focused only on organic chemistry, a particularly fertile area at the time, not least in the field of polymer chemistry.[64]

Toward the end of 1936 Ludwig Hermann decided to unite all branches of research and development that touched on "physical areas" into a new department for "process engineering." He entrusted the management of the new department to Karl Winnacker, who had already successfully worked on viscous carbon substances together with Siegfried Kiesskalt during his stint in the Dyes Department and whose ideas and projects were being resisted there.[65] The new department was placed under the general control of Friedrich Jähne, who transferred this control to Kiesskalt on whose initiative the new department originally had been set up. This department included the three groups coloristics and physics, the Physical Laboratory,

[61] HA, TEA 95c, Niederschrift über die Maingau-Direktions-Sitzung, May 30, 1934, and October 18, 1934.
[62] HA, TEA 95d, Niederschrift über die technische Direktionssitzung, October 18, 1937, and November 29, 1937.
[63] HA, Pol 18, Circular No. 70, Lautenschläger to Betriebsführer und Laboratoriumsleiter dated July 14, 1938.
[64] Cf. Carsten Reinhardt, "Basic Research in Industry: Two Case Studies at I.G. Farbenindustrie AG in the 1920s and 1930s," in Anthony S. Travis et al. (eds.), *Determinants in the Evolution of the European Chemical Industry, 1900–1939. New Technologies, Political Frameworks, Markets and Companies* (Dordrecht: Kluwer Academic Publishers, 1998), pp. 67–89.
[65] HA, TEA 95c, Niederschrift über die technische Direktionssitzung, June 17, 1935; Winnacker, *Challenging Years*, pp. 74–6, 79–81.

and physicochemical processes.[66] Jähne, Kiesskalt, and Winnacker now worked closely together in this new department.

This new department in Hoechst formed the nucleus of what at a later date was to be known and taught as process engineering in West Germany. Both Kiesskalt and Franz Patat, the head of the Physical Laboratory and later successor to Winnacker as head of the department, became professors for process engineering at the Technical Universities in Aachen and Munich. The work of the team of Kiesskalt, Patat, and Winnacker was exceptionally progressive. The hiring of Patat in particular also turned out to be extremely profitable for Hoechst. Patat had studied chemistry, physics, and mathematics in Vienna and worked as an assistant to Professor Hermann Mark, who had previously worked in the main laboratory in Ludwigshafen. In 1934 Patat switched to the University of Göttingen, where he became an assistant to Professor Arnold Eucken, a physical chemist and cofounder of process engineering. In 1936 Patat qualified to become a university professor; as he was not given a position at the university he accepted a job in Hoechst in the same year, where he took on the management of the Physical Laboratory and, as of 1938, of the entire Process Engineering Department.[67]

A high-speed centrifuge, which was first designed in 1935 under Kiesskalt by Dr. Gnann, one of the engineers, was continually being improved on in the department. The centrifuge was primarily used by the Dyes Department, but was also used to determine molecular weights.[68] On Winnacker's suggestion, Hermann also bought in 1938 the first electron microscope ever to be used in industry for the process engineering division: "It cost about 150,000 marks, a considerable amount of money for those days."[69] As one of the key departments of the whole plant the Process Engineering Department carried out various finishing tasks for the Dyes Department. It also investigated the properties of fibers and their finishing and the effect on fibers of the latest products of the detergents division. In later years the department began to work on "extensive and basic investigations into polymerization reactions, work which continued up until 1945." In

---

[66] HA, TEA 95d, Niederschrift über die technische Direktionssitzung, November 30, 1936; PA Abteilung VI, Kiesskalt, preliminary reference for Kiesskalt from Jähne, July 10, 1945.

[67] HA, PA Patat (Mikrofilm), form filled out on joining the company, personal data sheet of Patat and Eucken's reference letter for Patat, August 1936.

[68] For more on the ultracentrifuge and Staudinger's interest in and criticism of the one constructed by Hoechst, see HA, PA Staudinger, vol. 3, memo from the Chefingenieur-Büro (Kiesskalt) for Patentabteilung dated March 20, 1935; Staudinger to Kränzlein, April 3, 1935; Staudinger to Kränzlein, May 2, 1935, and Staudinger to Direktion IG Werk Hoechst, May 2, 1935 (copy); Kränzlein to Staudinger, May 9, 1935; Staudinger to Direktion der IG in Hoechst, May 16, 1935; Brüning and Kränzlein to Staudinger, May 22, 1935.

[69] Winnacker, *Challenging Years*, pp. 87–8.

addition to work on a de-ashing procedure for brown coal, investigations were also carried out on the possibility of utilizing recovered paper by the precipitation of soot. And the department also worked on a new form of hydrolyzing aniline into phenol under high pressure and cooperated in the investigation of the extraction of ethylene and coal gas by thermodynamic adsorption.[70]

The establishment of the Process Engineering Department is a good example of the spirit of innovation that prevailed in Hoechst in the 1930s. In 1939 a chemist wrote about the importance of the two branches, chemical technology and process engineering: "Chemical technology takes as its starting point substances which are produced. The field can be divided according to the individual methods and emphasis focuses on the chemical aspects. In the new field of process engineering the starting point is the apparatus in which and with which the substances are processed, and the focus is on the constructive elements, and the physical and physico-chemical bases for the apparatus; those aspects of different proceedings which have the same function are combined, e.g. filter facilities, evaporators and so on; this leads to a division based on physical and physicochemical processes." For chemists this parallel coexistence of the two fields of study was "very pleasant." For if one wanted to improve a process, the first step was to research the literature of chemical technology to see how such processes were carried out up to then. In contrast, the literature on process engineering covered all methods of operation that were used or could be utilized in similar cases: "From the point of view of the specialist, these two branches of science, chemical technology and process engineering, complement one another extremely well because the specialist is interested in translating planned improvements and new processes into practice as quickly as possible."[71] In Hoechst all areas cooperated closely with one another: Kiesskalt, the instigator of the new department, was an engineer and one of the main proponents in the Association of German Engineers, *Verein Deutscher Ingenieure* (VDI), of setting up a division for process engineering in the VDI in 1934. Winnacker, who had been appointed to head the Process Engineering Department in Hoechst at Kiesskalt's instigation, had a background in chemical technology and had been trained by Berl, its leading representative. And Patat had completed his training as a physical chemist under Professor Eucken, the leading scientist in this new field at the time. The close interdisciplinary cooperation between the three men yielded excellent successes for Hoechst – this was an area that was still wide open and unregulated, in which innovations could prosper.

[70] HA, PA Patat (Mikrofilm), reference for Patat from Bockmühl dated July 27, 1945.
[71] HA, PA Kiesskalt, Dohse, "Verfahrenstechnik vom Standpunkt des Chemikers" (1939),
     p. 3.

Assessing the activities Hermann had initiated to put the plant back on its feet and his drive to continually bring new products and production lines to Hoechst, Lautenschläger concluded that "with respect to its range of activities and its fields of research" Hoechst was "probably the most heterogeneous plant" of the IG Farben group in 1938.[72] But Hermann's interest in new products and processes was more than just an agenda, it was a simple necessity. Within the corporate structure Hoechst was forced to fight for every product and every product line. This also applied to synthetic materials and plastics, a field that saw a remarkable cooperation between industry and science in research and development.

Around 1942 Lautenschläger wrote that the importance of research into synthetic materials and plastics had increased most within the Four-Year Plan.[73] Synthetic materials were more than mere "substitutes," they also served as the basis for the development of completely new substances and led to completely new production methods. Their development also resulted in both incremental and radical innovations. According to Gottfried Plumpe, polymer chemistry, which created new materials for the production of synthetics and fibers, was "probably the most important innovation in chemical technology in the interwar years."[74] But he also noted that under the Nazi regime the Synthetic Materials Department was not particularly profitable and was not an area in which IG Farben was particularly interested in investing.[75]

When IG Farben set up the Synthetic Materials Commission in 1930, neither the number of scientists nor the expenditure on synthetic materials was particularly high. In August 1930 only thirty-eight chemists in the whole of IG Farben were working there; in March 1931 their numbers had dwindled to a mere twenty-five. Costs for research and development were estimated at around 1.3 million RM per year in 1931, that is, a mere 3.3 percent of sales.[76] In October 1931 only 18.2 chemists (according to the company's calculations) worked in the field of synthetic materials, while the laboratory costs were estimated to be around 170,000 RM per quarter, in other words, not more than 680,000 RM per annum.[77] But the IG Farben management regarded even these relatively low expenditures as too high. Thus at the end of April 1932 the minutes of a meeting of the Synthetic Materials Commission recorded that the Technical Committee "in its resolution of 1/28/32 has expressed its interest in the Kuko area [=Kunststoff-Kommission,

---

[72] HA, PA Lautenschläger, Lautenschläger, "Erinnerungen," vol. 3, [p. 57].

[73] Ibid., vol. 3, [pp. 10–21].

[74] Plumpe, *I.G. Farbenindustrie*, p. 154.

[75] Ibid., pp. 337–8, 552.

[76] HA, TEA 42, Niederschrift über die 5. Kuko-Sitzung am 20. März 1931 dated March 30, 1931; Plumpe, *I.G. Farbenindustrie*, pp. 330–1, gives slightly different numbers.

[77] HA, TEA 42, Niederschrift über die 7. Kuko-Sitzung am 16. Oktober 1931 dated October 28, 1931.

i.e., Synthetic Materials Commission]." Fritz ter Meer, who took part in the meeting, added "that the work is appreciated as being quite valuable insofar as it is of economic interest; however, he warned against a too broad approach and against areas whose economic success is doubtful."[78] These were not particularly cheering remarks, and it was the Nazi regime's policy of autarky, the Four-Year Plan, and then the war that triggered the first significant upswing for synthetic materials and made the area lucrative for IG Farben.[79]

Within IG Farben Hoechst contributed importantly to the research, development, and production of synthetic materials. The main research departments were located in Hoechst, Ludwigshafen, and Bitterfeld. With respect to production, however, Hoechst trailed behind the other two and even behind Leverkusen: with over 47 percent of all its investments in plants and equipment in this area in the years from 1926 to 1943 Ludwigshafen was the most important and the largest site, followed by Leverkusen with 20.5 percent, Bitterfeld with 12.9 percent, and finally Hoechst with 12.4 percent.[80]

According to the commemorative brochure published for the company's centennial in 1963, the man to whom Hoechst owed its first synthetic materials operation and "a fair share" of the production of synthetic materials within IG Farben was Georg Kränzlein. Under his direction, production of polyvinyl acetate was started up in a building of the Alizarin Department in 1928, where it went under the trade name of Mowilith.[81] Hermann promoted Kränzlein, making him manager for organic chemistry research and considered him to be one of "IG's best scientists."[82] He was not alone in this opinion. After his visit to the IG Farben plant in Hoechst, Bolton of DuPont wrote very positively about Kränzlein. Bolton was most impressed by the work in the Alizarin and Vat Dyeing Laboratory under Kränzlein's management and described him as an "outstanding chemist" in his field. Bolton also remarked on the fact that the head of the laboratory assigned different journals to different chemists; every eight days one of the members of staff was expected to present and discuss any interesting developments reported in the journal assigned to him. Bolton felt that this was an extremely efficient means of keeping staff informed of the most recent literature on scientific topics and patents: "In this laboratory, like the Central Laboratory in Leverkusen, a great deal of time is spent keeping abreast of every new development reported in the scientific or patent literature relating to the fields in which the laboratory is engaged." In addition, Bolton wrote that Leverkusen maintained contacts to former

[78] HA, TEA 42, Niederschrift über die 8. Kuko-Sitzung am 20. April 1932 dated May 7, 1932.
[79] Cf. Plumpe, *I.G. Farbenindustrie*, pp. 330 and 337.
[80] Ibid., pp. 330, 338.
[81] Bäumler, *Century*, p. 165.
[82] HA, PA Kränzlein, speech of Hermann's on the occasion of the celebration of Kränzlein's twenty-five years' employment by the company, November 9, 1933.

laboratory employees, many of whom had done very good work in certain fields and who were able to pass on their experience to younger staff in the form of lectures.[83] Kränzlein, who was endowed with a boundless optimism, was also capable of inspiring his employees with an equal enthusiasm.[84]

Professor Hermann Staudinger, with whom Kränzlein cooperated closely from the late 1920s onwards, was one of the "fathers" of polymer chemistry; nevertheless he had been able to gain acceptance of his concept of macromolecules only after years of controversy.[85] Staudinger had a "freelance contract" with IG Farben for the Leverkusen plant.[86] However, he cooperated particularly closely with Hoechst in the field of synthetic materials, apparently because Kränzlein was one of the first to accept his theory of macromolecules – ever since attending a conference in Düsseldorf organized by Professor Richard Willstätter on macromolecules in organic chemistry in September 1926.[87] Since then Kränzlein and Staudinger had cooperated whenever possible, and Kränzlein tried to hire organic chemists who had trained under Staudinger to work in Hoechst.[88] Staudinger was much interested in working together with Kränzlein, not least because his "main opponents" in the dispute about macromolecules, Professors Kurt Hans Meyer and Hermann Mark, carried out their research in the main laboratory in Ludwigshafen, and the former was also a member of the IG Farben Management Board. However, both Meyer and Mark left IG Farben in 1932 – Meyer because of differences with Carl Duisberg and probably also because of the increasing anti-Semitism in Germany to which he felt himself exposed, while Mark reported that Director Gaus had even explicitly suggested to him that as a "half Jew" he should leave the company.[89]

---

[83] Hagley Museum and Library, DuPont Company Records, Jasper Crane Papers, Series II, Part 2, Box 1038, Bolton to Crane, June 11, 1936.

[84] Cf. Winnacker, *Challenging Years*, pp. 70–1; HA, PA Kränzlein, speech of Hermann's on the occasion of the celebration of Kränzlein's twenty-five years' employment by the company, November 9, 1933.

[85] See Claudia Krüll, "Hermann Staudinger. Aufbruch ins Zeitalter der Makromoleküle," *Kultur & Technik*, issue 3, 1978, pp. 44–9; Yasu Furukawa, *Inventing Polymer Science. Staudinger, Carothers, and the Emergence of Macromolecular Chemistry* (Philadelphia: University of Pennsylvania Press, 1998), pp. 69–91, 166–75; Claus Priesner, *H. Staudinger, H. Mark und K. H. Meyer: Thesen zur Größe und Struktur der Makromoleküle. Ursachen und Hintergründe eines akademischen Disputes* (Weinheim: Chemie, 1980).

[86] Plumpe, *I.G. Farbenindustrie*, p. 319.

[87] HA, PA Staudinger, vol. 4, Kränzlein to Staudinger, March 21, 1941; Staudinger to Kränzlein, April 3, 1941; on the meeting in Düsseldorf organized by Willstätter in September 1926, see Furukawa, *Inventing Polymer Science*, pp. 72–4.

[88] HA, PA Staudinger, vol. 1, Kränzlein to Staudinger, November 21, 1929; Deichmann, *Flüchten*, pp. 399–400.

[89] Cf. Stephan H. Lindner, "Die IG Farben und ihre jüdischen und als Juden geltenden Mitarbeiter in leitenden Positionen während des 'Dritten Reichs' – Das Beispiel des IG Werks Hoechst," in Haus der Geschichte Baden-Württemberg (ed.), *Jüdische Unternehmer und*

Over and beyond their scientific partnership, a personal relationship, a very real friendship, developed between Kränzlein and Staudinger. Staudinger held lectures in Hoechst, and Kränzlein came alone or together with employees to Freiburg. Together with their wives, Staudinger and Kränzlein visited one another and went on trips together.[90] This close cooperation between the two men, which was further underpinned by the fact that Hoechst provided Staudinger with money and chemicals, prompted Kränzlein to consider a transfer of Staudinger's contract with IG Farben from Leverkusen to Hoechst. At the end of 1933 Kränzlein wrote to Leverkusen that Staudinger had already approached him some time before with the request that the contract "which he had with IG for Le. [Leverkusen] should be transferred to the Hoechst plant." Ter Meer was also of Kränzlein's opinion, and it required only the consent of Leverkusen, which Kränzlein now considered to be imperative "in order that the flow of correspondence initiated by Prof. Staudinger with the IG plants becomes the responsibility of a single department which would in future prevent unpleasant correspondence within the IG plants."[91]

He was referring to a delivery of styrene from Ludwigshafen to Staudinger, which the Patents Division of Leverkusen had objected to. Kränzlein subsequently emphasized in a letter to Leverkusen how important it was "to continue to provide our partners in universities with basic materials in order to support their work." He added that Hoechst had been "supplying monomers and polymers to Staudinger for years," providing him with cellulose ester, polyvinyl alcohols, and ethylene oxides. "In our relationship with our associate Staudinger in the field of synthetic materials, we are today approaching the ideal which used to exist in the field of dyes between German technology and university science at the time when our IG plants still worked individually and were not yet filled with the spirit of systematizing everything. Regrettably, our lack of consideration and support of valuable and competent German university specialists has contributed to the condition which has quite properly been described as the excessive atomization of German chemical science." Kränzlein concluded by stating that the close cooperation with Staudinger prompted him to request that Staudinger's contract with IG Farben be transferred to

Führungskräfte in Südwestdeutschland 1800–1950. Die Herausbildung einer Wirtschaftselite und ihre Zerstörung durch die Nationalsozialisten (Berlin: Philo, 2004), pp. 196–7; Deichmann, *Flüchten*, pp. 181–3; Herman F. Mark, *From Small Organic Molecules to Large. A Century of Progress* (Washington, DC: American Chemical Society, 1993), pp. 61–3.

[90] HA, PA Staudinger, vol. 2: see, e.g., Kränzlein to Staudinger, February 9, 1932, May 24, 1932 and August 1, 1932; Staudinger to Kränzlein, October 19, 1932, February 1, 1933 and December 23, 1933.

[91] HA, PA Staudinger, vol. 2, Kränzlein to Direktor Brüggemann, Leverkusen, December 12, 1933.

Hoechst.[92] Kränzlein informed Staudinger of his request and reported "in confidence" on the disputes surrounding the delivery of styrene; he added that he had applied for Staudinger to become "a partner of Hoechst as of next year," so that there would be no further necessity to waste "valuable time with long useless IG letters."[93] However, Kränzlein withdrew his request after consultation with Leverkusen because, as he wrote to Staudinger, he "did not wish to create the feeling that I want to poach a valuable contractor from the Leverkusen plant." However, this would in no way preclude a continuation of their previous cooperation.[94]

At the end of April 1934 after Kränzlein had agreed that Staudinger should continue to work for Leverkusen, Heinrich Kühne, a member of the IG Farben Management Board based there, wrote to Kränzlein about a visit of Staudinger to that plant. Staudinger wished to extend his contract but "under somewhat changed conditions." The state's austerity measures were forcing him, unless he cut back his research, to augment his additional income. On the subject of his contract with Leverkusen, Staudinger had remarked that in the past he had the impression that "with the exception of yourself [i.e., Kränzlein] no particular importance was attached to his work in Leverkusen and perhaps also in the rest of IG. For this reason he had mainly communicated with you." Leverkusen thereupon emphasized that, in the future, every effort would be made to work more closely with Staudinger again, particularly since he was increasingly devoting himself to the field of cellulose, which held great interest for Leverkusen. Kühne felt that Staudinger's demand to be given an annual remuneration of 10,000 RM in the future instead of the 6,000 RM he had previously received was "acceptable."[95] Kränzlein, who could only feel flattered by so much attention and encouragement, replied to Kühne that he had recommended Staudinger to continue working with Leverkusen, even if Staudinger would have "preferred to work with Hoechst" because of the "existing personal relationship between him and myself." Kränzlein considered the increase of Staudinger's remuneration to 10,000 RM "very much justified, because St. has used his emoluments from IG and a large proportion of his earnings for the benefit of his students in a very idealistic manner. Furthermore Leverkusen, Hoechst, and *Geheimrat* Haeuser should provide Staudinger with certain monies promised him directly or indirectly so that he would be able to purchase a high-speed centrifuge that currently cost around 20,000 RM. Staudinger accounted for his wish for a high-speed centrifuge by the fact that "foreign researchers were considerably better equipped than he was." Signer, one of his employees, had carried out "measurements into

---

[92] Ibid., vol. 2, Kränzlein to IG Farben Patentabteilung Leverkusen, December 13, 1933.
[93] Ibid., vol. 4, Kränzlein to Staudinger, December 11, 1933.
[94] Ibid., vol. 3, Kränzlein to Staudinger, February 12, 1934.
[95] Ibid., vol. 3, Kühne (Leverkusen) to Kränzlein, April 20, 1934, and April 23, 1934.

the size of molecules using a high-speed centrifuge" at The Svedberg's institute in Uppsala, which was "not possible" in the Freiburg laboratory. Kränzlein therefore concluded: "It is very regrettable that foreign institutes are much better equipped to carry out new scientific research, particularly in the field of high polymers, than our university laboratories, which means that German research must inevitably fall behind if matters continue like this. Perhaps it will be possible to find a way out if IG acquires a high-speed centrifuge and makes it available to St. on loan."[96]

Kränzlein's support of Staudinger began to waver only when the two men came into conflict. Kränzlein had wished to act as a mediator in the dispute on macromolecules between Staudinger and Kurt Hess, who, next to Meyer and Mark, was one of Staudinger's most determined opponents. Hess, one of IG Farben's most highly paid contractors at the Kaiser Wilhelm Institute for Chemistry in Berlin, was – like Kränzlein – a fanatical Nazi.[97] Kränzlein offered to organize a meeting between the two men to allow them to settle the dispute. An opportunity for this presented itself in the context of a lecture that Hess was to hold in Hoechst. Staudinger replied sharply to this suggestion, writing that he understood that IG Farben wished to inform itself about the work of Hess "as it [IG Farben] had for years spent large sums of money on the furtherance of the same [i.e., of Hess's work]." The reason for Staudinger's refusal to come to Hoechst was not because Hess had "wrong ideas about the constitution of cellulose," but because of his "disloyal behavior toward the work carried out here." The opinions of Hess, who worked "in the Kaiser Wilhelm Institute under the most favorable conditions and was in a position to dedicate his time fully to scientific research" on cellulose, "were no longer recognized today by the scientific world." Staudinger added that he believed that "no other researcher would have been able to afford a similar number of mistakes without rendering his position as a scientist untenable." Now that the results from Staudinger's laboratory had gained acceptance despite having had to work "under far worse conditions," Hess was attempting "to discredit these results, which were supported by experiments, by making disparaging remarks but without being able to disprove them experimentally." In short, Staudinger accused Hess of nothing less than of "disparaging my work against his better knowledge to cover up his scientific failures."[98]

[96] Ibid., vol. 3, Kränzlein to Kühne (Leverkusen), April 25, 1934.

[97] Cf. Furukawa, *Inventing Polymer Science*, pp. 166–7; Susanne Heim, *Kalorien, Kautschuk, Karrieren. Pflanzenzüchtung und landwirtschaftliche Forschung in Kaiser-Wilhelm-Instituten 1933–1945* (Göttingen: Wallstein, 2003), pp. 139–40; Hess was not simply a staunch National Socialist, he was also the person responsible for denouncing Lise Meitner in the 1930s, who had been allowed to continue working at the KWI in Berlin, despite being a "Jewess," until her native country of Austria was annexed.

[98] HA, PA Staudinger, vol. 3, Kränzlein to Staudinger, October 18, 1934; Staudinger to Kränzlein, November 10, 1934.

Kränzlein replied that he inferred from Staudinger's letter "with deep regret" that he would not be able to bring about a reconciliation between Staudinger and Hess. And yet "today more than ever German researchers should get along with one another," even more so if both were partners of IG Farben. Kränzlein added: "If Hess is at the Kaiser Wilhelm Institute, this is not something to reproach him for; you must also be aware that as one of the very few Aryans there he has been battling against Jewry in the K.W.I. for years; this too must be taken into consideration because these circumstances mean that he was by no means well off."[99] After having received Kränzlein's letter Staudinger showed himself to be slightly mollified as he apparently hoped that through Kränzlein's offices he would finally be accorded the recognition in Germany and abroad that was due him. Staudinger agreed to talk with Hess but did not change his opinion of him, since he believed that Hess was continuing to write "distorting reports about the results of the work carried out here" and that one of his assistants had taken the liberty of carrying out "an outrageous attack against me in the most respected German journal." "If furthermore you are appealing to the fact that in the past few years as one of the few Aryans in the Kaiser Wilhelm Institute K. Hess has been in a very difficult position, then I believe that he can ascribe this unfavorable position essentially to his failures in his scientific work, not just in the field of cellulose but also his earlier work on alkaloids. If he now complains about the bad treatment meted out to him by non-Aryans, then this is in sharp contrast to the fact that he attacks me in his publications citing works of K. H. Meyer and H. Mark which I have long since disproved."[100]

The conflict surrounding Hess led to a cooling off of the friendship between Kränzlein and Staudinger, as an extensive correspondence shows. Kränzlein had less and less patience with Staudinger's quarrels. Kränzlein defended the National Socialist Hess to Staudinger and could not comprehend Staudinger's attacks on Meyer and Mark, who as "Jews" were unworthy of being the subject of any conflict: "In my opinion you are making the mistake of constantly arguing with Jews. This results in your self-justifying publications, which appear to many of us and now also to the editorial staff of the *Berichte* as enough of a good thing. You have no need to get mixed up in polemical discussions with Jews because that is to accord them too much honor. You should avoid and ignore that lot because otherwise you repeatedly offer them the chance to have the last word and this will time and again rebound on yourself." And Kränzlein added: "We are systematically dissociating ourselves from the Jews, the Nuremberg Laws prove that. By these means we will push them back to where they came from. Why don't you distance yourself in your scientific work? They will

[99] Ibid., vol. 3, Kränzlein to Staudinger, November 14, 1934.
[100] Ibid., Staudinger to Kränzlein, November 23, 1934.

have to return to their intellectual ghetto again, whence they came, and go back to their Talmud from which they never manage to get away." Kränzlein went even further: "It is now your duty not to mention Jews at all any more, let alone to get yourself mixed up in polemical discussions with them. Our German ideological concepts are important here."[101]

That not everyone in IG Farben shared Kränzlein's anti-Semitism is demonstrated by the fact that almost simultaneously Heinrich Hörlein offered to arrange a talk between Meyer and Staudinger in order "to finally conclude the polemics." However, Staudinger did not want this: "Furthermore my misgivings have increased after having received just recently a longer letter from another department within IG, reproaching me that I am making the mistake of constantly battling Jews and which points out that we should dissociate ourselves from the Jews through the Nuremberg Laws."[102]

Staudinger now even began to behave as though he was "a victim" of "Jews"; all the evidence in both official and private letters testifies to this. He complained to a friend that he continued to have problems with Meyer and Mark, that he found himself "facing a completely closed group, which had developed before 1933 and which continued to stick together today." In his own words: "It is very hard to prevail against these Jews abroad and the comrades of these Jews inside the country."[103] But in addition to such private tirades against an alleged "Jewish" conspiracy, he also attempted to present himself to the National Socialist rulers as the victim of such a conspiracy. After he had been prohibited by the Nazi regime from traveling to the international congress of chemists in Rome, Staudinger wrote to the Reich Ministry of Science in June 1938 that his position as a chemist within Germany was "influenced very unfavorably" by "a scientific struggle I was obliged to wage against mainly Jewish circles concerning the recognition for and preservation of the work of my laboratory." In 1926 he experienced rejection at the conference in Düsseldorf "since many Jewish scientists had

[101] Ibid., Kränzlein to Staudinger, June 3, 1936 (copy); Deutsches Museum, NL 88 Staudinger-Archiv, D II 15.12, Kränzlein to Staudinger, June 3, 1936.

[102] Deutsches Museum, NL 88 Staudinger-Archiv, D II 15.13, Heinrich Hörlein (IG plant Elberfeld) to Staudinger, June 9, 1936; D II 15.14, response from Staudinger to Hörlein, June 12, 1936. On this topic, see also Deichmann, *Flüchten*, pp. 402–6: Deichmann oddly enough writes on p. 404 that Kränzlein and Hörlein had attempted to mediate in the dispute between Staudinger and Meyer. However, on pp. 405–6 she writes correctly that Kränzlein did not wish to do so – he attempted only to mediate in Staudinger's quarrel with Hess.

[103] Deutsches Museum, NL 88 Staudinger-Archiv, D II 15.21, Staudinger to Helmut W. Klever, June 9, 1941; Klever wrote a letter on February 19, 1941, to Staudinger (D II 15.19) in which he indulged in anti-Semitic ranting; "So National Socialism is smashing up Jewry. But this is like using a broomstick against mosquitoes. The organization is shaken up and shortly afterwards the same old swarms form again. ... What use was the reign of cudgels in the *Reichskrystallnacht*? Jewish high finance continues to be firmly ingrained in industry."

completely opposing views at that time." After 1928 the same men, spearheaded by Kurt Hans Meyer, had then adopted his results without citing him, as would have been customary usage in scientific circles. As Staudinger, of course, could not put up with such behavior, this had resulted in the polemical infighting, which had been continuing for a long time and had been "very detrimental" for him, since his main opponent, Meyer, was a member of the IG Farben Management Board and a director in Ludwigshafen and held "a very influential position in German chemistry." "The success of Jewish circles in science is based on the same methods they also employ in other fields: highlighting their own achievements and corrosive criticism of others." The fact that he was not permitted to travel to Rome would give them particular pleasure: "I only regret that my battle over many decades to break the Jewish influence in this important field of chemistry was in vain."[104]

Staudinger thus tried to make capital out of the Jewish origin of some of his opponents for his own purposes but met with decided opposition from his old friend Kränzlein. Staudinger had given notice at the end of December 1936 that he wished to terminate his contract with IG Farben on April 1, 1937, since, as he wrote, he did not know how to reconcile the duties laid on him by the Four-Year Plan with his contractual obligations toward the company.[105] Prior to his giving notice he had had a dispute with IG Farben concerning the filing of a patent for polystyrene. Staudinger terminated his contract and demanded a 30 percent share of the profits. While the Patents Department regarded this sum as "high but acceptable," Kränzlein raged and declared it "impossible" to agree to such a large sum. Because Staudinger already had received a lot of money from IG Farben and had been supplied with the necessary materials, additional claims seemed "unusual." However, the Patents Departments of the plants in Ludwigshafen and Leverkusen were initially ready to sign a contract with Staudinger. But IG Farben then decided that it had no prospect of beginning production of the plastic anyway.[106] Summing up the whole affair, Kränzlein wrote that Staudinger believed that he had "invented a particularly valuable synthetic material by the polymerization of styrene in the

[104] HA, PA Staudinger, vol. 4, Der Reichs- und Preussische Minister für Wissenschaft, Erziehung und Volksbildung to Kränzlein, June 29, 1938; in the letter, notarized excerpt from Staudinger's letter to the Ministry for Research, evidently dated June 1938.

[105] Ibid., vol. 3, Staudinger to Direktion der IG, Werk Leverkusen, c/o Dr. Kühne, December 23, 1936.

[106] Ibid., vol. 4, Kränzlein to IG Patentabteilung im Werk Leverkusen, August 2, 1934, with enclosures on Staudinger's proposed patent; Patentabteilung Ludwigshafen to Patentabteilung Leverkusen, June 3, 1936; Patentabteilung Leverkusen to Patentabteilung Ludwigshafen, November 5, 1936; Memo from Kränzlein to Patentabteilung, November 9, 1936; Patentabteilung Ludwigshafen to Patentabteilung Leverkusen, November 13, 1936; IG Farben to Staudinger, February 16, 1937.

presence of divinyl benzol." He had then registered the process as a patent at home and abroad and terminated his contract with IG Farben, relinquishing any further cash compensation. He had then offered the patent to the company: "It turned out, however, that although the polystyrene produced from styrene by the admixture of divinyl benzol represents a very interesting product scientifically, it has no technical importance." IG Farben had therefore abstained from acquiring the invention.[107] The relationship between Staudinger and Kränzlein cooled even more in consequence.[108]

In a letter to the Minister of Science concerning Staudinger, Kränzlein made this very clear. Kränzlein wrote that the "most acrimonious conflict of opinions" had raged between Staudinger and "the Aryan Hess." The quarrel between the two scientists, both of whom "were amply remunerated by IG," had assumed such proportions "that IG resolved to end it." An offer by Kränzlein to mediate between the two men had been declined by Staudinger, while Hess had been prepared to accept. Nor had Staudinger been discriminated against because of Meyer's position in the company. Quite the contrary: he had been supplied not only with money but also with the newest preparations and high polymers that he required for his work. Kränzlein therefore concluded: "Staudinger's theoretical ideas concerning the constitution of high polymers were fully accepted by circles within IG. The undersigned has repeatedly referred to the obvious truth of Staudinger's interpretations, both in discussions and in writing. Staudinger always gratefully appreciated this. Why has Staudinger taken pains to hide this active and moral support by IG and why did he put it about that he was a victim of Jews when his bitterest quarrel was with the Aryan Kurt Hess? Moreover, Staudinger as a highly remunerated employee of IG has enjoyed its full indirect support against K. H. Meyer in their dispute. It is therefore most regrettable that Staudinger is giving completely one-sided accounts and depicting the matter as though IG had used K. H. Meyer to embitter his life."[109]

Whether Staudinger ever learned about this letter is not clear. The two men continued to write one another and extend mutual invitations, but appear to have avoided accepting them. Nevertheless they remained in contact until the war. But by the late 1930s Kränzlein was increasingly occupied with armaments assignments and jobs for the SS,[110] which had

---

[107] Ibid., vol. 4, draft of Kränzlein's letter to the Minister for Research dated July 7, 1938, which was not sent (in the letter that was finally sent this story is missing).

[108] Ibid., vol. 4, Kränzlein to Gajewski (Wolfen), December 8, 1939.

[109] Ibid., vol. 4, Kränzlein to Wissenschaftsminister, July 11, 1938. In the first draft of the letter many formulations were even sharper and the story of polystyrene was reported in detail. The draft, dated July 7, 1938, was passed on to ter Meer on the same day, who clearly revised the text of the letter.

[110] Cf. HA, PA Kränzlein, eulogies given at the memorial service for Kränzlein, November 10, 1943.

consequences for research in Hoechst. The field of plastics shifted more and more into the hands of Otto Ambros and Ludwigshafen.[111]

Two aspects are particularly interesting when we look at the initially close and later more selective cooperation between Staudinger and Kränzlein. Kränzlein recognized the potential of plastics and synthetic fibers early on, not only for the war effort but also in times of peace. Therefore he felt his association with Staudinger, whose theory on macromolecules he accepted at a very early stage, to be so important. In a memorandum written in 1941 Kränzlein explicitly stated that plastics were important for both a wartime and a peacetime economy; to use a modern expression they were "dual use items." This was what distinguished research and development in the 1930s from similar activities during the First World War. During the First World War, according to Kränzlein, Hoechst had had to devote "all its energies to national defense," and in the "laboratories almost everything focused on the war." In this war things were different, particularly in the field of plastics: "in Hoechst we have long since agreed to take an active part in the scientific and technical developments in the areas of plastics and rubber, especially since today the conditions in the war, unlike those during the [First] World War, offer us far more opportunity to carry out research which will provide a basis for future peacetime developments." Kränzlein considered it particularly convenient that Hoechst was even able to "get young chemists fit for active service back from the troops and employ them to carry out scientific work" in the interests of the Four-Year Plan and future autarky.[112] The cooperation with Staudinger and the hiring of some of his best students led to good research and development results in the field of polymers in Hoechst.

Yet despite Kränzlein's realistic assessment of the situation, which is very obvious here, his political attitude and his support of Hess cost him Staudinger's friendship and cooperation. The conflict between the two men escalated due to Kränzlein's keen National Socialism, which was also underpinning his support of Hess. That the conflict arose because Staudinger attacked IG Farben, as the historian of science Ute Deichmann writes,[113] does not appear plausible since Kränzlein supported Staudinger in his dispute with Meyer and Mark at a time when they were still working for the company. Kränzlein's ideological leanings, his support of the "National Socialist comrade" Hess, and Staudinger's habit of dividing the world into "friends" and "foes" seem to have been decisive for the quarrel.[114]

---

[111] Cf. Morris, "Ambros"; Stokes, "IG Farben Fusion," pp. 286–7, 308–9.
[112] HA, PA Kränzlein, memo by Kränzlein dated January 6, 1941.
[113] Deichmann, *Flüchten*, p. 403.
[114] Cf. oral communication by Dr. Gerhard Bier on Staudinger, his former teacher, according to which Staudinger categorized people either as "friends" or "foes."

Staudinger himself, as he said several times, was very interested in continuing to work together with Hoechst or IG Farben. Thus, in 1936 he emphasized the importance of cooperation between universities and industry: "In essence, we owe the successes in the field of plastics to industry." The universities had played a "conspicuously small" part in the work on plastics: "The reason for this is because developments in the field of plastics require a considerably greater outlay of material and mechanical resources. The universities are absolutely not equipped adequately for this, and there are no funds available for such facilities. For the same reason it is not possible to retain competent employees at the universities."[115] So it comes as no surprise that in 1943 Staudinger again concluded a freelance contract with IG Farben – but this time he cooperated with Leverkusen and Ludwigshafen, while Hoechst was excluded.[116]

We can summarize the developments as follows: the efforts and funds Hoechst spent on researching and developing polymers and the focus on their use as "substitutes" was in line with the program of IG Farben as formulated in the memorandum of 1934 outlined above. Research and development in Hoechst overlapped with, and often competed against, that of other company plants, as can be seen by the example of the cooperation with Staudinger. Hoechst drew on the know-how and recruited competent people from other IG Farben works, assigning them senior positions in Hoechst. And Hoechst attempted to extend its cooperation with out-standing scientists. It still followed the strategy of "mass attack" criticized by a manager from DuPont in 1930, that is, it continued to assign many chemists and other staff to work on a particular subject without much regard for cost. However, in contrast to the manager from DuPont, I would argue that these were not second- or third-rate people. Hoechst tried to get the best employees for work on plastics, albeit with one important exception: they could not be "Jewish" or be considered "Jewish." Hoechst sent Franz Henle, one of its top men in research on azo dyes and artificial wax, into early retirement as early as 1935 because he was a "Jew by decree." And Hoechst opposed Meyer and Mark despite the fact their results in polymer research had been effusively praised in 1930 – but at the time they had not yet been regarded as Jews.[117]

---

[115] HA, TEA 191b, Niederschrift der 2. Sitzung der "Fachgruppe für Chemie der Kunststoffe im V. D. CH." in Berlin am 3.12.1936, dated December 18, 1936; see also PA Staudinger, vol. 3, Staudinger to Kränzlein, June 9, 1934; cf. also Furukawa, *Inventing Polymer Science*, pp. 91–2, 174–5.

[116] HA, PSW 22, according to these documents Staudinger was employed by Ludwigshafen from January 1, 1943, to December 31, 1944, with an annual salary of 12,000 RM.

[117] HA, TEA 42, Niederschrift über die 2. Kunststoff-Sitzung am 16.10.1930 in Ludwigshafen dated November 30, 1930, praise by Kränzlein for Meyer and Mark.

4.3. PRODUCTION AND ITS IMPORTANCE FOR REARMING AND WAR

In his history on Hoechst during the time it formed part of IG Farben, Otto Hirschel, who as former head of the Directorate T could be expected to know most about figures and statistics, wrote: "Determining the business volume of the Hoechst plant is a difficult task because there are hardly any documents available." However, he added that "on the basis of how overall turnover of IG developed" it was possible to "gauge the situation in specific years," an assessment that "by and large also applies to Hoechst."[118] Indeed, with the exception of the figures on the volume of costs, only very few figures could be found in the archives on the development of Hoechst during the Third Reich – and these figures do not permit us to learn anything about developments in turnover and profits.[119] The fact that the former head of the Directorate T who had been responsible for cost accounting could not come up with much in the way of figures or statistics leads us to the conclusion that profits were not Hoechst's top priority. In the 1930s and early 1940s Hoechst was no "profit center" within the IG Farben corporate group but merely a production facility that produced certain products and product lines as part of the war economy of the Third Reich and its policy of self-sufficiency. Hoechst not only had no sales outlets of its own any longer, it clearly no longer even maintained proper accounts.

However, some figures are available that allow us more or less to follow how production developed in Hoechst – with the qualification that, due to the extraordinary heterogeneity of the plant, it is difficult to convey any comprehensive pattern.[120]

Let us begin by looking at developments in the volume of costs in Hoechst (Table 4.2). This term includes all costs or expenses accruing in the course of production, primarily therefore all labor, energy, and freight costs together with the cost of servicing debts.[121] But these figures give no indication of the returns, in other words, of how profitable Hoechst was. What they do show, however, is that during the Third Reich Hoechst was not doing quite as well as Ludwigshafen and Leverkusen. Both Hoechst and Ludwigshafen had been hit badly by the Great Depression, and their total outlays had decreased accordingly. After the seizure of power by the National Socialists total costs in Ludwigshafen increased again considerably. After the mid-1930s, but particularly during the Second World War,

---

[118] HA, 12, Hirschel, "Das Werk Hoechst im Verbande der I.G.," p. 40.
[119] Written and oral communications from the archives of Hoechst, Bayer, and BASF.
[120] Cf., e.g., the report HA, 12, Hirschel, "Das Werk Hoechst im Verbande der I.G.," which mentions all the divisions but does not give any indication of the importance of individual divisions for the plant.
[121] See ibid., p. 27.

TABLE 4.2. *Volume of Costs for the Major Plants in Hoechst (Including Nitrogen), Leverkusen (Including Titanium and Photography), and Ludwigshafen (in Million RM)*

Year	Hoechst	Leverkusen	Ludwigshafen
1927	32.1	27.8	34.0
1928	35.0	32.7	38.1
1929	35.5	35.7	38.6
1930	26.8	31.8	30.3
1931	22.9	30.7	26.7
1932	20.1	25.2	20.9
1933	19.9	26.9	21.8
1934	21.3	33.5	26.9
1935	22.9	35.5	34.1
1936	25.7	39.3	36.6
1937	32.3	46.4	44.7
1938	33.8	48.2	46.5
1939	39.4	55.6	55.2
1940	43.2	57.4	58.9
1941	46.8	62.8	69.6
1942	49.5	63.3	69.3
1943	55.7	67.3	84.4

Source: HA, RFL 78, Statistisches der Farbwerke 1927–1939/48, volume of costs of the plants at Hoechst, Leverkusen, and Ludwigshafen from 1927 to 1943.

Ludwigshafen experienced extensive growth. Leverkusen had not suffered quite so much during the Great Depression as had the two other former major works of IG Farben. Yet in the second half of the 1930s Leverkusen too was able to record extensive growth, even if it was not quite as spectacular as that of Ludwigshafen. In contrast, Hoechst achieved the same level of expenditure as prior to the Great Depression only in 1939, and during the war growth was considerably lower than at Leverkusen and Ludwigshafen.

According to Gottfried Plumpe, dyes, including dyeing aids and raw materials for washing agents, were IG Farben's most profitable areas for the years 1932–44; with a return on sales of 30.8 percent for those years this sector enjoyed above-average returns. In second position were pharmaceuticals/pesticides with a profit-earning capacity of 24.5 percent. And in third place was photography with a return on sales of 19.8 percent. These three areas, according to Plumpe, were all areas that were "least affected by the typical economic policy of the Third Reich" and were able to maintain the greatest share of exports and were the most important "profit earners" for IG Farben in those years. With a share of overall turnover of 34 percent, the share they contributed to profits stood at 54.9 percent between 1932 and 1944. The share of turnover of those areas that profited most from the government's industrial policies and its policy of self-sufficiency – mineral oil, chemicals, and synthetic fibers – was 52.7 percent, but their share of profits was only 33 percent.[122] Plumpe was of the opinion that this was due to the fact that public orders and projects had only "relatively low profit margins."[123] Yet pharmaceuticals and dyes for uniforms were also sold to the state at discount prices.[124]

Hoechst was well represented in the sectors Plumpe considered to be particularly lucrative, which might indicate that the plant was fairly profitable. There is even a calculation of profit for the year 1935 for Hoechst that itemizes the most important areas of production (Table 4.3). The calculation shows that dyes, dyeing agents, and pharmaceuticals were very profitable, just as they were for the whole of IG Farben. Organic intermediate products were also very lucrative. In fact, net income had increased considerably compared to 1934, but the growth in profits was generated almost exclusively by the Pharmaceutical Department. The chemicals sector was also profitable. After 1932 it became the largest area in IG Farben and the most diversified: "It was in this sector that internal diversification in IG took place, it included the areas polymers and light metals, which played an increasingly important role in the development of the chemical industry worldwide."[125] But, as was already noted in the fact-finding memorandum commissioned by Hermann in 1934, pride of place at Hoechst was held by the Pharmaceutical Department.

Carl Ludwig Lautenschläger's memoirs include a list showing his estimates of annual turnover and profits for the years 1940 through 1944. However, it is not clear when he drew up these figures, where he got the numbers from, and what changes occurred during this period. The figures he gives for the Pharmaceutical Department are clearly too high, as the figures that Hirschel used, given in Table 4.4, are considerably lower. Lautenschläger's figures are therefore not particularly accurate, but they are the best ones available and better than pure guesswork.

An attempt will be made here, based as far as possible on the account given by Otto Hirschel, to trace the development of the most important lines of production in Hoechst and their significance for rearmament and the war against the backdrop of developments in IG Farben in these years.

The Inorganics Department experienced enormous growth during the Third Reich. After the enforced retirement of Rohmer, Staib, who came from the IG Farben plant at Rheinfelden, took over the Inorganics Department. He succeeded in setting a number of projects in motion and increasing the profits of his department. But there were a number of initial difficulties that needed to be overcome.

---

[122] Plumpe, *I.G. Farbenindustrie*, p. 550. [123] Ibid., pp. 550–1.
[124] Cf. Bartmann, *Tradition*, pp. 152–3. [125] Plumpe, *I.G. Farbenindustrie*, p. 551.

TABLE 4.3. *Calculation of Profit for the Hoechst Plant in 1935 (in 1,000 RM)*

Sector	Gross Profit Sales	Gross Profit	% of Sales	Net Income before Deduction of Interest	Net Income after Deduction of Interest	Net Income 1934 (Comparison)
Chemicals, inorganics group	4,059	745	18	388	348	295
Organic intermediate products	2,965	995	34	741	689	750
Organic tanning agents	262	53	21	16	12	33
Vulcanization accel. activated carbon	49	8	19	5	4	2
Synthetic materials/ solvents	9,897	1,306	13	430	293	525
Total chemicals	17,226	3,107	18	1,580	1,346	1,605
Textile additives	5,333	2,524	47	1,855	1,686	1,843
Dyes	44,250	14,846	34	10,870	9,568	8,714
Pharmaceuticals	39,702	12,263	31	7,928	6,909	3,671
TOTAL	106,511	32,740	31	22,233	19,509	15,833

Source: HA, TEA 975a, Hoechst (Hermann) to ter Meer, August 20, 1936, in the appendix a summary of the development and profit and loss account of the Hoechst plant for 1935. It is not clear whether the figures given for the profit in 1934 refer to profit before or after deductions.

The Nitrogen Department had been expanded during the First World War to cope with the enormous demand for explosives. A computation of costs for the different IG Farben plants at the end of 1923 showed that Hoechst had the highest production costs.[126] Nevertheless Hoechst also profited from the enormous demand for nitrogen products and from the positive economic situation until 1929. The Great Depression and a new

---

[126] HA, 12, Hirschel, "Das Werk Hoechst im Verbande der I.G.," p. 46–64.

TABLE 4.4. *Annual Turnover and Net Profits (in 1,000 RM) at Hoechst for 1940–1944 According to Lautenschläger's Estimates*

Category	Turnover	Net Profit
Inorganics	35,000	5,000
Intermediate products	10,000	Mainly to IG or Hoechst
Dyes	60,000	15,000
Textile additives	25,000	5,000
Solvents	35,000	6,000
Synthetic materials	10,000	3,000
Pharmaceuticals	85,000	17,000
Other products	20,000	2,000
TOTAL	280,000	53,000

Source: HA, PA Lautenschläger, Lautenschläger, "Erinnerungen," vol. 8, [pp. 177–8].

process developed by the Norwegian group Norsk Hydro for the production of calcium nitrate put paid to the boom. The nitrogen syndicate to which IG Farben belonged was forced to conclude an agreement with Norsk Hydro and to make considerable concessions regarding its sales quotas. This resulted in an abrupt reduction in the production of nitrogen products by IG Farben. Hoechst was obliged to restrict production to highly concentrated nitric acid, Chile saltpeter, and potassium nitrate. Production by ammonia oxidation in 1931 sank to less than one-fifth of the levels of 1928.[127]

When Hermann took over management of the plant the future of the Hoechst Nitrogen Department looked anything but rosy. At the end of May 1933 Fritz ter Meer informed Carl Krauch that the works managers Pistor (from Bitterfeld/Wolfen) and Hermann (from Hoechst) had approached him to express their worries concerning the restrictions on the processing of nitrogen. He suggested carrying out the restrictions in proportion to the capacities available for the processing of primary nitrogen into nitric acid or fertilizer. For the three works this was "of crucial importance, since after the negative growth in the dyes and intermediate products trade in the last decade and the resulting consolidations, production volumes had decreased to such an extent, particularly in Hoechst, that any reduction in the volume of production would considerably compromise the profitability of these works."[128]

The situation in Hoechst became so bad that prior to his departure to the United States, ter Meer wrote an urgent letter to Krauch in which he again raised the subject of nitrogen production in Hoechst, describing the

[127] Ibid., p. 65; cf. also Plumpe, *I.G. Farbenindustrie*, pp. 223–37.
[128] BASF UA, IG Archiv C121/11, ter Meer to Krauch, May 31, 1933.

situation as presenting "a quite unedifying picture." After terminating production of Chile saltpeter Hoechst's share in ammonia oxidation within IG Farben had fallen to a mere 4 percent in July 1933 as a result of the assignment of quotas for saltpeter production: "The loss of income from the Nitrogen Department is quite significant and results in a heavy extra burden being laid on the other Hoechst production facilities." As Oppau had just recently begun production of Chile saltpeter, ter Meer considered that under those circumstances "the sacrifice of ceding the current production of calcium nitrate in Oppau to Hoechst might be tolerable." It would also be an opportunity of "paying proper tribute" to "the achievements of the gentlemen in Hoechst," who had helped develop the new low-cost method of Chile saltpeter production used in Oppau.[129] But Krauch was not prepared to agree, and he pointed out that the facilities in Oppau that made use of the Hoechst method were not that large and that one might even have to be closed down. It was therefore not possible to effect an exchange in the production of Chile saltpeter and calcium nitrate between Oppau and Hoechst, even if Krauch was "by no means" unaware of "the difficult situation of Hoechst." Yet even if Krauch emphasized that any other distribution of production than the one currently existing was not in the interest of IG Farben, he did wish to appear conciliatory: "In order to take the interests of Hoechst as far as possible into account, I have arranged that allowances should be made for Hoechst, when possible, when allotting the quotas for saltpeter production."[130]

The Nitrogen Department in Hoechst did indeed experience an enormous upturn in the years to come (Table 4.5). Fertilizers played the most important role in IG Farben's nitrogen production; their share between 1929 and 1936 was never less than 80 percent. By 1942 this share had decreased to a little below 60 percent, however, by 1944 it had fallen to 19 percent. Thus, the fertilizer industry had profited from the focus of the Nazi regime on self-sufficiency and its support of agriculture, as had the Pesticide Department. While to begin with, in 1931 and 1932, the nitrogen business was running at a loss, in the 1930s it became a profit earner with a return on sales of more than 20 percent in the years between 1936 and 1942.[131]

As a result of this development and of Krauch's obligingness production in Hoechst began to increase again after 1935, and after 1937 it even rose steeply. The main part of Hoechst's production went into agriculture, mainly into the manufacture of fertilizers; a not insignificant share, primarily of highly concentrated nitric acid (referred to as Hoko acid, from *hoch konzentrierte*), went to the armed forces for use as an intermediate product in the manufacture of explosives – an area in which IG Farben

[129] Ibid., ter Meer to Krauch, September 14, 1933.
[130] Ibid., Krauch to ter Meer, September 20, 1933.
[131] Plumpe, *I.G. Farbenindustrie*, pp. 241–3.

TABLE 4.5. *Main Production of the Hoechst Nitrogen Department 1928–1944 (in Tons)*

Year	Ammonia Oxidation (into N)	Concentrated Nitric Acid (into HNO₃ 100%)	Calcium Nitrate (into Goods)
1928	33,842	27,377	102,941
1931	6,332	20,081	–
1933	6,012	17,766	–
1935	10,702	42,803	–
1936	14,701	54,960	6,013
1937	41,415	75,277	115,910
1939	48,684	93,841	138,056
1941	51,526	102,776	131,498
1943	51,860	110,329	112,920
1944	41,876	105,199	43,990

Source: HA, 12, Hirschel, "Das Werk Hoechst im Verbande der I.G.," p. 67.

carried out some development work. The overall growth of this department in Hoechst was phenomenal, particularly if one takes the years 1931 through 1933 as a starting point. But IG Farben's leadership clearly was not of the opinion that this department had a very promising future because, despite the enormous increases in production, no substantial investments were made in Hoechst in this area.

Winnacker, head of the Inorganics Department since 1938, complained in a draft of a letter written at the end of October 1943 to Krauch as the *Gebechem* (Plenipotentiary General for Special Questions of Chemical Production in the Four-Year Plan) that Hoechst faced great obstacles in its production of the nitric acid used for explosives. Winnacker wrote that the production of highly concentrated nitric acid in Hoechst had "only been maintained at the projected levels up to now with the greatest difficulties." There were a number of different reasons for the drop in production during the past few weeks, "the remedying of which is for the most part beyond the capacity of the work." The nitric acid operations are "technically very outdated" and worked "in part with facilities which still dated from the last world war." Attempts to overhaul the facilities had already failed prior to the war because of problems in procuring the necessary materials, and even improvements made during the war had not been able to eliminate the deficiencies. The *Hoko*acid production suffered particularly from the fact that although it was termed a "special operation" of the High Command of the Armed Forces (*Oberkommando der Wehrmacht*, OKW), otherwise it was "in no way privileged with regard to being given priority over the large number of chemical operations." "Even if the special importance for the war of this manufacture was as far as possible always taken into account within the plant, the lack of priority has had an adverse effect on obtaining spare parts and auxiliary materials."[132]

Another important product for Hoechst during the war was hexogen, which combined a "strong explosive force and shattering power with an excellent chemical stability and relative insensitiveness to mechanical stress." Hexogen was still regarded as "the most important explosive for special military purposes" in the 1950s. And it was found that the admixture of equal quantities of trinitrotoluene (TNT) and hexogen was highly safe under fire, despite the increased explosive force and effect.[133]

An important development of the Inorganics Department was the electrolysis of sodium chlorine. Originally Hoechst had a Billiter electrolysis facility, which Hermann had helped set up. At the beginning of the 1930s the demand for chlorine increased again, primarily because of the rise in methane chlorination. Hoechst applied to expand its annual capacity from 3,000 to 5,000 tons annually, and the expansion was approved. In the original facility sixty-three mercury cathode cells were set up, following the model of the cells developed in Leverkusen, so that, as Hirschel wrote, "Hoechst is the first IG plant" to adopt "mercury electrolysis for large-scale production." However, the original pumps that transported the mercury from the decomposers to the baths had to be replaced by pumps of a newer type. The costs of the new electrolysis method were considerably lower, and it produced very pure caustic soda lye without any sodium chloride content, which, according to Hirschel, was particularly suitable for use in artificial silk and staple fiber factories. The original Billiter facility, which at first had been allowed to continue production, was pulled down in April 1935, and the new facilities were expanded. In 1936/37 mercury electrolysis was expanded to include 147 cells, and during the war this increased to 210 cells to satisfy the growing demand for chlorine and caustic soda solution. This resulted in a capacity of no less than 32,000 tons annually.[134]

Chlorine was used for the production of intermediate products, dyes, and pharmaceuticals, for the production of solvents, and for the chemical production of varnishes and plastics.[135] The production of *Nebelsäure* or "acid fog" was considered important for the war effort, and production increased enormously in the 1930s and during the war.[136] Overall, the Inorganics

[132] HA, PA Lautenschläger, Sekretariat Lautenschläger, Vertrauliche Korrespondenz, draft by Winnacker for Lautenschläger dated October 25, 1943, in response to a letter from Krauch dated October 7, 1943.

[133] Karl Winnacker and Ernst Weingaertner (eds.), *Chemische Technologie*, 4 vols. (Munich: Hanser, 1950–4), vol. 4 (1954), pp. 405–6, and vol. 2 (1950), pp. 203–4.

[134] HA, 12, Hirschel, "Das Werk Hoechst im Verbande der I.G.," p. 43.

[135] Winnacker and Weingaertner (eds.), *Chemische Technologie*, vol. 1 (1950), pp. 437ff.

[136] HA, IG/50, IG-Prozessakten, Jähne document 9, Affidavit by Dr. Fritz Bachran dated January 7, 1948, on the production of acid fog; see also Winnacker and Weingaertner (eds.), *Chemische Technologie*, vol. 2 (1950), pp. 65–7.

Department expanded greatly in the Third Reich. Some of the older facilities were running at full capacity, strongly increasing their wear and tear,
while in other areas production was carried out using the most modern
technology available.

This development was also noted for the Acetone and Acetic Acid
Department, referred to after the war as the Acetic Acid and Solvents
Department (ELKA or *Essigsäure- und Lösungsmittelfabrik*). At a special
meeting of the Technical Committee on January 13, 1933, the decision was
taken to reassess acetic acid production in Hoechst. Hoechst, which in
Hirschel's words, had been the first "together with Wacker to introduce
acetylene chemistry into technology," was "annoyed by this decision to the
highest degree" and "pulled out all stops to forestall any undermining of the
Acetic Acid and Solvents Department." At first its efforts were successful.
But only two years later the department in Hoechst was once again under
threat. After an explosion in the aldol workshop a fire broke out, which not
only killed several workers but also largely put an end to production of
ethyl acetate, crotonaldehyde, unprocessed acetic acid, and all acetic acid
derivatives. Hoechst managed to resume production and simultaneously
applied for a new building for the Acetic Acid and Solvents Department, an
application that was initially approved. But in 1935 plans were made to set
up a new factory for Buna in the middle of Germany: "This work was to be
built on a grand scale in accordance with the wish of Dr. Bosch, and it
would form the nucleus for a solvents factory for IG."[137] But according to
Hirschel, Hoechst opposed Bosch's plans, presenting its own calculations to
the effect that a new building would be considerably more economical in
Hoechst than in the IG Farben plant in Schkopau. In the meeting of the
Division's Technical Commission in February 1936 the calculation was
approved, and it was agreed "to leave the solvents in Hoechst for the time
being and to grant the above loan." Hoechst was able to start constructing
the new building, and the modern, open-style factory for the production of
acetic acid and solvents marked the beginning of "a new era for the Solvents
Department in Hoechst."[138] According to a report on Paul Roth, the head
of the department, the facilities for the production of solvents in Hoechst
were designed between 1937 and 1940 "according to modern technical
considerations for new workshops" and Hoechst, subsequently became
"the biggest and most modern producer of solvents on the continent."[139]

Before the war, and even during the war, several new products went into
production: one of them was glycerogen, begun in 1938, which was produced from sugar using high-pressure hydrogenation and used to compensate for the lack of glycerol – which was widely used in explosives and

[137] HA, 12, Hirschel, "Das Werk Hoechst im Verbande der I.G.," pp. 91–3.
[138] Ibid., pp. 93–4.
[139] HA, PA Abteilung VI, Roth, Report on Dr. Paul B. Roth dated February 7, 1952.

TABLE 4.6. *The Most Important Products of the Hoechst Acetic Acid and Solvents Department 1929–1944 (in Tons, with the Exception of Synthetic Materials)*

Year	Acetic Acid	Ethyl Acetate	Butanol	Glycerogen	Vinyl Acetate
1929	9,634	1,991	1,104	–	–
1931	3,725	863	575	–	–
1933	4,153	1,942	643	–	–
1935	4,719	3,855	1,337	–	–
1937	8,697	4,177	2,745	–	203
1939	10,480	7,026	2,910	837	1,581
1941	13,057	10,174	2,842	1,845	5,373
1943	15,646	8,782	3,847	2,074	8,516
1944	14,027	8,794	4,024	1,530	7,944

Source: HA, 12, Hirschel, "Das Werk Hoechst im Verbande der I.G.," p. 96.

blasting agents, dyes production, and as a coolant. In 1940 production of Uresin B was started, a soft resin used in the production of high-gloss combination varnishes. The plastics workshop, which began production in the spring of 1940, was particularly important.[140] As part of the Four-Year Plan, the department produced Mowilith, Novital, Hostalon, and a number of other plastics (Table 4.6).[141]

Dyes, textile additives together with pharmaceuticals, and pesticides were among the company's most profitable lines of business – and these were all sectors, part of IG's Division 2 under Fritz ter Meer, in which Hoechst was well represented.

As long as Hoechst was an independent company its dyes production was very decentralized, to the extent that even dyes of the same class were produced in different places in the factory. Hermann, who wished to reorganize the plant to conform to the framework set by IG Farben and ease work in its committees and commissions, combined the four dyes departments into a single dyes factory in 1934. With the outbreak of the Second World War the "mob[ilization] plan" that had been set up during the "peace years" came into effect. Accordingly the production of dyes in Hoechst was to be reduced to 50 percent of production levels in 1936. The halving of production did not uniformly apply to all dyes. The aim was to reduce production of less successful brands, like indigo or sulfur black, more while the more valuable brands would be allowed to continue production at almost their former levels or without any reductions. The sulfur black and indigo operations were

[140] HA, 12, Hirschel, "Das Werk Hoechst im Verbande der I.G.," pp. 94–5.
[141] HA, PA Lautenschläger, Lautenschläger, "Erinnerungen," vol. 3, [unpaginated].

finally shut down in 1940. The overall production and sale of dyes by IG Farben and Hoechst continued to do quite well until the beginning of the war.[142] While the Alizarin Department had been under threat at the beginning of the 1920s, under Georg Kränzlein's management it became "again a strong and economically viable part of the plant in Hoechst," according to Hirschel, as is borne out by the department's sales figures and profits. Thus the net proceeds of the department increased from 7.3 million RM in 1932 to 11.5 million RM in 1936 and 20.2 million RM in 1939. At the same time gross profits increased from 3.5 million RM in 1932 to 6.9 million RM 1936 and 12.2 million RM in 1939.[143]

The helindon dyes, used for the uniforms of the *Wehrmacht*, were important money producers during the war. Georg Kränzlein had already recognized the importance of these uniform dyes in September 1932, even prior to the seizure of power. He had decidedly opposed suggestions made by Leverkusen at the time to give up the production of helidon dyes, stating that "with the coming rearmament of Germany increased sales of helindon dyes can be anticipated" and Hoechst should therefore continue to maintain proper operations: "we are the only manufacturer and IG should not interfere with our responsibility to bring order to these operations."[144] The Dyes Department was fully occupied during the war, producing brown, khaki, gray, and blue dyes for use by the army, air force, and navy.[145]

In the Third Reich the Dyes Department was enlarged to include some new buildings. The ranges of indigosol and indanthrene were considerably expanded, "which meant a considerable increase in production." Similarly the production of important brand name varnishes such as Hansa yellow and the paranitraniline reds was expanded as sales continued to grow. Hansa yellow took over from the inorganic chrome and cadmium dyes, which were no longer available. And red varnishes were in strong demand because mail vans and mail boxes, formerly yellow, were now being painted red.[146]

An important impulse for research in the field of dyes was the development of synthetic fibers, particularly viscose, which confronted chemists working with dyes "with new problems" in the 1930s: "It was necessary to search for dyes and dyeing processes which will dye the so-called Wollstra fibers and textiles, produced from a mixture of wool and viscose, evenly and ensure that the colors will remain fast." The Central Laboratory, the Coloristic Department, and the laboratory of the Azo Department cooperated

[142] HA, 12, Hirschel, "Das Werk Hoechst im Verbande der I.G.," pp. 127–8, on indigo, p. 152.
[143] Ibid., p. 165.
[144] HA, PA Abteilung VI, Greune, Kränzlein to Roemer and Greune, September 8, 1932 (copy).
[145] HA, PA Lautenschläger, Lautenschläger, "Erinnerungen," vol. 3, [unpaginated].
[146] Ibid.

TABLE 4.7. *Turnover of the Igepon and Igepal Products 1931–1943 (in Million RM)*

Product Group	1931	1934	1937	1940	1943
Igepon	2.2	5.1	6.0	3.6	3.4
Igepal	–	–	0.5	2.8	7.9
TOTAL	2.2	5.1	6.5	6.4	11.3

Source: HA, 12, Hirschel, "Das Werk Hoechst im Verbande der I.G.," p. 171.

closely.[147] According to Lautenschläger, Hoechst managed to find "dyes with completely new properties" – so-called autazoles – and develop them until ready for production: "These self-coupling dyes allow fast dyeing of viscose and natural fibers, such as Vistra and cotton, to be carried out simultaneously in a single operation."[148]

Textile additives were another very important area for Hoechst, particularly during the war. The products Igepon A and Igepon T had already been developed during the Great Depression. They were detergents developed in Hoechst, Leverkusen, and Ludwigshafen and then produced in Hoechst: "Due to their excellent properties the two Igepons very quickly became major products despite the fact that they were launched on the market at the time of the Great Depression. Besides their resistance to calcium, their capacity to remove oil, and their wetting action, they have the additional advantage that when used to wash woolens they do not create the same sort of felting effect as occurs with soap. They were increasingly utilized in the textiles sector, in raw materials for washing, and in cosmetics. After the outbreak of World War II production declined as, because of the rationing of fats, only inferior fats were available."[149] But after the beginning of the Four-Year Plan "interest focused on work carried out on textile additives and raw materials for detergents on a fat-free basis in order to become independent of foreign countries in the matter of fat." Hoechst developed different "Igepal" products as detergents for cotton, wool, and feathers, which, as Table 4.7 shows, were very successful. In 1939 a separate factory was set up for the production of Igepal products.[150]

Hoechst did not content itself with these successes but continued to carry out research and development in the field of textile additives and raw materials for detergents. With the development of synthetic fibers this proved to by a very rewarding field of work, not least because of the autarky policy.

[147] HA, 12, Hirschel, "Das Werk Hoechst im Verbande der I.G.," p. 134.
[148] HA, PA Lautenschläger, Lautenschläger, "Erinnerungen," vol. 3, [unpaginated].
[149] HA, 12, Hirschel, "Das Werk Hoechst im Verbande der I.G.," p. 170.
[150] Ibid., pp. 170–1.

In 1935 Professor Ludwig Orthner, who had headed the commission for textile additives since 1934, was transferred to Hoechst, and in 1937 he became head of the laboratory for textile additives in Hoechst.[151] In the years to come, he worked very successfully in this field together with researchers from Hoechst and other IG Farben plants. The hydrophobing of staple fibers, which were becoming increasingly important, was one of Hoechst's research areas. The aim was to prevent water absorption by staple fibers, particularly in blended fabrics containing wool. One important product was Persistol VC, which came out in 1940 and which was introduced to the market as a "wash-fast softening agent for natural and artificial cellulose fibers." During the war the main focus was on products requiring little or no grease, and in 1940 this resulted in the development of an emulsifier, which was used as an oil for drilling, cutting, and milling, as an anticorrosive in fuels, and as an antifreeze agent. It was a prime example of what chemistry could accomplish.[152]

The Pharmaceutical Department under Lautenschläger's management was apparently the most profitable area, and Lautenschläger refused to relinquish his position as head of the department even after he became works manager. The department had been negatively affected by the merger in 1925. Hoechst had been particularly hard hit by the merging of the sales divisions and their subsequent transfer to Leverkusen, which resulted in the department's becoming much smaller. However, Hoechst did receive some compensation for these losses. The gaps in the Pharmaceutical Department created by the merger were "partly compensated" by products from Cassella and Kalle. With the takeover of "some drugs discovered there and already well established, such as Trypaflavin, Panflavin, Tonophosphan, and Omnadin, our Pharmaceutical Department in Hoechst had a substantial increase, in terms of both production facilities and staff."[153]

In 1942 Lautenschläger retrospectively reviewed the successes of the Pharmaceutical Department since the introduction of the Four-Year Plan (Table 4.8): "In addition to work on therapeutic agents against parasitic diseases, drugs for the treatment of congolense infection, a sleeping sickness in animals in Africa, agents for chemotherapeutical internal antisepsis, for the treatment of flu and rheumatism, research was continued in the fields of vitamins, enzymes, and hormones. A vital hormone from the adrenal cortex, a lactation hormone from the pituitary gland, E and K vitamins, all found their way to the pharmaceuticals market." A cancer research team that included "important clinicians" was put together to study preparations

[151] Cf. ibid., pp. 171–3.
[152] Ibid., 172–3; HA, PA Abteilung VI, Orthner, application by Lautenschläger to increase Orthner's salary as a reward for his achievements, May 6, 1942.
[153] HA, PA Lautenschläger, Lautenschläger, Erinnerungen, vol. 2, [unpaginated]; 12, Hirschel, "Das Werk Hoechst im Verbande der I.G.," p. 230.

TABLE 4.8. *Turnover of the Hoechst Pharmaceutical Department 1925–1942 (in Million RM)*

Year	Turnover
1925	19.0
1935	38.9
1937	45.6
1939	52.9
1942	67.4

Source: HA, 12, Hirschel, "Das Werk Hoechst im Verbande der I.G.," p. 234.

developed in Hoechst together with staff in Hoechst's own research laboratories. Progress was also made in the field of immunology with new therapeutic antitoxins and vaccines and technical improvements to existing sera. Great progress was made in two particular areas. One of them was insulin, where systematic investigations into pancreatic hormones led to "the discovery of the genuine incretion of the islets of Langerhans in the pancreas," "a native insulin which represented an important breakthrough in the treatment of diabetes and which found great acceptance in international scientific circles. This islet hormone and the previously used crystalline insulin were used to develop deposit insulins with a much longer action duration than that of already available insulin preparations and which made it possible for severe diabetics to get by with one injection every 1–2 days instead of repeated daily injections as had previously been the case." And for the first time, Hoechst was successful in "developing an alkaloid with a similar effect as regards pain relief and antispasmolytic action as preparations from the opium family such as morphine and papaverin. The synthetic product, coming as it did at the start of the war, was just in time to make us independent of the opium market." Sales of this preparation, called Dolantin, rose rapidly and brought Hoechst "the highest recognition" on the part of the Reich government, particularly from the Inspectorate for Military Hygiene.[154]

The Pharmaceutical Department not only had a very good reputation, it had also grown quite considerably in the 1930s and, as already mentioned, was a very lucrative part of IG Farben's manufacturing range. Between 1932 and 1944 the pharmaceuticals and pesticides sector became IG Farben's "second most important profit earner."[155] And Hoechst played an

[154] HA, PA Lautenschläger, Lautenschläger, "Erinnerungen," vol. 3, [unpaginated]; Bartmann, *Tradition*, pp. 171–3.
[155] Plumpe, *I.G. Farbenindustrie*, p. 554.

important role, even if its share decreased during the Third Reich compared to Leverkusen/Elberfeld. Under Gerhard Domagk's leadership Elberfeld developed sulfonamides, highly efficient antibacterial chemotherapeutic agents and real "blockbusters" – to use a modern term; chemotherapy became Elberfeld's domain.[156] Indeed, the majority of the company's top profit earners in pharmaceuticals came from the Bayer laboratories in Elberfeld and Leverkusen and not from Hoechst at all: three out of four preparations in 1938 and four out of five in 1943. Hoechst's share of the entire pharmaceuticals business of IG Farben correspondingly decreased from 41.7 percent in 1935 to 34.7 percent by 1942.[157]

Nevertheless the position of the Hoechst Pharmaceutical Department was quite strong, and this was due not least to Lautenschläger. A number of new preparations came out, and in-house research was considerably expanded under his management.[158] As a trained pharmacist Lautenschläger was very interested in all biochemical procedures and not such an advocate of pure chemical synthesis as others in IG Farben were. Like Lautenschläger, Max Bockmühl, the new head of research, had also studied pharmacy and chemistry after first completing primary training as a pharmacist. There was therefore great interest in the work on insulin, vitamins, and, later, penicillin in Hoechst. Largely as a result of Lautenschläger's interests, the Pharmaceutical Department in Hoechst never focused merely on chemical synthesis but always intensively pursued biochemical work as well. There was never a complete path dependency on the chemical synthesis of drugs in the Pharmaceutical Department in Hoechst.[159]

In addition to the work in these fields, the Pharmaceutical Department also had a range of special projects as part of the Four-Year Plan "which had the goal of making us independent of fats, foreign drugs, spices and luxury foodstuffs such as caffeine, cinnamon, pepper, cocoa butter and the like." During work on the production of artificial pepper Hoechst began cooperating with Staudinger again, on whose work the new development was based.[160] Lautenschläger drew a very positive conclusion: "All these newly created products which even during the war quickly developed into important commercial preparations for domestic and foreign markets involved a considerable increase in our previous pharmaceuticals production by expansions to old building and the setting up of new ones."[161]

---

[156] Bartmann, *Tradition*, pp. 147, 168–71.

[157] Ibid., pp. 150–3, 175–7.   [158] Ibid., p. 125.

[159] Ibid., pp. 127, 181–5; cf. also Luitgard Marschall, *Im Schatten der chemischen Synthese. Industrielle Biotechnologie in Deutschland (1900–1970)* (Frankfurt am Main: Campus, 2000).

[160] HA, PA Lautenschläger, Lautenschläger, "Erinnerungen," vol. 3 [unpaginated]; cf. also Deutsches Museum, NL 88, Staudinger Archiv, D II 18.1–18.33, correspondence of Lautenschläger and Staudinger between 1933 and 1943 on Staudinger's development of artificial pepper, which became quite profitable for both parties.

[161] HA, PA Lautenschläger, Lautenschläger, "Erinnerungen," vol. 3, [unpaginated].

TABLE 4.9. *Turnover of the Hoechst Pesticides Department 1926–1942 (in Million RM)*

Year	Turnover
1926	1.2
1929	2.6
1935	1.2
1937	2.2
1939	3.1
1941	6.3
1942	7.7

Source: HA, 12, Hirschel, "Das Werk Hoechst im Verbande der I.G.," p. 242.

Pesticides had formed part of the Pharmaceutical Department since the beginning of the 1930s, and a number of new preparations were developed (Table 4.9). Foremost among them was "Nirosan," largely developed by Michael Erlenbach, which became the most important insecticide for viniculture in Germany. Previously arsenic had been much in use for viniculture in Germany, but Nirosan (1,3,6,8-tetranitrocarbazole) was nontoxic for humans and contained no arsenic. After it came on the market and was used against pests on vines and fruit trees, arsenic-bearing agents were finally forbidden.[162] Thus production in Hoechst, particularly in IG Farben's most lucrative sectors, was on the increase – and overall Hoechst seems to have been quite profitable. Above all, Hoechst expanded those sectors that looked particularly promising in a future peacetime economy.

But Hoechst also manufactured products for the war – and not just so-called dual use items such as synthetic materials or dyes for uniforms. In Hoechst some work was carried out on chemical weapons, and small amounts of them were probably produced – although apparently not of the desired quality. According to the historian Olaf Groehler, during the war chemists in the company's plants at Hoechst and Mainkur (Cassella) worked on "preparations for the mass production of the chemical weapon 'Exelsior,' a poisonous gas (*Blaukreuzgas*) with a considerably stronger action than those available up to now." Data from the Army Weapons Office in the summer 1942 indicate, in Groehler's words, that laboratory tests had been "concluded and a pilot plant with five tons per month began operations."[163] Whether this plant was in Hoechst or in Mainkur is not

---

[162] Bäumler, *Century*, pp. 314–15; Winnacker and Weingaertner (eds.), *Chemische Technologie*, vol. 4 (1954), pp. 980–3.

[163] Olaf Groehler, *Der lautlose Tod*, 5. überarb. Auflage (Berlin: Verlag der Nation, 1990), p. 215.

clear. However, confirmation that work on warfare agents was being carried out in Hoechst is also available from another source. A letter dated August 6, 1942, from the Army Weapons Office to Armaments Minister Albert Speer concerning the urgency of the chemical weapons project noted that work in Hoechst was being carried out on "intermediate products for a new C-weapon."[164] But overall, Hoechst neither led the way in the development of chemical weapons nor produced important research and developments.[165]

After the war, Lautenschläger and his defense gave very different accounts of Hoechst's importance for armaments production. When he was questioned in Nuremberg on March 20, 1947, about production in the plant during the war, Lautenschläger admitted on the record that Hoechst had produced chemical products, nitrogen, and nitric acid – some of which had been used for the hydrogenation of dyes, some for fertilizer, and "the rest was used for explosives," which were supplied to the army. When asked how large the percentage of production had been that he would describe "as pure armaments production," Lautenschläger said that this had varied. When the question was put to him more precisely, suggesting that he give an example for May 1942, Lautenschläger replied: "About 40%, this probably is somewhat overstated. Drugs were a big item. I cannot be sure what was for civilian use and what was for the war production and what was passed on from the civilian sector to the hospitals."[166] In a statement under oath Lautenschläger formulated the matter as follows: "Approximately 40% of the production during the war of the IG Farben plant Hoechst was purely armaments production and was in fact production of intermediate products for explosives, acid fog, dyes for uniforms and for camouflage."[167]

In his memoirs Lautenschläger included even more items under the category of armaments production. He stated that the annual value of production for 1940 had amounted to 150 million RM, and in the years 1940–4 50–60 percent of this was for the war, the rest for the civilian sector.[168] Not only IG Farben itself but Hoechst as well "had carried out a lot of war work in its workshops." Research not connected to the war "unfortunately had had to take very much a back seat." But as Lautenschläger used a very broad interpretation of which products were relevant for the war, this apparently only affected small areas: "In the inorganic workshops all

---

[164] Hans Günter Brauch and Rolf-Dieter Müller (eds.), *Chemische Kriegführung – Chemische Abrüstung. Dokumente und Kommentare* (Berlin: Berlin-Verlag-Spitz, 1985), pp. 176–7, document 46; here p. 284: reference to Groehler and the work on "Excelsior" in Hoechst.

[165] Cf. Robert Harris and Jeremy Paxman, *A Higher Form of Killing: The Secret History of Chemical and Biological Warfare* (London: Arrow Books, 2002), pp. 53–67; Hayes, *Industry and Ideology*, pp. 188, 333.

[166] BA, Film 44839, Lautenschläger's testimony to Cooper dated March 20, 1947.

[167] NI-6415, Affidavit by Lautenschläger dated March 26, 1947.

[168] HA, PA Lautenschläger, Lautenschläger, "Erinnerungen," vol. 8, [p. 177].

capacities were used to produce sulfuric acid and nitric acid for nitration products, for the production of explosives such as di- and trinitrobenzol and -toluene, as well as special explosives such as hexogen, likewise for the electrolysis of caustic soda solution for the textiles industry (military textiles) and of chlorine for chlorinated products and chemical weapons, furthermore for special carbon for electrodes and gasmasks." In the organic workshops "the dyes factories for the dyeing of military textiles, for example, indanthrene and helindone for blues, grays, and khaki, dyes for tarpaulins and red flags, topcoat enamels for automobile and airplane varnishes, photosensitizers for color films, and triphenylmethane and sulfur colors, particularly methyl violet and methylene blue for drugs in veterinary medicine, were running at full speed." The Drugs Departments in Hoechst and Marburg as well as in Neuhausen and Vienna, "which worked under the supervision of the Inspectorate for Military Hygiene," were also extremely busy with the manufacture of a range of products: "in addition to the most important drugs for war surgery, particularly the sera and vaccination workshops were operating at full capacity to continually produce vaccines against typhoid, dysentery, cholera, and bacteriophages and as well sera against tetanus, gas gangrene, and dysentery." Lautenschläger additionally noted: "new workshops for the production of special propellants to increase the engine performance of airplanes and allow quicker warming-up and better performance at high altitudes, and workshops for the production of illuminants for signaling on airfields and battlefields were set up at our facilities within a very short time. Large amounts of camouflage colors for buildings and runways and fire retardants had to be produced. The solvents workshops producing varnishes and films were working to full capacity. Synthetic materials as substitutes for wood and metals, and softeners and textile additives as well as substitute substances for soap were produced in ever greater amounts." The extensive expansion of production meant that the workshops for intermediate and auxiliary products had to be expanded together with the factory for construction materials.[169]

Lautenschläger's defense counsel in Nuremberg saw this quite differently. In the trial brief of his lawyer Hans Pribilla, Lautenschläger's statement "that approximately 40% of the production of the Hoechst plant had been purely armaments production" was referred to as a "false report on the production in Hoechst" and was withdrawn as being "objectively false."[170] This view was later perpetuated. According to a memorandum by the Personnel Department at the beginning of 1947, during the war the plant at Hoechst had catered to "predominantly extremely important civilian needs (curative drugs, pesticides, detergents, fertilizers, etc.)." This

[169] Ibid., vol. 3, [pp. 118–19].
[170] HA IG/54, IG-Prozessakten, trial brief for Lautenschläger from defense lawyer Hans Pribilla, p. 5.

could also be deduced by the fact that Hoechst "was not supervised by the controlling agencies of the Third Reich as an armament factory."[171] Winnacker made a similar statement in his memoirs: "Moreover, Hoechst had not taken part in certain types of particularly important war work."[172]

Lieutenant-Colonel Herbert Moulton of the U.S. Army, who wrote a report on the main plant in Hoechst on July 30, 1945, during the first months of the Allied occupation, agreed with this statement: "Besides reemphasizing the conclusions of the Production Control Agency that there was virtually no bombing damage, Moulton noted that the factory produced few materials directly related to the war effort."[173]

Indeed, Hoechst does not appear to have been a plant that was essential to the war effort, as far as this can be said of any chemical plant at the time. As Winnacker already noted: "Basically, all that was produced in Hoechst, and indeed throughout the I.G. Farbenindustrie, was necessary for the general economy both in peace and in war."[174] But a substantial part of production, such as blasting agents, explosives, or acid fog, went directly into the war machinery. Another part consisted of the production of so-called dual-use items, products such as medicines, synthetic materials, or dyes that had both civilian and military uses. And one part served the production of consumer goods, particularly many of the dyes and the textile additives. However, because of the attempts at self-sufficiency and the shortage of raw materials many of these products could also be numbered among those considered "essential to the war effort." During the war it was important that such products counted as "essential to the war effort" as this secured labor and raw materials that would otherwise have been diverted to other industries considered more important.

Thus it was vital to the corporate logic of a company wishing to continue to exist and produce that Hoechst received a letter from the Reich Office for Chemistry written at the end of April 1941 confirming that the "products detergents, textile additives, leather additives, crop protection and pesticides, dyes and intermediate products, pharmaceuticals" of Hoechst were "essential for the war."[175] In addition, strategically important research and development for the war was being carried out, for example, in the Process Engineering Department where Kiesskalt and Patat worked on camouflage paints and similar projects to prevent submarines from being detected by radar.[176]

---

[171] HA, file Pensel-Entnazifizierung, Note Frank dated March 10, 1947, on "Besprechung mit Herrn Krepp, Spruchkammer Frankfurt am 7.3.1947."

[172] Winnacker, *Challenging Years*, p. 114.

[173] According to Stokes, *Divide and Prosper*, p. 52.

[174] Winnacker, *Challenging Years*, p. 96.

[175] HA, Pol 15, Reichsstelle Chemie, Bestätigung für I.G Farbenindustrie, Werk Hoechst, April 26, 1941 (copy).

[176] Winnacker, *Challenging Years*, p. 97.

Yet just how unimportant Hoechst was for the war production in IG Farben during the Second World War is clear from the letter, already referred to above, drafted by Karl Winnacker for Lautenschläger to send to Krauch at the end of October 1943, for it complained in detail about Hoechst's current position. After the reorganization of the chemical industry "with regard to its strategic importance for the war" Hoechst had been "removed from the ranks of companies supervised by the *Gebechem*" and was now supervised by the Reich Ministry of Economics: "With this any claims to preferential treatment from our local supervisory offices such as armaments and *Gebechem* companies are entitled to receive are denied us. This circumstance has a particularly adverse effect on the allocation of labor." As Hoechst did not benefit "from the protection accorded *Gebechem* companies, either with regard to conscription or with regard to the allocation of labor," the number of its employees continuously fell, so that in view of the "disastrous labor situation" it was no longer possible to give preference to individual areas of production, lest other areas "which were likewise of great importance for the army would immediately grind to a halt."[177] It is not clear whether this letter from Lautenschläger to Krauch was ever sent or whether an attempt was made to solve the problem during face-to-face talks. At all events, the draft makes very clear that Hoechst was not one of IG Farben's foremost plants.

The operations working to full capacity were only one aspect of the plant; there were also units that had been shut down or were working at reduced capacity, primarily because of the shortage of workers and raw material. Lautenschläger wrote on this subject: "At the time it was often difficult for the chemical industry, which was primarily geared to exports, to obtain material for substitute production quickly enough, and a lot of worry and efforts went into finding the right way to carry out our work of construction and to maintain the workshops, laboratories, and offices in their entirety because of the lack of workers, coal, and raw materials." After 1943 particular difficulties for industry arose due to the austerity measures and the lack of important raw materials that prevented even important repairs from being carried out: "This led to the workshops gradually becoming dilapidated and to a corresponding reduction in production."[178]

Production in all IG Farben plants became increasingly difficult once the Allies had won air supremacy in Germany in 1943, due to the many interruptions of production by air raids. According to Lautenschläger, by the end of 1944 production in IG Farben plants "had decreased on average

[177] HA, PA Lautenschläger, Sekretariat Lautenschläger, Vertrauliche Korrespondenz, draft by Winnacker for Lautenschläger dated October 25, 1943, on letter from Krauch dated October 7, 1943.

[178] HA, PA Lautenschläger, Lautenschläger, "Erinnerungen," vol. 3, [p. 122].

to 50% of the output in 1943." This apparently also applied to Hoechst: "Our Hoechst plant recorded many air-raid warnings until the end of the war; however, it suffered only limited damage."[179] Little changed thereafter, and at the beginning of January 1945 the plant was largely closed down and production was discontinued because of the lack of raw materials.

In conclusion it can be stated that production in Hoechst did not focus on "large-scale chemistry," but it did include some of IG Farben's most profitable lines of business: pharmaceuticals, textile auxiliaries, and dyes – most of which did well. For the Pharmaceutical Department it was important to note that it did not exclusively center on chemical synthesis but was also very open to biochemical processes as demonstrated by its insulin production. It is therefore hardly surprising that after the war the first large-scale production of penicillin started up in Hoechst – after all, Hoechst had already been working in that area during the war.[180]

It is impossible to discover a clear policy in Hoechst with respect to its product range because heterogeneity of production had been Hermann's program. Hermann's openness to almost all aspects and sectors of the chemical industry encouraged an impressive capacity for innovation that is most obvious in the Department for Process Engineering. Hoechst was not a preeminent plant, but its foundations were sound, and it included important innovations and profitable areas, and it was constantly attempting to obtain new products or product lines without giving up any existing ones.

4.4. DRUGS AND EXPERIMENTS ON HUMAN SUBJECTS – THE PHARMACEUTICAL DEPARTMENT DURING THE WAR

When Carl Ludwig Lautenschläger took over the management of the Pharmaceutical Research Department at the beginning of the 1920s, it consisted of only four chemists, Max Bockmühl and "three younger gentlemen."[181] After Lautenschläger was appointed manager of the entire Pharmaceutical Department, Bockmühl became head of pharmaceutical research and was thus responsible for all synthetic and biochemical work in the drug laboratories. Bockmühl celebrated his sixtieth birthday in the summer of 1942. To mark the occasion, an article on his life and work was published in the *Frankfurter General-Anzeiger* under the heading "Battling to Save Lives." The article listed the drugs to which Bockmühl had contributed or that had been developed under his aegis, particularly the

---

[179] Ibid., vol. 8, [pp. 14–15].

[180] Cf. Ingrid Pieroth, *Penicillinherstellung. Von den Anfängen bis zur Grossproduktion* (Stuttgart: Wissenschaftliche Verlagsgesellschaft, 1992), pp. 82–6, 126–7; Bartmann, *Tradition*, pp. 183–5, 232–3, 241–3.

[181] HA, PA Lautenschläger, Lautenschläger, "Erinnerungen," vol. 2, [unpaginated].

anti-rheumatic and analgesic preparations Novalgin and Gardan. His achievements were summed up by the words "Millions of people have subsequently been delivered from terrible suffering and excruciating pain by research carried out in the Hoechst works."[182]

Indeed, before and during the war, the Pharmaceutical Department in Hoechst carried out research on beneficial drugs. Dolantin, a highly effective pain killer, was not just important for the war, although it did benefit the war effort. Researchers in Hoechst also worked on penicillin and Amidon (methadone); both would become very important after the war.

At the time when the celebratory article was published, however, the Pharmaceutical Department had already sought out and made contact with the SS. Instead of solely benefiting mankind and working toward the preservation of human life, the department also colluded in monstrous crimes, in experiments and tests carried out on inmates in different concentration camps. Primarily on charges of collaborating in the "medical experiments" carried out in concentration camps, Lautenschläger and his colleagues Wilhelm Mann and Heinrich Hörlein from the Management Board were tried in Nuremberg during the IG Farben trial. All three men stood accused of providing the SS with pharmaceutical preparations and vaccines "for the purpose of having them tested, knowing that the tests would be conducted by medical experimentations upon concentration-camp inmates without their consent." The prosecution claimed "that each of said defendants took the initiative in getting [IG] Farben products tested by the SS through the means of criminal medical experiments," leading to the death or grievous permanent injury of a number of persons.[183]

The accusations focused mainly on the development and testing of drugs for the prophylaxis and treatment of typhus – also called spotted fever – during the Second World War. The Behring Works in Marburg and the Pharmaceutical Departments of Hoechst and of Elberfeld/ Leverkusen had been working on vaccines and drugs, some of which had not yet been tested and were not yet officially approved for medical use. According to the charges, various people in IG Farben had pushed for tests and clinical trials to be carried out so as to be able to market the products as soon as possible. It was claimed that "the circumstances surrounding the testing of [IG] Farben's vaccine, as well as with respect to its acridine, rutenol, and methylene blue, in combating typhus discloses that the defendants Hoerlein, Lautenschlaeger, and Mann, in particular, well knew that concentration-camp inmates were being criminally infected

---

[182] HA, PA Abteilung VI, Bockmühl, article "Im Kampf um die Erhaltung des Lebens. Max Bockmühl-Hoechst feiert seinen 60. Geburtstag" from *Frankfurter General-Anzeiger*, August 29/30, 1942.

[183] *Trials of War Criminals before the Nuernberg Military Tribunals*, vol. VIII, p. 1170.

with the typhus virus by SS doctors for the deliberate purpose of conducting experiments with these Farben products." The charges established that "criminal experiments" had been carried out by SS doctors on concentration camp inmates to test various preparations of IG Farben, and that the type of experiments carried out could be deduced from the reports of the doctors.[184]

Neither Lautenschläger's connections to the Behring Works nor the role of Elberfeld/Leverkusen will be examined here, only that of Hoechst and its preparations. We will be looking only at tests and experiments carried out using the preparations "nitroacridine compound 3582" – also referred to as "acridine" or preparation 3582 for short – and "rutenol." Both preparations were administered to camp inmates from the Buchenwald concentration camp who had previously been deliberately infected with typhus under the direction of SS doctor Erwin Ding,[185] and were also tested on inmates from the concentration camps Auschwitz and Gusen/Mauthausen under the direction of Hellmuth Vetter,[186] an SS doctor and former employee of IG Farben in Leverkusen.

Before looking at the experiments carried out in the different concentration camps it is necessary to clarify who the key players in Hoechst were in this chapter of the plant's history and who bore responsibility. In the "trial brief" written for Lautenschläger, his lawyer Hans Pribilla named Julius Weber as the person responsible. Pribilla wrote that part of Weber's duties was to arrange clinical trials of medicinal substances developed in the research laboratories, in accordance with the guidelines given him by Lautenschläger. As works manager Lautenschläger had the obligation to exercise due care in his choice of subordinates, and Weber had "proven himself to be a reliable and irreproachable employee since 1927," whom Lautenschläger could fully trust: "If therefore something actionable occurred during the testing of the Hoechst drugs against typhus for which Hoechst must be held responsible, then the person responsible under criminal law was not Dr. Lautenschläger but Dr. Weber." But according to the evidence, "both Prof. Lautenschläger and Dr. Weber proceeded correctly in their choice and supervision of the investigators carrying out the

---

[184] Ibid., pp. 1170–2. The German text of the verdict correctly reads "Flecktyphus-Bazillus" (typhus bacillus) instead of the erroneous "typhus virus": *Das Urteil im I.G.-Farben-Prozess. Der vollständige Wortlaut mit Dokumentenanhang* (Offenbach: Bollwerk, 1948), pp. 110–12.

[185] Dr. Erwin Ding was born as the illegitimate son of a doctor named Schuler and was later adopted by Ding. He applied for permission to use the name Schuler, and his application was granted by Himmler. In the literature he is referred to either as Ding or as Ding-Schuler.

[186] According to Weindling, *Epidemics and Genocide*, p. 359, experiments on "deliberately infected patients" were also carried out in the concentration camps of Auschwitz and Gusen in addition to Buchenwald.

tests." Furthermore, in clinical trials the investigating doctor was "solely responsible," and he was provided with the Hoechst guidelines governing such trials. From this it followed "that Hoechst cannot be held responsible for crimes which were committed by the persons investigating its nitroacridine preparations."[187]

Julius Weber, an employee of Hoechst, reported directly to Lautenschläger. Born in 1896, Weber served four years in the army during the First World War, by the end of which he held the rank of lieutenant. After the war Weber studied medicine and chemistry in Göttingen, Gießen, Freiburg, and Frankfurt. In 1923 he was licensed to practice medicine. He subsequently completed a doctorate in medicine in 1927 in Frankfurt under the direction of Professor Gustav Embden, whom he served as an assistant, and a doctorate in chemistry in 1928 in Gießen. After brief hesitation he accepted a position in Hoechst offered by works manager Paul Duden in 1927. Weber was already given commercial power of attorney in 1928, and in 1935, when he was head of the Pharmaceutical Research Department, he became an authorized signatory. His duties included setting up initial clinical series with doctors and clinics to test preparations developed in Hoechst, monitoring the trials and finally compiling the results of the tests for broader clinical trials and a later market launch of the tested preparations – all of which corresponds to the description of his duties given by Pribilla. The extent of his responsibilities and the degree of independence allowed him in the discharge of his duties became clear at the beginning of 1938 when Hoechst arranged for him to be exempt from military duties, as in his position as head of the Pharmaceutical Research Department he was expected to "deputize for Lautenschläger and relieve him of many duties in his work."[188]

In addition to Julius Weber and Lautenschläger, who had maintained his influence on the Pharmaceutical Department even after becoming works manager, Max Bockmühl, the head of the entire pharmaceutical research, also played an important role in clinical trials, and the three men were jointly responsible for deciding which substances should be released for clinical testing. Born in 1882, Bockmühl first trained as a pharmacist. After passing his state examinations, he studied chemistry in Munich under Professor Adolf von Baeyer and completed his doctorate under Professor Alfred Einhorn, the inventor of Novocain. Bockmühl joined Hoechst in 1910, where he became part of the group that founded the Pharmaceutical

[187] HA, IG/54, IG-Prozessakten, trial brief for Lautenschläger from defense lawyer Hans Pribilla, pp. 59–60.
[188] HA, PA Abteilung VI, Julius Weber, personal data sheet; Weber to Ammelburg, January 12, 1927; Weber to Duden, January 12, 1927; references for Embden dated December 19, 1927, and for Rosemann dated October 7, 1926 (copies); application for his appointment to become authorized signatory dated March 11, 1935; Hoechst to *Wehrbezirkskommando* Frankfurt a.M., January 29, 1938.

Laboratory. He was given commercial power of attorney in 1927; in 1930 he became head of all pharmaceutical research. Because of this position, he became a deputy director in 1937 and a full director in 1938.[189] Bockmühl's co-worker Rudolf Fußgänger, born in 1901, also played a not insignificant role in the clinical testing of the two above-mentioned preparations. Fußgänger had studied in Freiburg im Breisgau, Jena, Tübingen, and Frankfurt, where he completed his doctorate in chemistry. In 1929 he completed a second doctorate, this time in medicine, in Tübingen. Apparently at the suggestion of Lautenschläger, who was looking for another pharmacologist for Hoechst, Fußgänger completed his training as a doctor and was licensed to practice medicine in 1931 despite having already joined Hoechst in 1930, where he began work in the Pharmacological Laboratory. In 1938 Fußgänger moved to the Chemotherapeutical Laboratory and took over its management after Robert Julius Schnitzer had been dismissed because he was Jewish.[190]

As a defense witness at the Nuremberg trial, Weber stated that clinical testing had been governed by the medical precept *nil nocere*, never to do harm: "Only after thorough testing in the laboratory, in which all available biological methods of examination were employed, was a newly developed remedy allowed to be used on patients."[191] Such conduct corresponded to the "Final Draft of the Guidelines for New Therapies and for the Performing of Scientific Experiments on Human Subjects" of the Reich Health Council, published in a circular of the Reich Interior Ministry at the end of February 1931. This expressly stated that scientific experiments on human subjects were indispensable "as otherwise progress in knowledge, healing and the prevention of diseases would be impeded or even excluded." But the guidelines also emphasized the doctor's responsibility to the persons subjected to new methods of treatment or other scientific tests. In both cases the rule applied that the persons or their legal representatives "must unequivocally agree to the carrying out of the test on the basis of adequate information given previously." For scientific tests the guidelines even stated: "Every experiment on human subjects must be rejected if it can be replaced by animal experiments. An experiment on a human subject may only be carried out if previously all information has been obtained which can be gained for clarification and protection by means of the biological methods available to medical science such as laboratory tests and animal experiments. Under these conditions any unnecessary and haphazard

---

[189] HA, PA Abteilung VI, Bockmühl, application to appoint him director dated May 22, 1936, by Hermann; detailed curriculum vitae (no author or date).

[190] HA, PA Abteilung VI, Fußgänger, Personalblatt; Fußgänger to Bockmühl, June 30, 1943 (copy).

[191] HA IG/54, IG-Prozessakten, Lautenschläger document 58, affidavit by Dr. Julius Weber, dated March 20, 1948.

experimentation on human subjects is out of the question."[192] This statement must be qualified by noting that the guidelines of the Reich Health Council were apparently not very well known.[193]

A report written in 1936 by Bolton, an employee from the DuPont company, also appears to confirm Weber's statement. Bolton visited several laboratories in the Pharmaceutical Department and noted: "Many of these experiments were set up specially for the purpose of demonstrating to us the methods that have been developed in this laboratory for use in connection with their study of the effects of chemical compounds." For, according to Bolton: "Prior to clinical examination upon human beings a very great deal of vivisection is carried on, and for the experiments mice, rats, cats, rabbits, dogs, salamanders and fish are used. Only at the end of an investigation are monkeys used because of the cost and difficulty of getting healthy ones."[194]

Thus any new active agent was only to be tested clinically, that is, on human subjects, after it had first been tested extensively in animal experiments. IG Farben did not just propagate this attitude to its American visitor but also to its own employees and other readers of a self-published 1938 brochure "Products of Our Work." In a short summation of the development of pain-relieving drugs as well as of poisons based on the analysis of opium, the brochure stated: "Approximately two dozen such alkaloids have been obtained to date from opium alone, for example, the soothing cough medicine codeine and the convulsant agent Thebain. But how was it that the dangerous properties of the latter were recognized? For people will not allow themselves to be used for such tests! Here we made use of a very young science, 'pharmacology,' which experimentally studies the effect of individual drugs (pharmaca) using principally frogs, mice, and rats."[195]

This public statement given in 1938 and Bolton's report from 1936 are mentioned here since they show that already before the Second World War the company was considering how best to test the effects, side-effects, and tolerance of new drugs before using them on humans. When the company

[192] Circular from Reichsminister des Inneren dated February 28, 1931: "Endgültiger Entwurf von Richtlinien für neuartige Heilbehandlung und für die Vornahme wissenschaftlicher Versuche am Menschen" of the Reichsgesundheitsrat, excerpts printed in Alexander Mitscherlich and Fred Mielke (eds.), *Medizin ohne Menschlichkeit. Dokumente des Nürnberger Ärzteprozesses* (Frankfurt a.M.: Fischer, 1960), pp. 269–71. Cf. Rolf Winau, "Der Menschenversuch in der Medizin," in Angelika Ebbinghaus and Klaus Dörner (eds.), *Vernichten und Heilen. Der Nürnberger Ärzteprozess und seine Folgen* (Berlin: Aufbau, 2001), pp. 93–109; Ludger Weß, "Menschenversuche und Seuchenpolitik – Zwei unbekannte Kapitel aus der Geschichte der deutschen Tropenmedizin," *1999. Zeitschrift für Sozialgeschichte des 20. und 21. Jahrhunderts* 8 (1993), issue 2, pp. 10–50.
[193] Cf. Winau, "Der Menschenversuch," pp. 108–9; Wess, "Menschenversuche," p. 29.
[194] Hagley Museum and Library, DuPont Company Records, Jasper Crane Papers, Series II, Part 2, Box 1038, Bolton to Crane, June 11, 1936.
[195] *I.G. Farbenindustrie Aktiengesellschaft, Erzeugnisse unserer Arbeit* (Frankfurt a.M.: I.G. Farbenindustrie A.G., 1938), pp. 108–9.

emphasized that it wished to test new active agents clinically only after having tested them thoroughly and extensively on animals, it indicated a certain respect for human beings.

The reality was different. IG Farben, and thus also Hoechst, did not use only volunteers for clinical tests. The journalist and historian Ernst Klee writes that already in the 1920s tests using pharmaceutical substances from Hoechst were carried out on patients in the medical institution Eichberg at Kiedrich in the Rheingau (near Wiesbaden), and these clearly were done without the consent of either the patients or their guardians.[196] In an article the historian Ludger Weß draws attention to antimalaria tests carried out with the company's drugs in German medical institutions in the interwar years.[197] That clinical testing on human subjects without their consent by IG Farben appears to have been routine, rather than the exception, is borne out by the testimony of a witness to the defense – not a witness for the prosecution – in the IG Farben trial in Nuremberg. Aloys Auer, senior consultant and director of the Municipal Hospital Frankfurt-Höchst and a friend of Lautenschläger, who had "tested around 100 new preparations of Hoechst during their more than 20 years' collaboration," stated: "The doctor is solely responsible for clinical tests; with a conscientious consideration of the indication he selects the patients to whom he wishes to administer the new therapeutic drug. It is generally not considered common practice to obtain the patient's express consent – irrespective of which treatment class the patient is in – because information concerning the preparation being tested will often interfere with an objective assessment of its effect by the doctor." Auer did, however, add: "On the other hand I have repeatedly drawn patients' attention to the use of new drugs, without ever having encountered opposition from them."[198] Nonetheless, even according to a statement by a witness for the defense in the Nuremberg trial, "normal" clinical testing on human subjects was carried out without the patients' knowledge or express consent. That this statement, which throws a telling light on Hoechst's practice of clinical testing, has remained unnoticed until now is probably due to the fact that it was an unlikely statement to be found as part of the defense.

So what was the concrete procedure required for preparations of IG Farben to pass from the laboratory into clinical testing? According to director Anton Mertens, the head of the "Bayer" Pharmaceutical Research Department in the IG Farben plant at Leverkusen, preparations came from laboratories in Leverkusen, Elberfeld, or Hoechst. Hörlein was informed about new preparations from Elberfeld and Leverkusen, and Lautenschläger about similar

[196] Klee, *Auschwitz*, pp. 301–4.
[197] See Wess, "Menschenversuche," pp. 15–16.
[198] HA IG/54, IG-Prozessakten, Lautenschläger document 51, affidavit by Aloys Auer dated March 15, 1948.

new preparations from Hoechst by the respective heads of the laboratories during internal meetings. In Hoechst, where Julius Weber also participated in the internal meetings in addition to the heads of the laboratories, Lautenschläger decided whether a preparation would be approved for "clinical therapeutic application" after the laboratory stage. Once this had been decided, the head of the laboratory wrote a synopsis that included the possible indications and information about the preparation's effects and possible areas of application. The synopses from both Elberfeld/Leverkusen and Hoechst were then sent on to Mertens in Leverkusen with instructions that the new preparation should "be tested in this or that medical field."[199]

Weber also indicated with respect to these procedures that "experiments on oneself" almost always preceded the first clinical tests, that is, researchers tested the active agents on themselves first.[200] But experiments were admittedly also being carried out on test persons who had not volunteered, as Mertens stated in the Nuremberg trial: "If the internal tests in human beings in Elberfeld or Hoechst did not lead to satisfactory results, tolerance tests of IG drugs were also carried out in mental institutions and lunatic asylums."[201] It was not merely the therapeutic effect that was in question. General tolerance to the preparation was of prime importance during the testing stage, and when "self-experiments" to test tolerance were not sufficiently convincing, then tests were performed on people outside the works – and they were carried out prior to beginning official clinical testing. Weber described the practice as follows: "According to the guidelines of the Bayer organization I was allowed to carry out the first preliminary clinical tests of newly developed drugs independently; this usually took place in hospitals in the immediate vicinity of Frankfurt. After these so-called tentative tests [*Tastprüfung*] I was not permitted to initiate any more testing of drugs without informing the head office in Leverkusen and involving the local Bayer offices."[202] Mertens, for his part, testified that he was unable to say anything about tests in psychiatric wards as these had already been underway in 1935, the year in which he began working there, and had been treated as "very confidential." They had been the "business of the laboratories" that were subordinate to Hörlein and Lautenschläger. While in Elberfeld the heads of the laboratories had contacted the clinics themselves, in Hoechst this was Weber's duty.[203]

Contradicting Weber's statement that after the first "tentative tests" nothing was undertaken without informing Leverkusen, Mertens stated that

[199] NI-13823, affidavit by Anton Mertens dated September 30, 1947.
[200] HA IG/54, IG-Prozessakten, Lautenschläger document 58, affidavit by Julius Weber dated March 20, 1948.
[201] NI-13823, affidavit by Anton Mertens dated September 30, 1947.
[202] HA IG/54, IG-Prozessakten, Lautenschläger document 58, affidavit by Julius Weber dated March 20, 1948.
[203] NI-13823, affidavit by Anton Mertens dated September 30, 1947.

Hoechst had not always adhered to this principle. Although in Elberfeld "as a rule everything found there" was passed on to Mertens, this did not apply to Hoechst – because of the "perennial efforts of IG Hoechst to be independent." "In IG Hoechst it could happen that – independently of the Leverkusen tests – Hoechst let its preparations be tested by its own test centers, which happened with preparation 3582."[204] In contrast to his own statement quoted above, Weber admitted – albeit in a somewhat roundabout fashion – that this had been the case with preparation 3582: "As in all other cases, in the matter of the clinical testing of our preparations against epidemic typhus I proceeded by contacting all clinics or centers which appeared suitable to me, either by letter, through our Bayer offices, or personally, informing them of our therapeutic drugs and urging them to request the drugs if necessary."[205] In this way Weber made contact with Joachim Mrugowsky from the SS-Hygiene Institute, a connection that was to pave the way for experiments on inmates of the Buchenwald concentration camp using preparations from Hoechst.

After the war Mertens claimed that he had not been informed about the tests of preparation 3582 and rutenol by SS doctor Ding in Buchenwald. He had learned about them only after the war from Eugen Kogon's book. The cooperation between Hoechst and Ding had "very much surprised" him as in his view it "was not in any way sanctioned." "The silence on the part of IG Hoechst in this matter is quite noticeable, particularly since this was not a preparation which interested only Hoechst, but Leverkusen similarly had an interest in it." Lautenschläger, "from whom the instructions must have come to maintain silence about the collaboration between Hoechst and Dr. Ding," must – Mertens believed – have been "conscious of the possibility of being involved in criminal actions" and apparently "considered keeping the matter to himself as the only proper thing" to do. Mertens added that he did not know whether Lautenschläger had informed Hörlein about the cooperation with Dr. Ding.[206] This statement by Mertens was almost certainly false. On January 8, 1943, Leverkusen wrote Hoechst to enquire whether it was supplying Mrugowsky with the preparation, as had been inferred from a letter to the Bayer office in Berlin. Hoechst replied by letter and emphasized that the cooperation with Mrugowsky already existed for some time; the letter went on to mention that Mrugowsky had previously received smaller quantities of the preparation from the Behring Works from Professor Richard Bieling, and after September 1942 the supplies had come from Hoechst.[207] As the Bayer office was also fully

[204] Ibid.
[205] HA, IG/54, IG-Prozessakten, Lautenschläger document 58, affidavit by Julius Weber dated March 20, 1948.
[206] NI-13823, affidavit by Anton Mertens dated September 30, 1947.
[207] HA, file Fleckfieberversuche, Bayer, Leverkusen (Mertens, Koenig) to Hoechst, January 8, 1943; NI-11415, response Hoechst (Bockmühl and Weber) to Bayer, Leverkusen, January 13, 1943.

acquainted with the connections to Ding – the office even having organized some of the contacts itself – and as Ding's name crops up in a letter to Mertens, it is almost impossible to draw any other conclusion than that Mertens must have been aware of Ding's experiments.[208] But it is impossible to state whether Mertens knew more or even wanted to know more.

The only thing that is certain is that Hoechst took steps that exceeded the authority allowed to the plant and that it employed every means at its disposal to have preparation 3582 tested – and was even prepared to circumvent the normal channels and bypass Leverkusen. This is a good illustration of IG Farben's polycratic nature, whereby individual works competed against one another and were more likely to discuss and initiate some form of cooperation with important Nazi figures like Mrugowsky than with the company's own decision makers.

According to Weber's postwar statement, the two preparations to be tested, 3582 ("acridine") and rutenol, "had already proved their medical worth under other names." Preparation 3582 had been clinically tested at an earlier date under the name "3043 B"; moreover it was commercially available in the drug "Entozon." Rutenol had been tested under the name "Balkanol" and was regarded as a "highly effective remedy."[209] Fußgänger also emphasized that the two preparations had been clinically tested in the years between 1930 and 1935 and that both had been "recognized as very remarkable chemotherapeutic agents particularly against various streptococcal septic processes."[210] What neither Weber nor Bockmühl said was that the preparations had never got beyond the initial testing stages. With the discovery of sulfonamides by Gerhard Domagk in IG Farben's Elberfeld laboratories considerably more effective remedies against bacterial infections were on the market than these two acridine preparations.[211]

This did not mean that Hoechst gave up its search for an indication against which the two preparations might, under certain circumstances, be effective. At the beginning of September 1938 Hoechst applied to Leverkusen through Weber for permission to start clinical tests into the potential use of preparation 3582 as a drug against Lamblia in the small intestine. The preparation had previously been tested, primarily in mice but also in other animals, with some success, and according to Hoechst's application it had been well tolerated. Furthermore, in vitro tests had

[208] HA, file Fleckfieberversuche, Memo from Leverkusen re. "Besprechung von Direktor Mertens in Hoechst," May 31, 1943.
[209] HA, I.G./54, IG-Prozessakten, Lautenschläger document 58, affidavit by Julius Weber dated March 20, 1948.
[210] Ibid., Lautenschläger document 59, affidavit by Rudolf Fußgänger dated February 27, 1948.
[211] See Ekkehard Grundmann, *Gerhard Domagk. Der erste Sieger über die Infektionskrankheiten* (Münster: Lit, 2001), pp. 41–65; and Lesch, *Chemistry*.

shown that the preparation was effective against bacterial infections.[212] Leverkusen temporarily rejected the application, primarily because IG Farben already had two unbeatable drugs against lambliasis, Atebrin and Sostol, both of which were also used to treat other things, namely, malaria and as a vermifuge, respectively.[213] But Hoechst had no intention of taking no for an answer, and Lautenschläger personally lobbied for preparation 3582 to be tested, describing it as considerably more effective against Lamblia than Sostol.[214] Leverkusen thereupon requested an opinion from the head of its own Chemotherapeutic Laboratory, Professor Walter Kikuth. Kikuth emphasized the difference between mice and people and between the Lamblia in their respective small intestines. While the Lamblia existing in the human small intestine could be effectively combated in a very short time using Sostol and Atebrin in low and harmless doses, the Lamblia found in mice were more tenacious, and these two drugs were less effective there. Systematic chemotherapeutic investigations would therefore "probably within a very short time" find preparations that were superior in mice. Kikuth added: "As we are convinced of the clinical efficacy of our agents in humans, we did not continue our chemotherapeutic work, for if one of our preparations shows a 100 percent success rate and is further distinguished by being extremely well tolerated, and the particulars of its toxicology are fully known, then at present we consider further chemo-therapeutic work in this area to be pointless." After receiving the report Mertens in Leverkusen decided that no clinical testing should be carried out with preparation 3582 and asked Lautenschläger to agree.[215]

The preparation remained in the laboratory at Hoechst, to the discontent of Lautenschläger and his employees, until Germany attacked the Soviet Union in 1941. New dangers to German soldiers meant new opportunities for the pharmaceutical industry, and one of the most potent dangers was epidemic typhus, against which there was as yet no effective therapeutic remedy at the time. At Lautenschläger's behest, Fußgänger began looking for an effective chemotherapeutic agent against typhus. He infected mice and tested a large number of active substances in accordance with the systematic methods used in IG Farben.[216] Fußgänger found that "the che-motherapeutic agents of the acridine series preparation 3582 and its arsenic acid salt Balkanol are the optimal means up to now of influencing typhus

---

[212] HA, file Fleckfieberversuche, Hoechst (Weber, Fehrle) to Leverkusen, September 2, 1938.
[213] Ibid., Leverkusen (Mertens, Koenig) to Hoechst, September 15, 1938.
[214] Ibid., Hoechst (Lautenschläger, Weber) to Leverkusen, September 21, 1938.
[215] Ibid., Mertens (Leverkusen) to Lautenschläger (Hoechst), November 3, 1938, where Kikuth's statement is quoted.
[216] NI-9811, affidavit by Carl Ludwig Lautenschläger dated May 2, 1947; HA, IG/54, IG-Prozessakten, Lautenschläger document 59, affidavit by Rudolf Fußgänger dated February 27, 1948.

infection in mice." After these animal experiments, clinical testing of the preparation was begun in Frankfurt.[217] Thus in summer 1942, once again Professor Lehmann-Facius from the Frankfurt Clinic for Emotionally Disturbed Persons was encouraged to test patients' tolerance to the drug – he had allegedly already "successfully" tested the preparation against Lamblia in 1938. The results of these experiments on patients from the clinic, which were carried out in 1942 and almost certainly without the patients' consent, were apparently also positive.[218]

While these tests could be termed "tentative tests," to use Weber's expression, Weber was by no means satisfied – nor had he been so earlier. Weber began to take all possible steps to start proper clinical testing.[219] In August 1942 Hoechst was already writing to a certain Professor Otto that, in cooperation with Frankfurt, plans were now being made to use preparation 3582 "in clinical tests carried out on persons suffering from epidemic typhus, for which there is currently an opportunity in the East."[220] One of Weber's activities was to approach Mrugowsky. Weber himself did not adhere to the prescribed path of organizing clinical tests through Leverkusen, which he had described as compulsory. As Mertens quite rightly complained, he took the initiative himself.

Hoechst was already in communication with Joachim Mrugowsky, an army doctor and head of the Hygiene Institute of the Waffen-SS since the summer of 1941.[221] On September 10, 1942, Weber and a representative of the Behring Works held a discussion with Mrugowsky on vaccines and the "typhus fever preparation 3582" – for that was how it was already being described. Mrugowsky was sent a review of the preparation and agreed "to carry out tests with 3582 in suitable cases very soon." For his part, Weber promised to supply the preparation and noted "Dr. M. expressly declared himself to be very interested in the tests and has promised to expedite them to the best of his ability. In our presence he called in his deputy, a senior army physician, and gave him the necessary instructions concerning the testing of the preparations and instructed us in his absence or his being

[217] HA, IG/54, IG-Prozessakten, Lautenschläger document 59, affidavit by Rudolf Fußgänger dated February 27, 1948.
[218] Ibid., Lautenschläger document 58, affidavit by Julius Weber dated March 20, 1948.
[219] Ibid.
[220] NI-14711, Hoechst (Lautenschläger, Weber) to Geheimrat Prof. Dr. Otto, Forschungsinstitut für Chemotherapie Frankfurt, August 25, 1942.
[221] See NI-11415, Hoechst (Bockmühl, Weber) to Bayer, Leverkusen, January 13, 1943; NI-13581, Mertens and von Engelhardt to Hoechst, May 15, 1941; NI-14708, Hoechst to Behringwerke Gruppe Leverkusen, March 2, 1942 on deliveries to Mrugowsky on August 4, 1941; NI-14709, Hoechst to Behringwerke Gruppe Leverkusen, March 12, 1942; NI-14710, IG Pharma-Büro Berlin, Abteilung Behringwerke to Hoechst, July 20, 1942; for a detailed account on Mrugowsky and his role in experiments on human subjects and in the Holocaust see Weindling, *Epidemics and Genocide*, pp. 246–59.

otherwise occupied to get in touch directly with the deputy and the official in charge. (Sturmbannführer Dr. Murthum)."²²²

While Weber was seeking on his own initiative to establish links to Mrugowsky for the potential clinical testing of preparation 3582, Leverkusen changed its mind and no longer objected to clinical tests being carried out. Mertens wrote to Weber at the beginning of November 1942 that Hoechst had sent "substantial amounts" of preparation 3582 to the Bayer office in Berlin "for testing against Volhynian fever and perhaps also against typhus." As Leverkusen was very much of the opinion that there would be a demand for remedies against typhus in the course of the winter, it wished to know "how far preparations for the clinical testing of the substance had progressed."²²³ Hoechst immediately answered that there had been "no problems with compatibility in any tests to date" when preparation 3582 had been administered in tablet form, and therefore "no reservations" were felt about a "release of the preparation for more general testing." However, Hoechst added that, as a precaution, the preparation could also be administered in the form of granules. But it should only be given in this form "if, at a later date, complaints concerning the gastric tolerance of 3582 are voiced."²²⁴ This time Hoechst was successful – Leverkusen began organizing tests of the preparation.

On December 2, 1942, Leverkusen informed Hoechst that 1,000 tablets of preparation 3582 had been sent to *SS-Obersturmführer* Hellmuth Vetter for testing. "Like last year Dr. Vetter has a large typhus fever station in Auschwitz, where he has the opportunity of thoroughly testing new drugs." Hoechst "very much" welcomed the fact that Vetter was carrying out tests in Auschwitz and "looked forward very much" to the results.²²⁵ At the same time tests were also being carried out in the army, but by December 1942, after episodes of vomiting by individual patients, it was decided that the preparation should be administered only in granular rather than tablet form.²²⁶ At the end of January 1943, Hoechst also informed Vetter in Auschwitz that administration in tablet form could result in extreme intolerance of the preparation, particularly in patients whose digestive system was weakened and who suffered from high fever and diarrhea. Some researchers had completely discontinued testing with 3582 because of the intolerance manifested in patients. It was hoped that tolerance to the granular form

---

²²² NI-13588, Bayer Pharma Büro Berlin (Berg) to Behringwerke Leverkusen, August 26, 1942, on the planned talk with Mrugowsky in the first week of September; HA, file Fleckfieberversuche, Niederschrift von Dr. W/T [Weber] über die Besprechung mit Dr. Mrugowsky, Berlin, in Gegenwart von Dr. Kohlhaas, Behringwerke, Berlin, am 10.9.1942, dated September 18, 1942 (copy) [=NI-9701].

²²³ HA, file Fleckfieberversuche, Bayer (Mertens, Koenig) to Hoechst, November 7, 1942.

²²⁴ Ibid., Hoechst (Bockmühl, Weber) to Leverkusen, November 9, 1942.

²²⁵ Ibid., Leverkusen (Mertens, Koenig) to Hoechst, December 2, 1942; Hoechst (Bockmühl, Weber) to Bayer Gruppe Leverkusen, December 3, 1942 [=NI-11413].

²²⁶ Ibid., Hoechst (Bockmühl, Weber) to Bayer Gruppe Leverkusen, December 15, 1942.

would be better. Hoechst therefore sent Vetter some trial doses that the latter was requested to test in further patients as quickly as possible.[227]

Vetter reported to Leverkusen at the end of February 1943 that he judged tolerance of the preparation when administered in tablet form to be "generally very bad." "Vomiting occurred, considerably weakening the patients; [administration of the preparation by] clysma resulted in violent diarrhea with tenesmus and defecation of up to 15 times daily. Overall, vomiting occurred in 78% of cases. The patients additionally complained of a strong burning sensation in the mouth and pharynx after taking the tablets, which often continued for some time." Vetter therefore regarded the preparation as "worthless" against typhus fever. According to Vetter, mortality after administration of the preparation was 30 percent, almost equal to that of untreated cases, which had stood at 34 percent the year before. Commenting on these results to Hoechst Leverkusen referred to them as "not very agreeable." According to Leverkusen, Vetter was even of the opinion that the effect of 3582 on reducing fever was harmful for the organism rather than otherwise and tended to have "an unfavorable" effect on mortality. Leverkusen commented further that Vetter, "who really does have sufficient experience in the clinical treatment of epidemic typhus[,] is at all events very disappointed in preparation 3582." As he had "unfortunately been transferred from Auschwitz," it was unclear whether and to what extent he would continue to be available to carry out further clinical tests in future.[228]

Hoechst found itself in a very unpleasant position, for other doctors who had also administered the substance had likewise come to the conclusion that tolerance was poor and found the effects of the preparation to be unconvincing.[229] But Hoechst was not prepared to accept Vetter's criticism, indeed, Fußgänger became quite irate. In a note sent to Weber he wrote that Vetter is "apparently impressed" by the "vomit-inducing effect of the preparation." Nor was it true that mortality was negatively influenced by the preparation's antipyretic effect, as after all mortality had sunk from 34 to 30 percent. As no other chemotherapeutic agent was available yet and according to Kikuth animal testing of preparation 3582 had shown it to be more effective than the methylene blue that Leverkusen had released for testing, Fußgänger believed that "enough other investigators can be found, who will be similarly prepared to support large-scale clinical testing."[230]

---

[227] NI-11417, IG Hoechst (Weber, Koenig) to Vetter, KL Auschwitz, January 27, 1943.

[228] HA, file Fleckfieberversuche, Leverkusen (Bayer) to Hoechst, March 9, 1943, in the appendix: Aktennotiz Leverkusen dated March 24, 1943 on a discussion with Vetter on February 24, 1943.

[229] On experiments in Lemberg, see NI-9711, Richard Haas to Hoechst, February 12, 1943; Hoechst (Lautenschläger, Weber) to Haas, March 19, 1943; Haas to Hoechst, March 21, 1943.

[230] NI-12445, Note by Fußgänger for Weber dated March 16, 1943, re. "Besprechung Dr. Vetter in Lev."

Fußgänger was completely unmoved by the fact that preparation 3582 had not significantly decreased mortality and that patients additionally suffered from the preparation's considerable side-effects; he demanded further testing despite its "vomit-inducing effect." Nevertheless Hoechst decided to abandon testing with tablets and only distribute the granules, and it was conceded that "As regards the curative results reported by Dr. Vetter, unfortunately they cannot be called particularly overwhelming." But it was felt that the reference to a higher mortality occurring after administration of the preparation was incorrect: since mortality had dropped from 34 to 30 percent, it would be more proper to speak "only of its not very convincing efficacy."[231]

Around the same time Vetter was carrying out his experiments in Auschwitz with preparation 3582, testing was also begun on inmates from the concentration camp at Buchenwald, evidently making use of the preparations that had been sent to Mrugowsky. At the beginning of January Hoechst asked Mrugowsky what his experience had been with the experimental quantities he had been provided with, since nothing had been heard from him. Hoechst reported initial successes, but added that these were individual cases that they had "happened to have seen." Moreover, as regards the tolerance of the tablets it had been proven "that the tablets should not be taken on a completely empty stomach, but together with plenty of liquid or soup." But Hoechst was forced to admit that lengthier courses of treatment with the tablets resulted "now and then in an irritation of the stomach, which manifested itself in short spells of vomiting."[232] Hoechst wrote to Mrugowsky again at the beginning of February 1943 to inform him that the latest test results appeared encouraging and that the company was very interested in large-scale testing. Under these circumstances Weber would be interested in calling on him in person.[233]

In the meantime Mrugowsky had apparently had the opportunity to have the tablets tested. In a talk with Weber in February 1943 he criticized the poor tolerance of the tablets.[234] Among the tests organized by Mrugowsky must have been the first series of experiments carried out with the preparation in Buchenwald; this can be inferred from the fact that they began on January 26, 1943, under Ding's supervision. Prisoners were artificially infected with viruses previously cultivated in chicken yolk sacs – a clear indication that these were experiments carried out on human guinea pigs. Once again it was found that tolerance of the preparation was "extremely bad": almost all persons treated with preparation 3582 suffered from strong

[231] HA, file Fleckfieberversuche, Hoechst (Fehrle und Weber) to Leverkusen, March 17, 1943.
[232] Ibid., Hoechst (Bockmühl, Weber) to *SS-Standartenführer* Mrugowsky, *SS-Sanitätsdienstamt*, January 8, 1943 (copy) [=NI-9576].
[233] NI-11418, Hoechst (Bockmühl, Weber) to Mrugowsky, February 2, 1943.
[234] HA, IG/54, IG-Prozessakten, Lautenschläger document 54, affidavit by Joachim Mrugowsky dated September 3, 1947: according to his statement the preparation was sent to the SS military hospitals in Berlin and apparently also in Prague.

nausea and vomiting. Moreover, no positive effect on the course of the disease was noted.[235]

Because of the problems experienced with tolerance Hoechst offered Mrugowsky two different granulate forms of preparation 3582 at the end of February 1943 for further experiments, even if this was considered an "inconvenient form of application." "But after your cases have several times vomited the tablets, and we have also heard similar complaints from other parties we consider it to be advisable for the time being to use the granulate preferentially." Testing should begin with a 5 percent granulate material, which had already been tested elsewhere, however, "not in patients ill with fever." The other test persons had "tolerated the granulate with no difficulties." Although Hoechst was well aware that "tolerance tests carried out in persons not ill with typhus do not allow any conclusions to be drawn," nevertheless it was believed that the granulate "could also be tolerated without difficulty by sick persons." Hoechst also offered a 10 percent granulate that had not yet been clinically tested, writing "our patient material is unsuitable for the carrying out of tolerance tests." It was hoped that it would be possible for Mrugowsky to use this granulate immediately on typhus patients "without preliminary time-consuming tests on our part."[236]

At the same time, or only a short time previously, Ding and Weber had managed to meet up; according to Weber it was a "chance" meeting in the Bayer office in Berlin that took place sometime in the middle of February 1943.[237] At all events, Hoechst sent test batches of 3582 granulate together with the relevant sheet of instructions to Waldemar Hoven, "the resident doctor of the Waffen-SS" in Weimar on February 19, 1943, requesting him to pass them on to Ding. Additional quantities of the preparation "were easily available on request."[238] At the same time Mrugowsky was also sent the granulate, as it was believed that this would be tolerated better.[239]

On March 25, 1943, Ding and Weber held a telephone conversation to discuss the experiments carried out until then. During the conversation it

[235] National Archives, NO-265 "Ding diary"; NI-9710, Vorläufiger Bericht über den Fleckfieber-Therapieversuch mit Nitroakridin-Präparat 3582 bis zum Abend des 9.2.1943; NI-9709, Weiterer Bericht bis zum Abend des 19.2.1943; Mrugowsky declared after the war (HA, IG/54, Lautenschläger document 54, affidavit by Joachim Mrugowsky dated September 3, 1947) that he had not passed on the preparation to Ding – but based on the chronology of events which is discussed below, this appears implausible, despite certain reservations respecting the "Ding diary." Yet another possibility is that Ding received the preparation from Bieling.

[236] HA, file Fleckfieberversuche, Hoechst (Lautenschläger und Weber) to Mrugowsky, February 27, 1943.

[237] HA, IG/54, IG-Prozessakten, Lautenschläger document 58, affidavit by Julius Weber dated March 20, 1948.

[238] HA, file Fleckfieberversuche, Hoechst (Lautenschläger und Weber) to Hoven, February 19, 1943.

[239] NI-9712, Hoechst (Lautenschläger, Weber) to Mrugowsky.

was noted that the "therapeutic results had not been particularly good," but this might be due to the fact that, in contrast to the animal experiments carried out in Hoechst, treatment had only been initiated on the third day of illness. However, Weber noted contentedly: "Ding does not appear to have experienced any difficulties with tolerance of the granulate." Furthermore, at Ding's request, the two men had agreed during the conversation that Ding, who wished to have a look at the experimental arrangements in Hoechst, would be invited through Mrugowsky.[240]

In a second series of experiments in Buchenwald that began on March 31, 1943, forty prisoners were infected with Rickettsia from eggs, but none of the patients fell ill.[241] Hoechst kept Buchenwald liberally supplied with the preparation, also supplying it in the form of granules, and generally showed itself to be pleased that experiments were being carried out. At one point, when difficulties temporarily halted delivery, Hoechst immediately sent replacements so that testing could continue.[242] As requested by Ding in his telephone call with Weber on March 25, Hoechst wrote to Mrugowsky that in view of the problems concerning the correct dosage of the preparation it would be helpful if one of his associates could come to Hoechst to study the laboratory tests.[243]

Ding visited Hoechst on April 14, 1943, and made an extremely bad impression. In the Nuremberg trial Lautenschläger testified that the course of disease in Ding's experimental series had been "mostly fatal" and the results had been "considerably less favorable" than in other clinics. When Ding had talked about "regulated infections" in his conversation with Lautenschläger, Lautenschläger had realized that Ding was not carrying out clinical experiments on soldiers who had fallen ill with typhus fever "but on artificially infected people." Lautenschläger had therefore declined to let Ding be supplied further and had come to an agreement with Weber that they could not "be party to" such experiments as those carried out by Dr. Ding. Lautenschläger added that "from an immunological point of

[240] HA, file Fleckfieberversuche, memo by Weber, dated March 30, 1943, on "Gespräch mit Dr. Ding, Weimar, am Donnerstag, den 25. März 43" [–NI-9727]. Based on a statement by the Kapo Arthur Dietzsch, various authors – e.g., Klee, *Auschwitz*, p. 308 – write that Weber had stayed in Weimar/Buchenwald at the time. Weber, however, maintained that there had only been a telephone call between Ding and himself. This version is confirmed in NI-9725, telegram from Ding to Weber or representative, March 23, 1943, in which Ding requests a call from Weber on Thursday, March 25, 1943, in the morning between 10 and 11 o'clock to Weimar telephone 6311.

[241] National Archives, NO-265, "Ding diary."

[242] NI-11404, Weber to Bieling, Behring-Werke Marburg, March 17, 1943; see also, e.g., NI-11405, internal Hoechst memo from Jeck for Weber concerning a telephone conversation with Dr. Kern, March 16, 1943; NI-9717, Hoechst (Fehrle, Weber) to Standortarzt der Waffen-SS Weimar (Hoven), March 8, 1943; NI-11408, Hoechst (Lautenschläger, Weber) to Ding, Weimar, March 31, 1943.

[243] NI-10260, Hoechst (Lautenschläger, Weber) to Mrugowsky, March 27, 1943.

view" he considered "such tests to be worthless and inexpedient (experiments on unsuitable objects)." It was decided to discontinue correspondence with Ding and not to provide him with any new preparations.[244] Fußgänger also emphasized that in the course of the meeting in Hoechst Ding did "not give the impression of being a serious scientist."[245] Weber likewise declared after the war that at the time Ding had given "confused replies" to questions and had not shown himself to be "particularly competent." At the meeting in Hoechst with Lautenschläger, Fußgänger, and Weber, Ding made an "unfavorable impression" on all three men: "He revealed himself to be an inexperienced, ambitious careerist without the necessary professional qualifications required for an investigator of curative drugs." Weber emphasized, however, that at this meeting it had not become obvious that people were being artificially infected in Buchenwald. He himself had used the expression "regulated infections" in his conversation with Dr. Ding, who in his turn had immediately made use of the term in his talk with Lautenschläger. After Ding's visit Lautenschläger had made it clear that he did not "wish for further clinical testing of our preparations by Dr. Ding"; according to Weber, Lautenschläger wished him "to get Dr. Ding off his back." Weber, however, was of the opinion that "that before pronouncing judgment it would be necessary to study the graphs and tables carefully, and according to the information provided by Dr. Ding they would only be available in 4 weeks time at the earliest." "Prior to this I thought it would not be possible to definitively judge either Ding's mode of operation or his results."[246]

Thus, all three persons who had participated in the meeting in Hoechst agreed that Ding was incompetent and that his test series would very probably not yield much of value. Nevertheless, it is not logical to suggest that Lautenschläger declined any further cooperation because of the high rates of mortality in Ding's experiments, for the test series that resulted in a death rate of more than 50 percent of the artificially infected camp inmates began only on April 24, 1943, that is, ten days after the meeting in Hoechst. In his interrogation by the Limburg public prosecutor's office in 1960 Fußgänger similarly emphasized that Lautenschläger must have been mistaken about the high death rates since these occurred only in the third series of experiments.[247]

[244] NI-9811, affidavit by Carl Ludwig Lautenschläger dated May 2, 1947.
[245] HA, IG/54, IG-Prozessakten, Lautenschläger document 59, affidavit by Rudolf Fußgänger dated February 27, 1948.
[246] Ibid., Lautenschläger document 58, affidavit by Julius Weber dated March 20, 1948.
[247] BA Aussenstelle Ludwigsburg, Fußgängerverfahren, Einstellung des Ermittlungsverfahrens, pp. 40–1; Klee, *Auschwitz*, pp. 309–10, claims that the experimental series commenced on April 11. But according to the "Ding diary," National Archives, NO-265, that was when the "preliminary experiments" started, i.e., inmates were infected intravenously with fresh blood from typhus patients, but no attempts were made to treat infected persons.

In the course of these so-called therapeutic experiments thirty-nine inmates were "infected by intravenous injection of 2 cc fresh blood each from typhus patients" on April 24, 1943, as the previous attempt at artificial infection using Rickettsia from eggs had not gone as planned. The effectiveness of this means of infection had been successfully tested in "preliminary experiments" on April 11 and 13. Accordingly, all infected persons fell ill with "very serious typhus." Twenty-one of the thirty-nine forcibly infected inmates died: fifteen prisoners were treated with preparation 3582, and eight of them died; fifteen prisoners were treated with rutenol, and eight of them likewise died; and five of the nine infected inmates from the control group also died. The mortality in all three groups was therefore over 50 percent with no significant differences between groups. In addition, the side-effects of the two Hoechst preparations experienced by the people being experimented on were extremely unpleasant.[248]

According to the statement of one of his colleagues, Hoven, Ding was "most indignant" about the rutenol preparation. In a report on the experiments intended for publication and written in 1944, Ding wrote that the overall results of the clinical tests of preparation 3582 and rutenol showed that neither of the preparations had any febrifugal or palliative effect on the disease, and tolerance had been very bad: "patients" who had been administered 3582 had vomited up to seven times per day. The final mortality rate had been "about as high as that of the controls who had not been treated with these preparations." His results therefore agreed with those given to him by Vetter – and completely contradicted those given in an article by Professor Holler, published in 1944, in which the substance was extolled.[249] But Ding did not reflect on the value of his experiments, which both Eugen Kogon and Lautenschläger, as previously mentioned, considered to be very low, if not completely worthless. Kogon wrote that infection using fresh blood from a sick person had "bordered on lunacy" and had "smashed through all immunization measures, resulting disastrously in nearly every case."[250]

Whether Lautenschläger, as he later declared, really considered Ding's reference to "regulated infections" to be indicating that experiments were being carried out on artificially infected persons and thereupon disapproved of such experiments is possible. It is likewise possible that in a subsequent

[248] National Archives, NO-265, "Ding diary"; NI-9742, Results of the therapeutic experiments with acridine (=preparation 3582) and rutenol, dated June 1, 1943. This document is printed in Klee, *Auschwitz*, pp. 312–13.
[249] NI-12182, Testimony by Waldemar Hoven dated October 3, 1947; NI-9752, Report by Ding "Zur Fleckfieberbehandlung mit Acridinderivaten," September 29, 1944.
[250] Eugen Kogon, *The Theory and Practice of Hell: The German Concentration Camps and the System Behind Them*, translated by Heinz Norden (New York: Berkley, 1950), pp. 141–2.

meeting Weber informed Lautenschläger that the expression used by Ding had originally come from Weber himself. Deliveries to Ding ceased, as is confirmed by a letter from Ding, dated July 1944, in which he complained that he had not received any more supplies since the meeting, a termination that is additionally confirmed by Kogon.[251] But a thoroughly plausible explanation for the termination of deliveries is that Lautenschläger considered Ding to be so incompetent that he wanted to have nothing more to do with him.

Although Lautenschläger, Fußgänger, and Weber had received a very negative impression of Ding, Weber nevertheless said to Bieling at the Behring Works that Hoechst was very curious to hear about the results of Ding's experiments. And Hoechst wrote to Mrugowsky, thanking him for sending Ding to Hoechst, as the discussion with him had been "very valuable."[252] Both statements indicate that despite their poor estimate of Ding, Hoechst, and above all Weber, wished to leave no stone unturned in finding a use for the preparations.

At the end of May Mertens came from Leverkusen to Hoechst to discuss preparation 3582 and rutenol. Despite the poor tolerance it was agreed that further testing of the two preparations should be carried out in Auschwitz in order "to try everything." During the meeting it also was mentioned that Vetter had heard that another SS doctor, Dr. Ding, had also formed a poor opinion of the preparation, both with regard to its therapeutic efficacy and with regard to tolerance – an indication that Leverkusen must have been aware of Ding's experiments.[253]

Tests had been carried out not only in concentration camps but also at various other locations, and the results were discouraging. In a memorandum written for Weber at the beginning of June Fußgänger noted that the latest reports – whereby he also referred to those of Ding – would have led him to doubt the effectiveness of the nitroacridine preparations "if all applications of the preparation had not been carried out on patient material in which the powers of resistance had been already diminished from the start." To prevent the preparation from being completely withdrawn and since the Rickettsia causing Volhynian fever were of a

[251] NI-9747, Ding to Lautenschläger dated July 11, 1944, in which Ding deplored the fact that he had heard nothing more from Lautenschläger since their last meeting in April 1943. This was confirmed by Kogon, *Theory and Practice*, pp.141–2, n. 4.

[252] NI-11425, Weber to Bieling, Behring-Werke Marburg, April 19, 1943; NI-11424, Hoechst (Lautenschläger, Weber) to Mrugowsky, April 17, 1943; HA, IG/54, IG-Prozessakten, Lautenschläger document 58, affidavit by Julius Weber dated March 20, 1948, in which Weber attempted to interpret these letters according to how he perceived them, namely that it would not have been possible to simply break off the experiments and sever connections to the SS.

[253] HA, file Fleckfieberversuche, Memo Leverkusen re "Besprechung von Direktor Mertens in Hoechst," May 31, 1943.

less virulent strain, he recommended using the preparation to treat this disease.[254]

Hoechst accordingly wrote to Mrugowsky that clinical tests of preparation 3582 had "generally not turned out the way we had expected." The main reason for this was poor gastric tolerance, even of the granulate, which had come as a surprise. Hoechst was therefore considering whether to continue testing at all, and was consequently particularly interested in his results. At all events 3582 was considered to be an effective agent for the treatment of Volhynian fever.[255]

Hoechst was thus interested in maintaining its connection to Mrugowsky and did not intend to give it up quite yet. Weber also kept in touch with Ding, for whom he did several favors, even after the meeting in Hoechst.[256] In mid-June the medical inspection office (*Sanitätswesen*) of the Waffen-SS in the SS Main Operational Headquarters (*SS-Führungshauptamt*) asked Hoechst to meet with Ding to discuss his results. Lautenschläger and Weber promised that Weber would come to Berlin for a meeting on June 30. Almost concurrently, Weber received a letter from Professor Bieling from Marburg, who gave him an account of Ding's experiments. After the poor results that had been achieved there, it was "understandable" that testing would not be continued in the same manner. However, Bieling had the impression that the matter was not yet settled: "After all, the conditions which prevail here are quite special and I think they should be disregarded and one should see what can be achieved under other, better conditions." The allusion to the conditions prevailing in a concentration camp spoke for itself. Bieling added that Ding would be staying in Weimar between June 21 and June 26 and would be happy to show Weber his experiments in detail. But Bieling was of the opinion that "after what I have seen and can describe to you" it would probably not be worth spending the time.[257]

It was not possible to ascertain whether Weber met Ding in Berlin or in Weimar at the end of June 1943. In the Buchenwald trial, Arthur Dietzsch, the kapo of the Typhus Department in Buchenwald, declared that either Weber or an employee from Hoechst had visited Buchenwald. Weber, on the other hand, claimed in a statement written for the investigators in the Nuremberg trial in 1947, apparently with reference to the appointment on June 30, that he had "never entered" the Buchenwald concentration camp and that he had not met Ding at any time other than the two meetings on record. "My efforts to achieve some degree of clarity concerning the nature

[254] NI-12448, Fußgänger to Weber, June 5, 1943.

[255] NI-9743, Hoechst (Lautenschläger, Weber) to Mrugowsky, June 8, 1943.

[256] HA, file Fleckfieberversuche, Ding to Weber, May 5, 1943; NI-11497, Weber to Ding, June 15, 1943.

[257] NI-11498, Hoechst (Lautenschläger, Weber) to Ding, June 17, 1943: promise that Weber would come to Berlin; NI-11499, Weber to Kern, Bayer office Berlin, June 17, 1943, on Weber's meeting with Ding; NI-9824, Bieling to Weber, June 18, 1943, on Ding.

of his tests by visiting Dr. Ding failed: in Berlin I received the information
that Dr. Ding was currently in Weimar. In Weimar I did not manage to get
hold of Dr. Ding using the telephone number given me in Berlin so that I left
Weimar again without having achieved anything and without setting foot
outside the station in Weimar."[258]

But in all probability it was not even necessary for Weber to visit the
concentration camp and see the conditions for himself. It seems very
probable that at least Weber and Lautenschläger were aware of the fact that
Auschwitz and Buchenwald were concentration camps. After all, Hoechst
had already been sending letters in March 1941 addressed to the "K.L. [for
Konzentrationslager, i.e., concentration camp] Buchenwald 'camp doctor' "
in Weimar/Buchenwald. Hoven's statement that, following requests from
suppliers, after the middle of 1941 people preferred to use the address
"Resident doctor of the Waffen-SS Weimar" seems quite plausible.[259]
Fußgänger's memorandum for Weber on the "patient material" does seem
to indicate that he too had his suspicions concerning the type of experimental subjects being used, even if he had no proof.

At any rate, Hoechst learned about the Buchenwald results, either in
Berlin, Weimar, or Buchenwald or from Bieling, who apparently looked at
the documentation of the experiments in Berlin.[260] Ding's letter to
Lautenschläger, written in July 1944 – the most important document
exonerating Lautenschläger – contained more than merely the statement
that Ding had heard nothing from Hoechst and the acridine preparations
"since our last meeting." The letter also gave the exact number of people
who had died during the experiment – and it may be inferred from
Lautenschläger's statement in Nuremberg that he was confused, perhaps
even that when confronted with Ding's letter he made a mistake. For in his
letter Ding wrote that the results of his experiments in Buchenwald from the
beginning of January until the end of April 1943 had been discussed in
Hoechst – which is impossible since the meeting took place on April 14 and
the last of Ding's experimental series started on April 24. The letter is also
important since the venue for the experiments is given as the Clinical Ward
of the "Department for Typhus and Virus Research" of the SS-Hygiene
Institute in "Weimar Buchenwald."[261]

Typically – and this was another indication of the initiative Hoechst took
to get its preparations tested – Hoechst replied to Ding that Professor

[258] HA, PA Abteilung VI, Julius Weber, Weber to Benvenuto von Halle, February 15, 1947:
     including affidavit.
[259] NI-12179, Lagerarzt des K.L. Buchenwald to IG Farbenindustrie AG, Frankfurt-Hoechst
     dated March 11, 1941, on the storage life of the typhus-paratyphoid vaccine; NI-12182,
     Affidavit by Waldemar Hoven dated October 3, 1947.
[260] According to NI-9747, Ding to Lautenschläger, July 11, 1944, and NI-9824, Bieling to
     Weber, June 18, 1943, on Ding.
[261] NI-9747, Ding to Lautenschläger, July 11, 1944.

Holler's initially negative experiences with the acridine preparations 3582 and rutenol "induced us to ask you to re-check the chemotherapeutic effectiveness of our drugs." After Ding had obtained similarly negative results, rutenol was primarily used to combat Volhynian fever and abdominal typhus. Holler was the first to achieve some measure of success when using the preparations against typhus, but only after he had completely changed the dosage.[262]

Even if Weber subsequently declared that the experiments in Buchenwald were carried out without Hoechst's knowledge or approval and not at Hoechst's instigation,[263] the circumstantial evidence tells a different story. It must be emphasized that Hoechst really wanted preparation 3582 and rutenol to reach the clinical testing stage and put much effort into achieving its goal. In addition to the tests organized though the agency of Leverkusen, quite a number of tests were also carried out that Weber had initiated independently. Only one of these tests demonstrated any positive results, the series of Professor Holler's that was subsequently published. Therefore when Hoven stated after the war that "the initiative for the tests in concentration camps" had lain not with the SS but with IG Farben,[264] he was not entirely wrong. Weber had touted his preparations everywhere. Nor is the statement in the so-called Ding diary according to which the anti-typhus agents preparation 3582 from Hoechst and methylene blue from Elberfeld were tested "at the instigation of I.G. Farbenindustrie A.G." completely unfounded.[265] Weber must therefore have known, either from Bieling but perhaps also from Ding, by June 1943 that these monstrous experiments were being carried out on inmates from the Buchenwald concentration camp. By July 1944 at the latest after the receipt of Ding's letter, but – on the basis of his close collaboration with Weber – probably already in 1943 Lautenschläger must also have been aware of what was going on.

This did not prevent either Weber or Lautenschläger from continuing to cooperate with the SS. Once again through the instrumentality of Bayer, Vetter carried out a new series of tests with rutenol and 3582. This time Vetter wished to test whether the remedy might be effective against tuberculosis. He had reported to Leverkusen that rutenol had a beneficial effect on open pulmonary tuberculosis. The recoveries he had observed had been "so striking" that "at all costs he wished to continue clinical work" and additionally requested that experiments be carried out in the laboratories at Hoechst. Leverkusen therefore asked Hoechst to send supplies to Vetter, who was by then camp doctor in the concentration camp at

---

[262] NI-9748, Hoechst (Lautenschläger and Weber) to Ding, July 13, 1944.
[263] HA, PA Abteilung VI, Julius Weber, Weber to Benvenuto von Halle, February 15, 1947, including affidavit.
[264] NI-12182, affidavit by Waldemar Hoven dated October 3, 1947.
[265] National Archives, NO-265, "Ding diary."

Gusen.[266] Hoechst was astonished to hear of Vetter's observations. While 3582 was a "very polyvalent nitroacridine preparation" with wide-reaching effects, in animal experiments its effect against tuberculosis bacilli had "not been the best." Although Hoechst had considerable reservations concerning the possible effectiveness of the preparation against tuberculosis, Vetter was encouraged to continue his experiments.[267] Hoechst consequently sent supplies of the preparation to Vetter before the year was out. Both Hoechst and Leverkusen claimed not to attach "much importance" to the tests – however, at the end of January 1944 it was agreed that testing should continue.[268]

Experimental testing continued even when supplies began to run low. Bayer requested Hoechst to maintain its deliveries to Vetter to prevent him having to cut short his experiments, and Hoechst agreed that supplies to Vetter would be given priority.[269]

When Leverkusen (Bayer) wrote to ask whether Vetter could publish his results on the use of 3582 and rutenol against typhus and tuberculosis, Hoechst asked that it should not be "emphasized" that "his patients are only subjected to the treatment to determine tolerance."[270] Thus, in all experiments, Hoechst was now less interested in indications than in the question of tolerance, and at this point, as the request shows, this was no room for large-scale clinical testing. Nevertheless, following Leverkusen's initiative, Hoechst continued preferentially to supply Vetter.[271] Leverkusen had higher expectations of what could be achieved with the preparation than Hoechst did, for at the time no effective remedy against tuberculosis existed in Germany. Bayer asked Hoechst to speed up deliveries to Vetter, since it was felt that with his experiments he was "pursuing an interesting topic and a priori it is not possible to say whether, when the largely war-related importance of typhus begins to decline, this area may one day be more significant for this preparation."[272] Thus Hoechst continued sending material to Vetter in Gusen until February 1945.[273] But there was likewise no medical justification for the experiments carried out in Gusen. Like Ding, Vetter considered the camp inmates to be objects, who might or might not

[266] NI-9424, Bayer (Mertens) to Hoechst, December 15, 1943.
[267] NI-9425, Hoechst (Bockmühl und Weber) to Bayer, Leverkusen, December 23, 1943.
[268] NI-9427, Hoechst (Bockmühl, Weber) to Vetter, December 29, 1943; NI-12449, Niederschrift über die Besprechung mit Mertens am 28. Januar 1944 in Hoechst.
[269] NI-14007, Bayer (Koenig) to Hoechst, June 10, 1944; NI-9428; Hoechst (Bockmühl, Weber) to Bayer, June 19, 1944.
[270] NI-9435, Bayer (Mertens, Koenig) to Hoechst, May 31, 1944; NI-9429, Hoechst (Bockmühl, Weber) to Bayer, June 28, 1944.
[271] NI-9431, Hoechst (Bockmühl, Weber) to Vetter, June 29, 1944.
[272] NI-9432, Bayer (Mertens, Koenig) to Hoechst, July 7, 1944.
[273] NI-9440, Vetter to Hoechst, December 14, 1944; NI-9441, Hoechst (Bockmühl, Weber) to Vetter, December 18, 1944; NI-9442, Hoechst (Bockmühl Weber) to Vetter, January 17, 1945; NI-9443; Hoechst (Bockmühl, Weber) to Vetter, February 7, 1945.

die in the course of his "experiments." According to a survivor from Gusen, Vetter together with another SS doctor continued murdering camp inmates "until almost the last day of the camp" – and Hoechst assisted him.[274]

While Ding opted for suicide, the other SS doctors with whom Hoechst had cooperated, namely Hoven, Vetter, and Mrugowsky, were sentenced to death and executed. Nobody from Hoechst was sentenced, since during the trial the judges were of the opinion that the available evidence against Lautenschläger militated both for and against his innocence – and they subsequently acquitted him in accordance with the principle of *in dubio per reo*. But the responsible persons in Hoechst – Lautenschläger, Bockmühl, Fußgänger, and above all Weber – knew very well with whom they had collaborated for so long.[275]

At all events we can be sure that since the beginning of 1942 Weber was acquainted with the conditions prevailing in German concentration camps, as he was an active supporter of the Roman Catholic resistance to the regime. Although he himself was a Protestant, Weber was a friend of Hans Carls, the head of Caritas in Wuppertal; their friendship started in 1936, and Carls, a diabetic, was treated by Weber. When the Gestapo arrested Carls in November 1941, Weber did his best to have Carls declared unfit for detention because of his diabetes. He failed in this and was himself summoned and interrogated by the Gestapo.[276] After Carls was transferred to the Dachau concentration camp in March 1942, Weber regularly supplied him with insulin as well as other medicines. Weber received reports about the concentration camp and the conditions there from Carls via an intermediary or from Carls's secretary, Maria Husemann. Apparently in the beginning Carls did not tell him everything. In a letter to Maria Husemann, written at the beginning of July 1943, Carls described the malaria experiments of Professor Claus Carl Schilling, which were mainly carried out on Polish priests: "What I wrote about Prof. Sch. is not just true but the unscrupulousness goes much further, it is such that Dr. Weber would be horrified, if he knew everything, and would seize every opportunity as a

---

[274] Statement of Hans Marsalak, as quoted in Klee, *Auschwitz*, p. 320; on Vetter see Klee, *Auschwitz*, pp. 315–21, in which the reference letter from Bayer (Mertens, Mentzel) for Vetter dated October 30, 1945, is printed on p. 321.

[275] For more on the poor quality of the SS doctors, cf. NI-12182, affidavit by Waldemar Hoven dated October 3, 1947, and Kogon, *Theory and Practice*, pp. 133–6 on Hoven.

[276] HHSTAW, Abt. 520 F (A–Z), Julius Weber, Weber to Mayor of Frankfurt, December 20, 1944; Statement by Hans Carls dated November 1, 1945; Weber to Vorstand des Caritas-Verbands, July 7, 1941 (certified copy of a copy). On Carls see also his book on his concentration camp experience: Hans Carls, *Dachau. Erinnerungen eines katholischen Geistlichen aus der Zeit seiner Gefangenschaft 1941–1945* (Cologne: Bachem, 1946); on Carls and Weber see also Maria Husemann, *Mein Widerstandskampf gegen die Verbrechen der Hitlerdiktatur. Bericht von Karl Sommer 1964* (Wuppertal: Stadtdekanat und Katholikenrat, 1983), pp. 31–3.

doctor to put a stop to this."[277] But Weber was given quite detailed information about the conditions in the camp. He even took steps to ensure that these reports found a wider audience. After the war a priest reported that Weber had given him a copy of a letter from Carls "in which he described the barbarities which he had witnessed in the concentration camp in grim – or as I felt at the time – desperate detail, with the explicit request to preserve it."[278] Carls himself said during Weber's denazification trial that he had written to him in secret. In one of his letters he had reported that "Prof. Shilling is carrying out experiments with pyogenic bacteria." Weber had subsequently written to him "that he was unable to believe it." Carls had replied to Weber "that nonetheless the experiments were being carried out by Dr. Schilling."[279]

Weber thus knew about the conditions in the concentration camps and about the inhuman experiments to which prisoners were subjected even by well-known doctors such as Schilling, of whom Ernst Klee was later to write that "he was the eldest of the medical perpetrators and it was his passion."[280] Despite his knowledge of the criminal nature of experiments carried out on concentration camp inmates, Weber was involved in similar experiments by virtue of his position at Hoechst. Weber himself invited Ding to test the Hoechst preparations. Whether he already knew at the time what went on in the medical experiments on human subjects in concentration camps cannot be ascertained, but in the course of the experiments carried out in Buchenwald he must have realized. He finally refrained from further collaborations with Ding on Lautenschläger's instructions, but he was the person who maintained the connection longest. We cannot know whether – as he declared after the war – this was for tactical reasons, as it was simply not done to sever connections to the SS, or for professional reasons, in that he wished to leave nothing undone. Thus it is not actually important whether Weber actually visited Buchenwald in person or not. His story that he never left the station in Weimar may be factually correct – but he did not have to visit Buchenwald to know what was going on there.[281]

[277] HHSTAW, Abt. 520 F (A-Z), Julius Weber, Affidavit by Maria Husemann dated January 11, 1949; Protokoll der öffentlichen Sitzung der Spruchkammer, January 10, 1949, especially testimonies by Carls and Husemann; on Schilling cf. Klee, *Auschwitz*, pp. 117–25.
[278] HHSTAW, Abt. 520 F (A-Z), Julius Weber, statement by Maria Travers, January 14, 1949; statement by priest Rudolphi, January 13, 1949.
[279] Ibid., Protokoll der öffentlichen Sitzung der Spruchkammer, January 10, 1949 – testimony by Carls.
[280] Klee, *Auschwitz*, pp. 117–25.
[281] Compare this, however, to Klee, *Auschwitz*, p. 308; Ulrich Schneider and Harry Stein, *IG-Farben AG, Abt. Behringwerke Marburg- KZ Buchenwald Menschenversuche. Ein dokumentarischer Bericht* (Kassel: Brüder-Grimm-Verlag, 1986), p. 54.

The schizophrenic situation in which he found himself – feeling obliged for professional reasons to do something deeply abhorrent to his conscience – made Weber a broken man. After the war he became seriously ill, physically and mentally. In 1946/47 he hid in order to avoid having to testify in Nuremberg – and he left Hoechst.[282]

But Lautenschläger, Bockmühl, and Fußgänger had initiated and supported the clinical tests just as much – they too were aware that the experiments were carried out on people who were very much weakened, as Fußgänger's notes concerning the "patient material" attest. And Lautenschläger at least knew what testing in Auschwitz meant. By the end of the war they no longer bothered to give any other address for Vetter than that of the Gusen concentration camp, where the preparation continued to be tested until 1945. If Weber was as talkative as Fußgänger claimed after the war,[283] all persons involved must have known exactly what they were letting themselves in for. And the request not to mention that the experiments were primarily designed to assess tolerance makes a mockery of all principles, including their own – namely, that clinical testing should be begun only very late and that experiments should not be carried on human subjects.

But right from the beginning Hoechst had never had any problems with experiments being carried out on people who had not given their consent – whether these were "tentative tests" carried out in the Clinic for Emotionally Disturbed Persons in Frankfurt or Auer's experiments in Hoechst. From there it was no great step to regard concentration camp inmates as "patient material," even if their physical constitution was considerably worse than that of the patients in Hoechst or in psychiatric clinics.

The judges in Nuremberg refused to follow the line of argumentation put forward by the prosecution. The evidence produced there did not convince them that the behavior of the defendants was actionable: "The inference that the defendants connived with SS doctors in their criminal practices is dispelled by the fact that [IG] Farben discontinued forwarding drugs to these physicians as soon as their improper conduct was suspected. We find nothing culpable in the circumstances under which quantities of vaccines were shipped by Farben to concentration camps, since it was reasonable to suppose that there was a legitimate need for such drugs in these institutions." And the question as to whether the responsible persons in IG Farben could have concluded from the reports of the doctors that their drugs were being used for illegal experiments hinged on the dispute about whether the German word *Versuch* was properly translated by the English

[282] HA, PA Abteilung VI, Weber, correspondence and notes, especially from 1947.
[283] BA Aussenstelle Ludwigsburg, 403 AR-Z 12/60, Abschlussbericht, dated July 17, 1961, of the Staatsanwaltschaft Limburg/Lahn in the preliminary proceedings against Fußgänger, statement of Fußgänger, quoted on p. 42 of that report.

word "experiment," as the prosecution claimed, or by the word "test," as the defendants claimed. For the defendants insisted that within the scope of proper precautions tests would not only have been only permissible but even expedient. In the end, the court applied the rule "that where from credible evidence two reasonable inferences may be drawn, one of guilt and the other of innocence, the latter must prevail." The judges therefore ruled in accordance with the principle "in doubt in favor of the accused."[284] Lautenschläger was acquitted.

At the beginning of 1960, on the instructions of the chief public prosecutor in Frankfurt, the public prosecutor's office in Limburg instituted preliminary proceedings "against Dr. Rudolf Fußgänger in Frankfurt (M) and others for murder (typhus experiments in the concentration camp Buchenwald)." However, the proceedings against him and the others under investigation were dropped in July 1961 on the grounds that "a considerable number of the persons involved in the typhus experiments in the concentration camp Buchenwald are dead, others have already been called to account by the courts of the former occupying powers." In the meantime Julius Weber died in August 1960 before he could be questioned. Lautenschläger had been acquitted in Nuremberg, and the questioning of Fußgänger was felt to confirm the fact that the Hoechst employees responsible for the typhus experiments, that is, Lautenschläger, Weber, and Fußgänger, could not be prosecuted.[285]

This is not a view that has been generally shared in literature. In his book *The Theory and Practice of Hell* (in German: *Der SS-Staat*) Eugen Kogon gave a very critical assessment of Hoechst, but ultimately he confirmed the statement that contacts to Ding-Schuler had been broken off. Nevertheless he did feel that this called for some comment. In the second edition of his book, published in 1946, Kogon added a footnote to his reporting of the case: he had been informed by the Hoechst employees Julius Weber and Rudolf Fußgänger that they had believed the medicine was being tested on soldiers who were ill with typhus. When they subsequently found out that it was being tested on concentration camp inmates, they had immediately ended all cooperation. Kogon confirmed this but added that he had been surprised by the medical and scientific code of ethics that permitted persons to be given chemotherapeutic drugs that had previously only been tested on animals without the persons' express consent.[286]

[284] *Trials of War Criminals before the Nuernberg Military Tribunals*, vol. VIII, pp. 1171–2.
[285] BA Aussenstelle Ludwigsburg, 403 AR-Z 12/60, Oberstaatsanwalt beim Landgericht Frankfurt a.M. to Hessischer Minister der Justiz durch den Generalstaatsanwalt, January 26, 1960 (copy); Oberstaatsanwalt Limburg/Lahn to Hessischer Minister der Justiz durch den Generalstaatsanwalt, April 13, 1960; Abschlussbericht, dated July 17, 1961, of the Staatsanwaltschaft Limburg/Lahn in the preliminary proceedings against Fußgänger. On Fußgänger and Weber see also HA, PA Abt. VI, Weber and Fußgänger.
[286] Kogon, *Theory and Practice*, pp. 140–3, see especially n. 4. on pp. 141–2.

The first publication that specifically attacked Hoechst for its involvement was a "documentary report" by Ulrich Schneider and Harry Stein. They wrote that in the trial the responsible persons from IG Farben were able to weasel their way "out of their responsibility." In the case of Lautenschläger and Hoechst it could be proved that the company had been "the driving force behind the typhus therapy experiments in the concentration camp Buchenwald." Hoechst had "unscrupulously" taken advantage of "the career ambitions and craving for recognition of a person like Ding-Schuler for its own interests" without "paying the least attention to the victims."[287] In a study on typhus experiments in human subjects, the historian Thomas Werther came to the conclusion that with its "experience in carrying out tests on human subjects and its personal and political contacts to the Wehrmacht, the SS and state institutions," IG Farben had been "particularly good" at pushing through its own agenda: "These tests, which were mostly ineffective, resulted in many deaths, of which the number cannot be determined exactly, and in an even larger number of serious and less severe typhus cases as well as continued attempts to get the 'typhus problem' under control with the help of modified or new methods, which in turn led to new victims."[288]

It was above all Ernst Klee who made a detailed investigation into IG Farben's and Hoechst's collaboration in the medical experiments carried out on concentration camp inmates. His conclusion was highly unflattering for Hoechst. In the trial the judges had believed Lautenschläger when he stated that he had broken off all ties to Ding after his visit to Hoechst. However, this had not been true; moreover, in the person of Vetter Hoechst had a second member of the SS willing to test its preparations in concentration camps: "The concentration camps served as a laboratory for the pharmaceutical industry and the Wehrmacht. Inmates were tortured to death like laboratory rats or suffered permanent damage. None of the persons responsible in the pharmaceuticals industry was sentenced after 1945."[289]

In contrast, the last study on this subject accepted the verdict of the Nuremberg trial. In his work the historian Wilhelm Bartmann came to a considerably more favorable assessment of Lautenschläger. Bartmann describes Lautenschläger as a devout Christian who was involved in the criminal experiments on concentration camps inmates but was at the same time practically a victim of the regime. Bartmann felt that Klee was demanding "a considerable degree of moral courage" if he expected Lautenschläger to pit himself against such experiments, for "a general

[287] Schneider and Stein, *IG-Farben*, pp. 51–4.
[288] Thomas Werther, "Menschenversuche in der Fleckfieberforschung," in Angelika Ebbinghaus and Klaus Dörner (eds.), *Vernichten und Heilen. Der Nürnberger Ärzteprozess und seine Folgen* (Berlin: Aufbau, 2001), pp. 152–73, here pp. 168–9, 172–3.
[289] Klee, *Auschwitz*, pp. 304–5, 341.

refusal by Hoechst to participate in the typhus experiments could have been classified as sabotage." He concluded this passage on Hoechst by repeatedly emphasizing that Lautenschläger was acquitted in Nuremberg and had subsequently not been prosecuted again.[290]

But this is a question that cannot be simply reduced to its legal aspects. The Hoechst Pharmaceutical Department would surely not have aided and abetted such crimes in other circumstances. But at the time, Bockmühl, Fußgänger, Lautenschläger, and Weber jettisoned their moral values and misgivings to be able to carry out clinical tests of a preparation that for a long time had not been released for testing. Whether they were aware at the beginning of the series of experiments carried out in Buchenwald and Auschwitz of what was happening is not clear. But we can assume with near certainty that, over time, they did know. A number of facts, such as the reports of Vetter and Ding and, not least, Weber's own activities in support of the Roman Catholic resistance, support this assumption. Nevertheless, Hoechst collaborated closely with the SS for a long time. Up until February 1945, Hoechst continued to supply SS doctor Vetter in the Gusen concentration camp with a substance that was primarily known for its poor tolerance – and this despite the fact that Hoechst did not expect much from Vetter's experiments. But Hoechst wanted this preparation to be successful, whatever the cost. It was not Hoechst's ideological orientation that was responsible for the darkest chapter in the plant's history during the Nazi period, as Weber, the Hoechst employee chiefly responsible, was active in the Roman Catholic resistance. Naked ambition made Hoechst complicit in the crimes committed: its wish to turn the preparations, which had been so long in the making, to good economic account and to score a success over its competitor, Elberfeld. The proximity of humanity and monstrosity becomes clear when we look at Weber. He was the person in Hoechst responsible for providing the necessary materials for medical experiments on concentration camp inmates in Buchenwald, and at the same time he was illegally supplying the inmates of the Dachau camp with medicines – it would be difficult to find a plainer example of mistaken professionalism.

---

[290] Bartmann, *Tradition*, pp. 206–11, especially pp. 209–11.

5

The Postwar Years: Dealing with the Past

Around the middle of March 1945 Lautenschläger wrote a "personal" and "strictly confidential" letter to Karl Weber, the works manager in Gersthofen. In his letter he gave an account of the last board meetings of IG Farben, which had been "completely dominated by the serious situation of the war." Production had strongly declined due to "the loss of the plants in the east" and heavy damage in many other works. In the Maingau plants "currently only salaries and wages were being paid without any productive work being carried out." "Even the best funded company" would "not be able to stand this for any length of time." Plans were made to reduce the salaries primarily of the white-collar employees and to review the entire workforce with an eye to reducing their numbers where possible. Weber was enjoined to think about which "of the useless employees" he could "get rid of" and which employees aged 60 and above or "sickly" he could pension off. Lautenschläger added: "I would also like to ask you to prepare for a timely offloading of the foreign workers. A call for workers for trench digging has gone out, at least in our area, and this offers the opportunity to get rid of the foreigners." While he appeared to feel few qualms concerning the future of these people, he was all the more worried about the welfare of his family, requesting Weber to send foodstuffs to them, and about his own creature comforts: "And finally I would like to ask if possible to have someone grow a little tobacco for me this year so that in these bad times I will at least have something to smoke in my pipe."[1]

---

[1] Clariant WA, file 125, letter from Lautenschläger to Weber, March 16, 1945; cf. Hayes, *Industry and Ideology*, p. 376: "As the Reich caved in, the pressing concern before Farben's leaders was the survival not of their firm but of their persons."

According to Lautenschläger, by the end of 1944 production in the IG Farben plants had been reduced by the air raids of the Allies "to 50% of the output of 1943."[2] But despite many air-raid warnings, Hoechst had suffered only little damage. Lautenschläger reported that in the West the large Hoechst plant "alone still retained its full productive capacity until the end of the war and often had to take over manufacturing or alternative production for other plants in the West, as far as possible." But Hoechst too began shifting the manufacture of a large number of its products to the Black Forest, the Allgäu, or Vorarlberg.[3]

The plant was finally largely shut down at the end of January 1945, apparently "because of the lack of coal." A directive of Lautenschläger's on January 22, 1945, decreed that by January 26th "manufacturing and laboratories, etc., will be shut down for the time being," and energy consumption was to be immediately reduced. Only the workshops, those auxiliary and companion plants "which do not require energy," and the production of metal treatment agents, of butanol, alkali-chlorine electrolysis, and parts of the Nitrogen Department, were exempt from this directive.[4] Only a few weeks later, on March 29, American troops occupied the works. For Hoechst the war had ended.

Many people had lost their lives in the war, many had been persecuted, and many had suffered. However, not a few had profited from the regime and had adapted, or had even become perpetrators themselves. The prosecution of such persons and their removal from positions of influence was an important goal of the American Military Government in its zone of occupation – which became the "Land of the *Fragebogen*" in which Germans were obliged to fill out questionnaires concerning their past under the Nazi regime. With the American occupation came "denazification," as it was called.[5]

Denazification was also influenced by the general unemployment at the time. Employees in Hoechst who had not been members of the NSDAP or of one of its organizations expected to be given preference over those persons who had been active in them. After the Americans had placed the works under forced administration, the Workers' Representation of the former IG Farben plant at Hoechst wrote to Colonel Percival on July 9, 1945, that it expected the "still existing pernicious influence of the Nazi regime in the plant" to be "cleared away." It was rumored that only 1,500

[2] HA, PA Lautenschläger, Lautenschläger, Erinnerungen, vol. 8, [p. 14].

[3] Ibid., [pp. 15, 18–19].

[4] HA, Rundschreiben 00003 (interne Rundschreiben), Lautenschläger to Abteilungs- und Betriebsleiter, January 22, 1945, re. "Betriebseinschränkung"; PA Lautenschläger, Lautenschläger, Erinnerungen, vol. 8, [p. 39]; *Chronik der Hoechst AG*, p. 182; cf. also Winnacker, *Challenging Years*, pp. 106–7.

[5] John Dos Passos, "Land of the Fragebogen," in Dos Passos, *Tour of Duty* (Boston: Houghton Mifflin, 1946), pp. 243–74.

persons would be employed while currently around 4,200 people worked in the plant. According to the estimate of the Workers' Representation, an average of around one-third of the employees were members and supporters of the former NSDAP, in individual departments perhaps even one-half of all employees: "in the antifascist camp of the plant some degree of indignation is understandably felt about this, and this has resulted in repeated disturbances of law and order in different factories." It would "be in keeping with the principles of justice and would promote peaceful reconstruction" if jobs would initially be reserved for those persons who had kept their distance from the Nazi regime. The Workers' Representation was of the opinion that a preferential treatment of opponents to the former regime via the allocation of jobs was feasible with the exception of some technical specialists.[6]

In the first phase directly after the war Lautenschläger took on the task of denazification in the plant himself, since he regarded himself as having a completely clean record. In April 1945 he began complying with directives from the Frankfurt City Council and ordered the summary dismissal of leading representatives and early supporters of the Nazi regime. Ferdinand Pensel was charged with the task of carrying out denazification, and in June he accordingly ordered that as the current situation in the plants required extensive staff cuts, "in the first instance those Party members who had been politically active" would be dismissed; indeed, when choosing between two employees with the same aptitude and performance, Party membership was the single decisive factor. By the end of June 101 persons had been suspended; Winnacker, described by some as a staunch National Socialist, was likewise threatened by dismissal; however, he was able to prevent this.[7]

On July 1 Lautenschläger finally decreed that the guidelines of the Chamber of Commerce and Industry concerning the "purging of industry" would be adopted in Hoechst. A new element in these guidelines was that it was no longer necessary to have been a member of the NSDAP or one of its organizations to be suspended. The new guidelines placed a greater emphasis on the extent of a person's active support of National Socialism. Thus, not only all Party members who had joined the NSDAP before April 1, 1933, were to be dismissed. Likewise all members of the SS, all members and representatives of the Gestapo, all leaders of the SA, the NSKK, and the

[6] HA, file Entnazifizierung (Vorstandssafe), Arbeitnehmervertretung des früheren IG Werkes Hoechst (Weber, Bassing, Becker, Frank, Theis, Schenkelberg, Dietz, Schmitt, Habicht) to Colonel Percival, July 9, 1945 (in an English and a German version, in the latter version the name of the recipient has been misspelled).
[7] HA, file Pensel-Entnazifizierung, Circular from Pensel, June 20, 1945; Clariant WA, file 586, Pensel to Abteilungsleiter, June 9, 1945; circular from Lautenschläger to Abteilungsleiter und Betriebsführer, July 1, 1945; private files Winnacker, Winnacker to Lautenschläger, June 4, 1945; declaration of fellow employees of Winnacker according to which Winnacker had not been "active in a Party political sense," June 5, 1945 (copy).

NSFK "from *Sturmführer* upwards" were to be dismissed. The heads of various other Nazi organizations were also to be dismissed. The guidelines began with an even more sweeping statement on who should be dismissed: "All persons particularly active in support of National Socialism even if they had not been Pgs [*Parteigenossen*, i.e., Party comrades]. This also includes such persons who have behaved as enemies of their people [*volksfeindlich*]."[8]

While this wording offered some leeway for a fairer judgment of people's behavior in everyday life, as purely formal reasons such as membership or rank no longer solely determined whether someone would be dismissed, it also provided an opportunity for denunciations and for resentment. Quite often, in a reversal of the normal procedure for taking evidence, the accused were expected to prove their innocence.

Thus Siegfried Kiesskalt, one of the senior engineers in the works, emphasized at the beginning of May 1945 that he not only had never been a member of the NSDAP but he was even of partly Jewish descent, a fact that he had been able to hide successfully. This "personally hard fate," by which he presumably meant the necessity of concealment, had made him see many things "much more sharply"; he had taken pains "in conflicts concerning employees to act as a counterweight," although, as he added "without of course being able to outwardly expose myself."[9] According to him, prior to the arrival of the Americans he had not only taken steps to ensure that the plant was not destroyed as *Gauleiter* Sprenger had ordered – an act of "heroism" that incidently both Winnacker and Lautenschläger also claimed as their own – but had also put together a "self-protection unit against looting Eastern workers which I subsequently transferred to the MG police."[10] The story given by the union representatives reads a little differently. Both Kiesskalt and his colleague Franz Patat were listed under the category of "sympathizers who were not Party members," and a lengthy explanatory paragraph was devoted to both men: "After occupation by the American Army Dr. Kiesskalt and Dr. Patat became members of the Military Police. Without going into the details of the Party membership of these two people which, for several reasons, requires some clarification, the

---

[8] HA, file Pensel-Entnazifizierung, "Grundlagen für Suspendierungen – Bereinigung der gewerblichen Wirtschaft," dated June 12, 1945 (copy); memo dated August 20, 1945, with attachments; Clariant WA, file 586, Pensel to Abteilungsleiter, June 9, 1945, and Lautenschläger to Abteilungsleiter und Betriebsführer, July 1, 1945, but which does not include the words "even if they were not Party members"!

[9] HA, PA Michalson (microfilm). Thus, possibly Kiesskalt had not personally broached the subject of dismissing the "Jew by decree" Michalson to Michalson himself in 1938, as was originally planned, but had left the task to senior engineer Berger.

[10] HA, file Entnazifizierung (Vorstandssafe), Kiesskalt, Politische Erklärungen zu meinem Lebenslauf, May 5, 1945 (copy); PA Lautenschläger, Lautenschläger, Erinnerungen, vol. 8, [pp. 40–41]; Winnacker, *Challenging Years*, p. 107.

following can be said: Kiesskalt and Patat were heads of the research laboratory for submarine camouflage. When filling important jobs whose correspondence was marked 'secret command matter' it was the practice of the Nazi government to ensure that the persons appointed had been previously judged absolutely reliable by the Gestapo. These persons had to give a statement declaring their loyalty to the Nazi government or several persons well known to the Gestapo had to guarantee for them."[11] Kiesskalt was neither a "victim" nor the "hero" he made himself out to be, although he apparently had experienced some difficulties during the Nazi period. Patat also appears to have experienced political difficulties at the University of Göttingen and had therefore switched to work in industry.[12] But it appeared that they had demonstrated themselves to be "good Germans" sufficiently during their time in the plant for them to become targets. No proof was provided; the statement of the union representative about both men quoted above consists only of suspicions and assumptions. However, this was enough – both Patat and Kiesskalt were dismissed in July 1945.[13]

Two former employees of the Hoechst factory administrator and Defense Commissar Hans Pöhn were also dismissed. The *Betriebsobmann* of Hoechst, Hermann Zeh, who was in custody at the time, had apparently described them as Gestapo informers, which resulted in their immediate dismissal. But the works management vehemently defended them, accusing Zeh and "other ill-disposed people" of informing on them and of false testimony.[14] The records do not show whether the works management was successful, nor is it known whether the accusations were correct. But the accusations and their subsequent repudiation are typical, in both form and content, of the climate of denunciation, mutual suspicion, and distrust in the works at the time.

On July 5, 1945, the American Military Government confiscated the assets of IG Farben in the American sector on the basis of General Order No. 2 of the Military Government Law No. 52. Lautenschläger was dismissed from his position as works manager two days later; his dismissal was

---

[11] HA, file Entnazifizierung (Vorstandssafe), Liedtke, Freier Deutscher Gewerkschaftsbund, Industriegruppe Chemie to Military Government, Frankfurt, re. "I.G. Farbenindustrie, Werk Hoechst," dated June 30, 1945, in the appendix a report by Becker, Bassing and Weber dated June 28, 1945, on the attitude of individual persons in Hoechst toward the Nazi regime (copy only).

[12] Private files Winnacker, certificate from Patat for Winnacker, March 22, 1947.

[13] HA, PA Abteilung VI, Kiesskalt, personal data sheet, "leaving the company: July 27, 1945" by the Military Government; Kiesskalt to Pensel, April 20, 1946. Request for a certificate that he was "sent into retirement and dismissed by the US control officer because of my technical defense departments and not because of Party political incriminations"; PA Patat (microfilm), reference from Bockmühl for Patat dated July 27, 1945.

[14] HA, file Pensel-Entnazifizierung, IG Farben (Roth) to Headquarters USFET [?] C.I.C. Detachment 970, Regional Office VII, Hoechst Team, dated January 29, 1946, concerning Wilhelm Görz and dated February 18, 1946, concerning Peter Kullmann.

subsequently replayed for American news cameras.[15] In the following days a number of other senior employees, including Winnacker and Jähne, were also dismissed.[16] Colonel Percival, the American officer responsible, appointed Max Bockmühl, previously manager of pharmaceutical research, interim works manager in mid-July. However, Bockmühl was subsequently removed at the end of November and was accused of leaving too many former Nazis in office. He was replaced by Paul Roth, the head of the Acetone and Acetic Acid Department. During Bockmühl's period as works manager, denazification continued in accordance with Pensel's specifications; by the middle of August 1945 the number of suspended persons had risen to 161.[17]

Law no. 8, passed on September 26, 1945, resulted in further dismissals, since now even the most insignificant formal charge required the removal of such a person from any senior position, and on the orders of the control officer the concept of what constituted a senior position in Hoechst was a broad one, as it included even the foremen. In itself the law provided only for the removal of persons from senior and top positions, so that persons could in fact be kept on in the plant if they took on lower positions. A total of 42 persons were subsequently dismissed while 52 were demoted to "ordinary labor." All persons concerned were entitled to file a protest and endeavor to prove that they should be classified as "worthy of being employed in every position" or alternately "worthy of being employed in a nonexecutive position."[18] A final wave of dismissals – the so-called Patterson action – was carried out at the beginning of 1946. At the beginning of April 1946 Captain Patterson ordered that all persons particularly identified on the lists of the politically incriminated should be summarily and immediately dismissed. This measure affected another 182 persons. According to the records of the Personnel Department, the last politically incriminated person in Hoechst, with the exception of blue-collar workers, was obliged to leave the plant. Thus, for Hoechst, the Law for the Liberation from National Socialism and Militarism (known in Germany more tersely under the name *Befreiungsgesetz*) passed on March 5, 1946, resulted only in the dismissal of a Swiss and an Austrian national.[19]

On February 5, 1946, Michael Erlenbach was appointed "Property Custodian" and took over the functions that had been previously held by the Sub-Control Officer. He rapidly came into conflict with works manager Roth, who quite patently had no intention of allowing Erlenbach to

[15] HA, PA Lautenschläger, Lautenschläger, Erinnerungen, vol. 8, [pp. 51–7]; Winnacker, *Challenging Years*, p. 118.
[16] HA, file Pensel-Entnazifizierung, List of dismissals dated November 12, 1945 (copy).
[17] Ibid., memo dated August 20, 1945 with enclosures; cf. Winnacker, *Challenging Years*, pp. 118–19; Bäumler, *Rotfabriker*, pp. 315–16.
[18] HA, file Pensel-Entnazifizierung, Note Frank, Personalabteilung, dated December 13, 1948.
[19] Ibid.

interfere in the management of the works. Roth's high-handednesses and his attitude toward Erlenbach finally forced Erlenbach to turn to the American Army for help. His report on the problems began by stating that the senior executives of the plant believed after his appointment as Custodian that "the time has come for them to seize power again; they are only prepared to grant the Custodian a subordinate role." Roth had "not shown much good will" in cooperating with him. A "clique headed" by Roth was trying to undermine his position with the help of German government authorities. Senior executives were hoping for a "change of circumstances" and tried to portray the Custodian and his representative as working "only for the interests of the Americans." The report went on to state that before transferring power to such people who hindered the process of democratization more than they supported it, it would be better to wait. At all events: "The dualism between Custodian and works manager can only result in success if both are working toward the same end."[20]

How justified Erlenbach's complaint in this matter was is borne out by Roth's reply. Roth began by pointing out that "right from the beginning a clear demarcation between the responsibilities of the works manager and those of the Custodian" had been lacking. However, the Custodian had "done nothing to prevent such difficulties by clearly demarcating the area of operations of the Custodian from that of the factory management." Indeed, Roth clearly implied that Erlenbach was incompetent and incapable of managing the works, and went on to emphasize that, unlike Erlenbach, he himself had experience both as a researcher and as an executive: "Even if the duties of the Custodian were restricted to the handling of financial matters and the maintaining of the materials of the plant, it would be hardly possible constantly to provide him with enough information that he will be able to fill out the gaps in his knowledge concerning the affairs of the plant within a short space of time. In reality the Custodian's wishes go beyond mere information on these matters. Of course, this is not intended as a reproach against the Custodian, Dr. Erlenbach, who worked in the plant as lab. chemist [*sic*] for 20 years but had no opportunity to go beyond his scientific work and deal with general questions concerning the plant." But, of course, this was precisely what Erlenbach was being reproached for. The reason given for the lack of information provided to Erlenbach was his alleged inefficiency. Roth moreover emphasized that it was not true to say that senior executives "were still caught up in the way of thinking of the Nazi period": "All persons who had any contact to the NSDAP have disappeared from the works management."[21] After two consultations with

[20] HA, Winnacker Archiv, Sekretariat Wi. Vorlesungen 1967–9, excerpted German translation from the summary report of the IG Hoechst Custodian, undated [Juli 1946] and long English version (both copies only).

[21] Ibid., statement by Roth on the Summary Report, July 24, 1946 (copy only).

the American military authorities, however, it became clear to Roth that he would not be kept in office. On July 30 Colonel Percival instructed Erlenbach to dismiss Roth immediately.[22]

After a brief transition phase Erlenbach took over the management of the works himself. At the end of April 1947 he was appointed "Trustee" of the plant by the American military authorities; in June a new works management was appointed, headed by Erlenbach.

Pensel correspondingly summarized the time of "denazification" in Hoechst: "The political resentments of the two opposing political camps were for a long time a considerable psychological strain for the people in the Personnel Department and the active members of the Works Council, who worked together in a commendable spirit of unanimity and cooperation with the different succeeding works managements to overcome personal and social difficulties during the first years after the war."[23]

In summer 1947, shortly after Erlenbach was appointed Trustee of Hoechst, the IG Farben trial, in which top managers of the former IG Farben corporation faced charges concerning their participation in Nazi crimes, began in Nuremberg. This did nothing to lighten the prevailing climate of mutual distrust in Hoechst – after all, its former works manager Lautenschläger and his deputy, chief engineer Jähne, were also among the accused. The new works management of Hoechst apparently offered no help to either of the men during the trial; in all likelihood it did not want to be implicated by association.[24] However, Bockmühl did help Lautenschläger with his exonerating statements, and Wagenheimer provided both Lautenschläger and Jähne with cigarettes and alcohol during their time in prison.[25] Arrest, imprisonment, and the death of his beloved son while he remained in custody in Nuremberg so that he did not manage to see him alive once more were too much for Lautenschläger. He suffered a complete collapse. After initially being very cooperative and working together with the prosecutors, he subsequently refused to give any statement during the trial. He was finally acquitted for lack of evidence. The denazification tribunal even classified

[22] Ibid., order by Percival dated July 30, 1946 (copy); see also minutes by Roth dated July 26, 1946, on the second negotitiation with U.S. military autorities on July 25, 1946 (copy only).

[23] HA, file Pensel-Entnazifizierung, memo by Pensel dated April 25, 1955, re. "Die soziale Situation des Werkes nach dem Zusammenbruch 1945."

[24] Cf. HA, Hirschel, "Aus meiner I.G. Zeit," pp. 33–6, pushed the blame for this on to Erlenbach, and Erlenbach seems to have heard about this from a third party; PA Abteilung VI, Hirschel, memos dated December 14, 1967, January 10, and February 15, 1968; this was apparently one of the reasons why, on the orders of the management of Hoechst, Hirschel was not allowed to publish his memoirs.

[25] HA, file Fleckfieberversuche, Lautenschläger (or his lawyer) to Bockmühl, March 2, 1948, asking what he should write in two affidavits (copy); two affidavits by Bockmühl dated March 15, 1948; Hirschel, "Aus meiner I.G. Zeit," p. 35.

him as "exonerated."[26] This was definitely a judicial misjudgment. Lautenschläger should at the very least have been regarded as a "fellow traveler," for he was an opportunist of the first order, a spineless conformist who had handed over the plant and its employees to the *Gauleitung* and the Gestapo.[27]

Lautenschläger was a broken man after the war, believing himself to have been cheated of the fruits of his labor and of his life's work. Full of self-pity and constantly complaining, he considered himself to be a "victim," and in his memoirs he did not refrain from anti-Semitic tirades. Apparently it never occurred to him to reflect critically on his own inglorious role during the Third Reich or on his behavior toward Jewish employees, political victims of persecution, or "foreign workers."[28]

After his dismissal in July 1945 Lautenschläger wished to have nothing more to do with Hoechst – he never set foot in the plant again. As early as 1946, Haberland arranged for him to switch to Elberfeld, his old competitor in IG Farben, where Lautenschläger began working in a pharmaceutical laboratory. He was working there when he was arrested, at first as a witness, then as a defendant. After he was acquitted and released from custody, he returned to Elberfeld.[29] His attitude toward Hoechst may have been fueled by the fact that he was not accorded the support he had hoped for during the trial, and that after his acquittal Hoechst continued to ignore him. After he was acquitted it was the management of the Behring Works that sent a car to pick him up in Nuremberg.[30] Lautenschläger retired at the beginning of the 1950s and returned to his hometown, Karlsruhe, where he died, a rather isolated man, at the end of 1962.[31]

His deputy in the plant, Friedrich Jähne, was not acquitted in Nuremberg as Lautenschläger was, but was sentenced to a short term of imprisonment. Fritz ter Meer, on the other hand, was sentenced to seven years'

[26] HA, PA Lautenschläger, Lautenschläger, "Erinnerungen," vol. 8, [pp. 77–136], which included very unpleasant antisemitic tirades directed against the trial; HHStAW, Abt. 520 F (A–Z), Lautenschläger; in the judgment given by the denazification tribunal on December 29, 1948, Lautenschläger was initially classified as "Group 4 (fellow travelers)" *(Mitläufer)*, but in the judgment given on June 17, 1949, after his appeal in which he had been represented by the lawyer Heinrich von Rospatt, he was classified as Group 5, i.e., "exonerated" *(Entlastete)*.
[27] Cf. Feldman, *Allianz*, p. 537–8.
[28] Cf. HA, PA Lautenschläger, Lautenschläger, Erinnerungen, vol. 6, [pp. 224–30] anti-Semitic rants of the worst sort about the "Jewish question"; vol. 8, [pp. 57, 59, 79, 121, 268–71, 297–303] on his role as a "victim," both volumes were written around 1952; cf. also Bartmann, *Tradition*, pp. 83, 199–201.
[29] Bayer WA, PA Lautenschläger.
[30] HA, PA Lautenschläger, Lautenschläger, Erinnerungen, vol. 8, [p. 135]; Hirschel, "Aus meiner I.G.-Zeit," pp. 42–3.
[31] Hoechst Human Resources, PA Abteilung VI, Lautenschläger, cf. correspondence from the 1950s; see also Heine, *Verstand*, pp. 161–3.

imprisonment. Throughout the trial he had organized the concerted defense of the Management Board, "coordinated" their statements, and thus also ensured the silence of Lautenschläger.[32] Ter Meer subsequently collapsed. Karl Bornemann, ter Meer's defense counsel, reported to Struss, formerly the managing director of the Technical Committee office, that ter Meer, "who held up brilliantly in Nuremberg, has suffered from a complete nervous breakdown because of the unexpectedly hard sentence (6 years) [*sic*] in Landsberg." Ter Meer had given the following statement that Bornemann wished to discuss with Struss: "He, ter Meer, had been overridden by the younger members of his colleagues on the Management Board in the matter of Auschwitz, particularly by Ambros. He had been against the building of Auschwitz and had even noted this in a document which must be in his files." For this reason ter Meer wished to "force a revision of his trial." But a search had not revealed the document, and Bornemann and Struss agreed that "an appeal on this basis would be completely impracticable." Bornemann informed ter Meer of this, "who, in the meantime, had got a grip on himself again and recognized the impossibility of an appeal."[33]

Fritz ter Meer and a number of other senior management figures of West German industry later waged a journalistic battle through the media against the verdict of Nuremberg. Like Tilo von Wilmowsky from the Krupp corporation, they attempted to defame the trials of German industrialists "as a political act, which arose directly out of the war psychosis" and which "outraged any sense of justice."[34]

5.2. THE REESTABLISHMENT OF HOECHST AND THE "IG FAMILY"

After the denazification tribunals took up their work in summer 1946, many former National Socialists, who had been dismissed from Hoechst in the immediate aftermath of the war, were classified as "exonerated" or "fellow travelers" and consequently demanded their reinstatement. A large number of them were indeed taken on again, although they were not always reinstated in the position they had hoped for or given their previous jobs. By the end of the 1940s many were again in office or holding positions of responsibility, and those who had not managed to achieve this by the

[32] Wagner, *IG Auschwitz*, pp. 302–3.
[33] HA, WIN 653, Büro Winnacker, Files on Struss, which includes a memorandum by Struss about the IG Farben trial in Nuremberg and the attempt by ter Meer to appeal, dated May 31, 1952.
[34] Tilo Freiherr von Wilmowsky, *Warum wurde Krupp verurteilt? Legende und Justizirrtum*, 2nd edition (Stuttgart: Vorwerk, 1950), pp. 7–8 (preface); cf. on this topic in detail Jonathan Wiesen, *West German Industry and the Challenge of the Nazi Past, 1945–1955* (Chapel Hill: University of North Carolina Press, 2001).

beginning of the 1950s complained of their heavy lot and regarded themselves as "victims."[35]

By the end of 1951 every one of the managers sentenced during the IG Farben trial had been released.[36] The IG Farbenindustrie AG was dissolved, and the three "big sisters" were founded anew or reestablished – including Hoechst. After the breaking up of the old IG Farben group in the Western zones, one of the questions that needed to be settled was that of the future chairman of the Management Board of Hoechst. Michael Erlenbach had been the most important man in Hoechst since 1946, first as "Property Custodian," then as "Trustee" of the "Independent Unit." But he was only responsible for Hoechst itself, as the individual plants of the former Maingau Works Group were now independent companies.

Karl Winnacker, the "crown prince" of Hoechst during its IG Farben period, had been dismissed in summer 1945 shortly after Lautenschläger, who had previously even made an attempt himself to dismiss Winnacker for being a "National Socialist." After Winnacker's dismissal he worked according to his own report as a gardener for almost two years. At the same time, as he wrote himself, he often felt drawn to Leverkusen and maintained close ties to his old friends from IG Farben, particularly Haberland. In addition he edited a multivolume compendium on "chemical technology." Winnacker continued to put his time to good use until he once again found an employment as a chemist in the Duisberger Kupferhütte. He then switched to Knapsack near Cologne, where he was appointed a member of the Management Board.[37]

Once it was clear that dissolution of the old IG Farben was pending in the early years of the young Federal Republic of Germany, Winnacker as the former crown prince of Hoechst once again came under discussion as its potential chairman. He himself wrote about this period as follows: "There was fierce competition for the top job in Hoechst. A decision was urgent because in both Leverkusen and Ludwigshafen the boards of management had already crystallized."[38] There the crown princes Ulrich Haberland and Karl Wurster had already established themselves. In Hoechst, however, at first it appeared that Winnacker did not stand much chance, but subsequently his name increasingly began to be mentioned.

Winnacker held his first discussion concerning his possible appointment to the management board of Hoechst with Randolph Newman, the American control officer for IG Farben, on September 18, 1951. Shortly before their conversation Winnacker received a letter from Heinrich von

---

[35] HA, file Pensel-Entnazifizierung, Memo dated April 18, 1947, on discussion of the same date; Memo dated March 31, 1948, re. "Mitläufer-Problem"; Aktennotiz dated March 13, 1950.

[36] Cf. Wagner, *IG Auschwitz*, pp. 310–11.

[37] Cf. Winnacker, *Challenging Years*, pp. 118–40.   [38] Ibid., p. 146.

Rospatt, a representative of the Federal Ministry of Economics, who had
defended the IG Farben managers in Nuremberg prior to moving to a job in
the Ministry of Economics. After talking with Newman, Friedbert Ritter,
the Trustee of Knapsack, and Konrad Weil, the Trustee of Griesheim,
Rospatt advised Winnacker "not to insist" on becoming chairman of the
Management Board of Hoechst: "I am convinced that with such a demand
you would prevent your appointment to the Management Board of
Hoechst. But in my opinion, you would serve the purpose more if at least
you were on the Management Board, even if you were not its chairman,
than if you were not on the board at all. Mr. Abs [a leading executive of the
Deutsche Bank during the Third Reich and later chairman of the Deutsche
Bank in the Federal Republic of Germany] sends you the same advice."
According to the planned statutes the chairman of the management board
would "merely be primus inter pares." The Allies had as yet no intention of
appointing any chairman to the Management Board but intended to leave
this matter to the Supervisory Board. Rospatt even advised Winnacker to
show himself prepared to be even more accommodating: "Furthermore you
might combat possible resistance on the part of Dr. Erlenbach and
Mr. Newman by not taking over the management of the Hoechst plant but
instead you could take on the management of the production office and
leave the internal works management of Hoechst to Dr. Erlenbach."[39]

In a talk with Newman on the same day, Newman informed Winnacker
that he had heard "from different sides" about the suggestion to appoint
Winnacker to the Management Board of the soon to be reestablished
Hoechst Farbwerke. Newman also appeared "inclined to follow this
suggestion" and requested Winnacker's comments. Contrary to all advice
Winnacker answered that he would "only join the Hoechst Management
Board as chairman." He had not attempted to return to Hoechst and was of
the opinion that it would be "associated with many difficulties, above all
with resentment on the part of the gentlemen who performed the office of
trustees up until now." If he went to Hoechst, then he would only go under
the condition that "he would hold the senior position, as had been decided
by the Management Board of the old IG Farbenindustrie." Newman replied
that he "could not agree to this condition" since there were no plans for a
chairman, and it likewise represented "an affront" to the gentlemen currently in office. On being asked at the conclusion of their talk by Newman
whether or not he would be prepared to join Hoechst, even if he did not

[39] HA, Winnacker Archiv, Sekretariat Winnacker Vorlesungen, Letter from Rospatt, BWMin
(i.H.), to Winnacker, September 18, 1951. On Rospatt see Bernhard Löffler, *Soziale
Marktwirtschaft und administrative Praxis. Das Bundeswirtschaftsministerium unter
Ludwig Erhard* (Stuttgart: Steiner, 2002), pp. 169–70, and Norbert Frei,
Vergangenheitspolitik. Die Anfänge der Bundesrepublik und die NS-Vergangenheit
(Munich: dtv, 1999), p. 164.

become chairman, Winnacker was not prepared to commit himself. The decision was adjourned until the following day. Winnacker then proposed a compromise, which he had obviously agreed on with Ritter, who wanted him back in Hoechst as its chairman. Instead of the position of chairman, it was suggested that a position of "technical manager" could be created, who would be put in charge of the entire chemical production and research – and this was the position Winnacker wanted to have.[40]

After this proposal met with the approval of the American authorities, Winnacker inquired of his former mentor and sponsor Fritz ter Meer what he thought of it – and here the old hierarchy of IG Farben, which obviously still existed in people's minds, becomes very obvious. In a private memorandum Winnacker noted: "First of all ter Meer emphasized that I am fully and completely entitled to demand the chairmanship in Hoechst on the basis of the decisions of the former Management Board. However, in view of the attitude of the Americans he believes the right thing to do is not do demand anything impossible and to pursue the suggestion made by Ritter."[41] But now Gustav Pistor, the former plant manager of Bitterfeld, and Paul Duden, the old plant manager of Hoechst, opposed Winnacker's appointment, and in a letter to the Federal Minister for Economic Affairs, Ludwig Erhard, a copy of which was sent to German Chancellor Konrad Adenauer, they proposed that Ritter be appointed chairman. Winnacker was deeply outraged by this and typically declared to ter Meer that he "would not continue the negotiations which were now underway a moment longer" if he could not be "absolutely sure that I am acting in accordance [with the wishes of] all of the old IG." Ter Meer, who had also been annoyed by the letter, intervened immediately and contacted Pistor, who claimed it had all been a misunderstanding.[42]

At this point Winnacker's position was still far from secure. In a meeting with Ritter and Winnacker, Weil, the Trustee of Griesheim, and Otto Schulz, the Trustee of Offenbach, voiced their "considerable doubts regarding the consent of Dr. Erlenbach who has informed them by telephone that he will oppose a return of my person to Hoechst." Because of this "difficulty" they suggested to Winnacker that he should postpone his return to Hoechst "for one or two years." Whether Schulz and Weil were using Erlenbach as an excuse – which Schulz's suggestion of appointing Weil to be chairman of Hoechst seems to indicate – or whether Erlenbach

---

[40] HA, Winnacker Archiv, Tagebuch 1951–4, Aktennotiz by Winnacker dated September 24, 1951, on discussion with Newman and Thiernay on September 18, 1951, and on discussion with Newman et al. on September 19, 1951.

[41] Ibid., Aktennotiz by Winnacker dated September 24, 1951, on conversation with ter Meer on September 20, 1951.

[42] Ibid., Aktennotiz by Winnacker dated September 24, 1951, on telephone conversation with ter Meer on September 22, 1951, and on telephone call from Dr. Pistor on September 23, 1951.

was quoted correctly is not clear, but it was evident that there would be considerable problems with the existing Trustees.[43] The Government of the *Land* Hesse came to the aid of the Trustees Schulz and Weil, declaring that the return of Winnacker to Hoechst would mean "a resurgence of the old IG coterie."[44]

But Winnacker's appointment was also rejected by some of the former members of the IG Farben Management Board. Pistor and Duden had apparently favored Fritz Gajewski or Ritter as the new chairman of Hoechst. Wurster was apparently also opposed to Winnacker, who had "rushed through the IG nursery with the speed of an express train" and did not have any comprehensive training. Wurster favored Weil, although he would have preferred Erlenbach if he "had the capacity to undertake this," as he wrote in a memorandum. This seemed to be the problem. For although Newman was apparently quite satisfied with his choice, that is, Winnacker, he clearly would also have preferred Erlenbach. Newman is quoted as having said that he was "somewhat disappointed that in Germany Erlenbach was not regarded as suitable in any way to be chairman of the Management Board since he was apparently considered to be too stupid."[45]

On December 7, 1951, the Farbwerke Hoechst AG were finally formed, followed on December 19 by the Farbenfabriken Bayer AG, and on January 30, 1952, by the Badische Anilin & Soda-Fabrik AG (BASF). As technical manager of the new Hoechst, Winnacker was appointed to the Management Board, as was Michael Erlenbach, now the manager of the Pharmaceutical and Inorganic Division.[46] By the following year Winnacker had already achieved his ambition – to become chairman of the Management Board in Hoechst.

Once he had firmly established himself as chairman at Hoechst, Winnacker quickly proceeded to help the "old boys" from the old IG Farben and to obtain positions for them within Hoechst where possible, irrespective of their political incrimination. Thus the unease expressed by the government of the *Land* Hesse, which considered that he would fetch the "old IG coterie" back, turned out to be more than justified.

Already in June 1952, on the occasion of a talk with Adenauer concerning the new Hoechst corporation, Winnacker began enquiring into the possibility of cooperating with those members of the IG Farben Management Board who had been sentenced in Nuremberg. According to Winnacker's

---

[43] Ibid., Aktennotiz by Winnacker dated September 26, 1951, re "Besprechung in Knapsack, 25.9.1951" (copy).

[44] HA, Winnacker Archiv, Sekretariat, Vorlesungen, Memos on the filling of management positions in Hoechst September 28–October 20, 1951 (apparently a transcript made of recordings of Mr. Thies).

[45] Ibid.

[46] See Winnacker, *Challenging Years*, pp. 146–50, in which Winnacker's appointment appears less controversial.

notes, Adenauer answered that "after the ratification of the General Treaty" Hoechst would be in a position "to cooperate more closely with these gentlemen in accordance with our wishes." Adenauer had then added: "only the assumption of official positions on the Management Board or Supervisory Board is still restricted."[47] This statement by Adenauer mirrored the policies decided on in the Federal Cabinet in May of that year concerning former members of the IG Farben Management Board who had been sentenced as war criminals.[48] Accordingly, only a few days later Winnacker took steps to set up a form of cooperation with his former superior Jähne, whereby it was planned that Jähne would start working as a technical consultant to the Bobingen plant, now part of Hoechst. He would be paid DM 5,000 for his work. Jähne would receive a further DM 10,000 "after the disincorporation of the company and as soon as we have the right to freely dispose of our capital and our assets" as "a contribution toward expenses" that would allow him to "cover his financial obligations which he incurred as a result of being a defendant in Nuremberg."[49] This was a considerable financial sum spent to aid a former member of the Management Board, particularly in view of the fact that Winnacker opposed the idea of Hoechst being IG Farben's legal successor when the matter concerned the reimbursement of victims of National Socialism.

Walther Dürrfeld also benefited from Winnacker's solidarity. Concerning Dürrfeld, one of the three heads of the IG Farben plant at Auschwitz, the historian Bernd Wagner judged that as the dominating figure of the Auschwitz plant Dürrfeld had been a particularly terrible accomplice of the SS, nay more: "In particular the stepping up of disciplinary measures, the systematic surveillance and the exchange of weakened prisoners for 'new' inmates which was to prove fatal to thousands, all this can be traced back to the personal interventions of Dürrfeld. An increasing lack of scruples in dealings with prisoners is observable over time. The example of the murder of Jews on a massive scale within view of the factory premises,

---

[47] HA, Winnacker Archiv, Tagebuch 1951–1954, Aktennotiz by Winnacker dated June 21, 1952, re "Besprechung mit Bundeskanzler Adenauer am 19.6.52 in Bonn."

[48] Die Kabinettsprotokolle der Bundesregierung, vol. 5, 1952, p. 255, n. 32, Draft of Ludwig Erhard dated April 24, 1952, and pp. 322–3: special session held on May 14, 1952, on the Transition Agreement which also addressed the "problem of war criminals": with respect to the convicted IG Farben managers it was noted that according to the provisions of Law No. 35 they would not be permitted to participate in the management of the successor companies of IG Farben; but the following note was added: "It is not possible to dispense with the contribution of these persons. The Allies – with the exception of Mr. v. Schnitzler – have no objection to these persons working in some position, irrespective of what those positions might be, as long as they do not officially appear as members of the board of management. This will apply for the duration of Law No. 35. This would objectively suffice to satisfy German demands."

[49] HA, Winnacker Archiv, Tagebuch 1951–1954, Aktennotiz ("Confidential!") by Winnacker dated June 28, 1952, re "Besuch von Jähne am 25.6.52 (Königstein)."

the unsatisfactory level of productivity despite all countermeasures taken, and the situation of the war which was turning against Germany made the orders he gave become increasingly radical."[50]

On October 8, 1952, Winnacker wrote to Dürrfeld that after holding numerous talks he could now firmly promise him a job in Hoechst, adding: "We would like to ask you to give us a little time before we make out the contract since we are looking for a suitable way to do this, due to circumstances with which you are familiar."[51] On October 10, 1952, Winnacker discussed several matters with Ritter, the chairman of Knapsack, including the case of Dürrfeld. The two men came to an agreement that Knapsack would take on Dürrfeld: "Once everything has been definitely concluded concerning the employment of Dr. Dürrfeld, the contract will be transferred to Hoechst. But it is already agreed that the contract will be with Hoechst."[52]

After he had settled this matter, Winnacker informed Hans Bassing, chairman of the Works Council in Hoechst, and Erlenbach, his colleague on the Management Board, the following day that he wanted to hire Dürrfeld for the Engineering Division and would shortly be concluding a contract with Knapsack. Winnacker's memorandum on this matter is worth being quoted in full: "Initially the name of Dürrfeld did not mean anything to Bassing. I emphasized to him that D. had been the builder of the IG plant Auschwitz and had been sentenced to a longer term of imprisonment in Nuremberg. Bassing was of the opinion that on account of the limited contact, the workforce would hardly take any notice of Dr. Dürrfeld. But there was a danger that somebody might feel slighted when he was taken on by the engineering division and this would cause some disturbance. B. agrees to D. being hired in Hoechst, but he recommended that this should be done carefully and at the beginning no special external emphasis should be accorded him. I discussed the matter with Dr. Erlenbach who was of the same opinion."[53] The significance of this is that Winnacker and the men he talked with were aware whom they were hiring, and Bassing's only worry was to avoid Dürrfeld being given preferential treatment too early.

Dürrfeld signed the contract in October 1952, and it even awarded him a pension without requiring a minimum length of service with the company "in consideration of your long-standing years of service in IG," as Winnacker wrote.[54] But Dürrfeld never took up his duties either in Knapsack or in Hoechst. The reason for this was that the Wollheim trial

[50] Wagner, *IG Auschwitz*, p. 291.
[51] HA, PA Abteilung VI, Dürrfeld, Winnacker to Dürrfeld, October 8, 1952.
[52] HA, Winnacker Archiv, Tagebuch 1951–1954, Aktennotiz by Winnacker dated October 15, 1952, re. "Besprechung mit Dr. Ritter, Knapsack, am 10.10.52."
[53] Ibid., Aktennotiz by Winnacker dated October 11, 1952, re. "Unterhaltung mit dem Betriebsratsvorsitzenden Bassing bezüglich Einstellung Dr. Dürrfeld (11.10.52)."
[54] HA, WIN 653, Büro Winnacker, Personalia, Dürrfeld to Winnacker, November 18, 1952; PA Abteilung VI, Dürrfeld, Winnacker to Dürrfeld, October 20, 1952.

started its proceedings in Frankfurt, the "great importance" of which was immediately obvious to Winnacker.[55] In a memo written some time later, Winnacker noted rather sourly: "shortly after we had signed the contract it became known to us that the Wollheim trial was about to begin in Frankfurt and that Dr. Dürrfeld would appear in this trial as a prominent witness. While we were not aware of these facts, I think I can safely assume that Dr. Dürrfeld had already been apprised of the situation."[56]

Indeed, Wollheim's statement of claim against IG Farbenindustrie AG in Liquidation had been filed at the civil division of the district court of Frankfurt upon Main on November 3, 1951, by Wollheim's lawyer Henry Ormond.[57] Dürrfeld was summoned to appear as a witness on November 20, 1952, and claimed that he had only learned of this fact on November 14 in a letter from one of the liquidators of IG Farben dated November 13. Dürrfeld described the chronology of the events to Winnacker at Winnacker's direct request. Dürrfeld claimed to have only learned from the aforementioned letter that he was expected to appear as a witness and had then immediately contacted the IG Farben liquidators and the lawyers. On November 17 he had received the summons to appear of the court and had then consulted Rospatt, the last lawyer to represent him during his time in custody in Landsberg. On November 18 he was certified as too ill to appear by his doctors and informed the district court of his inability to appear.[58]

In a meeting between Dürrfeld and Winnacker on December 1, Winnacker told Dürrfeld that he would not be able to begin work yet. First, a delay had occurred in the lifting of American control, and Winnacker had no desire to seek permission. Second, Winnacker pointed out that "the trial currently underway in Frankfurt will have repercussions on the workforce in Hoechst which makes an appearance during the trial seem inadvisable."[59] According to a note by Dürrfeld on their conversation: "It was felt to be a lucky coincidence and in our mutual interest that it had not been possible for me to appear on the appointed day in front of the district court in response to the summons. Furthermore it will be expedient for the time being if I do not become the focus of attention."[60]

---

[55] HA, Winnacker Archiv, Tagebuch 1951–1954, Aktennotiz by Winnacker dated December 1, 1952, re. "Unterredung mit Dr. Brinckmann (Liq. A.) am November 29, 1952."

[56] HA, WIN 653, Büro Winnacker, Personalia, Aktennotiz by Winnacker dated March 25, 1954, on Dr. Dürrfeld.

[57] IfZ, ED 422, vol. 1, statement of claim filed by the lawyer Mr. Ormond at the Landgericht Frankfurt am Main dated November 3, 1951; cf. Wagner, *IG Auschwitz*, pp. 311–15.

[58] HA, PA Abteilung VI, Dürrfeld, Dürrfeld to Winnacker, December 3, 1952 – with attachments.

[59] HA, WIN 653, Büro Winnacker, Personalia, Aktennotiz by Winnacker dated December 1, 1952, re. "Unterredung mit Dr. Dürrfeld am 1.12.52 in Hoechst (Zeitw. anwesend: Dr. Wengler)."

[60] HA, PA Abteilung VI, Dürrfeld, Aktennotiz by Dürrfeld ("Confidential!") dated December 3, 1952, re. "Besprechung mit Direktor Dr. Winnacker und Direktor Dr. Wengler am

Winnacker even sent an employee to the trial to report on its progress to him and Erlenbach. The employee, Reinhold Bertrams, a trained commercial clerk from the Personnel and Welfare Department, gave a more or less detailed account of the individual days of the trial and the witnesses and occasionally enclosed newspaper articles.[61] As the press, particularly the *Frankfurter Rundschau*, was considered to be "hostile to IG," Winnacker was not unhappy when Dürrfeld received an offer from Ambros.[62] Ambros, who worked as a consultant for Friedrich Flick among others, offered Dürrfeld a "very attractive job in one of the large Ruhr plants."[63] To help Winnacker out, Haberland also offered to employ "Dürrfeld in RUHRBAU for an interim period in case that we in Hoechst meet with difficulties."[64]

However, Dürrfeld finally preferred to take up the job offered him by Ambros, since Winnacker made it clear that Hoechst wished to abide by the contract but was unable to employ him "in view of the ongoing trial in Frankfurt." It was suggested that he should work for Uhde for several months. However, Dürrfeld did not wish to do so and became a member of the management board of Scholven-Chemie, a subsidiary of Hibernia. But in a letter terminating his contract with Hoechst Dürrfeld was apparently inclined to be ungracious. The letter "included phrasing" that Winnacker "could not accept." In his memorandum on the conversation he held with Dürrfeld concerning the letter Winnacker showed himself to be most incensed: "I have drawn the attention of Mr. Dürrfeld to the fact that Hoechst was the only IG plant to give him an extraordinarily favorable contract immediately after he had asked for one, and one which was also approved by all the gentlemen in Hoechst. I find it unjust and ungrateful if the blame for the facts of the ongoing trial in Frankfurt and its consequences are now somehow placed ·at Hoechst's door. Dr. Dürrfeld thereupon asked to be allowed to retract his letter."[65] But Dürrfeld was by no means grateful, and a long spate of negotiations between Dürrfeld and Hoechst followed concerning eventual financial obligations on the part of Hoechst.[66]

Winnacker's experience with Dürrfeld and the Wollheim trial did not shake his resolve to show solidarity to the so-called victims of denazification;

1.12.1952 in Hoechst" (attachment to the letter from Dürrfeld to Winnacker, December 3, 1952).

[61] HA, PSW 1332/ Dez. 1951–Juni 1953 (Sozial-Ausschuss, Nachlass Dr. Erlenbach).

[62] Ibid., here note dated December 13, 1952.

[63] HA, WIN 653, Personalia, Dürrfeld to Winnacker, December 25, 1952.

[64] HA, PA Abteilung VI, Dürrfeld, Aktennotiz by Winnacker dated January 21, 1953, re. "Besprechung mit Wurster und Haberland am 19.1.1953 betr. Dürrfeld."

[65] Ibid., Aktennotiz by Winnacker dated January 27, 1953, re. "Besprechung mit Dürrfeld in Hoechst am 26.1.53 (mit Dr. Wengler)."

[66] Ibid., Dürrfeld died at the beginning of 1967. An obituary notice from Scholven-Chemie in the files is dated March 1, 1967.

he only became more careful.[67] Thus Winnacker expressed his regret in one case that he was unable to offer any more help. He explained to the person concerned that "in his case and that of many others we are in the difficult situation of having to redress and moderate the results of denazification and that certain limits are set us in our efforts to do so."[68] How much he was committed to helping persons he declared to be "victims" was particularly noticeable in the case of Max Ilgner, formerly a member of the IG Management Board. In a conversation with him in the spring of 1955 Winnacker emphasized: "A moral standpoint toward those who were sentenced in Nuremberg is a matter of course for us. We will continue to maintain humanitarian contacts to this circle as we have done up to now. Together with the successor companies we are currently considering how we can provide an economic basis for those persons who were affected most, particularly those who have an insufficient pension."[69] Winnacker had already contacted Wurster and Haberland, and both men were prepared to provide some form of assistance to the younger men of those sentenced in Nuremberg. But nothing was specifically decided as yet.[70]

Ilgner stressed in his conversation with Winnacker that he "gratefully appreciated the moral support of Hoechst," but he was now of the opinion that this "must also find expression in an economic form." He proposed concluding a freelance contract with Hoechst. While Winnacker declared that Hoechst was prepared to provide financial support even in the event that "contrary to expectations" the other two successor companies did not wish to contribute anything, nevertheless he wanted at all costs to ensure that "no affiliation between Dr. Ilgner and Farbwerke Hoechst is discernible in any form." Winnacker's rejection of the suggestion of Hoechst financing an institute of foreign trade by concluding a direct contract with Ilgner was

[67] Just how careful Winnacker became is not only evident from this story. There is also a nice example in the memoirs of Franz Josef Strauß, *Die Erinnerungen* (Berlin: Siedler, 1989), pp. 238–9: After Strauß had been appointed West German Defense Minster, he wished to appear once more at a meeting of the supervisory board of the Karlsruhe reactor construction company in his former position as Minister for Atomic Affairs. This was rejected, to his incomprehension, by his successor Balke and by Winnacker. Winnacker telephoned him: "'I was once already mixed up in a war criminal trial' – cue IG Farben – 'and I tell you that we who are in atomic research should not have the slightest thing to do with military considerations.'" Strauß continued: "This episode belongs to the sad chapter of the wounded German nation." A complete misjudgment, at least as far as Winnacker was concerned.

[68] HA, PA Abteilung VI, Hermann Holler, Winnacker to Direktor Ritter in Knapsack-Griesheim AG, February 26, 1953.

[69] HA, PA Abteilung VI, Ilgner, Aktennotiz by Winnacker re "Dr. Ilgner. Unterhaltung am 21.3.1955 (anwesend: Dr. Weil)."

[70] Ibid., excerpt from a memorandum of Winnacker concerning a talk with Haberland in Leverkusen dated January 10, 1955, and on a discussion with Wurster on January 12, 1953, in Ludwigshafen (dated January 13, 1955).

even blunter: "Hoechst cannot conclude any freelance contract. It is completely unthinkable for any political-economic department or even a loose cooperation to be created with a connection to Hoechst which would in any form remind people of NW 7 [Ilgner's former office as a lobbyist for IG Farben in the Third Reich] or its circle of employees. Irrespective of any personal esteem [in which he is held] it is completely out of the question for Hoechst to expose itself in any way in this dangerous area, particularly since these activities are being closely watched by foreign countries in the past and also now."[71] Although Ilgner had apparently been very much disliked by Wurster, Haberland, and Jähne because of "his arrogant behavior based on his close connection to Schmitz," whose nephew he was, "and because of Nuremberg," Winnacker was ready to help him – although Winnacker himself described Ilgner as "slightly crazy."[72] After Ilgner was unable to drum up sufficient support for his project of a research institute for foreign trade financed by German industry, Winnacker offered him, in addition to his pension from IG Farben, the choice of either receiving an annual payment of DM 12,000 through the offices of a friendly bank or of receiving a single payment of DM 100,000 – a more than handsome offer. Ilgner chose the latter option but was far from content, and later attempted to negotiate for still more money.[73]

Winnacker knew why he demanded strict confidentiality in the matter of any help offered to Ilgner and why he declined to rehire him. Quite apart from any personal problems with Ilgner, what was far more important was the fact that the appeal against the verdict of the Wollheim trial was still underway concerning the reimbursement of concentration camp inmates who had suffered in the IG Farben plant at Auschwitz.

The IG Farbenindustrie in Liquidation had been sentenced in the first instance by the district court of Frankfurt upon Main to pay Norbert Wollheim compensation amounting to DM 10,000 plus 4 percent interest as of July 1, 1951. Wollheim, a Jew, had been arrested by the Gestapo in Berlin on March 8, 1943, and deported to Auschwitz together with his wife and three-year-old son. There he was separated from his family, who were subsequently murdered in Auschwitz. Wollheim himself survived the concentration camp as the SS sent him to the Auschwitz plant, where he worked as a welder.[74] The court, which clearly indicated in its verdict that

---

[71] Ibid., Aktennotiz by Winnacker re. "Dr. Ilgner. Unterhaltung am 21.3.1955 (anwesend: Dr. Weil)."

[72] Ibid., handwitten note (not signed) dated May 20, 1955; see also Aktennotiz by Menne dated February 6, 1956.

[73] Ibid., Briefwechsel 1955–1962.

[74] IfZ, ED 422, vol. 9, Wollheim to Ormond, July 23, 1951, as attachment: "Tätigkeit als Häftling für die IG-Farben," July 20, 1951; vol. 1, Ormond to Landgericht Frankfurt am Main, Zivilkammer, Wollheim's claim against IG in Liquidation, November 3, 1951; final speech of Ormond at the hearing on May 11, 1953.

the witnesses from IG Farben had left a bad impression, found "an appalling indifference on the part of the defendants [i.e., of IG Farben] and its employees toward the plaintiff and the imprisoned Jews." IG Farben had not complied with its obligations as an employer to provide for the welfare of its Jewish employees in Auschwitz, and did not even attempt to argue that "it had undertaken any serious efforts to improve the lot of the plaintiff and of the Jewish prisoners." The court was of the opinion that as IG Farben "did not do everything it could or should have done for the plaintiff; it had infringed against its obligations as an employer to provide for the welfare of its employees, and at the very least it had acted negligently, that is, culpably." Since right from the beginning the case was clearly a test case, as the plaintiff's lawyer admitted, IG Farben in Liquidation now faced the threat of numerous subsequent prosecutions.[75]

Not least because of this, IG Farben in Liquidation appealed against the verdict. Already before the judgment was pronounced, IG Farben in Liquidation turned to W. Alexander Menne, a member of the Management Board of Hoechst and president of the Association of Chemical Industry, and sought to enlist his support in the Wollheim trial and the trial of Rudolf Wachsmann, which followed on the tail of the Wollheim trial. It was hoped that Menne could harness German and foreign media to the interests of the industry, and additionally get in touch "with reasonable Jewish organizations" so that "these circles would dissociate themselves in public from the trials."[76]

At the same time, IG Farben in Liquidation attempted to put pressure on the Federal Government and its successor companies by declaring that if it lost its case in front of the appellate court and did not receive sufficient support from the Federal Government the reserves of IG Farben in Liquidation would not be sufficient and it would be obliged to draw on the assets of its successor companies. A letter from IG Farben in Liquidation was sent to Wurster, Haberland, and Winnacker, which stated unequivocally: "In the event that, contrary to expectation, the measures we have undertaken in approaching various ministries are not successful within a very short time, we will be forced to discuss with you to what extent you will assume liability for the claims of former concentration camp inmates which are not covered by the residual assets of IG in order to enable you to immediately exchange share certificates."[77]

---

[75] IfZ, ED 422, vol. 1, witnesses for the plaintiff who had also suffered in Auschwitz, declared in the hearings that they also intended to file claims; final speech of Ormond at the hearing on May 11, 1953, and judgment of the *Landgericht* Frankfurt am Main dated June 10, 1953; cf. Wagner, *IG Auschwitz*, pp. 311–12.

[76] HA, Menne Nachlass, Schadensansprüche ehemaliger KZ-Häftlinge, IG Farben in Liquidation (Schmidt) to Menne, Hoechst, August 4, 1953.

[77] Ibid., IG Farben in Liquidation (Schmidt, Reuter) to Haberland, Wurster, Winnacker, August 14, 1953; idem to Allied High Commission, Tripartite Farben Control Group,

The judges in the second instance were interested in arranging a settlement, as were both parties, IG Farben in Liquidation and Wollheim and the Jewish organizations that in the meantime were supporting him. Despite their sympathy for Wollheim's claims, the large Jewish organizations such as the Claims Conference or the United Restitution Organization (URO) were in a critical position as they were in the midst of negotiations, initiated by Adenauer on behalf of the Federal Republic of Germany, concerning possible "restitution" payments. A trial against IG Farben in Liquidation was "therefore a risk in their eyes, since this could arouse resistance on the part of German industry against a general law regulating indemnification." On the other hand, appeal proceedings also posed a considerable risk for IG Farben in Liquidation, as previously mentioned.[78]

IG Farben in Liquidation faced two major problems. First, a settlement would entail "the recognition of responsibility for the conditions in Auschwitz," all the more so as "both the verdict in Nuremberg and the Wollheim trial in the first instance created an unfavorable picture of IG in the public mind." Second, IG Farben also very much opposed any settlement because this would raise the question of individual compensation: "The precondition to weaken this antagonism [against any settlement] is a proposal by the Jewish associations which will provide sufficient guarantees to ensure that no further claims against IG will be made after a global compensation has been paid out."[79]

The hiring of Max Ilgner by Hoechst at this critical point in time could have endangered the negotiations for a settlement. Likewise, Fritz ter Meer declined to accept a position on the Supervisory Board of one of the IG Farben successor companies before the end of the Wollheim trial.[80] Then, shortly before the beginning of the trial, Winnacker, as mentioned above, to his own annoyance hired no less a person than Dürrfeld, the former head of the company's plant at Auschwitz – but Winnacker was able to prevent his being taken on in Hoechst. Consequently Winnacker preferred to provide former members of IG Farben Management Boards with consultancy contracts that were not concluded directly with Hoechst. Thus, for example, a highly remunerated job was found for Heinrich Bütefisch, previously sentenced to six years' custody in Nuremberg, as a consultant to the Ruhrchemie, which was jointly financed by Hoechst and Mannesmann.[81]

August 14, 1953; idem to Bundeswirtschaftsminister Erhard, August 14, 1953; idem to Ministerialrat Kuschnitzki, Bundesfinanzministerium, August 13, 1953.

[78] Wagner, *IG Auschwitz*, p. 312.
[79] HA, Menne Nachlass, Schadensansprüche ehemaliger KZ-Häftlinge, Memo from Kaufmann for Winnacker and Menne dated March 18, 1954, re "Besprechung bei der I.G. i. L. am 17.3.1954."
[80] HA, Winnacker Archiv, Tagebuch 18.9.1951–31.12.1954, here: memo by Winnacker on meeting in Eltville on December 14, 1953 (excerpts).
[81] HA, PA Abteilung VI, Bütefisch.

Helping former Hoechst employees with Nazi pasts was less politically charged, since with Winnacker's support they were already "rehabilitated" before the Wollheim trial ended at the beginning of 1957. Jürgen von Klenck, a former Hoechst employee under Kränzlein, a member of the SS and senior executive of Ambros who was involved in the organization of the "entire chemical weapons industry" during the war, was taken on by Hoechst as manager of all applications technology in 1955 and appointed a director in 1958. He finally switched to the Management Board of Mannesmann in 1967.[82] In 1953 Hans Wagenheimer joined Anorgana in Gendorf and became a managing director there in 1955. After the merger with Hoechst at the end of 1955 he was given commercial power of attorney. At the beginning of 1958 he switched to Kalle, where first he became a deputy member and in 1961 a full member of the Management Board.[83] Otto Hirschel, who had been responsible for the treatment of the forced laborers and had worked together with the Gestapo as head of the Social Department during the war, was also rehired as an authorized signatory.[84]

And finally Hans Streeck was also back in Hoechst again, where he had been taken on in 1934 as an employee of Kränzlein and where he had joined the SS in 1935. Like von Klenck, Streeck worked for Ambros in the context of the "Krauch plan" during the war. Streeck was involved in the planning and erection of the IG Farben plant at Auschwitz, and at the beginning of 1944 he became head of the Department for Intermediate Products, which was just being set up. A very positive reference from IG Farben dated July 5, 1945, included the following encomium: "For his colleagues and subordinates he provides an example of the conscientious fulfillment of duties and diligence, while his willingness to help colleagues spread through his sphere of activity and made him popular with everybody." Leaving aside the question whether all "employees" would have agreed with this judgment, the end of the reference may strike the reader as particularly cynical: "Because of the changed situation and the resultant abandonment of our Auschwitz plant on 1/21/1945 Dr. Streeck lost his extensive and responsible field of activity so that he was not able to reap the fruits of his industry and his long labors at this place of activity."[85] Writing to Hoechst, Streeck applied directly to Winnacker and not to the Personnel Department in June 1953 since he wanted to avoid "a refusal resulting in a 'fait accompli' which might occur due to certain lingering resentments." This was not an erroneous idea, as in 1949 the Personnel Department had passed a resolution not to hire former members of the SS. Streeck and a certain Dr. Hampe were even referred to by name in a memorandum on the subject.

---

[82] HA, PA Abteilung VI, von Klenck. [83] HA, PA Abteilung VI, Wagenheimer.
[84] HA, PA Abteilung VI, Hirschel.
[85] HA, PA Abteilung VI, Streeck, IG reference dated July 5, 1945.

However, Winnacker was optimistic, writing back to Streeck: "I can perceive of no essential difficulties standing in the way of your return to Hoechst."[86] After joining Kalle, now again part of Hoechst, in January 1954 Streeck's career proceeded apace. He became an authorized signatory in the same year, and a director at the beginning of 1955. In the following year, he became a deputy and in 1957 a full member of the Management Board. When he died in 1963 he was chairman of the Management Board of Kalle.[87]

Werner Schultheiss was allowed to return to Hoechst in 1958, apparently as one of the last of the politically incriminated executives. In 1938 he had taken over the management of the Azo Division from Landers, and it was mooted that Schultheiss had been one of the three men responsible for almost getting Landers sent to a concentration camp, a fate he had avoided only by immediately stepping down from office. Schultheiss was dismissed from Hoechst as politically "incriminated" in 1945 – he had been a member of the NSDAP since 1932 and became a *Prominenter* of Hoechst in 1943. After the war he worked for many years for a small company in Bavaria. His relatively late return to Hoechst, only in 1958, may have been connected to the rumors linking him to Landers. However, after his return his career was as steep as that of Streeck. After his reinstatement in Hoechst he was appointed manager of the main Scientific Laboratory; at the beginning of 1961 he became head of research and consequently a full member of the Management Board, where he remained a member until his retirement in 1969.[88]

Winnacker was even prepared to be generous toward former employees of the *Gebechem* and the Four-Year Plan, despite the not infrequent problems IG Farben had experienced with that body. Councilor of state Walther Schieber, an "old fighter" of the NSDAP and former member of the SS who had first worked for the Four-Year Plan and later, during his work for the Armament Ministry, had been decisively involved in the use of concentration camp inmates in German industry, was given a monthly "financial subsidy which could be revoked at any time" by Winnacker and Haberland amounting to DM 500 from the 1950s until his death at the end of June 1960, since he was in financial difficulties;[89] after his death

[86] Ibid., Streeck to Winnacker, June 17, 1953; note by von Pensel dated March 16, 1949, on discussion with Sieglitz and Siebert dated March 14, 1949; Winnacker to Streeck, July 3, 1953.

[87] Ibid., Unterlagen (professional files).

[88] BA, BDC file on Schultheiss; HA, PA Abteilung VI, Schultheiss; HHStAW Abt. 520 F (A-Z), Schultheiss.

[89] Cf. Jan E. Schulte, *Zwangsarbeit und Vernichtung: das Wirtschaftsimperium der SS. Oswald Pohl, und das SS-Wirtschafts- und Verwaltungshauptamt* (Paderborn: Schöningh, 2001), pp. 214–15, 400.

Winnacker and Haberland continued to support his widow.[90] This at least throws a new light on the question to what extent the Four-Year Plan was a vehicle for the interests of IG Farben. Peter Hayes emphasizes that employees working for the Four-Year Plan were primarily loyal to that body and less so to IG Farben.[91] The payments to Schieber as well as other evidence in the archives appear to argue against this point of view. In a letter written to Hoechst in the middle of 1943 by Friedrich Baasch, head of a department on the staff of the *Gebechem*, Hoechst was requested to increase the salary of an employee provided by IG Farben to the *Gebechem* and currently deployed in Croatia and France, since the employee had "given a very good account of himself." Baasch did not know the salary of the person concerned and was therefore unable to offer a concrete suggestion: "I am asking you to take account of Mr. Fett, who by virtue of his work for GB-Chemie indirectly also works for I.G. Farbenindustrie AG, in any future arrangements."[92] It is not clear to what extent Fett considered himself a representative of IG Farben or of the *Gebechem* and acted accordingly, but his new employer clearly regarded him as an indirect employee of IG. The monthly payments to Schieber can be similarly interpreted.

After all, the men all knew each other from the war and respected each other's work. Serious political charges meant very little to them if they anticipated that the company would benefit from their work. Friedrich Dorn, who was hired by Kalle in 1953, is a good example of this. Dorn, born in 1906, completed his doctorate on the German cellulose industry in 1934 and was considered an expert on cellulose and paper. This was apparently the reason why he became the accredited expert on these questions for the Four-Year Plan. Dorn had been a member of the SS since 1934 with a break between 1935 and 1939 and a member of the NSDAP since 1937. He rose in the SS hierarchy to become an SS *Sturmbannführer* but was not appointed *Standartenführer* since he refused to move to the SS *Wirtschaftsverwaltungs-Hauptamt* (Main Office for Economic Administration) under Oswald Pohl. Dorn preferred to remain in the Reich Ministry of Trade and Commerce as one of the closest colleagues of Hans Kehrl. He became a *Ministerialdirigent* (assistant secretary) in the Ministry and Reich Commissioner for Paper and Packaging as well as head of the "monopoly corporation for textiles and paper for the occupied Eastern territories," a corporation set up by Reich Marshal Göring.[93] Kalle was interested in Dorn because of his achievements during the war, when he had been "assigned ever more and increasingly important tasks," all of which he "carried out to the entire satisfaction of his supervisors." In his report on the hiring of

[90] HA, PA Abteilung VI, Schieber.   [91] Hayes, *Industry and Ideology*, pp. 176–9.
[92] HA, PA Abteilung VI, W. Fett, Baasch, GB-Chemie, to IG Farben, April 5, 1943.
[93] BA, BDC file on Dorn.

Dorn, Director Todt from Kalle added that Dorn had a justifiable "drive to get on." For Dorn had had to "restrain himself very much in the last few years (in consideration of his intense involvement in the Four-Year Plan) to avoid getting into difficulties subsequently." These difficulties no longer existed, and he rose rapidly through the ranks in Kalle until switching to Zellstoff Waldhof at the end of 1956.[94]

The examples mentioned here not only give evidence of Winnacker's solidarity with these men, whom he often esteemed very highly, they also show that the best and most intelligent managers were also often to be found among the most unscrupulous and conformist individuals in the Third Reich – although some were convinced National Socialists and not merely opportunists. Winnacker hired these men on the basis of their previous performances or in expectation of their doing excellent work in future. Streeck, Schultheiss, and Dorn all were apparently outstanding employees. But they had also compromised themselves very much during the Third Reich, and Dürrfeld had even been actively involved in the committing of crimes.

Given the background of this highly efficient network of old IG Farben contacts, the solicitous care lavished on its members by the Hoechst corporation, which even partly reimbursed their expenses for the Nuremberg trial, it may strike onlookers as strange that Winnacker had no wish to take on the "inheritance" of the IG Farben group in other areas. In a conversation with Wurster on the attempts of the IG Farben liquidators to come to a settlement in the Wollheim trial, Winnacker expressed his "displeasure at the liquidators' method of proceeding and emphasized that Hoechst would under no circumstances be prepared to make any assurances or set aside reserves for the concentration camp trials." For, as Winnacker stated, "Hoechst does not regard itself as a legal successor to the IG Farbenindustrie and is additionally of the opinion that this is a matter which cannot be settled under civil law but which more properly belongs in the area of the politics. I expressed my incomprehension as to why the liquidators had allowed this problem to become the focus of attention and had even proffered reserves out of the capital of IG shareholders."[95]

Winnacker's obligations were primarily to the "victims" of denazification, but not to those who had suffered under the Nazi regime. Nazi victims should be reimbursed by means of government agreements or laws, since according to the general perception – not just that of Winnacker – the state had been responsible for the crimes of the Nazi regime. The concept that IG Farben had been complicit in Auschwitz and should there also be liable was and always remained foreign to him – and not only to him but to others as

---

[94] HA, PA Abteilung VI, Dorn.
[95] HA, Winnacker-Archiv, Tagebuch 18.9.1951–31.12.1954, Aktennotiz by Winnacker dated August 22, 1953, on conversation with Wurster (BASF, Ludwigshafen) on August 20, 1953.

well. He himself and his colleagues on the IG Farben Management Boards had only done "their duty" – and thus there existed a mutual obligation to help each other after the war and after denazification, and to ensure that former executives were given proper positions and supplied with sufficient money.[96] Nothing illustrates this attitude toward victims of the Nazi regime more than the treatment of Franz Michalson's widow. Michalson had been turned out of Hoechst as a so-called "Jew by decree" (*Geltungsjude*) in 1938 and had committed suicide in 1942 because of the treatment he and other Jews remaining in Germany experienced. After many years his widow was awarded a trifling sum in compensation, and she died soon afterwards.

In his memoirs Winnacker complained about the consequences of his dismissal in 1945: "As always at such times, those who could be manipulated, or who were prepared to fawn and court the favours of their new masters, rose to the top. Frequently you discovered that people with whom you had worked for many years no longer recognized you in the street."[97] The question cannot be avoided as to whether he and other leading IG Farben executives ever considered the feelings of the Jewish employees who were thrown out of work and robbed of their professional and social standing, or of the "foreign workers" who had frequently been forcibly torn from their native country and their families. But this was not a question Winnacker ever asked himself – and the same applies to many others of the economic elite.

A letter Winnacker wrote to Jähne on the occasion of his eightieth birthday recalled Jähne's outstanding work as chief engineer in Hoechst from 1931 until 1945, which had forcibly ended at the end of the war with the American occupation. Characteristically Winnacker wrote: "And then the year 1945 came, which saw the collapse of our proud company, the end of your active service, and the destruction of everything which had formed our way of life. For you this was probably the beginning of the most difficult time of your life which you were able to overcome with your admirable composure." Winnacker continued: "A few years later, when reason and order had returned and we began once more to refurbish our house, we were able to renew our old friendship."[98] Within only a few years, most of the "victims" of denazification, as Winnacker called them, were once more flourishing, now that "reason and order" had returned once more – to use Winnacker's words. Only Schilling or Zeh found the way back barred to them and were unable to return; by the mid-1950s at the latest Streeck, Schultheiss, or Wagenheimer were once again able to pursue their respective careers and found even the highest offices open to them. Even the former *Betriebsobmann* Retzinger was rehired and enjoyed a benevolent assessment in Bäumler's "family history" of Hoechst published in 1988.

---

[96] Cf. Wagner, *IG Auschwitz*, pp. 324–5. [97] Winnacker, *Challenging Years*, p. 120.
[98] HA, PA Abteilung VI, Jähne, Winnacker to Jähne, October 24, 1959 (copy).

The decision of the judges in the Wollheim trial, which finally resulted in IG Farben and thus probably also its successor companies paying 30 million DM to the victims through the Jewish Claims Committee, must be considered an important step in the acceptance of guilt and responsibility, since IG Farben in Liquidation was finally prepared to pay compensation. The victims can primarily thank the plaintiff Norbert Wollheim and his lawyer Henry Ormond, whose plea must be considered one of the great moments in the legal history of the Federal Republic of Germany.[99] In the following years, the Compensation Trust, which was specially founded for the administration and payment of compensation, thoroughly investigated the legitimacy of many thousands of applications. The unsalaried work of many volunteers also helped to disburse the sum of DM 27,841,500 to 5,855 former Jewish prisoners and workers in forty-two countries – a large sum of money, but it still only amounted to around DM 4,755 per person.[100]

There is much that could yet be written on the conflicts and collaborations between the successor companies of IG Farben in the years that were to follow. The cooperation between Haberland and Winnacker remained particularly close, as was demonstrated in the cases of Schieber and Dürrfeld. Nor was this cooperation limited to matters affecting former employees. As Kurt Hansen, formerly chairman of Bayer AG's Management Board, succinctly put it in a conversation about the cooperation between former IG Farben plants after the dissolution of the company at the beginning of the 1950s: "But after all, we were one family!"[101] This "family spirit" remained very much to the fore in the close cooperation between the three chairmen Haberland, Wurster, and Winnacker. The three men knew each other very well from their time at IG Farben, and collaboration between the new companies continued even after the company had been forcibly dissolved. It was manifest in joint celebrations of birthdays, such as that of Fritz ter Meer, but also extended to concerted regulations when dealing with former IG Farben employees, as well as the question of the treatment of former concentration camp inmates. Whether and to what extent they divided up the market between them is difficult to say. For despite all cooperation there was also competition and mutual distrust. But the social occasions referred to above also served as meeting places to discuss possible problems; moreover for many years the chairmen of the Management Boards of Bayer, BASF, and Hoechst continued to meet, apparently on a regular and very discrete basis, at the motorway restaurant Montabaur, accompanied only by their drivers.[102]

[99] IFZ, ED 422, vol. 1, final speech of Ormond at the hearing on May 11, 1953.
[100] Cf. Wagner, IG Auschwitz, pp. 313–14.
[101] Oral communication by Kurt Hansen.
[102] Cf. Winnacker, Challenging Years, pp. 101–3, 166–8; Stokes, "IG Farben Fusion", pp. 360–1; personal communication by Albrecht Winnacker.

In Hoechst, one manifestation of this "family spirit" was the retention of the old IG Farben colors of blue and orange – Hoechst was the only successor company to do so. Indeed, of all the old major companies of IG Farben Hoechst appears to have been the one most attached to its IG Farben history, which continued to live according to the old "IG culture." Winnacker's solidarity with "former IG employees" such as Ilgner, in which he even went against Haberland's and Wurster's wishes, is only one aspect of this, even if it is also the worst.

The economic and technical side of the company history of Hoechst in the 1950s and 1960s, years in which Winnacker remained the company's most important manager, is not the subject of this study. Randolph Newman had already noted that the choice of Winnacker "as an outstanding technical and economic authority"[103] to fill the position of senior member of the Management Board was the right one for the plant and, later, for the corporation as well. However, that is another story, worthy of being told and analyzed in a separate study.

---

[103] HA, Winnacker Archiv, Sekretariat, Vorlesungen, memos on the filling of management positions in Hoechst September 28–October 20, 1951 (apparently a transcript made of recordings of Mr. Thies).

Bibliography

ARCHIVAL SOURCES

Archiv der Friedrich-Ebert-Stiftung, Bonn
BASF Unternehmensarchiv, Ludwigshafen
Bayer Werksarchiv, Leverkusen
Bayerisches Hauptstaatsarchiv, Munich
Bundesarchiv, Berlin-Lichterfelde and Dahlwitz-Hoppegarten and Außenstelle
  Ludwigsburg
Bundesarchiv-Militärarchiv, Freiburg i. Br.
Caritas-Archiv, Wuppertal
Clariant Werksarchiv, Gersthofen
Deutsches Museum-Archiv, Munich
Hagley Museum and Library, Delaware
Hessisches Hauptstaatsarchiv, Wiesbaden
Hoechst Archives (HistoCom) and Hoechst Human Resources, Frankfurt a. M.
Institut für Stadtgeschichte, Frankfurt a. M.
Institut für Zeitgeschichte, Munich
National Archives, Washington, D.C.
Nordrhein-Westfälisches Entschädigungsamt, Bezirksregierung Düsseldorf
Nordrhein-Westfälisches Hauptstaatsarchiv Düsseldorf
Schweizer Bundesarchiv, Bern
Unternehmensarchiv der Wacker-Chemie, Burghausen

Private files Henle
Private files Hermann
Private files Lautenschläger
Private files Schnitzer/Mota
Private files Stoltzenberg
Private files Winnacker

INTERVIEWS AND INFORMATION

Gerhard Bier
Kurt Hansen (former Chairman of Bayer AG)

Oskar Henle
Elisabeth Lautenschläger (widow of Carl Ludwig Lautenschläger)
Wolfgang Metternich
Herbert Spahn
Albrecht Winnacker
Lore Wittmer (daughter of Ludwig Hermann)

LITERATURE

Abelshauser, Werner et al., *German Industry and Global Enterprise, BASF: The History of a Company* (Cambridge: Cambridge University Press, 2002)

Die Anfänge der Kunststoffwerkstätte in Hoechst, Dokumente aus Hoechster Archiven 40 (Frankfurt am Main–Hoechst: Farbwerke Hoechst AG, 1969)

Ausschuss zur Untersuchung der Erzeugungs- und Absatzbedingungen der deutschen Wirtschaft, *Verhandlungen und Berichte des Unterausschusses für allgemeine Wirtschaftsstruktur (I. Unterausschuss), 3. Arbeitsgruppe: Wandlungen in den wirtschaftlichen Organisationsformen. Erster Teil: Wandlungen in den Rechtsformen der Einzelunternehmungen und Konzerne* (Berlin: Mittler, 1928)

Baier, Helmut, *Die Deutschen Christen Bayerns im Rahmen des bayerischen Kirchenkampfes* (Nuremberg: Verein für Bayerische Kirchengeschichte, 1968)

Barera, Pilar, "The Evolution of Corporate Technological Capabilities: Du Pont and IG Farben in Comparative Perspective," *Zeitschrift für Unternehmensgeschichte* 39 (1994), pp. 31–45

Barkai, Avraham, *Das Wirtschaftssystem des Nationalsozialismus. Ideologie, Theorie, Politik 1933–1945* (Frankfurt am Main: Fischer, 1988)

Bartmann, Wilhelm, *Zwischen Tradition und Fortschritt. Aus der Geschichte der Pharmabereiche von Bayer, Hoechst und Schering von 1935 bis 1975* (Stuttgart: Steiner, 2003)

Bäumler, Ernst, *A Century of Chemistry* (Düsseldorf: Econ, 1963)

Farben, Formeln, Forscher. Hoechst und die Geschichte der industriellen Chemie in Deutschland (Munich: Piper, 1989)

Die Rotfabriker. Familiengeschichte eines Weltunternehmens (Munich: Piper, 1988)

Becht, Lutz, "Ausländische Arbeitskräfte und Arbeitseinsatz in Frankfurt am Main 1938–1945," *Archiv für Frankfurts Geschichte und Kunst* 65 (1999), pp. 422–72

Beck, Waltraud, Josef Fenzl, and Helga Krohn, *Juden in Höchst. Die vergessenen Nachbarn* (Frankfurt am Main: Jüdisches Museum, 1990)

Benz, Wolfgang, Hermann Graml, and Hermann Weiß (eds.), *Enzyklopädie des Nationalsozialismus*, 3rd edition (Munich: dtv, 1998)

Bergen, Doris L., *Twisted Cross. The German Christian Movement in the Third Reich* (Chapel Hill: University of North Carolina Press, 1996)

Borkin, Joseph, *The Crime and Punishment of the I.G. Farben* (New York: Free Press, 1978)

Bracher, Karl Dietrich, *The German Dictatorship. The Origins, Structure, and Effects of National Socialism*, translated from the German by Jean Steinberg, with an introduction by Peter Gay (Harmondsworth: Penguin, 1973)

(ed.), *Deutschland 1933–1945. Neue Studien zur nationalsozialistischen Herrschaft*, 2. ergänzte Auflage (Bonn: Bundeszentrale für Politische Bildung, 1993)

Bracher, Karl Dietrich, Manfred Funke, and Hans-Adolf Jacobsen (eds.), *Nationalsozialistische Diktatur 1933–1945. Eine Bilanz* (Bonn: Bundeszentrale für Politische Bildung, 1986)

Brauch, Hans Günter, and Rolf-Dieter Müller (eds.), *Chemische Kriegführung – Chemische Abrüstung. Dokumente und Kommentare* (Berlin: Berlin-Verlag-Spitz, 1985)

Braunbuch: Kriegs- und Naziverbrecher in der Bundesrepublik. Staat, Wirtschaft, Armee, Verwaltung, Justiz, Wissenschaft, herausgegeben vom Nationalrat der Nationalen Front des demokratischen Deutschland (Berlin [East]: Staatsverlag der DDR, 1965)

Broszat, Martin, *Der Staat Hitlers. Grundlegung und Entwicklung seiner inneren Verfassung*, 11. Aufl. (Munich: dtv, 1986)

Broszat, Martin, and Klaus Schwabe (eds.), *Die deutschen Eliten und der Weg in den Zweiten Weltkrieg* (Munich: Beck, 1989)

Browning, Christopher R., *Nazi Policy, Jewish Workers, German Killers* (Cambridge: Cambridge University Press, 2000)

Brüning, Gustav von, " 'Erwägungen' zu Carl Duisbergs Denkschrift 'Die Vereinigung der deutschen Farbenfabriken,' " *Tradition. Zeitschrift für Firmengeschichte und Unternehmerbiographie* 9 (1964), pp. 1–5

Buderath, Bernhard (ed.), *Peter Behrens. Umbautes Licht. Das Verwaltungsgebäude der Hoechst AG* (Frankfurt am Main: Prestel, 1999)

Budraß, Lutz, and Manfred Grieger, "Die Moral der Effizienz. Die Beschäftigung von KZ-Häftlingen am Beispiel des Volkswagenwerks und der Henschel Flugzeug-Werke," *Jahrbuch für Wirtschaftsgeschichte* 1993, pp. 89–136

Burleigh, Michael, *The Third Reich. A New History* (New York: Hill and Wang, 2000)

Büttner, Ursula, "The Persecution of Christian-Jewish Families in the Third Reich," *Leo Baeck Institute Yearbook* 34 (1989), pp. 267–89

Carls, Hans, *Dachau. Erinnerungen eines katholischen Geistlichen aus der Zeit seiner Gefangenschaft 1941–1945* (Cologne: Bachem, 1946)

Chandler, Alfred D., *Scale and Scope: The Dynamics of Industrial Capitalism* (Cambridge, MA: Belknap Press of Harvard University Press, 1990)

Chronik der Hoechst AG 1863–1988 (Frankfurt am Main: Hoechst-Aktiengesellschaft, 1990)

Coordination gegen Bayer-Gefahren e.V./CGB (ed.), *I.G. Farben: von Anilin bis Zwangsarbeit. Zur Geschichte von BASF, Bayer, Hoechst und anderen deutschen Chemie-Konzernen*, Bundesfachtagung der Chemie-Fachschaften/AK IG Farben (Stuttgart: Schmetterling, 1995)

Dahrendorf, Ralf, *Gesellschaft und Demokratie in Deutschland* (Munich: R. Piper, 1966)

Deichmann, Ute, *Flüchten, Mitmachen, Vergessen. Chemiker und Biochemiker in der NS-Zeit* (Weinheim: Wiley-VCH, 2001)

Deutschland-Bericht der Sopade (Sozialdemokratischen Partei Deutschlands), Zweiter Jahrgang 1935 (Frankfurt am Main: Zweitausendeins, 1980)

Didier, Friedrich, *Europa arbeitet in Deutschland. Sauckel mobilisiert die Leistungsreserven* (Berlin: Zentralverlag der NSDAP Franz Eher Nachf., 1943)

Dippel, John V. H., *Bound upon a Wheel of Fire: Why So Many German Jews Made the Tragic Decision to Remain in Nazi Germany* (New York: Basic Books, 1996)

Dos Passos, John, "Land of the Fragebogen," in Dos Passos, *Tour of Duty* (Boston: Houghton Mifflin, 1946), pp. 243–74

DuBois, Josiah E., *The Devil's Chemists. 24 Conspirators of the International Farben Cartel Who Manufacture Wars* (Boston: Beacon Press, 1952)

Duisberg, Carl, "Denkschrift über die Vereinigung der deutschen Farbenfabriken, Januar 1904," in *Abhandlungen, Vorträge und Reden aus den Jahren 1882–1921* (Berlin: Chemie, 1923), pp. 343–69

" 'Die Vereinigung der deutschen Farbenfabriken,' August 1915," in Wilhelm Treue, "Carl Duisbergs Denkschrift von 1915 zur Gründung der 'kleinen I.G.,' " *Tradition. Zeitschrift für Firmengeschichte und Unternehmerbiographie* 8 (1963), pp. 193–225

Ebbinghaus, Angelika, and Klaus Dörner, *Vernichten und Heilen. Der Nürnberger Ärzteprozess und seine Folgen* (Berlin: Aufbau, 2001)

Eichholtz, Dietrich, *Geschichte der deutschen Kriegswirtschaft 1939–1945*, vol. 1: *1939–41*, 3. durchgesehene Auflage (Berlin [East]: Akademie-Verlag, 1984)

"Die IG-Farben-'Friedensplanung.' Schlüsseldokumente der faschistischen 'Neuordnung des europäischen Grossraums,' " *Jahrbuch für Wirtschaftsgeschichte* 1966, part 3, pp. 271–332

d'Eramo, Luce, *Deviazione* (Milan: Mondadori, 1979); German edition: *Der Umweg* (Reinbek: Rowohlt, 1982)

Erker, Paul, and Toni Pierenkemper (eds.), *Deutsche Unternehmer zwischen Kriegswirtschaft und Wiederaufbau, Studien zur Erfahrungsbildung von Industrie-Eliten* (Munich: Oldenbourg, 1999)

Farbwerke, vorm. Meister Lucius & Brüning, 1863–1913 (Hoechst: Farbwerke Hoechst, [1913])

Feldman, Gerald D., *Allianz and the German Insurance Business, 1933–1945* (Cambridge: Cambridge University Press, 2001)

Fenzl, Josef, *Aus der Geschichte der Höchster Kaserne 1920–1945* (Frankfurt am Main: Kramer, 1998)

Fiedler, Martin, "Die 'Arisierung' der Wirtschaftselite. Ausmass und Verlauf der Verdrängung der jüdischen Vorstands- und Aufsichtsratmitglieder in deutschen Aktiengesellschaften (1933–1938)," in *'Arisierung' im Nationalsozialismus. Volksgemeinschaft, Raub und Gedächtnis, herausgegeben im Auftrag des Fritz Bauer Instituts von Irmtrud Wojak und Peter Hayes* (Frankfurt am Main: Campus, 2000), pp. 59–83

Fischer, Ernst, "Meister, Lucius und Brüning, die Gründer der Farbwerke Hoechst AG," *Tradition. Zeitschrift für Firmengeschichte und Unternehmerbiographie* 2 (1958), pp. 65–78

Fischer, Wolfram, "Dezentralisation oder Zentralisation – kollegiale oder autoritäre Führung? Die Auseinandersetzung um die Leitungsstruktur bei der Entstehung des I.G. Farben-Konzerns," in Norbert Horn and Jürgen Kocka (eds.), *Recht und Entwicklung der Grossunternehmen im 19. und frühen 20. Jahrhundert. Wirtschafts-, sozial- und rechtshistorische Untersuchungen zur Industrialisierung in Deutschland, Frankreich, England und den USA*, Kritische Studien zur Geschichtswissenschaft 40 (Göttingen: Vandenhoeck & Ruprecht, 1979), pp. 476–88

Flechtner, Hans-Joachim, *Carl Duisberg. Vom Chemiker zum Wirtschaftsführer* (Düsseldorf: Econ, 1959)

Freeman, Christopher, and Luc Soete, *The Economics of Industrial Innovation*, 3rd revised edition (Cambridge, MA: MIT Press, 1999)

Frei, Norbert, *Vergangenheitspolitik. Die Anfänge der Bundesrepublik und die NS-Vergangenheit* (Munich: dtv, 1999)

Frei, Norbert, Dirk van Laak, and Michael Stolleis (eds.), *Geschichte vor Gericht. Historiker, Richter und die Suche nach Gerechtigkeit* (Munich: Beck, 2000)

Frese, Matthias, *Betriebspolitik im 'Dritten Reich.' Deutsche Arbeitsfront, Unternehmer und Staatsbürokratie in der westdeutschen Grossindustrie 1933–1939* (Paderborn: Schöningh, 1991)

Friedländer, Saul, *Nazi Germany and the Jews, vol. 1: The Years of Persecution, 1933–1939* (New York: HarperCollins, 1997)

Fritzsche, Peter, *Germans into Nazis* (Cambridge, MA: Harvard University Press, 1998)

Furukawa, Yasu, *Inventing Polymer Science. Staudinger, Carothers, and the Emergence of Macromolecular Chemistry* (Philadelphia: University of Pennsylvania Press, 1998)

Gay, Peter, *My German Question: Growing Up in Nazi Berlin* (New Haven: Yale University Press, 1998)

Gellately, Robert, *Backing Hitler. Consent and Coercion in Nazi Germany* (Oxford: Oxford University Press, 2001)

Groehler, Olaf, *Der lautlose Tod*, 5. überarb. Auflage (Berlin: Verlag der Nation, 1990)

Gross, Hermann, *Further Facts and Figures Relating to the Deconcentration of the I.G. Farbenindustrie Aktiengesellschaft* (Kiel: Institut für Weltwirtschaft, 1950)

Grundmann, Ekkehard, *Gerhard Domagk. Der erste Sieger über die Infektionskrankheiten* (Münster: Lit, 2001)

Gruner, Wolf, *Der Geschlossene Arbeitseinsatz deutscher Juden. Zur Zwangsarbeit als Element der Verfolgung 1938–1943* (Berlin: Metropol, 1997)

Haber, Ludwig F., *The Chemical Industry 1900–1930: International Growth and Technological Change* (Oxford: Clarendon Press, 1971)

Hachtmann, Rüdiger, *Industriearbeit im 'Dritten Reich.' Untersuchungen zu den Lohn- und Arbeitsbedingungen 1933–1945* (Göttingen: Vandenhoek & Ruprecht, 1989)

Hahn, Otto, *My Life*, translated by Ernst Kaiser and Eithne Wilkins (London: Macdonald, 1970)

Hammermann, Gabriele, *Zwangsarbeit für den 'Verbündeten.' Die Arbeits- und Lebensbedingungen der italienischen Militärinternierten in Deutschland 1943–1945* (Tübingen: Niemeyer, 2002)

Hanslian, Rudolf (ed.), *Der chemische Krieg, vol. 1: Militärischer Teil*, 3. völlig neubearb. Aufl. (Berlin: Mittler, 1937)

Harris, Robert, and Jeremy Paxman, *A Higher Form of Killing: The Secret History of Chemical and Biological Warfare* (London: Arrow Books, 2002)

Hayes, Peter, "Big Business and 'Aryanization' in Germany, 1933–1939," *Jahrbuch für Antisemitismusforschung* 3 (1994), pp. 254–82

Hayes, Peter, *From Cooperation to Complicity. Degussa in the Third Reich* (Cambridge: Cambridge University Press, 2004)
  "Industrie und Ideologie: Die IG Farben in der Zeit des Nationalsozialismus," *Zeitschrift für Unternehmensgeschichte* 32 (1987), pp. 124–36
  Industry and Ideology. IG Farben in the Nazi Era, 2nd edition (Cambridge: Cambridge University Press, 2001)
  "Zur umstrittenen Geschichte der I.G. Farbenindustrie AG," *Geschichte und Gesellschaft* 18 (1992), pp. 405–17
Heim, Susanne, *Kalorien, Kautschuk, Karrieren. Pflanzenzüchtung und landwirtschaftliche Forschung in Kaiser-Wilhelm-Instituten 1933–1945* (Göttingen: Wallstein, 2003)
Heine, Jens Ulrich, *Verstand & Schicksal. Die Männer der I.G. Farbenindustrie A.G. (1925–1945) in 161 Kurzbiographien* (Weinheim: VCH, 1990)
Herbert, Ulrich, *Fremdarbeiter. Politik und Praxis des 'Ausländer-Einsatzes' in der Kriegswirtschaft des Dritten Reiches*, 2nd edition (Berlin: Dietz, 1986); abridged English edition: *Hitler's Foreign Workers. Enforced Foreign Labor in Germany under the Third Reich*, translated by William Templer (Cambridge: Cambridge University Press, 1997)
  (ed.), *National Socialist Extermination Policies. Contemporary German Perspectives and Controversies*, Studies on War and Genocide 2 (New York: Berghahn Books, 2000)
Herbst, Ludolf, *Das nationalsozialistische Deutschland 1933–1945. Die Entfesselung der Gewalt: Rassismus und Krieg* (Frankfurt am Main: Suhrkamp, 1996)
Heuer, Renate, and Siegbert Wolf (eds.), *Die Juden der Frankfurter Universität*, unter Mitarbeit von Holger Kiehnel und Barbara Seib, mit einem Vorwort von Notker Hammerstein (Frankfurt am Main: Campus, 1997)
Hilberg, Raul, *Perpetrators, Victims, Bystanders. The Jewish Catastrophe, 1933–1945* (New York: Aaron Asher, 1992)
Hoffmann, Friedrich, *Die Verfolgung der nationalsozialistischen Gewaltverbrechen in Hessen* (Baden-Baden: Nomos, 2001)
Holdermann, Karl, *Im Banne der Chemie. Carl Bosch, Leben und Werk. Bearb. von Walter Greiling* (Düsseldorf: Econ, 1953)
Hounshell, David A., and John K. Smith, *Science and Corporate Strategy. Du Pont R&D, 1902–1980* (Cambridge: Cambridge University Press, 1989)
Hromadka, Wolfgang, *Die Arbeitsordnung im Wandel der Zeit. Dargestellt am Beispiel der Hoechst AG* (Cologne: Heymann, 1979)
Husemann, Maria, *Mein Widerstandskampf gegen die Verbrechen der Hitlerdiktatur. Bericht von Karl Sommer 1964* (Wuppertal: Stadtdekanat und Katholikenrat, 1983)
I.G. Farbenindustrie Aktiengesellschaft, Erzeugnisse unserer Arbeit (Frankfurt am Main: I.G. Farbenindustrie A.G., 1938)
Jähne, Friedrich, *Der Ingenieur im Chemiebetrieb* (Weinheim: Chemie, 1951)
James, Harold, *The Deutsche Bank and the Nazi Economic War against the Jews: The Expropriation of Jewish-Owned Property* (Cambridge: Cambridge University Press, 2001)

Johnson, Jeffrey Allan, "The Power of Synthesis (1900–1925)," in Werner Abelshauser et al., *German Industry and Global Enterprise, BASF. The History of a Company* (Cambridge: Cambridge University Press, 2002), pp. 115–205

Jones, J. Steven, "Review of Werner Abelshauser, Wolfgang von Hippel, Jeffrey Allan Johnson, and Raymond G. Stokes, 'German Industry and Global Enterprise BASF: The History of A Company' Economic History Services, May 14, 2004," URL: *http://www.eh.net/bookreviews/library/0785.shtml* (last retrieved: July 1, 2006)

Die Kabinettsprotokolle der Bundesregierung, vol. 5: 1952, bearbeitet von Kai von Jena (Boppard am Rhein: Boldt, 1989)

Kershaw, Ian, *Hitler*, 2 vols. (London: Allan Lane, 1998–2000)
   The Nazi Dictatorship: Problems and Perspectives of Interpretation, 4th edition (London: Arnold, 2000)

Kingston, William, "Streptomycin, Schatz v. Waksman, and the Balance of Credit for Discovery," *Journal of the History of Medicine and Allied Sciences* 59 (2004), pp. 441–62

Klee, Ernst, *Auschwitz, die NS-Medizin und ihre Opfer* (Frankfurt am Main: S. Fischer, 1997)

Kobrak, Christopher, *National Cultures and International Competition. The Experience of Schering AG, 1851–1959* (Cambridge: Cambridge University Press, 2001)

Kogon, Eugen, *The Theory and Practice of Hell. The German Concentration Camps and the System behind Them*, translated by Heinz Norden (New York: Berkley, 1950)

Köhler, Otto, *... und heute die ganze Welt. Die Geschichte der IG Farben Bayer, BASF und Hoechst: Mit der Rede des Autors und BASF-Miteigentümers auf der BASF-Hauptversammlöung 1987 in Ludwigshafen* (Cologne: PapyRossa, 1990)

Kracauer, Siegfried, *Geschichte – Vor den letzten Dingen*, Schriften 4 (Frankfurt am Main: Suhrkamp, 1971)

Kroll, Gerhard, *Von der Weltwirtschaftskrise zur Staatskonjunktur* (Berlin: Duncker & Humblot, 1958)

Kropat, Wolf-Arno, *Kristallnacht in Hessen. Der Judenpogrom vom November 1938* (Wiesbaden: Kommission für die Geschichte der Juden in Hessen, 1988)
   'Reichskristallnacht.' Der Judenpogrom vom 7. bis 10. November 1938 – Urheber, Täter, Hintergründe (Wiesbaden: Kommission für die Geschichte der Juden in Hessen, 1997)

Krüll, Claudia, "Hermann Staudinger. Aufbruch ins Zeitalter der Makromoleküle," *Kultur & Technik* 3 (1978), pp. 44–9

Kwiet, Konrad, "The Ultimate Refuge. Suicide in the Jewish Community under the Nazis," *Leo Baeck Institute Yearbook* 29 (1984), pp. 135–67

Laqueur, Walter, and Richard Breitman, *Breaking the Silence* (New York: Simon & Schuster, 1986)

Lautenschläger, Carl Ludwig, *50 Jahre Arzneimittelforschung* (Stuttgart: Thieme, 1955)

Leitner, Gerit von, *Der Fall Clara Immerwahr. Leben für eine humane Wissenschaft*, 2. durchgesehene und verbesserte Auflage (Munich: Beck, 1994)

Lesch, John E., "Chemistry and Biomedicine in an Industrial Setting: The Invention of the Sulfa Drugs," in Seymour H. Mauskopf (ed.), *Chemical Sciences in the Modern World* (Philadelphia: University of Pennsylvania Press, 1993), pp. 158–215

Lindner, Stephan H., "Die IG Farben und ihre jüdischen und als Juden geltenden Mitarbeiter in leitenden Positionen während des 'Dritten Reichs' – Das Beispiel des IG Werks Höchst," in Haus der Geschichte Baden-Württemberg (ed.), *Jüdische Unternehmer und Führungskräfte in Südwestdeutschland 1800–1950. Die Herausbildung einer Wirtschaftselite und ihre Zerstörung durch die Nationalsozialisten* (Berlin: Philo, 2004)

Das Reichskommissariat für die Behandlung feindlichen Vermögens im Zweiten Weltkrieg. Eine Studie zur Verwaltungs-, Rechts-, und Wirtschaftsgeschichte des nationalsozialistischen Deutschlands (Stuttgart: Steiner, 1991)

Lindner, Stephan H., and Michael Lindner, "Dr. Robert Julius Schnitzer – ein führender Forscher auf dem Gebiet der Chemotherapie," in Albrecht Scholz und Caris-Petra Heidel (eds.), *Emigrantenschicksale. Einfluss der jüdischen Emigranten auf Sozialpolitik und Wissenschaft in den Aufnahmeländern* (Frankfurt am Main: Mabuse, 2004)

Lindner, Stephan H., and Michael Lindner, "Das Ende des 'Zauberbergs': Robert Julius Schnitzer und die erfolgreiche Bekämpfung der Tuberkulose," *Atemwegs- und Lungenkrankheiten. Zeitschrift für Diagnostik und Therapie* 30 (2004), issue 4, pp. 198–203

Lochner, Louis Paul, *Tycoons and Tyrant: German Industry from Hitler to Adenauer* (Chicago: Regnery, 1955)

Löffler, Bernhard, *Soziale Marktwirtschaft und administrative Praxis. Das Bundeswirtschaftsministerium unter Ludwig Erhard* (Stuttgart: Steiner, 2002)

Löhnert, Peter, and Manfred Gill, "The Relationship of I.G. Farben's Agfa Filmfabrik Wolfen to Its Jewish Scientists and to Scientists Married to Jews, 1933–1939," in John E. Lesch (ed.), *The German Chemical Industry in the Twentieth Century* (Dordrecht: Kluwer Academic Publishers, 2000), pp. 123–45

Lorentz, Bernhard, *Industrieelite und Wirtschaftspolitik 1928–1950. Heinrich Dräger und das Drägerwerk* (Paderborn: Schöningh, 2001)

Lotfi, Gabriele, *KZ der Gestapo. Arbeitserziehungslager im Dritten Reich* (Stuttgart: DVA, 2000)

Ludwig, Karl-Heinz, *Technik und Ingenieure im Dritten Reich* (Düsseldorf: Droste, 1979)

Mark, Herman F., *From Small Organic Molecules to Large. A Century of Progress* (Washington, DC: American Chemical Society, 1993)

Marsch, Ulrich, *Zwischen Wissenschaft und Wirtschaft. Industrieforschung in Deutschland und Grossbritannien 1880–1936* (Paderborn: Schöningh, 2000)

Marschall, Luitgard, *Im Schatten der chemischen Synthese. Industrielle Biotechnologie in Deutschland (1900–1970)* (Frankfurt am Main: Campus, 2000)

Martinetz, Dieter, *Der Gaskrieg 1914/18. Entwicklung, Herstellung und Einsatz chemischer Kampfstoffe. Das Zusammenwirken von militärischer Führung, Wissenschaft und Industrie* (Bonn: Bernard & Graefe, 1996)

Mason, Tim, "Der Primat der Politik – Politik und Wirtschaft im Nationalsozialismus," *Das Argument* 8 (1966), pp. 473–94

Metternich, Wolfgang, "Traditionsgebundene Baustrukturen," in Bernhard Buderath (ed.), *Peter Behrens. Umbautes Licht. Das Verwaltungsgebäude der Hoechst AG* (Frankfurt am Main: Prestel, 1999), pp. 139–47

Meyer, Beate, *'Jüdische Mischlinge,' Rassenpolitik, Verfolgungserfahrung 1933–1945* (Hamburg: Dölling & Galitz, 1999)

Milward, Alan S., *War, Economy, and Society, 1939–1945*, History of the World Economy in the Twentieth Century 5 (Berkeley: University of California Press, 1977)

Mitscherlich, Alexander, and Fred Mielke (eds.), *Medizin ohne Menschlichkeit. Dokumente des Nürnberger Ärzteprozesses* (Frankfurt an Main: Fischer, 1960)

Mokyr, Joel, *The Lever of Riches. Technological Creativity and Economic Progress* (Oxford: Oxford University Press, 1990)

Mollin, Gerhard Th., *Montankonzerne und 'Drittes Reich.' Der Gegensatz zwischen Monopolindustrie und Befehlswirtschaft in der deutschen Rüstung und Expansion 1936–1944* (Göttingen: Vandenhoeck & Ruprecht, 1988)

Mommsen, Hans, and Manfred Grieger, *Das Volkswagenwerk und seine Arbeiter im Dritten Reich* (Düsseldorf: Econ, 1996)

Morris, Peter J. T., "Ambros, Reppe, and the Emergence of Heavy Organic Chemicals in Germany, 1925–1945," in Anthony S. Travis et al. (eds.), *Determinants in the Evolution of the European Chemical Industry, 1900–1939. New Technologies, Political Frameworks, Markets and Companies* (Dordrecht: Kluwer Academic Publishers, 1998), pp. 89–123

Noakes, Jeremy, "The Development of Nazi Policy towards the German-Jewish 'Mischlinge' 1933–1945," *Leo Baeck Institute Yearbook* 34 (1989), pp. 291–354

Overy, Richard J., *War and Economy in the Third Reich* (Oxford: Claredndon Press, 1994)

Why the Allies Won (New York: Norton, 1995)

Penrose, Edith, *The Theory of the Growth of the Firm*, 3rd edition (Oxford: Oxford University Press, 1995)

Petzina, Dietmar, and Werner Plumpe, "Unternehmensethik – Unternehmenskultur: Herausforderungen für die Unternehmensgeschichtsschreibung?" *Jahrbuch für Wirtschaftsgeschichte* 1993, pp. 9–19

Pierenkemper, Toni, *Unternehmensgeschichte. Eine Einführung in ihre Methoden und Ergebnisse* (Stuttgart: Steiner, 2000)

Pieroth, Ingrid, *Penicillinherstellung. Von den Anfängen bis zur Großproduktion* (Stuttgart: Wissenschaftliche Verlagsgesellschaft, 1992)

Pinnow, Hermann, *75 Jahre Werksgeschichte Höchst. Zur Erinnerung an die 75. Wiederkehr des Gründungstages der Farbwerke vorm. Meister Lucius & Brüning, 1863–1938* (Munich: Bruckmann, 1938)

Plumpe, Gottfried, *Die I.G. Farbenindustrie AG. Wirtschaft, Technik, Politik 1904–1945* (Berlin: Duncker & Humblot, 1990)

Plumpe, Werner, "Unternehmen," in Gerhard Ambrosius, *Moderne Wirtschaftsgeschichte. Eine Einführung für Historiker und Ökonomen* (Munich: Oldenbourg, 1996), pp. 47–66

Priesner, Claus, *H. Staudinger, H. Mark und K. H. Meyer: Thesen zur Größe und Struktur der Makromoleküle. Ursachen und Hintergründe eines akademischen Disputes* (Weinheim: Chemie, 1980)

Pritzkoleit, Kurt, *Männer, Mächte, Monopole. Hinter den Türen der westdeutschen Wirtschaft* (Düsseldorf: Rauch, 1953)

Die neuen Herren. Die Mächtigen in Staat und Wirtschaft (Munich: Desch, 1955)

Rauh-Kühne, Cornelia, "Hans Constantin Paulssen: Sozialpartnerschaft aus dem Geiste der Kriegskameradschaft," in Paul Erker and Toni Pierenkemper (eds.), *Deutsche Unternehmer zwischen Kriegswirtschaft und Wiederaufbau. Studien zur Erfahrungsbildung von Industrie-Eliten* (Munich: Oldenbourg, 1999), pp. 109–92

Reader, William Joseph, *Imperial Chemical Industries. A History*, vol. 2: *The First Quarter-Century 1926–1952* (London: Oxford University Press, 1975)

Rebentisch, Dieter, *Führerstaat und Verwaltung im Zweiten Weltkrieg. Verfassungsentwicklung und Verwaltungspolitik 1939–1945* (Stuttgart: Steiner, 1989)

"Persönlichkeitsprofil und Karriereverlauf der nationalsozialistischen Führungskader in Hessen 1928–1945," *Hessisches Jahrbuch für Landesgeschichte* 33 (1983), pp. 293–331

"Zwei Beiträge zur Vorgeschichte und Machtergreifung des Nationalsozialismus in Frankfurt: Von der Splittergruppe zur Massenpartei," in Eike Hennnig (ed.), *Hessen unterm Hakenkreuz. Studien zur Durchsetzung der NSDAP in Hessen*, 2nd edition (Frankfurt am Main: Insel, 1984), pp. 279–97

Reibel, Carl-Wilhelm, *Das Fundament der Diktatur: Die NSDAP-Ortsgruppen 1932–1945* (Paderborn: Schöningh, 2002)

Reichelt, Werner-Otto, *Das Erbe der IG-Farben* (Düsseldorf: Econ, 1956)

Reinhardt, Carsten, "Basic Research in Industry: Two Case Studies at I.G. Farbenindustrie AG in the 1920s and 1930s," in Anthony S. Travis et al. (eds.), *Determinants in the Evolution of the European Chemical Industry, 1900–1939. New Technologies, Political Frameworks, Markets and Companies* (Dordrecht: Kluwer Academic Publishers, 1998), pp. 67–89

Forschung in der chemischen Industrie. Die Entwicklung synthetischer Farbstoffe bei BASF und Hoechst, 1863–1914 (Freiberg: TU Bergakademie, 1997)

Renneberg, Monika, and Mark Walker (eds.), *Science, Technology and National Socialism* (Cambridge: Cambridge University Press, 1994)

Rössler, Felix, *Der Führer des Betriebes (insbesondere: Die Rechtsnatur der Betriebsgemeinschaft und des Führeramts)*, Schriften des Instituts für Wirtschaftsrecht 13 (Jena: G. Fischer, 1935)

Sandkühler, Thomas, *'Endlösung' in Galizien. Der Judenmord in Ostpolen und die Rettungsinitiativen von Berthold Beitz* (Bonn: Dietz, 1996)

Schlumbohm, Jürgen, "Mikrogeschichte – Makrogeschichte: Zur Eröffnung einer Debatte," in Schlumbohm (ed.), *Mikrogeschichte – Makrogeschichte: komplementär oder inkommensurabel?*, 2nd edition (Göttingen: Wallstein, 2000), pp. 7–32

Schmidt, Albrecht, *Die industrielle Chemie in ihrer Bedeutung im Weltbild und Erinnerungen an ihren Aufbau* (Berlin: de Gruyter, 1934)

Schneider, Michael, *Unterm Hakenkreuz. Arbeiter und Arbeiterbewegung 1933 bis 1939*, Geschichte der Arbeiter und der Arbeiterbewegung in Deutschland seit dem Ende des 18. Jahrhunderts 12 (Bonn: Dietz, 1999)

Schneider, Ulrich, and Harry Stein, *IG-Farben AG, Abt. Behringwerke Marburg-KZ Buchenwald Menschenversuche. Ein dokumentarischer Bericht* (Kassel: Brüder-Grimm-Verlag, 1986)

Schoenbaum, David, *Hitler's Social Revolution. Class and Status in Nazi Germany, 1933–1939* (New York: Norton, 1980)

Schreiber, Peter Wolfram [pseudonym], *IG Farben. Die unschuldigen Kriegsplaner. Profit aus Krisen, Kriegen und KZs. Geschichte eines deutschen Monopols* (Stuttgart: Neuer Weg, 1978)

Schulte, Jan E., *Zwangsarbeit und Vernichtung: das Wirtschaftsimperium der SS. Oswald Pohl und das SS-Wirtschafts- und Verwaltungshauptamt* (Paderborn: Schöningh, 2001)

Smelser, Ronald, *Robert Ley: Hitler's Labor Front Leader* (Oxford: Berg, 1988)

Sohn, August W., *Das Manuskript. Roman* (Frankfurt am Main: Fischer, 1986)

Spoerer, Mark, *Zwangsarbeit unter dem Hakenkreuz. Ausländische Zivilarbeiter, Kriegsgefangene und Häftlinge im Deutschen Reich und im besetzten Europa 1939–1945* (Stuttgart: DVA, 2001)

Stefanski, Valentina Maria, *Zwangsarbeit in Leverkusen. Polnische Jugendliche im I.G. Farbenwerk* (Osnabrück: fibre, 2000)

Stokes, Raymond G., *Divide and Prosper. The Heirs of I.G. Farben under Allied Authority* (Berkeley: University of California Press, 1988)

"EH.Net Book Review: Author's Response," *http://www.eh.net/bookreviews/response/Stokes.htm* (last retrieved: July 1, 2006)

"From the IG Farben Fusion to the Establishment of BASF AG (1925–1952)," in Werner Abelshauser et al., *German Industry and Global Enterprise, BASF: The History of a Company* (Cambridge: Cambridge University Press, 2002), pp. 206–361

Stoltzenberg, Dietrich, *Fritz Haber. Chemiker, Nobelpreisträger, Deutscher, Jude* (Weinheim: VCH, 1994)

Straumann, Lukas, and Daniel Wildmann, *Schweizer Chemieunternehmen im 'Dritten Reich'* (Zurich: Chronos, 2001)

Strauß, Franz-Josef, *Die Erinnerungen* (Berlin: Siedler, 1989)

Streb, Jochen, Technologiepolitik im Zweiten Weltkrieg. Die staatliche Förderung der Synthesekautschukproduktion im deutsch-amerikanischen Vergleich, *Vierteljahrshefte für Zeitgeschichte* 50 (2002), pp. 367–97

Szöllösi-Janze, Margit, *Fritz Haber 1868–1934. Eine Biographie* (Munich: Beck, 1998)

(ed.), *Science in the Third Reich* (Oxford: Berg, 2001)

Tammen, Helmuth, *Die I.G. Farbenindustrie Aktiengesellschaft (1925–1933). Ein Chemiekonzern in der Weimarer Republik* (Berlin: H. Tammen, 1978)

Ter Meer, Fritz, *Die I.G. Farbenindustrie Aktiengesellschaft. Ihre Entstehung, Entwicklung und Bedeutung* (Düsseldorf: Econ, 1953)

Travis, Anthony S., et al. (eds.), *Determinants in the Evolution of the European Chemical Industry, 1900–1939. New Technologies, Political Frameworks, Markets and Companies* (Dordrecht: Kluwer Academic Publishers, 1998)

Trials of War Criminals before the Nuernberg Military Tribunals under Control Council Law No. 10, vols. 7–8: The Farben Case (Washington DC: U.S. Government Printing Office, 1952–53)

Unternehmen im Nationalsozialismus, herausgegeben und eingeleitet von Lothar Gall und Manfred Pohl (Munich: Beck, 1998)

Das Urteil im I.G.-Farben-Prozess. Der vollständige Wortlaut mit Dokumentenanhang (Offenbach: Bollwerk, 1948)

Vollert, Adalbert, *Nied – wie es einmal war. Historische Notizen eines Frankfurter Stadtteils* (Frankfurt am Main–Nied: Heimat- u. Geschichtsverein, 1989)

Die Vorbereitung des Zusammenschlusses der IG-Farbenindustrie im Jahre 1904, Dokumente aus Hoechster Archiven 9 (Frankfurt am Main–Hoechst: Farbwerke Hoechst AG, 1965)

Wagner, Bernd C., *IG Auschwitz. Zwangsarbeit und Vernichtung von Häftlingen des Lagers Monowitz 1941–1945* (Munich: Saur, 2000)

Wagner, Dieter, *Innovation und Standort. Geschichte und Unternehmensstrategien der Chemischen Fabrik Griesheim 1856–1925* (Darmstadt: Hessisches Wirtschaftsarchiv, 1999)

Weinberg, Gerhard L., *A World at Arms: A Global History of World War II*, 2nd edition (Cambridge: Cambridge University Press, 2005)

Weindling, Paul, *Epidemics and Genocide in Eastern Europe, 1890–1945* (Oxford: Oxford University Press, 2000)

Wengenroth, Ulrich, "Zwischen Aufruhr und Diktatur: Die Technische Hochschule 1918–1945," in Wengenroth (ed.), *Die Technische Universität München. Annäherungen an ihre Geschichte* (Munich: Technische Universität München, 1993), pp. 215–60

Werther, Thomas, "Menschenversuche in der Fleckfieberforschung," in Angelike Ebbinghaus and Klaus Dörner (eds.), *Vernichten und Heilen. Der Nürnberger Ärzteprozess und seine Folgen* (Berlin: Aufbau, 2001), pp. 152–73

Weß, Ludger, "Menschenversuche und Seuchenpolitik – Zwei unbekannte Kapitel aus der Geschichte der deutschen Tropenmedizin," *1999. Zeitschrift für Sozialgeschichte des 20. und 21. Jahrhunderts* 8 (1993), issue 2, pp. 10–50

Wickel, Helmut, *I.-G. Deutschland. Ein Staat im Staate* (Berlin: Der Bücherkreis, 1932)

Wiesen, Jonathan, *West German Industry and the Challenge of the Nazi Past, 1945–1955* (Chapel Hill: University of North Carolina Press, 2001)

Wilmowsky, Tilo Freiherr von, *Warum wurde Krupp verurteilt? Legende und Justizirrtum*, 2nd edition (Stuttgart: Vorwerk, 1950)

Wimmer, Wolfgang, *Wir haben fast immer was Neues. Gesundheitswesen und Innovationen der Pharma-Industrie in Deutschland, 1880–1935* (Berlin: Duncker & Humblot, 1994)

Winau, Rolf, "Der Menschenversuch in der Medizin," in Angelika Ebbinghaus and Klaus Dörner (eds.), *Vernichten und Heilen. Der Nürnberger Ärzteprozess und seine Folgen* (Berlin: Aufbau, 2001), pp. 93–109

Winkler, Heinrich A., *Der lange Weg nach Westen, vol. 2: Deutsche Geschichte vom 'Dritten Reich' bis zur Wiedervereinigung* (Munich: Beck, 2001)

Winnacker, Karl and Weingaertner, Ernst (eds.), *Chemische Technologie*, 4 vols. (Munich: Hanser, 1950–1954)

Winnacker, Karl, *Challenging Years. My Life in Chemistry* (London: Sidgwick & Jackson, 1972)

Zibell, Stephanie, "Der Gauleiter Jakob Sprenger und sein Streben nach staatlicher Macht im Gau Hessen-Nassau," *Zeitschrift für Geschichtswissenschaft* 49 (2001), pp. 389–408

Jakob Sprenger (1884–1945). NS-Gauleiter und Reichsstatthalter in Hessen (Darmstadt: Hessische Historische Kommission, 1999)

Zitelmann, Rainer, *Hitler. Selbstverständnis eines Revolutionärs* (Hamburg: Berg, 1987)

Zollitsch, Wolfgang, Arbeiter zwischen Weltwirtschaftskrise und Nationalsozialismus. Ein Beitrag zur Sozialgeschichte der Jahre 1928 bis 1936 (Göttingen: Vandenhoeck & Ruprecht, 1990)

Index

Aachen Technical University, 273
Abs, Hermann Josef, 348
Adenauer, Konrad, 349, 350, 351, 358
Agfa (Aktiengesellschaft für Anilinfabrikation) *See IG Farben plant Berlin*
Allgäu, 338
Allianz, 126
Alsace, 139, 159
Ambros, Otto, 1, 136, 261, 267, 285, 346, 354, 359
Ammelburg, Alfred, 29, 41, 101, 103
Anorgana GmbH, 359
Aschersleben, 77
Auer, Aloys, 313, 333
Augsburg, 11, 13, 76, 77, 80
Auschwitz (concentration camp), 211, 254, 309, 319, 326, 328, 333, 336, 346, 356, 358, 362
Austria, 200

Baasch, Friedrich, 361
Bad Orb, 238
Badoglio, Pietro, 223
Baeyer, Adolf von, 310
Baldus, Adolf, 162, 163, 248
Barell, Emil, 177
Barmen, 205
Basel, 172, 173
BASF (Badische Anilin- und Sodafabrik) *See IG Farben plant Ludwigshafen*
Bassing, Hans, 113, 352
Bavaria, 80, 89, 155, 159, 360
Bavarian Soviet Republic, 80

Bayer AG, *See IG Farben plant Elberfeld; IG Farben plant Leverkusen*
Behrens, Peter, 12
Behring Werke *See IG Farben plant Marburg*
Beil, Albert, 252
Beitz, Berthold, 5
Belgium, 176, 224, 228
Benda, Louis, 155, 156, 157, 159, 169
Berger, chief engineer, 162
Berl, Ernst, 136, 205, 206, 210, 274
Berlin, 39, 49, 139, 175, 224, 228, 260
Berlin University, 174
Bertrams, Reinhold, 354
Best, Charles, 177
Bieling, Richard, 315, 326, 327, 328, 329
Billiter, Jean, 77, 294
Binnewies, Wilhelm, 140, 141, 142, 143
Black Forest, 338
Blumrich, Karl Ferdinand, 135
Bockmühl, Max, 191, 301, 307, 310, 311, 316, 331, 333, 336, 342, 344
Boedecker, Mr., 206
Böker, Reinhard, 120
Bolton, Mr., 276, 312
Börgermoor, East Frisia, 142
Bormann, Mr., 97, 98
Bornemann, Karl, 346
Bosch, Carl, 17, 22, 24, 25, 26, 27, 28, 30, 31, 34, 35, 36, 42, 43, 44, 48, 49, 50, 57, 58, 65, 76, 81, 82, 154, 158, 295
Braunschweig (Brunswick) Technical University, 49, 129
Breloh, Lüneburg Heath, 78, 79, 81
Bremerhaven, 136

Breslau University, 93
Brüning, Adolf, 13
Brüning, Gustav von, 93, 94, 95, 96, 97, 98, 99, 106, 123, 127, 159, 207, 208, 214, 251
Brüning, Gustav von (the elder), 14, 15, 16
Bruyn, Johannes de, 219
Buchenwald (concentration camp), 176, 309, 315, 321, 323, 324, 327, 328, 329, 332, 334, 335, 336
Buhl, Bernhard, 49, 50
Bulgaria, 218
Burghausen, 183
Bütefisch, Heinrich, 358

Cairo, 176
Cambrai, 101
Canada, 178
Cannes, 49
Carl Weber AG, 173
Carls, Hans, 331, 332
Cassella (Leopold Cassella & Co. GmbH) *See IG Farben plant Mainkur*
Chamberlain, Arthur Neville, 146
Champagne, 78, 84
Chemische Fabrik Griesheim-Elektron *See IG Farben plant Griesheim*
Chemische Fabrik vorm. Weiler-ter Meer *See IG Farben plant Uerdingen upon Rhine*
China, 51
Cologne, 347
Crane, Jasper, 269
Croatia, 218, 227, 361

d'Eramo, Luce, 127, 131, 220, 231, 240, 248
Dachau (concentration camp), 331, 336
Daniel Sieff Research Institute (Weizmann Institute), 168, 177
Darmstadt, 164
Darmstadt Technical University, 35
Daumiller, Oscar, 88, 89
Denmark, 169, 218
Deutsche Bank, 164, 348
Dietzsch, Arthur, 327
Ding (-Schuler), Erwin, 309, 315, 316, 321, 322, 323, 324, 325, 326, 327, 328, 329, 330, 331, 332, 334, 335, 336
Domagk, Gerhard, 270, 301, 316
Dorn, Friedrich, 361, 362
Duden, Paul, 20, 28, 29, 36, 38, 39, 40, 41, 42, 43, 44, 48, 81, 155, 268, 349, 350

Duisberg, Carl, 13, 14, 15, 16, 17, 18, 21, 22, 23, 24, 25, 26, 35, 36, 44, 47, 48, 49, 50, 57, 58, 64, 253, 269, 277
Duisberger Kupferhütte, 347
DuPont de Nemours & Co., 11, 258, 259, 269, 276, 286, 312
Dürrfeld, Walter, 1, 351, 352, 353, 354, 358, 362, 364
Düsseldorf, 11, 277, 282

Eastern Europe, 3, 218, 227, 228
Eckelmann, Alfred, 118
Ehrlich, Paul, 155, 174, 270
Einhorn, Alfred, 310
Embden, Gustav, 88, 310
England, 13, 19, 168
Engler, Carl, 100
Epting, Max, 39
Erfurt, 184
Erhard, Ludwig, 349
Erlangen, 101
Erlangen University, 136, 182
Erlenbach, Arnold, 181
Erlenbach, Michael, 181, 182, 183, 302, 342, 343, 344, 347, 348, 349, 350, 352
Estonia, 222
Eucken, Arnold, 273, 274
Europe, 230, 264

Fehrle, Alfred, 103, 181, 251
Fett, Wolfgang, 361
Fischer, Hermann, 177
Flick, Friedrich, 354
France, 13, 19, 126, 176, 177, 222, 226, 228, 229, 239, 361
Frankfurt University, 100, 102, 105, 155, 189, 310, 311
Frantz, Dr., 79, 80
Freiburg, 184
Freiburg University, 101, 131, 133, 169, 182, 310, 311
Fries, Karl, 120, 129
Funke, Mr., 206
Fürth, 155
Fußgänger, Rudolf, 311, 316, 317, 320, 324, 326, 328, 331, 333, 334, 336

Gajewski, Fritz, 34, 35, 36, 350
Gärtner, Hugo, 145, 146, 147, 148, 149, 150, 151, 191, 199, 201, 202
Gau Hesse-Nassau *See Hesse-Nassau*
Gaus, Wilhelm, 262, 263, 266, 277

GB-Chemie *See Gebechem*
  (Generalbevollmächtigter Chemie)
Gebechem *(Generalbevollmächtigter*
  Chemie), 211, 226, 228, 229, 247, 293,
  306, 360, 361
Gendorf, 261, 359
Geneva, 217, 238
Geneva University, 28
German Association of Technicians,
  133, 134, 139
Gießen University, 133, 214, 310
Gimbel, Adalbert, 193
Gnann, Dr., 273
Goebbels, Joseph, 107
Goldschmidt, Stefan, 189
Göring, Hermann, 260, 261, 266, 361
Göttingen University, 101, 273, 310, 341
Graudenz, 161
Greif, Wilfrid, 29, 42, 51
Greifswald University, 101
Greune, Heinrich, 189, 206, 272
Griesheim-Elektron *See IG Farben plant*
  Griesheim
Grosch, Heinrich, 110, 111
Grossmann, Erna, 179, 180, 181
Grossmann, Karl, 179, 181
Grüneburg site, 43, 47, 48, 51, 110
Gusen (concentration camp), 309, 329, 330,
  331, 333, 336
Gutermuth, Dr., 215

Haber, Fritz, 77, 154
Haberland, Ulrich, 209, 345, 347, 354, 355,
  356, 357, 360, 361, 364, 365
Haeuser, Adolf, 16, 17, 19, 20, 22, 23, 24,
  26, 28, 39, 40, 44, 48, 49, 50, 101, 279
Hamburg, 184, 232
Hampe, Dr., 359
Hansen, Kurt, 364
Hardt, Albin, 179, 181
Hartmann, Bernhard, 144, 145
Heidelberg University, 101, 133, 150
Heisel, Paul, 182, 183
Hellmuth, Otto, 130
Henle, Franz, 140, 159, 160, 164, 165, 286
Henle, Karl, 160
Henle, widow of, 160, 163
Hermann, Ludwig, 43, 44, 51, 52, 54, 55, 56,
  57, 58, 66, 68, 70, 72, 76, 77, 78, 79,
  80, 81, 82, 83, 84, 85, 86, 87, 88, 89,
  90, 91, 92, 93, 94, 95, 96, 97, 98, 99,
  100, 102, 104, 105, 110, 113, 114, 115,
  117, 118, 120, 121, 122, 123, 127, 133,
  135, 137, 143, 155, 156, 157, 159, 160,
  186, 201, 204, 206, 207, 214, 251, 252,
  253, 254, 257, 259, 260, 261, 267, 268,
  269, 271, 272, 273, 275, 276, 289, 291,
  294, 296, 307
Herrmann, Walter, 43, 179, 181
Hess, Johannes, 78
Hess, Kurt, 280, 281, 284, 285
Hesse, 198, 350
Hesse-Nassau, 38, 106, 107, 121, 175, 194,
  195, 198
Hibernia, 354
Hilcken, Valentin, 165, 166, 251
Himmler, Heinrich, 38, 109, 110
Hindenburg, Paul von, 64, 66, 83
Hirsch, Josef, 148, 170, 199, 200, 201, 202,
  205, 239, 245
Hirschel, Otto, 94, 95, 104, 123, 125, 163,
  196, 203, 204, 213, 214, 220, 224, 228,
  229, 230, 232, 233, 234, 238, 241, 245,
  251, 258, 261, 268, 287, 289, 294, 295,
  296, 359
Hirschelmann, Walter, 66, 67, 108, 111, 128
Hissenauer, Georg, 187, 188
Hitler, Adolf, 3, 5, 64, 65, 66, 75, 82, 83, 84,
  85, 86, 87, 88, 89, 90, 91, 99, 107, 108,
  113, 114, 135, 146, 149, 152, 154, 180,
  186, 200, 223, 246, 259, 262, 266
Hoffa, Erwin, 157, 158, 159, 161
Hoffmann La Roche, 177, 178
Holländer, Mr., 194
Holler, Prof., 325, 328, 329
Hörlein, Heinrich, 156, 282, 308, 313,
  314, 315
Hoven, Waldemar, 322, 325, 328, 329, 331
Hungary, 218
Husemann, Maria, 331

I.G. Farbenindustrie AG *See IG Farben*
IG *See IG Farben*
IG Farben, 1, 2, 3, 4, 5, 6, 7, 8, 10, 11, 13,
  18, 26, 27, 28, 30, 33, 34, 35, 39, 43,
  45, 47, 48, 49, 50, 51, 52, 54, 55, 64,
  65, 69, 70, 72, 73, 74, 76, 81, 83, 88,
  92, 93, 94, 96, 103, 104, 105, 107, 109,
  110, 119, 121, 123, 127, 130, 132, 137,
  139, 150, 153, 154, 155, 156, 161, 180,
  181, 183, 187, 196, 199, 205, 215, 220,
  224, 226, 230, 233, 248, 249, 251, 252,
  261, 262, 263, 266, 267, 269, 275, 276,
  280, 287, 288, 289, 291, 292, 300, 301,

302, 303, 306, 307, 308, 309, 313, 316, 317, 335, 341, 344, 347, 349, 350, 357, 358, 359, 360, 362, 364, 365
IG Farben plant Auschwitz, 1, 7, 136, 210, 211, 247, 254, 259, 260, 261, 319, 320, 321, 346, 351, 352, 356, 357, 358, 359
IG Farben plant Autogen, 219
IG Farben plant Berlin, 15, 32, 33, 44, 46, 48, 50, 96, 152, 174, 227, 232, 263, 315, 319, 322, 327, 328
IG Farben plant Biebrich, 15, 32, 57
IG Farben plant Bitterfeld, 93, 276, 291, 349
IG Farben plant Bobingen, 351
IG Farben plant Elberfeld, 12, 32, 33, 34, 106, 156, 301, 308, 309, 313, 314, 315, 316, 329, 336, 345
IG Farben plant Gersthofen, 11, 13, 43, 76, 80, 81, 82, 88, 89, 182, 183, 242, 243, 247, 251, 268
IG Farben plant Griesheim, 1, 17, 19, 32, 48, 57, 72, 93, 94, 107, 117, 138, 163, 348, 349
IG Farben plant Heydebreck, 196, 260
IG Farben plant Leuna, 210, 259
IG Farben plant Leverkusen, 11, 12, 17, 27, 31, 34, 41, 43, 54, 56, 57, 62, 68, 105, 106, 107, 119, 227, 233, 235, 244, 251, 252, 253, 260, 261, 271, 276, 277, 278, 279, 283, 286, 287, 288, 294, 296, 297, 299, 301, 308, 309, 313, 314, 315, 316, 317, 318, 319, 320, 326, 329, 330, 347
IG Farben plant Ludwigshafen, 1, 11, 12, 17, 32, 34, 41, 54, 68, 82, 94, 188, 226, 244, 251, 252, 259, 260, 261, 262, 268, 271, 272, 273, 276, 277, 278, 283, 285, 286, 287, 288, 297, 347
IG Farben plant Mainkur, 15, 30, 31, 32, 33, 57, 72, 117, 302
IG Farben plant Marburg, 55, 57, 72, 103, 117, 304, 327
IG Farben plant Merseburg, 244
IG Farben plant Offenbach, 72, 117, 349
IG Farben plant Oppau, 53, 259, 260, 292
IG Farben plant Rheinfelden, 93, 252, 289
IG Farben plant Schkopau, 209, 259, 260, 295
IG Farben plant Uerdingen upon Rhine, 17, 32, 42, 209
IG Farben plant Wolfen, 33, 34, 154, 181, 183, 291

IG Farbenindustrie AG in Liquidation, 248, 353, 356, 357, 358, 364
Ilgner, Max, 263, 355, 356, 358, 365
Israel, 177
Italy, 218, 224, 229, 246

Jähne, Friedrich, 1, 76, 92, 95, 98, 103, 104, 107, 127, 136, 162, 186, 192, 207, 224, 230, 234, 235, 244, 252, 253, 254, 255, 256, 257, 272, 273, 342, 344, 345, 351, 356, 363
Jakobi, Constantin, 107
Japan, 158
Jena University, 28, 136, 214, 311
Jonas, Dr., 163
Joseph, Dr., 81

Kaiser Wilhelm Institute for Chemistry, Berlin, 280, 281
Kalle & Co. AG *See IG Farben plant Biebrich*
Karlsbad, 47
Karlsruhe, 99, 106, 345
Karlsruhe University, 101
Kassel, 141, 142, 144
Kehrl, Hans, 361
Keller, Gottfried, 85
Kelsterbach, 228
Kiedrich in the Rheingau, 313
Kiesskalt, Siegfried, 162, 206, 207, 255, 256, 257, 259, 272, 273, 274, 305, 340, 341
Kikuth, Walter, 317, 320
Klenck, Jürgen von, 359
Knapsack, 347, 348, 352
Knorr, Ludwig, 28
Koblenz, 106
Kossel, Albrecht, 101
Kostheim, 111
Kränzlein, Georg, 41, 104, 122, 123, 125, 128, 129, 130, 131, 132, 133, 136, 157, 158, 159, 160, 165, 169, 171, 172, 173, 189, 206, 226, 266, 271, 272, 276, 277, 278, 279, 280, 281, 282, 283, 284, 285, 296, 359
Kränzlein, Paul, 131
Krauch, Carl, 36, 96, 138, 196, 209, 226, 291, 292, 293, 306, 359
Krekeler, Karl, 29, 34
Krell, Ernst, 187
Krupp, 62, 346
Kühne, Heinrich, 279

L., Elisabeth, 202, 203, 204, 214, 215
Landers, Hermann, 129, 135, 139, 140, 160, 360
Landmann, Mr., 162, 164, 186
Landsberg, 346, 353
Lang, Karl, 114, 115
Langensteinbach, 106
Latvia, 222
Lausanne University, 164
Lautenschläger, Carl Ludwig, 1, 2, 11, 43, 51, 57, 73, 94, 95, 97, 98, 99, 100, 101, 102, 103, 104, 105, 106, 107, 108, 112, 121, 123, 125, 127, 133, 136, 137, 139, 145, 146, 147, 149, 155, 157, 161, 165, 166, 167, 168, 169, 170, 171, 172, 175, 177, 184, 192, 193, 194, 196, 197, 198, 199, 201, 204, 205, 208, 209, 210, 211, 212, 213, 214, 215, 218, 223, 224, 226, 228, 229, 230, 234, 238, 240, 241, 242, 246, 249, 251, 252, 254, 261, 267, 270, 271, 272, 275, 289, 297, 299, 301, 303, 304, 306, 307, 308, 309, 310, 311, 313, 314, 315, 317, 323, 324, 325, 326, 327, 328, 329, 331, 332, 333, 334, 335, 336, 338, 339, 340, 341, 344, 345, 346, 347
Lautenschläger, Erwin, 100
Leclercq, 232
Lehmann-Facius, Prof., 318
Leipzig University, 136, 182
Leopold, Prince of Prussia, 123
Leupold, Ernst Otto, 131
Ley, Robert, 108
Limburg, 324, 334
Liverpool University, 177
Lorraine, 182
Lower Franconia, 130
Lucius, Eugen, 13
Lummitsch, Otto, 77

Maingau Works Group, 13, 28, 35, 68, 72, 73, 76, 82, 85, 94, 95, 98, 117, 147, 187, 191, 194, 225, 228, 232, 234, 243, 260, 347
Mainz, 59
Mangione, Luce *See* d'Eramo, Luce
Mann, Wilhelm Rudolf, 64, 308
Mannesmann, 359
Marburg University, 28
Mark, Hermann, 272, 273, 277, 280, 281, 282, 285, 286
Mauthausen (concentration camp), 309
May, Richard, 183

Meiser, Bishop, 89
Meister, C.F. Wilhelm, 13
Memmingen, 77, 88
Mendelssohn-Bartholdy, Otto, 154
Menne, W. Alexander, 357
Mertens, Anton, 313, 314, 315, 316, 317, 318, 319, 326
Merton, Richard, 153, 154
Meyer, Kurt Hans, 41, 189, 272, 277, 280, 281, 282, 283, 284, 285, 286
Michalso(h)n, Franz, 161, 162, 163, 164, 363
Michalso(h)n, widow of, 164, 363
Michel, Oscar, 20
Middle Rhine Works Group *See Maingau Works Group*
Montabaur, 364
Montfaucon, 84
Montreux, 166
Morgenroth, Julius, 174, 175, 177
Moscow, 185
Moulton, Herbert, 305
Mount Sinai School of Medicine, 178
Mrugowsky, Joachim, 315, 316, 318, 319, 321, 322, 323, 326, 327, 331
Müller, August, 13
Müller, Georg, 198, 199, 205
Müller, Mayor, 48
Munich, 11, 81, 89, 131, 157, 158, 159
Munich Technical University, 77, 208, 273
Munich University, 136, 140, 159, 182, 310
Murthum, Dr., 319
Mussolini, Benito, 246, 247

Neuhausen, 304
Neuss, 253
New York, 51, 178
Newman, Randolph, 347, 348, 350, 365
Nicodemus, Otto, 133, 134, 135, 139, 140
Niihama, 158
Nobbe, Fritz, 49
Norsk Hydro, 291
Nuremberg (trial), 1, 2, 9, 10, 11, 91, 99, 103, 114, 125, 152, 153, 155, 157, 161, 182, 183, 186, 216, 219, 220, 224, 233, 240, 249, 281, 282, 303, 304, 308, 311, 313, 314, 323, 327, 328, 333, 334, 335, 344, 345, 346, 348, 350, 352, 355, 356, 358, 362
Nüsslein, Josef, 188, 252
Nutley, New Jersey, 178

Ormond, Henry, 353, 364
Orth, Gustav, 47, 48
Orthner, Ludwig, 189, 252, 298
Otto, Mr., Prof., 318

Palestine, 168
Papen, Franz von, 64, 200
Paris, 138, 176, 228, 229
Patat, Franz, 207, 208, 273, 274, 305,
    340, 341
Patterson, Captain, 342
Pensel, Ferdinand, 220, 339, 342, 344
Percival, Colonel, 338, 342, 344
Peterson, Max, 77
Pfaff, Kaspar, 182
Pfaffendorf, Wilhelm, 251
Pistor, Gustav, 291, 349, 350
Plato, Wilhelm, 251
Pohl, Oswald, 361
Pöhn, Hans, 121, 124, 125, 140, 141, 143,
    202, 203, 204, 341
Poland, 227
Popp, Bernhard, 130, 131
Popp, Walter, 186, 187
Prague, 138
Pribilla, Hans, 224, 304, 309, 310
Prussia, 89, 161

Rath, Walther vom, 48, 49, 50
Ref, Carl, 44, 45, 46, 47, 50, 51, 58, 252
Ref, widow of, 45
Reichenberg, Bohemia, 135
Reithinger, Arthur, 263
Reppe, Walter, 268
Retzinger, Ludwig, 67, 91, 103, 104, 105,
    108, 109, 111, 112, 113, 114, 115, 117,
    118, 120, 121, 124, 126, 127, 128, 133,
    144, 148, 185, 187, 204, 363
Rheinfelden, 170
Rheinisch-Westfälische
    Elektrizitätsgesellschaft (RWE), 256
Rhineland, 90, 91, 200
Rhine-Main area, 188
Rhône Poulenc, 177
Ritter, Friedbert, 348, 349, 350, 352
Robert Koch Institute, 174, 177
Röhm, Ernst Julius, 85, 200
Rohmer, Martin, 135, 139, 251, 289
Romania, 218
Rome, 282, 283
Roser, Wilhelm, 37, 102, 128
Rospatt, Heinrich von, 347, 348, 353

Roth, Leonhard, 66, 67
Roth, Paul, 191, 232, 295, 342, 343, 344
Ruhrbau, 354
Ruhrchemie, 358
Russia, 13, 19, 133, 138, 169

Saar area, 90
Saarland, 126
Safran, Jakob, 44, 45, 46, 47, 48
Sander, Ernst Ludwig, 164, 165, 166, 167,
    168, 171
Sander, Paul Richard, 164, 166
Sauckel, Fritz, 230, 231, 236, 237, 243, 245
Schacke, Bernhard, 61, 109, 110, 114, 118,
    120, 128, 136
Schering AG, 39
Schieber, Walther, 360, 361, 364
Schilling, Claus Carl, 331, 332
Schilling, Walter, 193, 197, 198, 199, 201,
    205, 363
Schindler, Oskar, 5
Schirmacher, Werner, 203, 204
Schleicher, Kurt von, 64
Schlichenmaier, Hans, 111, 118, 128
Schlick, Heinrich, 192, 193, 194, 195, 196,
    197, 198, 199, 205, 208, 214, 252
Schmidt, Albrecht, 29, 35, 37, 38, 39, 40, 41,
    42, 44, 51, 103, 182
Schmidt, Ernst, 118, 201, 202
Schmitz, Hermann, 104, 356
Schneider, Christian, 196
Schnitzer, Eva, 177
Schnitzer, Johanna, 174
Schnitzer, Manuel, 174
Schnitzer, Robert Julius, 174, 175, 176, 177,
    178, 183, 311
Schnitzler, Georg von, 19, 20, 29, 30, 36, 51
Scholl, Franz, 103
Scholven-Chemie, 354
Schulte, Eduard, 5
Schultheiss, Werner, 129, 140, 360, 362, 363
Schulz, Otto, 349
Schwamborn, Wilhelm, 68, 69, 70, 122, 123,
    141, 146, 147, 149, 151, 162, 167, 169,
    170, 171, 172, 175, 177, 191, 192, 193,
    195, 197, 201, 202, 214, 252
Selck, Erwin, 20, 49, 50, 51, 97, 138
Seldte, Franz, 200
Simon, Johann, 229, 230
Simson, Ernst von, 154
Slovakia, 218
Sorkin, Max, 169, 170, 171, 172, 173, 174

Sorkin, Nikolaus, 169
Sossenheim, 143
Southeast Europe, 227
Soviet Union, 226, 240, 317
Spain, 218
Speer, Albert, 303
Spiess, Franz, 237
Sprenger, Anneliese, 184
Sprenger, Jakob, 83, 97, 99, 103, 106, 121,
    132, 133, 184, 185, 193, 194, 196, 197,
    198, 244, 260, 340
Staib, Karl, 93, 94, 95, 96, 165, 207,
    252, 289
Staudinger, Hermann, 131, 132, 169, 277,
    278, 279, 280, 281, 182, 283, 284, 285,
    286, 301
Steindorff, Adolf, 138, 252
Stellmann, Wilhelm, 137, 138, 229
Strasbourg, 159
Strasbourg University, 140, 160
Strasser, Gregor, 128
Straub, Walther, 101
Streeck, Hans, 136, 359, 360, 362, 363
Struss, Ernst, 207, 346
Stuttgart, 84, 99
Swiss Federal Institute of Technology,
    Zurich, 155
Swiss Materials Science and Research
    Institute for Industry, Building, and
    Construction, St. Gall, 173
Switzerland, 156, 157, 166, 167, 168, 172

Tampke, Hans, 206, 207
Taunus, 141, 225
Technical College, Berlin-Charlottenburg, 253
Technical College, Brunswick, 205
Technical College, Darmstadt, 136, 164,
    185, 205
Technical College, Karlsruhe, 99, 161
Technical College, Mittweida, 161
Technical College, Stuttgart, 93
Tempelhofer Feld, 87
ter Meer, Fritz, 3, 30, 35, 36, 42, 43, 81, 82,
    98, 158, 183, 193, 207, 208, 209, 210,
    261, 262, 267, 268, 176, 278, 291, 292,
    296, 345, 346, 349, 358, 364
the Hague, 238
Svedberg's Institute, Uppsala, 280
Thiele, Johannes, 140, 159
Thoennissen, Emilie, 150, 151, 184
Thoennissen, Max, 150
Thuringia, 230

Tiedtke, Richard, 51
Toronto University, 177
tripartite agreement (*Dreiverband*) (includes
    Hoechst) *See IG Farben plant Biebrich;
    IG Farben plant Mainkur*
triple alliance (*Dreibund*) *See IG Farben
    plant Berlin; IG Farben plant Elberfeld;
    IG Farben plant Leverkusen; IG Farben
    plant Ludwigshafen*
Trost, Karl, 220, 231
Tübingen University, 311

Uhde GmbH, 354
Ukraine, 222
United States of America, 152, 205, 269, 291
Unterliederbach, 61

Versailles, 18, 26, 90
Vesper, Heinrich, 219
Vetter, Hellmuth, 309, 319, 320, 321, 325,
    326, 329, 330, 331, 333, 335, 336
Vienna, 273, 304
Vilnius, 169
Vogt, Helene, 159
Vorarlberg, 338
Voss, Arthur, 171
Vries, Albert de, 233

Wachsmann, Rudolf, 357
Wacker Chemie, 78, 81, 183, 295
Wagenheimer, Hans, 92, 99, 106, 111, 115,
    121, 122, 123, 124, 125, 127, 147, 149,
    150, 167, 172, 179, 181, 184, 185, 190,
    191, 193, 194, 195, 196, 197, 198, 199,
    205, 252, 260, 261, 344, 359, 363
Wagner, Hermann, 51
Waksman, Selman, 269
Waldenburg (Silesia), 131
Wangenheim, Colonel von, 79, 80
Weber, Johann, 66, 67
Weber, Julius, 189, 190, 309, 310, 311, 312,
    314, 315, 316, 318, 319, 320, 321, 322,
    323, 324, 326, 327, 328, 329, 331, 332,
    333, 334, 336
Weber, Karl, 88, 89, 242, 268
Weber, William, 51
Weidlich, Marianne, 49
Weidlich, Richard, 18, 29, 40, 44, 45, 46, 47,
    48, 49, 50, 51, 57, 58, 252
Weil, Bettina, 164
Weil, Konrad, 348, 349, 350
Weilbach, 110

Weiler-ter Meer *See IG Farben plant Uerdingen upon Rhine*

Weimar, 322, 327, 328, 332

Weimar Republic, 8, 64, 82, 83, 140, 200, 214

Weinberg, Arthur von, 15, 17, 154, 155, 156

Weinberg, Carl von, 15, 17, 88, 154, 155, 156

Weizmann, Chaim, 177

Western Europe, 218, 240

White Ruthenia, 222

Wieland, Heinrich, 182

Wiesbaden, 11, 201, 202, 228, 247, 255, 313

Willstätter, Richard, 189, 277

Wilmowsky, Tilo von, 346

Winnacker, Karl, 2, 9, 29, 35, 92, 93, 94, 104, 105, 108, 127, 135, 136, 199, 205, 206, 207, 208, 209, 210, 211, 230, 252, 253, 259, 272, 273, 274, 293, 305, 306, 339, 340, 342, 347, 348, 349, 350, 351, 352, 353, 354, 355, 356, 357, 358, 359, 360, 362, 363, 364, 365

Wolff, Hans Eduard, 51

Wollheim, Norbert, 211, 352, 353, 354, 356, 357, 358, 359, 362, 364

Wuppertal, 150, 331

Wurster, Karl, 193, 347, 350, 355, 356, 357, 362, 364, 365

Würzburg, 130

Würzburg University, 28, 101, 131, 132

Zeh, Hermann, 126, 127, 131, 185, 186, 187, 188, 193, 194, 195, 202, 205, 214, 341, 363

Zeilsheim, 61, 111

Zellstoff Waldhof, 362

Zurich, 49, 50, 174